"十三五"国家重点出版物出版规划项目 先进制造理论研究与工程技术系列
黑龙江省精品图书出版工程 / "双一流"建设精品出版工程

机械工程测试技术

MECHANICAL ENGINEERING MEASUREMENT TECHNOLOGY

（中英双语）

潘旭东　李跃峰　编著　王广林　主审

哈爾濱工業大學出版社
HARBIN INSTITUTE OF TECHNOLOGY PRESS

内 容 简 介

本书在以学生学习成效驱动的教学理念的指导下，在保证课程内容主线的基础上，梳理了课程内容之间的逻辑关系和衔接顺序，便于在主干内容学习完成的基础上，激发学生自主学习的兴趣，完善课程枝节，培养学生发散思维和创新能力。此外，本书还提供了中文全文的英文供对照参考，以期提升学生国际化视野。全书包括绪论、信号及其描述、测试系统的基本特性、传感器、信号分析与处理、工程应用实例等六章内容。

本书可作为高等学校机械类专业及相近专业本科生的教材，也可供大专、成人教育有关专业选用，还可作为有关专业高等学校教师、研究生和工程技术人员的参考书。

图书在版编目(CIP)数据

机械工程测试技术＝Mechanical Engineering Measurement Technology：汉、英/潘旭东，李跃峰编著. —哈尔滨：哈尔滨工业大学出版社，2022.3

ISBN 978-7-5603-9891-4

Ⅰ.①机… Ⅱ.①潘… ②李… Ⅲ.①机械工程-测试技术-高等学校-教材 Ⅳ.①TG806

中国版本图书馆 CIP 数据核字(2021)第 266709 号

策划编辑　张　荣　鹿　峰
责任编辑　周一瞳　孙连嵩
出版发行　哈尔滨工业大学出版社
社　　址　哈尔滨市南岗区复华四道街 10 号　邮编 150006
传　　真　0451-86414749
网　　址　http://hitpress.hit.edu.cn
印　　刷　黑龙江艺德印刷有限责任公司
开　　本　787 mm×1 092 mm　1/16　印张 32.5　字数 870 千字
版　　次　2022 年 3 月第 1 版　2022 年 3 月第 1 次印刷
书　　号　ISBN 978-7-5603-9891-4
定　　价　68.00 元

前　言

“机械工程测试技术”课程自 1978 年被正式列入机械类专业教学计划以来，一直受到各高等院校的重视。尤其是科学技术的高速发展及相关技术领域所取得的突破性成就，使得“机械工程测试技术”课程内容也得到了迅速发展和充实，已经成为机械类专业的主干课程之一，在机械类课程体系中具有越来越重要的地位。

随着新工科建设的深入，相关教学改革已经全面开展。在以学生学习成效驱动的教学理念的指导下，为满足培养在机械工程及相关领域引领未来发展的创新人才需求，作者梳理了“机械工程测试技术”课程内容，在保留主线内容的基础上，对课程内容进行了调整，加强了部分内容之间的衔接关系，更适应学生自学能力和发散思维的培养。同时，将作者自身完成的一些相关工程实例纳入课程内容，以工程实例为牵引，以期加深学生对课程内容的理解，培养其创新实践能力。此外，为了培养学生国际化视野，提升专业外语的理解能力，同时也为学生将来从事研究工作提供专业外语的支持，本书以中英文对照形式对全部内容予以展现。

本书由哈尔滨工业大学潘旭东、李跃峰编著。哈尔滨工业大学李金博和黑龙江大学王天予进行了中英文对照翻译工作。

本书由哈尔滨工业大学王广林教授主审。

由于时间仓促，本书在撰写过程中难免存在疏漏，恳请使用本书的教师、学生对书中的内容编排、取舍以及观点上偏误等提出批评、指正和修改意见。

作　者

2022 年 2 月

目　录

Contents

第1章　绪　　论

【本章学习目标】

1. 掌握测试系统的一般组成；
2. 了解测试系统的发展趋势；
3. 了解测试传感器的发展趋势。

Chapter 1　Introduction

【Learning Objectives】

1. To be able to grasp the general composition of measurement systems;
2. To be able to understand the development trend of measurement systems;
3. To be able to understand the development trend of sensors.

1.1 测试技术的重要性

测试的基本任务是获取有用的信息，首先检测出被测对象的有关信息，然后加以处理，最后将其结果提供给观察者或输入其他信息处理装置、控制系统。因此，测试技术属于信息科学范畴，是信息技术三大支柱（测试技术、计算技术和通信技术）之一。

测量是以确定被测物属性量值为目的的全部操作。测试是具有试验性质的测量，也可理解为测量和试验的综合。人类在从事社会生产、经济交往和科学研究活动中都与测试技术息息相关。

测试是人类认识客观世界的手段，是科学研究的基本方法。科学的基本目的在于客观地描述自然界。科学探索需要测试技术，用准确而简明的定量关系和数学语言来表述科学规律和理论需要测试技术，检验科学理论和规律的正确性同样需要测试技术。可以认为，精确的测试是科学的根基。

在工程技术领域中，工程研究、产品开发、生产监督、质量控制和性能试验等都离不开测试技术，特别是近代工程技术中广泛应用的自动控制技术已越来越多地运用测试技术，测试装置已成为控制系统的重要组成部分，甚至在日常生活用具，如汽车、家用电器等方面，也离不开测试技术。

总之，测试技术已广泛地应用于工农业生产、科学研究、国内外贸易、国防建设、交通运输、医疗卫生、环境保护和人民生活的各个方面，起着越来越重要的作用，成为国民经济发展和社会进步的一项必不可少的重要基础技术。因此，使用先进的测试技术也就成为经济高度发展和科技现代化的重要标志之一。

机械工业担负着装备国民经济各个部门的任务。在改革开放的过程中，机械工业面临着更新产品、革新生产技术、改善经营管理、提高产品质量、提高经济效益和参与国际市场竞争的挑战。测试技术将是机械工业应对上述挑战的基础技术之一。

1.1 The Importance of Measurement Technology

The basic task of measurement is to obtain useful information, which is firstly to detect the relevant information of the measured object, and then process it, and finally provide the result to the observer or input other information to process the treatment devices and control systems. Therefore, testing technology belongs to the category of information science and is one of the three pillars of information technology (testing technology, computing technology, and communication technology).

Measurement is the process of experimentally obtaining quantity values that can reasonably be attributed to a quantity. The quantity intended to be measured is called measurand. Testing is a procedure to determined characteristics of a given object and express them by qualitative and means. Human beings are closely related to testing technology in social production, economic communication and scientific research activities.

Testing, as a basic method of scientific research, is a means by which human beings understand the objective world. The basic purpose of science is to describe nature objectively. Scientific exploration requires measurement technology, which is also needed to express scientific laws and theories with accurate and concise quantitative relationships and mathematical language as well as testing the correctness of scientific theories and laws. Therefore, it can be considered that accurate testing is the foundation of science.

In the field of engineering technology, measurement technology is inseparable from engineering research, product development, production supervision, quality control and performance testing, especially applied in the automatic control technology which is widely used in modern engineering technology. Therefore, the measurement device has become an important part of the control system, and even measurement technology proves inseparable in daily life appliances, such as automobiles, household appliances, etc.

In short, measurement technology has been widely used in various aspects of industrial and agricultural production, scientific research, domestic and foreign trade, national defense construction, transportation, medical and health, environmental protection and people living, and it has played an increasingly crucial role and has become an essential and important basic technology for national economic development and social progress. Therefore, the use of advanced measurement technology has become one of the important signs of high economic development and modernization of science and technological.

The machinery industry shoulders the responsibility for all sectors of the national economy equipment. In the process of reform and opening up, the machinery industry is facing the challenges of products updates, production technology innovation, business management improvement, product quality improvement, economic benefit improvement, and the participation of international market competition. Measurement technology will be one of the basic technologies for the machinery industry to tackle the above-mentioned challenges.

1.2 测试过程和测试系统的一般组成

信息总是蕴涵在某些物理量之中，并依靠它们来传输，这些物理量就是信号。就具体物理性质而言，信号有电信号、光信号、力信号等。其中，电信号在变换、处理、传输和运用等方面都有明显的优点，因此成为目前应用最广泛的信号。各种非电信号也往往被转换成电信号，然后进行传输、处理和运用。

在测试工作的许多场合中，并不考虑信号的具体物理性质，而是将其抽象为变量之间的函数关系，特别是时间函数或空间函数，从数学层面加以分析研究，从中得出一些具有普遍意义的理论。这些理论极大地发展了测试技术，并成为测试技术的重要组成部分。

一般来说，测试工作的全过程包含许多环节：以适当的方式激励被测对象、信号的检测和转换、信号的调理、分析与处理、显示与记录，以及必要时以电量形式输出测量结果。测试系统框图如图 1.1 所示。

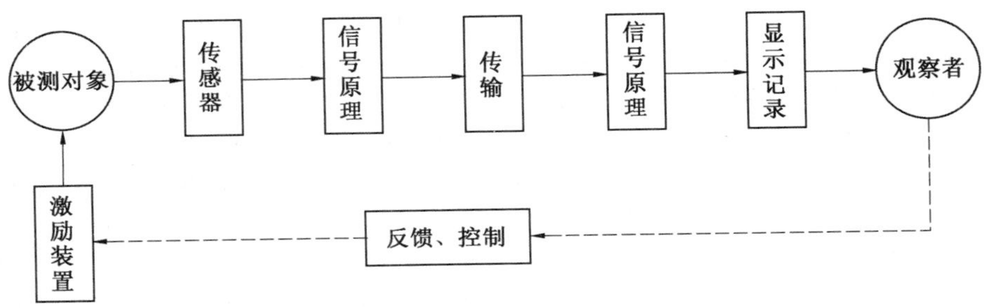

图 1.1　测试系统框图

客观事物是多样的。测试工作所希望获取的信息有可能已载于某种可检测的信号中，也有可能尚未载于可检测的信号中。对于后者，测试工作就包含着选用合适的方式激励被测对象，使其产生既能充分表征有关信息又便于检测的信号。事实上，许多系统的特性参量在系统的某些状态下可能充分地显示出来；而在另外一些状态下却可能没有显示出来或者显示得很不明显，以致难以检测出来。因此，在后一种情况下，要测量这些特性参量时，就需要激励该系统，使其处于能够充分显示这些参量特性的状态中，以便有效地检测载有这些信息的信号。

(1) 传感器直接作用于被测量，并能按一定规律将被测量转换成同种或别种量值输出，这种输出通常是电信号。

(2) 信号调理环节把来自传感器的信号转换成更适合于进一步传输和处理的形式。这时的信号转换在多数情况下是电信号之间的转换，如将阻抗的变化转换成电压的变化

1.2 The General Compositions of the Measurement Process and Measurement Systems

Information is always contained in certain physical quantities depended on for transmission, those which are signals, and in terms of specific physical properties, signals include electrical signals, light signals, force signals, and so on. Among them, electrical signals have so obvious advantages in conversion, processing, transmission, and application that it becomes the most widely used signal at present. And various non-electrical signals are often converted into electrical signals, and then they are transmitted, processed and used.

In many occasions of measurement work, the specific physical properties of signals are not being considered, but they are abstracted as functional relationships between variables, especially time functions or space functions. These theories which come from mathmatical analysis and research have greatly developed measurement technology and have become an important part of measurement technology.

Generally speaking, the whole process of the measurement work contains many links: excitation of the measured object in an appropriate manner, signal detection and conversion, signal conditioning, analysis and processing, display and recording, and output of measurement results in the form of electricity when necessary. The block diagram of the measurement systems is shown in Figure 1.1.

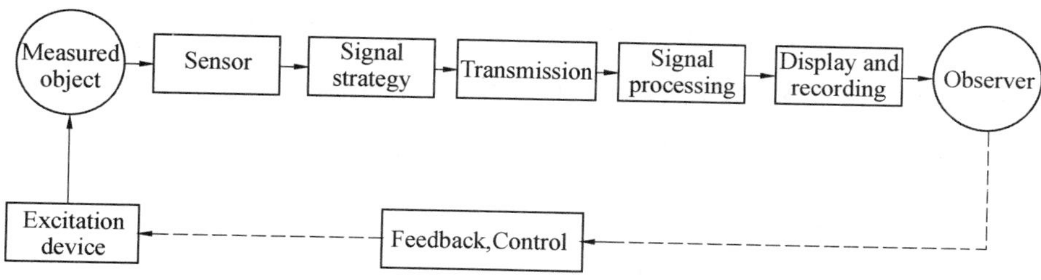

Figure 1.1 The block diagram of the measurement systems

As we all know, objective things are diverse. The information desired by the measurement work may have been or may not contained in a detectable signal. For the latter, measurement work involves in choosing an appropriate method to excite the measured object in order that it can generate the signal that can fully characterize the relevant information and is easy to detect. In fact, many characteristic parameters of the system may be fully displayed in some states of the system; in other states, they may not be displayed, or displayed very inconspicuously, making it difficult to detect. Therefore, in the latter case, when these characteristic parameters are to be measured, the system needs to be stimulated to be in a state that can fully display the characteristics of these parameters in order to effectively detect the signals carrying these information.

(1) The sensor directly acts on the measured quantity which can be converted into the same or another kind of value output according to a certain rule, as this output is usually an electrical signal.

(2) The signal conditioning link will convert the signal from a sensor into a more suitable

或将阻抗的变化转换成频率的变化等。

(3)信号处理环节接收来自调理环节的信号,并进行各种运算、滤波、分析,将结果输出到显示、记录或控制系统。

(4)信号显示、记录环节以观察者易于识别的形式来显示测量的结果,或者将测量结果存储,供必要时使用。

在所有这些环节中,必须遵循的基本原则是各个环节的输出量与输入量之间应保持一一对应和尽量不失真的关系,并尽可能地减小或消除各种干扰。

应当指出,并非所有的测试系统都具备图 1.1 中的所有环节,尤其是虚线连接的环节和传输环节。实际上,环节与环节之间都存在着传输,图 1.1 中的传输环节专指较远距离的通信传输。

测试技术是一种综合性技术,对新技术特别敏感。为了做好测试工作,需要及时将新技术运用在测试工作中。

1.3 测试技术的发展

现代测试技术既是促进科技发展的重要技术,又是科学技术发展的结果。现代科技的发展不断地向测试技术提出新的要求,推动测试技术的发展。与此同时,测试技术迅速吸取和综合各个科技领域(如物理、化学、生物学、材料科学、微电子学、计算机科学和工艺学等)的新成就,开发出新的方法和装置。

传感器在科学技术领域、工农业生产及日常生活中正发挥着越来越重要的作用。目前的传感器无论是在数量、质量还是在功能上都还远不能适应社会多方面发展的需要。人类社会对传感器提出的越来越高的要求是传感器技术发展的强大动力,而现代科学技术的突飞猛进则为传感器的发展提供了坚强的后盾。纵观几十年来传感技术领域的发展,不外乎两个方面:一是最大限度地提高现有传感器的性价比,二是寻求新原理、新材料、新工艺及新功能等。

当今传感器开发中,以下列几方面的发展最引人注目。

1. 研制新型传感器

传感器的工作机理遵从各种定律和效应。随着人们深入地认识自然,一些新的化学效应、物理效应、生物效应等会逐渐被发现。利用新发现的效应可研制出相应的新型传感器,从而使提高传感器性能和拓展传感器的应用范围成为可能。人们进一步探索具有新效应的感应功能材料,改变材料的结构、组成、添加物或采用各种工艺技术,利用材料形态

form for further transmission and processing. In most cases, the signal conversion is a kind of conversion between electrical signals at this time, for example, converting impedance change into voltage change, or into frequency change, etc.

(3) The signal processing link accepts the signal from the conditioning link, and performs various calculations, filtering, and analysis, and outputs the results to the display, recording or control system.

(4) The signal display and recording link will show the measurement results in an easy form for the observer to identify, or store the measurement results for use when necessary.

In all these links, the basic principle that must be followed is that the output and input of each link should maintain a one-to-one correspondence and try not to be distorted, and all kinds of interference must be reduced or eliminated as much as possible.

It should be pointed out that not all measurement systems have all the links in Figure 1.1, especially the links and transmission links connected by the dotted line. In fact, there is transmission between links. The transmission link in Figure 1.1 specifically refers to the long-distance communication transmission.

Since measurement technology is a comprehensive technology particularly sensitive to new technologies, it is necessary to comprehensively use knowledge of multiple disciplines and pay attention to the application of new technologies to do a good job of measurement.

1.3 The Development of Measurement Technology

Modern measurement technology is not only the important technology to promote the development of science and technology, but also the result of the development of science and technology. The development of modern science and technology continuously puts forward new requirements for testing technology and promotes the development of measurement technology. At the same time, it can quickly absorb and integrate new achievements in various scientific and technological fields (such as physics, chemistry, biology, materials science, microelectronics, computer science, and technology) to develop new methods and devices.

Sensors are playing an increasingly important role in the fields of science and technology, industrial and agricultural production and daily life. The current sensors, no matter in quantity, quality or function, are far from being able to meet the needs of the development of society in many aspects. Therefore, the strong driving force for the development of sensor technology is the higher and higher requirements of the human society on sensors, and the rapid advances in modern science and technology provide a strong backing for the development of them. Within a span of the decades of development in the field of sensing technology, there are no more than two aspects: one is to maximize and improve the performance price ratio of existing sensors, and the other is to seek new principles, new materials, new processes and new functions.

In today's sensor development, the most noticeable developments are in the following aspects.

1. The Development of New Sensors

As the working mechanism of sensors obeys various laws and effects, some new chemical

变化来提高材料对力、声、热、光、电、磁、分离载流子、吸附载流子、输送载流子，以及化学、生物等的感应功能，并以此研制出新型的传感器。

结构型传感器发展得较早，目前日趋成熟，但是它的结构复杂、体积较大、价格偏高。物性型传感器则与之相反，具有较多吸引人的优点，但其发展水平不足，世界各国都加大对于物性型传感器研究方面的投入，这成为一个值得注意的发展方向。其中，利用量子力学的效应研制的低灵敏阈传感器可用来检测微弱的信号，是发展新动向之一。例如，利用核磁共振吸收效应的磁敏传感器，可将灵敏阈提高到地磁强度的 10^{-6}；利用约瑟夫逊效应的热噪声温度传感器，可测 10^{-6} K 的超低温；利用光子滞后效应开发出响应速度极快的红外传感器等。此外，利用生物效应、化学效应开发的生物传感器和化学传感器更是有待开拓的新领域。

大自然是生物传感器的优秀设计师和工艺师，它通过漫长的岁月，不仅造就了集多种感官于一身的人类，而且还构造了许多功能奇特、性能高超的生物感官。例如，狗的嗅觉（灵敏阈为人的 10^6 倍），鸟的视觉（视力为人的 8～50 倍），蝙蝠、飞蛾、海豚的听觉（属于主动型生物雷达——超声波传感器）等，这些动物的感官性能是当今传感器技术所望尘莫及的，研究它们的机理、开发仿生传感器也是引人注目的方向。

2. 开发新材料

传感器材料是传感器技术的重要基础，材料科学的进步使得新型传感器的开发成为可能。近年来，对传感器材料的研究主要涉及以下几个方面：从单晶体到多晶体、非晶体；从单一型材料到复合材料；原子（分子）型材料的人工合成。利用复合材料来制造性能更加良好的传感器是今后的发展方向之一。

（1）半导体感应材料。

半导体感应材料在传感器技术中应用广泛，在相当长的时间内占据主导地位。半导体硅在力敏、磁敏、热敏、气敏、光敏、离子敏及其他敏感元件上具有广泛用途。

硅材料可分为单晶硅、多晶硅和非晶硅。目前，压力传感器仍以单晶硅为主，但有向多晶硅和非晶硅的薄膜方向发展的趋势。多晶硅传感器具有感温特性好、易小型化、易生产、成本低等优点。非晶硅主要应用于压力传感器、应变传感器、热电传感器、光传感器（如颜色传感器和摄像传感器）等。非晶硅因具有可用作薄膜光电器件、薄膜形成温度低、光吸收系数大、对整个可见光域敏感等良好的特性而获得迅速发展。

采用金属材料和非金属材料结合成化合物半导体是另一个思路。化合物半导体的发光效率高、抗辐射、耐高温、电子迁移率大，可制成高频率器件，预计在光敏、磁敏中会得到

effects, physical effects, and biological effects will gradually be discovered as people deeply understand nature. The newly discovered effects can be used to develop corresponding new sensors, which makes it possible to improve the performance of sensors and expand the application range of sensors. The new types of sensors are being developed by further exploring sensing functional materials with new effects, changing the structure, composition, addition, or adopting various process technologies, and making use of material morphology changes to improve the material's sensing functions for force, sound, heat, light, electricity, magnetism, separation carrier, adsorption carrier, transport carrier, and chemistry, biology, etc.

On the contrary to structure type sensors which developed earlier and now tends to be more mature, with complex structure, large volume, and relatively high price, physical property type sensors have more attractive advantages. But the development level of physical sensors is insufficient, and all over the countries should have increased their research investment, which has become a noteworthy development direction. Among them, the low sensitivity sensor developed by the effect of quantum mechanics can be used to detect weak signals, which is one of the new development trends. For example, the magnetic sensor using the NMR (Nuclear Magnetic Resonance) absorption effect can increase the sensitivity threshold to 10^{-6} of the geo-magnetic intensity; the thermal noise temperature sensor using the Josephson effect can measure ultra-low temperatures of 10^{-6} K; using the photon hysteresis effect to develop infrared sensors with extremely fast response speed, etc. In addition, biosensors and chemical sensors developed by using biological and chemical effects prove new areas to be explored.

Nature, like an excellent designer and craftsman of biosensors, has not only created a human with multiple senses, but also constructed many biological senses with peculiar functions and superb performance through the long years. For example, dogs' sense of smell (the sensitivity threshold is 10^{6} times that of humans), the birds' vision (the eyesight is 8–50 times that of humans), the bats', moths' and dolphins' hearing (belonging to the active bioradar—ultrasonic sensors), and so on, the sensory performance of these animals is beyond the reach of today's sensor technology, and studying their mechanism and developing bionic sensors are also eye-catching directions.

2. The Development of New Materials

Sensor materials are the significant foundation of sensor technology, and advances in material science have made it possible to develop new types of sensors. The recent research on sensor materials mainly involves the following aspects: from single crystals to polycrystalline and amorphous; from single-type materials to composite materials; artificial synthesis of atomic (molecular) materials. One of the future development directions is the use of composite materials to manufacture sensors with better performance.

(1) Semiconductor sensing materials.

Semiconductor sensing materials are widely used in sensor technology, occupying a dominant position for a long time. Semiconductor silicon has a wide range of applications in force sensitive, magnetic sensitive, heat sensitive, gas sensitive, photo sensitive, ion sensitive and other sensitive components.

Silicon materials can be divided into monocrystalline silicon, polycrystalline silicon and amorphous silicon. At present, pressure sensors are still dominated by monocrystalline silicon,

越来越多的应用。例如,用炉内合成生长的单晶,重复性、均匀性有较大提高,再用离子注入技术可制成性能优良的霍尔器件。

在半导体传感器中,场效应晶体管也是具有代表性的。在栅极上加一反向偏压,偏压的大小可控制漏极电流的大小。若用某种敏感材料将所要测量的参量以偏压的方式加到栅极上,就可以从漏极电流或电压的数值来确定该参量的大小,很容易系列化、集成化。

(2)磁性材料。

很多传感器应用的都是磁性材料。目前,磁性材料正在向薄膜化、非晶化等方向发展。非晶磁性材料具有电阻率高、磁导率高、耐腐蚀、矫顽力小、硬度大等特点,因此将获得越来越广泛的应用。

非晶体不具有磁的各向同性,因此是一种高磁导率和低损耗的材料,很容易得到旋转磁场,而且在各个方向都可得到高灵敏度的磁场,可用来制作磁通敏感元件、磁力计、高灵敏度的应力传感器,基于磁致伸缩效应的力敏元件也将得到发展。由于这类材料灵敏度比坡莫合金高几倍,因此可以大大降低涡流损耗,从而获得优良的磁特性,这对高频方面的应用来说更为可贵。利用这一特性,可以制造出用磁性晶体很难获得的快速响应型传感器。合成物可以在任意高于居里温度的温度下产生,这使得发展快速响应的温度传感器成为可能。

(3)陶瓷材料。

陶瓷材料在感应技术中具有较大的应用潜力。具有电功能的陶瓷又称电子陶瓷,可分为压电陶瓷、热电陶瓷、光电陶瓷、介电陶瓷、绝缘陶瓷和半导体陶瓷等,这些种类的陶瓷在工业测量方面均有应用,其中压电陶瓷和半导体陶瓷的应用最为广泛。半导体陶瓷是传感器常用的材料,尤以湿敏、气敏、热敏、电压敏最为突出。陶瓷感应材料的发展趋势是探索新材料,发展新品种,向高精度、高稳定性、小型化和长寿命、集成化、薄膜化和多功能化方向发展。

(4)智能材料。

智能材料是指通过控制和设计材料的机械、化学、物理、电学等参数,研制出具有生物体材料所有的特性或者优于生物体材料性能的人造材料。一般来说,具有以下功能的材料可称为智能材料:对环境判断的可自适应功能、自修复功能、自诊断功能、自增强功能(或称时基功能)。生物体材料的最突出特点是具有时基功能,它能够根据环境调节其自身的灵敏度。除生物体材料外,最值得关注的智能材料是形状记忆陶瓷、形状记忆合金和形状记忆聚合物。人类对智能材料的探索研究工作才刚刚开始,相信不久的将来会有很大的发展。

but there is a trend toward thin films of polycrystalline silicon and amorphous silicon. Polycrystalline silicon sensors have the advantages of good temperature sensing characteristics, easy miniaturization, easy production, and low cost. Amorphous silicon is mainly used in pressure sensors, strain sensors, thermoelectric sensors, light sensors (such as color sensors and camera sensors), etc. The rapid development of amorphous silicon is due to its good characteristics as thin-film optoelectronic devices, low film formation temperature, high light absorption coefficient, and sensitivity to the entire visible light range.

Another way of developing compound semiconductors is to combine metallic materials and non-metallic materials. Since compound semiconductors have high luminous efficiency, radiation resistance, high temperature resistance, high electron mobility, and can be made into high-frequency devices, it is expected to find more and more applications in photosensitive and magnetic sensitivity. For example, the repeatability and uniformity of single crystals synthesized in a furnace are greatly improved, and then ion beam implantation technique can be used to make Hall devices with excellent performance.

Among semiconductor sensors, field-effect transistors are also representative. A reverse bias voltage is applied to the grid, and the magnitude of the bias voltage can control the magnitude of the drain current. If a certain sensitive material is used to add the parameter to be measured to the grid in a biased manner, the magnitude of the parameter can be determined from the value of the drain current or voltage, which is easy to serialize and integrate.

(2) Magnetic materials.

Magnetic materials are applied in many sensors. At present, magnetic materials are developing in the direction of thin film and amorphization. The reason why amorphous magnetic materials are more and more popular is that they have the characteristics of high resistivity, high magnetic permeability, corrosion resistance, low coercivity and high hardness.

Not having magnetic isotropy, amorphous is a material with high magnetic permeability and low loss with easy to obtain a rotating magnetic field, and a high-sensitivity magnetic field obtained in all directions, so it can be used to make components of magnetic flux sensitive, magnetometers, high-sensitivity stress sensors and force-sensitive components based on the magnetostrictive effect will also be developed. Due to the sensitivity of this type of material that is several times higher than that of permalloy, it can greatly reduce the eddy current loss for obtaining excellent magnetic properties, which is more valuable for high-frequency applications. Using this feature, a fast response type sensor that is difficult to obtain with magnetic crystals can be manufactured. The composition can be produced at any temperature higher than the Curie temperature, making it possible to develop a quick-response temperature sensor.

(3) Ceramic materials.

Ceramic materials have great application potential in induction technology. Ceramics with electrical functions are also called electronic ceramics divided into piezoelectric ceramics and thermoelectric ceramics, photoelectric ceramics, dielectric ceramics, insulating ceramics and semiconducting ceramics, of which these types have applications in industrial measurement, with piezoelectric ceramics and semiconducting ceramics being the most widely used. Semiconducting ceramics are commonly used materials for sensors, especially humidity, gas, thermal, voltage-sensitive. The trend of ceramic induction materials is the exploration of new materials and the development of new varieties, with developing in the direction of high

3. 新工艺的采用

新工艺的采用对于发展新型传感器来说是不可缺少的。新工艺的含义很广,这里主要指微细加工技术(又称微机械加工技术),即分子束、电子束、离子束、激光束和化学刻蚀等用于微电子加工的技术,其目前已越来越多地用于传感器领域,如溅射、蒸镀、等离子体刻蚀、化学气体淀积(CVD)、扩散、外延、腐蚀、光刻等。迄今为止,已有大量采用上述工艺制成的传感器问世并且得到应用。

以应变式传感器为例,应变片可分为金属箔式应变片、体型应变片、扩散型应变片和薄膜应变片。薄膜应变片是今后的发展趋势,这主要是因为近年来薄膜工艺发展迅速,除采用高频溅射、真空淀积外,还发展了等离子体增强化学气相淀积、磁控溅射、分子束外延、金属有机化合物化学气相淀积、光 CVD 技术等,这些对传感器的发展起了很大推动作用。例如,目前常见的溅射型应变计是采用溅射技术直接在应变体即产生应变的柱(梁)、振动片等弹性体上形成的。这种应变计厚度很薄,不到箔式应变计的 1/10,故又称薄膜应变计,其优点是精度高、可靠性好,容易做成高阻抗的小型应变计,无迟滞和蠕变现象,有良好的冲击性能和耐热性等。

4. 多功能化与智能化、集成化

传感器集成化包括两种定义:一是在同一芯片上,将多个相同的敏感元件集成为一维、二维或三维阵列型传感器,如 CCD 图像传感器;二是多功能一体化,即将传感器与放大、运算及温度补偿等电路集成在一起,做在一块芯片上,使之具有校准、补偿、自诊断和网络通信的功能,可增强抗干扰能力,消除仪表带来的二次误差,具有很大的实用价值。固态功能材料——电介质、半导体、强磁体的进一步开发和集成技术的不断发展为传感器集成化开辟了广阔的前景。

通常情况下,一个传感器只能用来探测一种物理量。但是在许多场合中,为准确而完整地反映客观环境和事物,往往要同时测量多种不同的参数。可以把多个不同功能的敏感元件集成在一起(集成块)来同时测量多项参数,还可以对这些参数的测量结果进行综合处理和评价。传感器的多功能集成化不仅能够降低生产成本、减小体积,还可以提高传感器的可靠性和稳定性等性能指标。例如,美国某传感器研发中心推出的单片硅多维力传感器可以同时测量 3 个线速度、3 个角加速度和 3 个离心加速度(角速度),该传感器的主要元件包括 4 个安装在同一基板上的悬臂梁组成的单片硅结构和 9 个布置在各个悬臂梁上的压阻敏感元件。多功能传感器是当今传感器技术发展中一个全新的研究方向。将某些类型的传感器进行合理地搭配组合可成为新的传感器。

precision, high stability, miniaturization and long life, integration, thin film and multi-function.

(4) Smart materials.

Smart materials are designed materials that have all the characteristics of biological materials or are superior to biological materials by controlling and designing their mechanical, chemical, physical, electrical and other parameters. Generally, materials with the following functions can be called smart materials: self-adaptive function, self-repair function, self-diagnosis function, self-enhancement function (or time base function) for environmental judgment. Time base function, as the most prominent feature of biological materials can adjust their own sensitivity according to the environment. In addition to biological materials, the most noteworthy smart materials are shape memory ceramics, shape memory alloys and shape memory polymers. The exploration and research work on smart materials has just begun, and it is believed to have great development in the near future.

3. The Adoption of New Process

It is indispensable for new sensors to adopt new process. It covers a wide range of meaning, here mainly refers to the microfabrication technology (also known as micro-machining): molecular beam, electron beam, ion beam, laser beam and chemical etching, and other technologies used for microelectronic processing, which are now more and more used in the sensor field, such as sputtering, evaporation, plasma etching, chemical vapor deposition (CVD), diffusion, epitaxy, corrosion, photolithography, etc. And so far, a large number of sensors made by the above process have come out and applied.

Taking strain sensors as an example, strain gauges can be divided into metal foil strain gauges, bulk type strain gauges, diffusion strain gauges and thin film strain gauges. Thin film strain gauges are the future trend, which is mainly due to its rapid technology development recently, in addition to high-frequency sputtering and vacuum deposition, plasma enhanced chemical vapor deposition, magnetron sputtering, molecular beam epitaxy, metal organic chemical vapor deposition, and optical CVD technology, etc. have also been developed, all of which has played a big role in the development of sensors. For example, the current common sputtering strain gauges are formed by sputtering technology directly on the strain body, that is, the column (beam) that produces the strain, the vibrating plate and other elastic bodies. The thickness of this strain gauge is very thin, less than 1/10 of the foil strain gauge, so it is also called the thin film strain gauge with its advantages of high precision, good reliability, easy to be made into a small high-impedance strain gauge, no hysteresis and creep phenomenon, good impact performance and heat resistance, etc.

4. Multifunctionality, Intelligence and Integration

Sensor integration includes two definitions: one is to integrate multiple identical sensitive elements into one-, two-, or three-dimensional array sensors, such as CCD image sensors, on the same chip; the other is multifunctional integration, which means integrating the sensor with circuits such as amplification, calculation, and temperature compensation, and building it on a chip, so that it has the functions of calibration, compensation, self-diagnosis, and network communication, which can enhance the anti-interference ability and eliminate the secondary error caused by the instrument, with having great practical value. The further development of

传感器与微处理器相结合，在具有检测、转换功能的基础上，还能够具有记忆、存储、处理、逻辑思考、分析和结论判断等人工智能，称为传感器的智能化。智能传感器可以看作传感器与微处理器的结合体，其组成部分包括主传感器、辅助传感器及微处理器硬件系统。例如，智能压力传感器的主传感器是压力传感器，用来检测压力参数；辅助传感器通常为温度传感器（可以校正温度变化引起的测量误差）、环境压力传感器（可以测量工作环境的压力变化并对测定结果进行校正）；微处理器硬件系统除能够对传感器的弱输出信号进行放大、处理和存储外，还执行与计算机之间的通信联络。

借助于半导体集成化技术把传感器部分与信号预处理电路、输入输出接口、微处理器等制作在同一块芯片上，即成为集成智能传感器，它具有如下优点。

（1）自诊断功能。自诊断功能即接通电源时系统的自检和系统工作时的运行自检。当工作环境临近其极限条件时，将发出报警信号，并给出相关的诊断信息。当系统发生故障时，能够找出异常现象、确定故障的位置与部件等。

（2）自补偿功能。自补偿功能能够对信号检测过程中的非线性误差、温度变化及其导致的信号零点漂移和灵敏度漂移、响应时间延迟、噪声与交叉感应等进行补偿，改进了测试精度。

（3）自校正功能。自校正功能即系统中参数的设置与检查、测试中量程的自动转换和被测参量的自动运算等。

（4）通过数字式通信接口，传感器可以直接与计算机进行通信联络和信息交换，可以对检测系统进行远距离控制或在锁定方式下工作，也可以将测得的数据发送给远程用户。

（5）数据的自动存储、分析、处理与传输等能够很方便地实时处理探测到的大量数据。

目前，智能传感器技术正处于蓬勃发展时期。其代表性产品有德国斯特曼公司的二维加速度传感器、美国霍尼韦尔公司的 ST3000 系列智能变送器，以及含有微控制器（MCU）的具有多维检测能力的智能传感器和固体图像传感器（SSIS）、单片集成压力传感器等。此外，基于模糊理论的新型智能传感器和神经网络技术在智能传感器系统的研究和发展中的重要作用也日益受到研究人员的重视。

5. 操作简单化

在越来越多传感器新品出现的同时，传感器制造商也注重产品的操作舒适性。例如，作为专用于传感器和执行器之间联网通信的国际标准 AS-I（EN50295）就摒弃了传统接线中电源必须连接到每只传感器并且信号线必须连接到 I/O 模块中的限制。一个 AS-I

solid functional materials—dielectrics, semiconductors, and strong magnets, and integration technology have opened up broad prospects for sensor integration.

Generally, a sensor can only be used to detect one physical quantity, but in many occasions, in order to accurately and completely reflect the objective environment and things, it is often necessary to measure a variety of different parameters at the same time. We can integrate multiple sensitive components with different functions (integrated block) to measure multiple parameters at the same time, and can comprehensively process and evaluate the measurement results of these parameters. The multifunctional integration of the sensor can not only reduce the production cost and volume, but also improve the reliability and stability of the sensor and other performance indicators. For example, a monolithic silicon multi-dimensional force sensor launched by a sensor research and development center in the United States can simultaneously measure 3 linear velocities, 3 angular accelerations and 3 centrifugal accelerations (angular velocity). The main elements of the sensor include a monolithic silicon structure composed of 4 cantilever beams mounted on the same substrate, and 9 piezoresistive sensitive elements arranged on each cantilever beam. Multifunctional sensors are a brand-new research direction in the development of sensor technology today. A reasonable combination of certain types of sensors can become new sensors.

The combination of the sensor and the microprocessor, on the basis of the detection and conversion functions, can also have artificial intelligence such as memory, storage, processing, logical thinking, analysis and conclusion judgment, which can be called the intelligence of the sensor. An intelligent sensor can be regarded as a combination of a sensor and a microprocessor, and its components include a main sensor, an auxiliary sensor, and a microprocessor hardware system. For example, the main sensor of an intelligent pressure sensor is a pressure sensor, which is used to detect pressure parameters; an auxiliary sensor is usually a temperature sensor (which can correct measurement errors caused by temperature changes), or an environmental pressure sensor (which can measure pressure changes in the working environment and correct the measurement results); the microprocessor hardware system can not only amplify, process and store the weak output signal of the sensor, but also perform communication with the computer.

With the help of semiconductor integration technology, the sensor part, signal preprocessing circuit, input and output interface, microprocessor, etc. are made on the same chip, which becomes an integrated smart sensor. It has the following advantages.

(1) Self-diagnosis function. Self-diagnosis function refers to the self-check of a system when connecting the power and the running self-check when the system is working. When the working environment is close to its limit conditions, an alarm signal will be issued, and relevant diagnostic information will be given. When the system fails, it can find out abnormal phenomena, determine the location and parts of the fault, etc.

(2) Self-compensation function. Self-compensation function can compensate for non-linear errors, temperature changes and the signal zero drift and sensitivity drift, response time delay, noise and cross sensitivity in the signal detection process, and improve the measurement accuracy.

(3) Self-tuning function. Self-tuning function refers to the setting and checking of the parameters in a system, the automatic conversion of the measuring range during the

网络最多可包含124个传感器或31个可编程的AS-I传感器,用户可组合使用。

6. 微型化

各种控制仪器、设备的功能越来越多,部件数量也随之增多,这就要求各个部件的体积越小越好,因此传感器本身的体积也是越小越好。传统的体积较大、功能较弱的传感器将逐步被高性能、微型传感器取代。微米/纳米技术的进步和微机械加工技术的出现使3D工艺日渐成熟,为微型传感器的开发铺设了道路。微型传感器的特征是体积微小、质量较轻(体积、质量仅为传统传感器的几十分之一甚至几百分之一),其敏感元件的尺寸一般为微米级。

在当前技术水平下,微切削加工技术已经可以制作具有不同层次的3D微型结构,从而生产出体积非常小的微型传感器感应元件。例如,毒气传感器、离子传感器、光电探测器都装有极微小的感应元件。目前,利用硅材料制作的传感器体积已经很小。例如,传统的加速度传感器是由重力块和弹簧等制成的,体积较大,稳定性差,寿命短;而利用激光等各种微细加工技术制成的硅加速度传感器体积非常小,互换性和可靠性都较好。

1.4 本书的研究对象和性质

综上所述,本书所研究的对象是机械工程动态测试中常用的传感器、信号调理电路及记录仪器的工作原理,测量装置的评价方法,测试信号的分析和处理,以及常见物理量的测量方法。本书大体上分成两部分:第2~5章按信号的获取、调理、记录、分析的流程编写,讨论了测试技术的一般问题;第6章介绍了一些常见参量的测量。

对高等学校机械类的各相关专业而言,“机械工程测试技术”是一门技术基础课。通过本书的学习,可以培养学生合理地选用测试装置并初步掌握进行动态测试所需的基本知识和技能,为学生进一步学习、研究和处理机械工程技术问题打下基础。

从进行动态测试工作所必备的基本条件出发,学生在学完本书后应具有下列几方面的知识。

(1)掌握信号的时域和频域的描述方法,建立明确的信号频谱结构的概念;掌握频谱分析和相关分析的基本原理和方法;掌握数字信号分析中的一些基本概念。

(2)掌握测试装置基本特性的评价方法和不失真测试条件,并能正确地运用于测试装置的分析和选择;掌握一阶、二阶线性系统动态特性及其测定方法。

(3)了解常用传感器、常用信号调理电路和记录仪器的工作原理和性能,并能较合理地选用。

measurement, the automatic calculation of the measured parameters, etc.

(4) Through the digital communication interface, sensors can directly communicate and exchange information with the computer, remotely control the detection system or work in the locked mode, and also send the measured data to the remote user.

(5) The data function, including automatic storage, analysis, processing and transmission, etc., is convenient to process large amounts of data detected in real time.

At present, intelligent sensor technology is in a period of vigorous development. Representative products include two-dimensional acceleration sensor from Sterman in Germany, ST3000 series intelligent transmitter from Honeywell in the United States, as well as smart sensors with multi-dimensional detection capabilities containing MCU(Micro Control Unit) and solid-state image sensors (SSIS), monolithic integrated pressure sensors, etc. In addition, the important role of new intelligent sensors and neural network technologies based on fuzzy theory in the research and development of smart sensor systems has also received increasing attention from researchers.

5. The Simplification of Operations

Sensor manufacturers also pay attention to the operating comfort of their products with the emergence of more and more new sensor products. For example, as an international standard AS-I (EN50295) dedicated to networking communication between sensors and actuators, it eliminates the limitation that the power supply must be connected to each sensor and the signal line must be connected to the I/O module in the traditional wiring. An AS-I network can contain up to 124 sensors or 31 programmable AS-I sensors, which can be used in combination by users.

6. Miniaturization

The situation that various control instruments and equipment have more and more functions with the increasing number of components requires that each member smaller is better, and thus the volume of the sensor itself is as small as possible. Traditional sensors with larger volume and weaker functions will gradually be replaced by high-performance, miniature sensors. The advancement of micro/nano technology and the emergence of micro-machining technology have made the three-dimensional process more mature, and paved the way for the development of micro-sensors. Miniature sensors are characterized by small size and light weight (their volume and mass are only a few tenths or even a few hundredths of traditional sensors), and the size of their sensitive components is generally on the micron level.

At the current level of technology, micro-machining technology has been able to produce 3D micro structures with different levels, which can produce very small miniature sensor sensing elements. For example, agent gas sensors, ion sensors, and photodetectors are all equipped with tiny sensing elements. At present, the volume of sensors made of silicon materials is already very small. For example, traditional acceleration sensors are made of gravity blocks and springs, which are large in size, poor in stability, and short in life; however, the silicon acceleration sensor made by using various micro-processing technologies such as laser has a very small volume, good interchangeability and reliability.

(4)对动态测试工作的基本问题有一个比较完整的概念,并能初步运用于机械工程中某些参量的测试。

本书具有很强的实践性,只有在学习中密切联系实际、加强实验、注意物理概念,才能真正掌握有关理论。只有通过足够和必要的实验,学生才能得到应有的实验能力的训练,获得关于动态测试工作的比较完整的概念。也只有这样,学生才能初步具有处理实际测试工作的能力。

1.4 The Research Object and Nature of This Book

In conclusion, the objects studied in this book are the working principles of commonly used sensors, signal conditioning circuits and recording instruments in dynamic measurement of mechanical engineering, evaluation methods of measurement devices, analysis and processing of test signals, and test of common physical quantities method. The book is roughly divided into two parts: Chapter 2 to Chapter 5 are written according to the process of signal acquisition, conditioning, recording, and analysis, which discusses the general problems of testing technology; Chapter 6 introduces the test of some common parameters.

For all the related majors in machinery in institution of higher education, "Fundamentals of Mechanical Engineering Measurement Technology" is a basic technical course. Through this book, students will be able to select measurement devices reasonably and have a preliminary grasp of the basic knowledge and skills required for dynamic measurement, which will lay a foundation for students to further study, research and deal with mechanical engineering technical issues.

Starting from the basic conditions necessary for dynamic measurement, students should acquire the following knowledge after completing this book.

(1) To be able to master the description methods of the signal in the time domain and frequency domain, establish a clear concept of spectrum structure of signals; to master the basic principles and methods of spectrum analysis and related analysis; to master some basic concepts in digital signal analysis.

(2) To be able to grasp the evaluation method of the basic characteristics of the measurement device and the non-distortion measurement conditions, and to correctly apply it to the analysis and selection of the measurement device; to master the dynamic characteristics of first-order and second-order linear systems and their measurement methods.

(3) To be able to understand the working principles and performance of common sensors, common signal conditioning circuits and recording instruments, and to select them reasonably.

(4) To have a relatively complete concept of the basic problems of dynamic measurement, and can be initially applied to the measurement of certain parameters in mechanical engineering.

It is a very hands-on building book. Only by closely linking with reality, strengthening experiments, and paying attention to physics concepts can students truly master the relevant theories. Only by passing sufficient and necessary experiments can students receive the training of their due experimental ability and obtain a relatively complete concept of dynamic measurement work. It is in this way that students can have the initial ability to deal with actual testing work.

第2章 信号及其描述

【本章学习目标】

1. 了解信号的基本概念、种类及特点；

2. 掌握信号的描述方法，重点掌握信号的时域和频域描述方法；

3. 熟练掌握周期信号与非周期信号的频谱分析方法；

4. 了解典型信号的频谱；

5. 了解工程实际中信号在时域和频域下的特点。

Chapter 2 Signals and Their Descriptions

【Learning Objectives】

1. To be able to understand the basic concepts, types and characteristics of signals;

2. To be able to grasp the signal description methods, emphasis is placed on methods of the time domain and frequency domain description;

3. To be able to fully master the spectrum analysis methods of periodic signals and nonperiodic signals;

4. To be able to know the frequency spectrum of typical signals;

5. To be able to understand the characteristics of signals in the time domain and frequency domain in engineering practice.

2.1 信息和信号

信息是客观世界事物特征、状态、属性及其发展变化的直接或间接的反映，是客观事物之间相互联系和相互作用的表征，表现的是客观事物运动状态和变化的实质内容。测试是依靠一定的科学技术手段定量地获取某种研究对象原始信息的过程。通过对研究对象中有关信息做出客观、准确的描述，人们可以对其有一个合理、全面的认识，以达到进一步认知、改造和控制研究对象的目的。

信息的表达形式多种多样，如数字、文字、语言、声、光、电、图形等。信息的检测可以是直接的，也可以是需要对其进行加工处理才能获取的。信息只有通过信号才能实现传输、交换等功能，信号是携带信息的载体。

信号是可以察觉的带有信息的某种物理量，如光信号、声信号和电信号等。人们通过对光、声、电信号进行接收来获取信号所携带的信息。例如，道路交通信号灯通过红、黄、绿三种光信号来表示禁止通行、警示和允许通行的信息，用来指导车辆和行人安全有序地通过交叉路口；人们说话时，声信号通过声波传递到他人的耳朵，使别人了解我们的意图；各种频率空间中的无线电波、网线中的电流等都可以远距离传输各种信息，这里用的是电信号。

在机械工程中的测试方法中，通常大量采用非电量电测法。机械工程中经常涉及压力、流量、尺寸、位移、质量、力、速度、加速度、转速、温度等非电量参数的测试。与非电量对应的是电量，电量一般是指物理学中的电学量，如电压、电流、电阻、电感、电容和电功率等。非电量不能直接使用一般电工仪表、电子仪器测量，而在计算机测控系统中，更是要求输入的信号为电量。把被测非电量转换成与非电量有一定关系的电量再进行测量的方法就是非电量电测法，通常实现这种转换技术的器件称为传感器。现代测试技术的一大特点是采用非电量的电测法，其测量结果通常是随时间变化的电量，即电信号，本书中将其记为 $x(t)$ 。

测试技术与信号密切相关。在测试中除要观察大量的现象外，更重要的是还要测量大量的数据，这些数据常被转换为易于测量、传输、记录和分析的电信号。信号包含被测试对象的状态、特点等有用信息，是人们为研究客观事物而收集到的第一手素材。为收集、记录、整理、分析和使用好这些宝贵的素材，先要对信号的类型和特点等有一个清楚的

2.1 Information and Signals

Information is the direct or indirect reflection of the characteristics, states, attributes and development changes of things in the objective world, it is a representation of the interconnections and interactions between objective things, showing the actual content of the movement and changes of objective things. Measurement is a process of quantitatively obtaining the original information of a certain research object by relying on certain scientific and technology. Through an objective and accurate description of the relevant information in the research object, people will have a reasonable and comprehensive understanding of it, achieving the purpose of further cognition, transformation and control of the research object.

Information can be expressed in various forms, such as numbers, words, language, sound, light, electricity, graphics. The detection of information can be straightforward, or only be obtained through processing. Information can be transmitted and exchanged only through signals, carriers of carrying information.

A signal is a certain physical quantity that can be perceived with information, such as optical signals, acoustic signals, and electrical signals. People obtain the information carried by the signal through receiving above signals. For example, red, yellow, and green traffic lights are used to indicate prohibition, warning, and permission information, to guide vehicles and pedestrians to pass the intersection safely and orderly; when people speak, sound signals are transmitted through sound waves to other people ' s ears, making them understand our intentions; radio waves in various frequency spaces, currents in network cables and the like, can transmit various information overlong distances, signals are used here.

Non-electric quantity measurements are very popular among the measurement methods in mechanical engineering. Mechanical engineering often involves measurement of non-electrical parameters such as pressure, flow, size, displacement, mass, force, speed, acceleration, rotation speed, and temperature, etc. Electricity, corresponding to non-electricity, generally refers to electrical quantities in physics, such as voltage, current, resistance, inductance, capacitance, and electric power. Non-electricity cannot be directly measured with general electrical instruments and electronic devices. While the input signal is required to be electricity in the computer measurement and control system. Non-electric quantity measurement refers to the method of converting the measured non-electricity into a certain relationship with the non-electricity and then measuring, and the device realizing this conversion technology is usually called a sensor. One major feature of modern measurement technology is the use of non-electricity electrical measurement methods, of which the results are usually electrical quantities changing over time, that is, electrical signals, which are recorded in this book as $x(t)$.

The measurement technology is closely related to the signal. In addition to observing a large number of phenomena in the measurement, it is more important to measure a large amount

认识,以便对不同的信号采用合适的处理方法。

2.2 信号的分类

2.2.1 连续信号和离散信号

1. 连续信号

在时间的整个连续区间内都有定义或信号的数学表示式中的时间取值是连续的信号称为连续信号,它可以用一条曲线来表示信号随时间变化的特性,如图2.1(a)所示。正弦信号、直流信号、阶跃信号、锯齿波、矩形脉冲等都是连续信号。

连续信号的定义中未对幅值予以描述,因此连续信号的幅值可以是连续的,也可以是离散的。独立变量和幅值均取连续值的连续信号称为模拟信号。在实际应用中,连续信号与模拟信号不做区分。

2. 离散信号

在时间一定间隔内,只在时间轴上某些离散点上有函数值的信号称为离散信号,如图2.1(b)所示。与连续信号的定义类似,离散信号也未对幅值特征加以描述。离散信号中幅值也离散的信号称为数字信号。实际应用中,数字信号与离散信号也是通用的。二进制码就是一种数字信号,数字信号受噪声的干扰小,易于用数字电路进行处理,因此得到了广泛的应用。

离散信号中有一种特殊的信号——采样信号,即将连续信号转换为离散信号的一种重要信号。采样信号的主要特征和使用方法将贯穿本章内容。

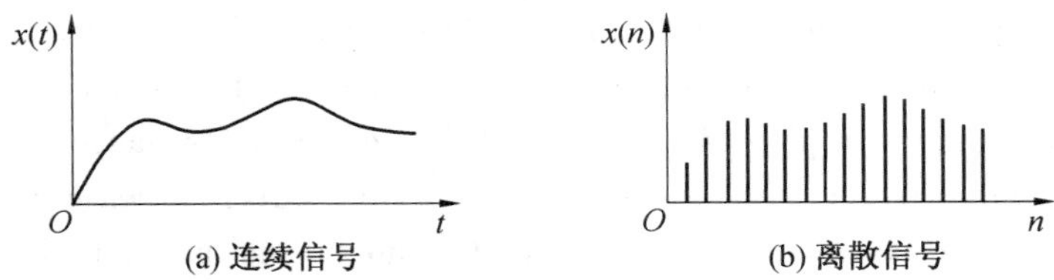

图2.1 连续信号和离散信号

2.2.2 静态信号和动态信号

1. 静态信号

静态信号是指在一定时间内,信号的幅值恒定,不随时间变化,可用静态检测手段来

of data, which is often converted into electrical signals that are easy to measure, transmit, record, and analyze. Signals includes the test object, features and other useful information which is the first-hand material that people collect for studying objective things. Before knowing how to collect, record, organize, analyze and use these valuable materials, we must first have a clear understanding of the types and characteristics of the signals in order to adopt appropriate processing methods for different signals.

2.2 The Classification of Signals

2.2.1 Continuous Signals and Discrete-Time Signals

1. Continuous Signals

A signal defined in the entire continuous interval of time, or a signal whose time value in the mathematical expression of a signal is continuous is called a continuous signal. It can be used a curve to represent the characteristics of the signal changing with time, as shown in Figure 2.1(a). Sine signals, DC signals, step signals, sawtooth waves, rectangular pulses, etc. are all continuous signals.

The definition of a continuous signal does not describe the amplitude. Therefore, the amplitude of a continuous signal can be continuous or discrete. An analog signal refers to a continuous signal whose independent variable and amplitude both take continuous values. In practical applications, we usually do not distinguish between continuous signals and analog signals.

2. Discrete-Time Signals

Discrete-time signals are signals that have function values at certain discrete points on the time axis within a certain interval of time, as shown in Figure 2.1(b). Similar to the definition of continuous signal, discrete-time signal fails to describe the amplitude characteristics. The signal with discrete amplitude in discrete-time signal is called a digital signal. In practical applications, digital-time signals and discrete-time signals are also common. Binary code is a kind of digital signal which has been widely used due to the less noise interference and easy to process with digital circuits.

There is a special kind of signal in the discrete-time signal—sampling signal, which is an important signal for converting a continuous signal into a discrete-time signal. The main characteristics and usage methods of the sampling signal will run through the content of this chapter.

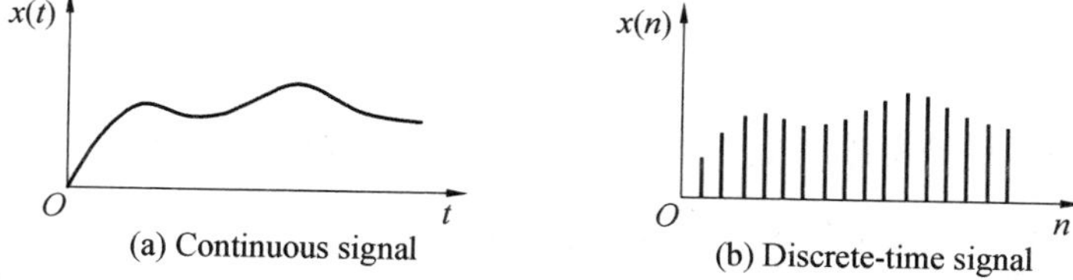

(a) Continuous signal (b) Discrete-time signal

Figure 2.1 Continuous signal and discrete-time signal

测量的信号。如果信号的幅值随时间变化非常缓慢,则称为缓变信号,缓变信号也可按照静态信号近似处理。值得注意的是,“缓慢”与否的判断是相对的,具体判断标准需要根据所关注的被测物理量的特征指标来确定。

2. 动态信号

动态信号是指瞬时值随时间变化的信号。值得一提的是,针对静态信号与动态信号开展的静态测试与动态测试有其各自关注的特征指标。

根据信号的取值是否确定,动态信号又可以分为确定性信号和非确定性信号两类,这两类信号还可根据特点进一步细分。动态信号的分类如图 2.2 所示。

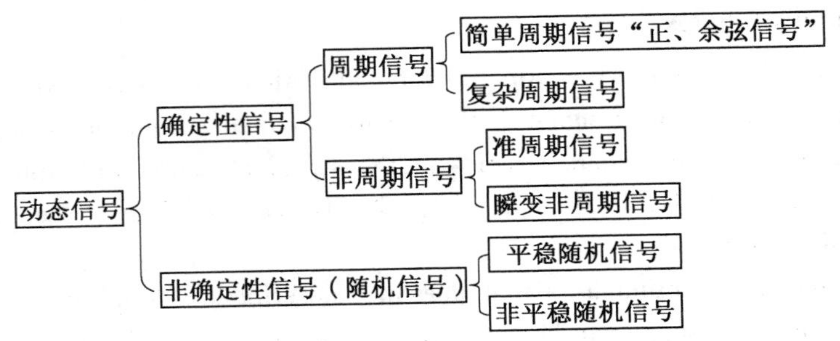

图 2.2　动态信号的分类

(1)确定性信号。

确定性信号是指信号可表示为一个确定的时间函数,因此可以确定其任何时刻的量值。例如,集中质量的单自由度振动系统如图 2.3 所示,具有集中质量的单自由度振动系统做无阻尼的自由振动时,其质点位移随时间的变化可表示为

$$x(t)=X_0\sin\left(\sqrt{\frac{k}{m}}t+\varphi_0\right)=X_0\sin(\omega t+\varphi_0) \tag{2.1}$$

式中　X_0、φ_0——振幅与初始相位,取决于初始条件;

m——质量块的质量;

k——弹簧刚度;

ω——系统的固有频率,$\omega=\sqrt{\frac{k}{m}}$。

如果确定性信号的波形按一定时间间隔周而复始重复出现,且在时间轴上无始无终,则称为周期信号,确定性信号中那些不具有周期重复性的信号称为非周期信号。

①周期信号。每隔一固定时间间隔周而复始重复出现的信号称为周期信号,可表示为

2.2.2 Static Signals and Dynamic Signals

1. Static Signals

A static signal refers to a signal that has a constant signal amplitude within a certain period of time, does not change with time, and can be measured by static detection means. When the amplitude of the signal changes very slowly with time, it is called a slowly varying signal, which can also be processed in a static signal. It is worth noting that the judgment of "slow" or not is relative, and the specific judgment standard needs to be determined according to the characteristic index of the concerned measured physical quantity.

2. Dynamic Signals

A dynamic signal is a signal whose instantaneous value changes with time. Particularly worth mentioning is that the static measurement and dynamic measurement carried out for static signals and dynamic signals have their own characteristic indicators.

According to whether the value of the signal is determined, dynamic signals can be divided into two types: deterministic signals and non-deterministic signals. These two types of signals can be further subdivided according to their characteristics. The classification of dynamic signals is shown in Figure 2.2.

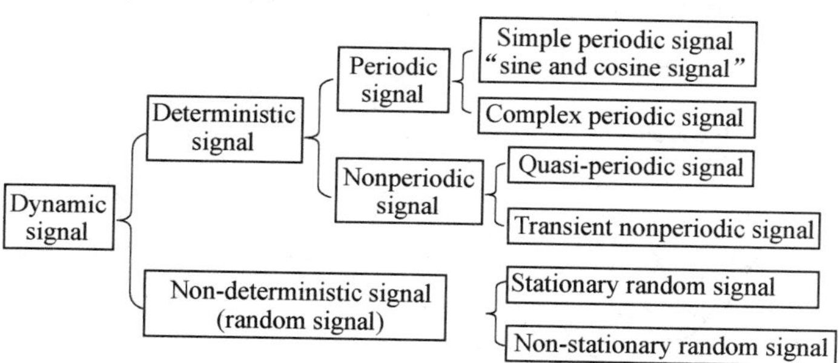

Figure 2.2 The classification of dynamic signals

(1) Deterministic signals.

A deterministic signal is a signal which can be expressed as a definite time function to determine its magnitude at any moment. For example, single-freedom vibration system with concentrated mass is shown in Figure 2.3, when a single-freedom vibration system with concentrated mass performs undamped free vibration, the change in the displacement of the mass point over time can be expressed as

$$x(t) = X_0 \sin\left(\sqrt{\frac{k}{m}}t + \varphi_0\right) = X_0 \sin(\omega t + \varphi_0) \tag{2.1}$$

where X_0, φ_0—the amplitude and the initial phase, depending on the initial conditions;

m —quality of mass block;

k — spring stiffness;

ω —natural frequency of system, $\omega = \sqrt{\frac{k}{m}}$.

A periodic signal refers to a signal that when the waveform of the deterministic signal

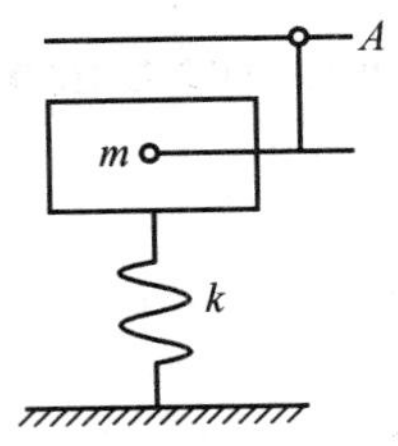

图 2.3　集中质量的单自由度振动系统

A—质点 m 的静态平衡位置

$$x(t)=x(t+nT),\quad n=\pm1,\pm2,\pm3,\cdots \tag{2.2}$$

$$T_0=2\pi/\omega_0$$

式中　T_0——周期；

ω_0——基频(圆频率,或称角频率)。

简谐波 $A\cos(\omega t+\varphi)$ 是最简单的周期信号。这种频率单一的正弦或余弦信号称为谐波信号(又称简谐信号、谐波)。由频率成整数倍关系的两个简谐波构成的周期信号如图 2.4所示,图中叠加后的信号数学表达式为 $x(t)=10\cos\omega t+5\cos 2\omega t$。

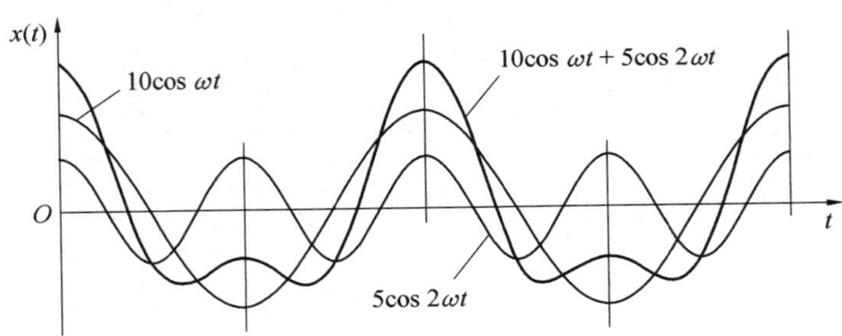

图 2.4　由频率成整数倍关系的两个简谐波构成的周期信号

②非周期信号。非周期信号可以用数学关系式描述,但是不会重复出现。非周期信号又可分为准周期信号和瞬变信号。

a. 准周期信号。准周期信号由两种以上的周期信号组成,但是其各组成分量不存在公共周期,各成分的频率不能都成整数倍关系,其中至少有两个成分频率的比是一个无理数。这样,无论经过多长时间,各成分的状态都不能恢复到刚刚发生时的初始状态,其合成信号不再为周期信号,因此准周期信号是非周期性的信号。

如果 $x(t)=\sin t+\sin\sqrt{2}t$ 是两个正弦信号的合成,则二者的频率分别为 $\omega_1=1$ 和 $\omega_2=\sqrt{2}$,其频率之比是无理数,两个周期信号的周期分别是 $T_1=2\pi$ 和 $T_2=\sqrt{2}\pi$。两个周期没有最小公倍数,说明两个信号分量没有公共周期。因此,信号 $x(t)$ 是非周期的,但是

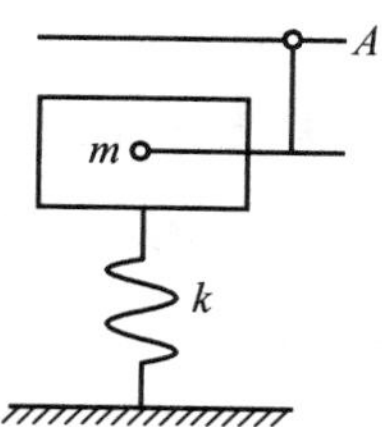

Figure 2.3　Single-freedom vibration system with concentrated mass
(A—static balance position of mass point m)

repeats itself at a certain time interval, and has no beginning and no ending on the time axis, while those signals that do not have periodical repetition in the deterministic signal are called a nonperiodic signal.

①Periodic signals. A signal that repeats itself at regular intervals is called a periodic signal, it can be expressed as

$$x(t) = x(t + nT) , \quad n = \pm 1, \pm 2, \pm 3, \cdots \tag{2.2}$$
$$T_0 = 2\pi / \omega_0$$

where　T_0 —period;

ω_0 —fundamental frequency (circular frequency, or angular frequency).

Simple harmonic wave $A\cos(\omega t + \varphi)$ is the simplest periodic signal. This single-frequency sine or cosine signal is called a harmonic signal (or simple harmonic signal, harmonic). The periodic signal composed of two harmonics whose frequencies are integer multiples is shown in Figure 2.4, the mathematical expression of the superimposed signal in the figure can be expressed by $x(t) = 10\cos\omega t + 5\cos 2\omega t$.

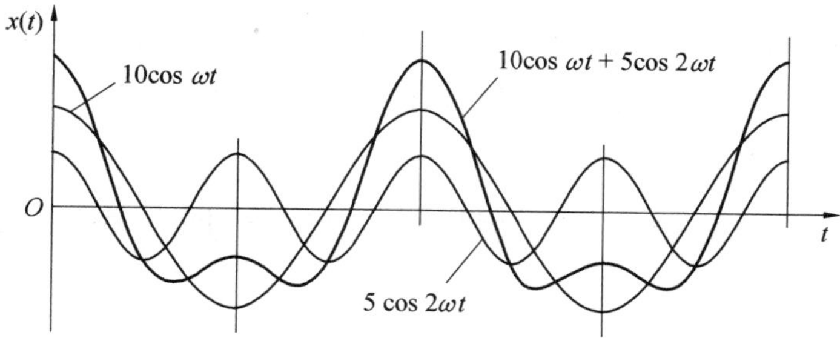

Figure 2.4　Periodic signal composed of two harmonics whose frequencies are integer multiples

②Nonperiodic signals. Nonperiodic signals can be described by mathematical relations without repeatedly appearing. They can be divided into quasi-periodic signals and transient signals.

a. Quasi-periodic signals. A quasi-periodic signal is composed of two or more periodic signals, but each component does not have a common period, and the frequency of each component cannot be an integer multiple relationship. There are at least two component frequencyratios is an irrational number, so no matter how long time elapses, the state of each component fails to be restored to the initial state in this way when it just occurred, and the synthesized signal is no longer a periodic signal, so the quasi-periodic signal proves a nonperiodic

其又是由周期信号合成的,故称为准周期信号。

准周期信号往往出现在通信、振动系统,应用于机械转子振动分析、齿轮噪声分析和语音分析等场合。当几个不相关的周期性现象混合作用时,常会产生准周期信号。多机组发动机不同步的振动信号属于准周期信号,多个独立振源激励起某个对象的振动往往是准周期信号。

b. 瞬变非周期信号(又称瞬态信号)。只在有限时间段内存在或随着时间的增长而幅值衰减到零的信号称为瞬变非周期信号,其特点是过程突然发生、时间短和能量大。持续时间较短的各种脉冲函数或者衰减函数都是瞬变非周期信号。例如,机械冲击信号、热源消失后的温度变化信号都是瞬变非周期信号。图 2.5 所示为典型的瞬变非周期信号。

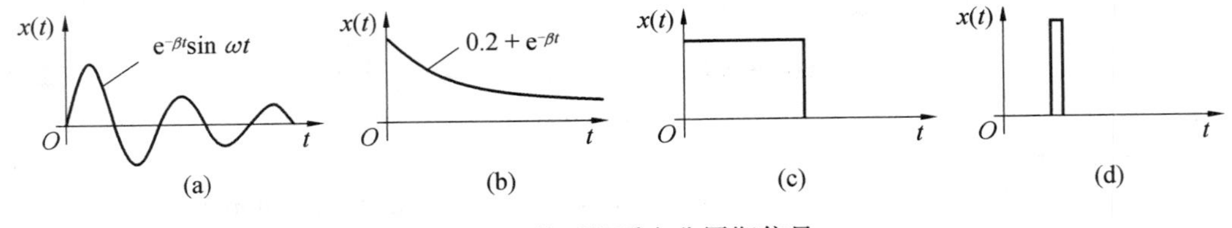

图 2.5　典型的瞬变非周期信号

(2)非确定性信号。

无法用数学式来表达,也无法预见未来任一时刻的瞬时值的信号称为非确定性信号。非确定性信号又称随机信号,其幅值、相位变化是不可预知的,所描述的物理现象是一种随机过程,如汽车奔驰时所产生的振动、飞机在大气流中的浮动、树叶随风飘荡、环境噪声、切削材质不均匀的工件时所产生的切削力等。在加工螺纹过程中,车床主轴受环境影响的振动信号波形就是一个非确定性信号。

由于随机信号具有某些统计特征,因此可以用概率统计的方法根据其过去来估计未来,但它只能近似地描述,存在误差。

2.3　信号的描述方式

2.3.1　信号的时域描述

直接观测或记录到的信号一般是以时间为独立变量的,称为信号的时域描述。信号时域描述能反映信号幅值随时间变化的关系,而不能明显揭示信号的频率组成情况。为研究信号的频率结构和各频率成分的幅值、相位关系,应对信号进行频谱分析,把信号的时域描述通过适当方法变成信号的频域描述,即以频率为独立变量来表示信号。

signal.

If $x(t) = \sin t + \sin\sqrt{2}t$ is the synthesis of two sinusoidal signals, the frequencies of the two are $\omega_1 = 1$ and $\omega_2 = \sqrt{2}$ respectively, and the ratio of the frequencies is an irrational number. The periods of the two periodic signals are $T_1 = 2\pi$ and $T_2 = \sqrt{2}\pi$. The two periods have no least common multiple, indicating that the two signal components do not have a common period. Therefore, the signal $x(t)$ is nonperiodic, but it is synthesized by periodic signals, so its name quasi-periodic signal.

Often appearing in communication and vibration systems, quasi-periodic signals are used in mechanical rotor vibration analysis, gear noise analysis, and speech analysis and other occasions. When several unrelated periodic phenomena are mixed together, quasi-periodic signals are often produced. The asynchronous vibration signal of multi-unit engines belongs to quasi-periodic signals, and the vibration of an object excited by multiple independent vibration sources is often a quasi-periodic signal.

b. Transient nonperiodic signals (also called transient signals). A transient nonperiodic signal refers to a signal, only existing in a limited period of time, or whose amplitude decaying to zero as time increases. Its characteristics are sudden occurrence of process, short time and large energy. Various pulse functions or attenuation functions with a short duration prove transient signals. For example, mechanical impact signals, and temperature change signals after the heat source disappears are all transient nonperiodic signals. Figure 2.5 is a typical transient nonperiodic signal.

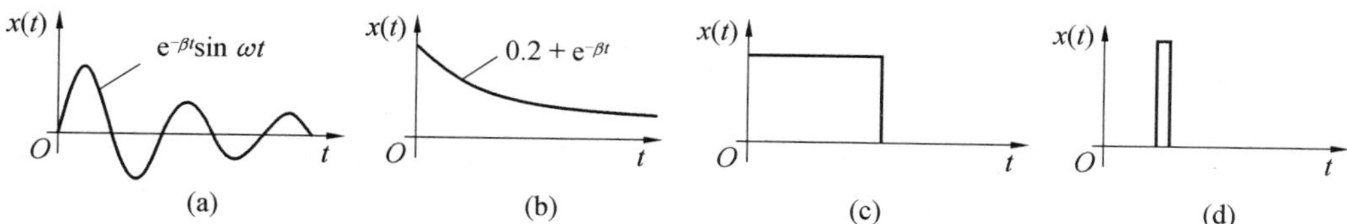

Figure 2.5　Typical transient nonperiodic signal

(2) Non-deterministic signals.

A signal that cannot be expressed mathematically, nor can it predict the instantaneous value at any time in the future, is called a non-deterministic signal. The non-deterministic signal is also called random signal, of which its amplitude and phase changes are unpredictable, and the physical phenomenon described is a random process, such as the generated vibration of a running car, the floating of an airplane in a large air current, the leaves drifting with the wind, the environmental noise, the generated cutting force during cutting workpieces with uneven cutting material. In the process of thread processing, the vibration signal waveform of the lathe spindle affected by the environment proves a non-deterministic signal.

Since random signals have certain statistical characteristics, probability statistics can be used to estimate the future based on their past, but it can only be described approximately with some errors.

图 2.6 所示为周期方波的一种时域描述,而下式则是其时域描述的另一种形式,即

$$\begin{cases} x(t)=x(t+nT_0) \\ x(t)=\begin{cases} A, & 0<t<\dfrac{T_0}{2} \\ -A, & -\dfrac{T_0}{2}<t<0 \end{cases} \end{cases}$$

图 2.6 周期方波的一种时域描述

若该周期方波应用傅里叶级数展开,则有

$$x(t)=\frac{4A}{\pi}\left(\sin \omega_0 t+\frac{1}{3}\sin 3\omega_0 t+\frac{1}{5}\sin 5\omega_0 t+\cdots\right)$$

式中

$$\omega_0=\frac{2\pi}{T_0}$$

该式表明,该周期方波是由一系列幅值和频率不等、初始相角为零的正弦信号叠加而成的。实际上,该式可改写成

$$x(t)=\frac{4A}{\pi}\left(\sum_{n=1}^{\infty}\frac{1}{n}\sin \omega t\right)$$

式中

$$\omega=n\omega_0,\quad n=1,3,5,\cdots$$

可见,该式除 t 外,还有另一变量 ω——各正弦成分的频率。若视 t 为参变量,以 ω 为独立变量,则此式为该周期方波的频域描述。

2.3.2 信号的频域描述

与时域描述相对应,以频率为独立变量对信号的描述称为信号的频域描述。在信号分析中,将组成信号的各频率成分找出来,按序排列,就可以得出信号的“频谱”。若以频率为横坐标,分别以幅值或初相位为纵坐标,便分别得到信号的幅频谱或相频谱,二者结合起来就是频谱。图 2.7 所示为周期方波的时域图形、幅频谱和相频谱的关系。

2.3 The Description of Signals

2.3.1 The Time Domain Description of Signals

The time domain description of the signal refers to a signal directly observed or recorded generally, with taking time as an independent variable. The signal time domain description can reflect the relationship of the signal amplitude over time, failing to clearly reveal the frequency composition of the signal. In order to study the frequency structure of the signal and the amplitude of each frequency component and phase relation, the signal should be analyzed in the frequency spectrum analysis, and the time domain description of the signal can be converted into the frequency domain description of the signal by an appropriate method, that is, the signal should be represented by frequency as an independent variable.

Figure 2.6 shows a time domain description of periodic rectangular, and the following formula is another form of its time domain description, which is

$$\begin{cases} x(t) = x(t + nT_0) \\ x(t) = \begin{cases} A, & 0 < t < \dfrac{T_0}{2} \\ -A, & -\dfrac{T_0}{2} < t < 0 \end{cases} \end{cases}$$

If this periodic rectangular is expanded by Fourier series, we get

$$x(t) = \frac{4A}{\pi}\left(\sin \omega_0 t + \frac{1}{3}\sin 3\omega_0 t + \frac{1}{5}\sin 5\omega_0 t + \cdots\right)$$

where

$$\omega_0 = \frac{2\pi}{T_0}$$

It indicates that the periodic rectangular is formed by the superposition of a series of unequal amplitude and frequency sin signals with an initial phase angle of zero. In fact, this formula can be rewritten as

$$x(t) = \frac{4A}{\pi}\left(\sum_{n=1}^{\infty} \frac{1}{n}\sin \omega t\right)$$

where

$$\omega = n\omega_0, \quad n = 1, 3, 5, \cdots$$

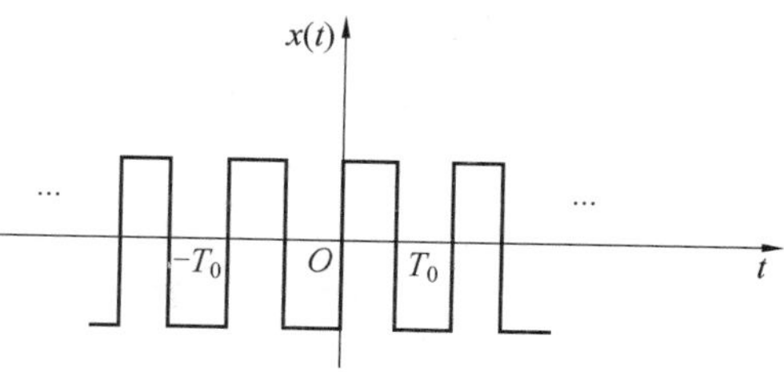

Figure 2.6　A time domain description of periodic rectangular

It can be seen that this formula has another variable ω besides t —the frequency of each sine component. If t is a parameter variable, ω is an independent variable, then this formula is the frequency domain description of the periodic rectangular.

2.3.2 The Frequency Domain Description of Signals

Corresponding to the time domain description, the description of the signal with frequency

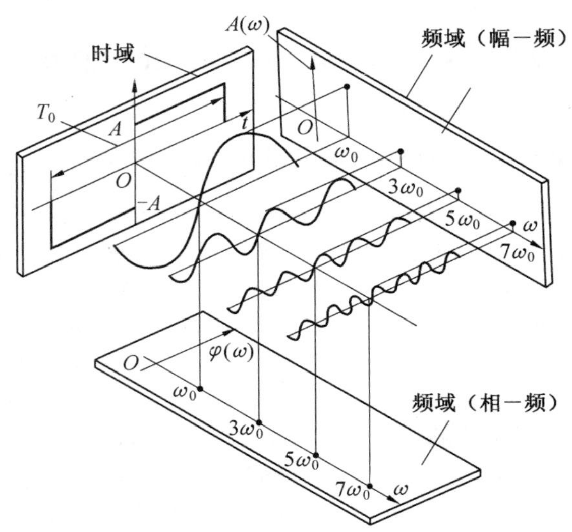

图 2.7　周期方波的时域图形、幅频谱和相频谱的关系

两个同周期方波及其幅频谱、相频谱见表 2.1。不难看出，在时域中，两方波除彼此相对平移 $T_0/4$ 外，其余完全一样。但二者的幅频谱虽相同，相频谱却不同。平移使各频率分量产生了 $n\pi/2$ 的相角，n 为谐波次数。总之，每个信号有其特有的幅频谱和相频谱，故频域中每个信号都需要同时用幅频谱和相频谱来描述。

表 2.1　两个同周期方波及其幅频谱、相频谱

时域波形	幅频谱	相频谱

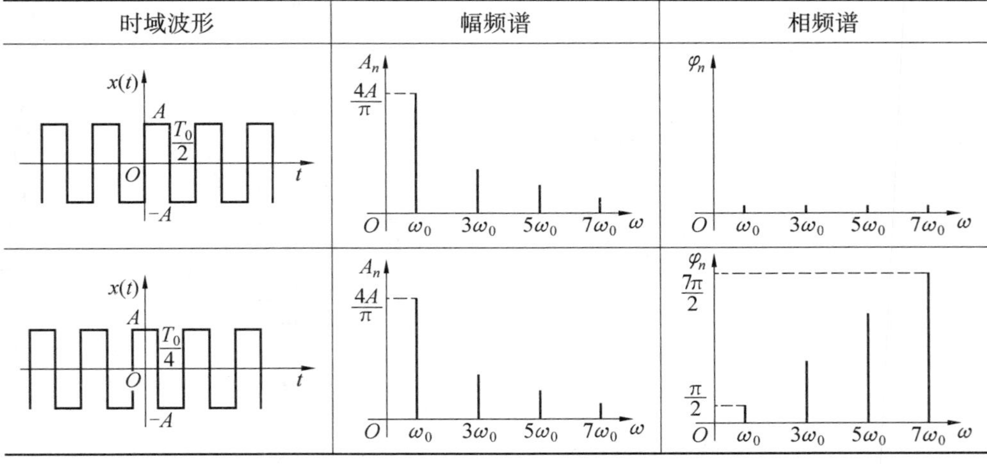

信号时域描述直观地反映信号瞬时值随时间变化的情况；频域描述则反映信号的频率组成及其幅值、初始相角的大小。为解决不同问题，往往需要掌握信号不同方面的特征，因此可采用不同的描述方式。例如，评定机器振动烈度，需用振动速度的均方根值来作为判据。若速度信号采用时域描述，就能很快求得均方根值。而在寻找振源时，需要掌

as an independent variable is called thefrequency domain description of the signal. In signal analysis, "frequency spectrum" of signals can be gotten by finding out the various frequency components that make up the signal with arrangement. Taking the frequency as the abscissa and the amplitude or initial phase as the ordinate respectively, amplitude frequency spectrum and phase frequency spectrum of the signal can be obtained respectively, two of which the combination is the spectrum. Figure 2.7 shows the relations between the time-domain graph, amplitude spectrum, and phase spectrum of the periodic rectangular.

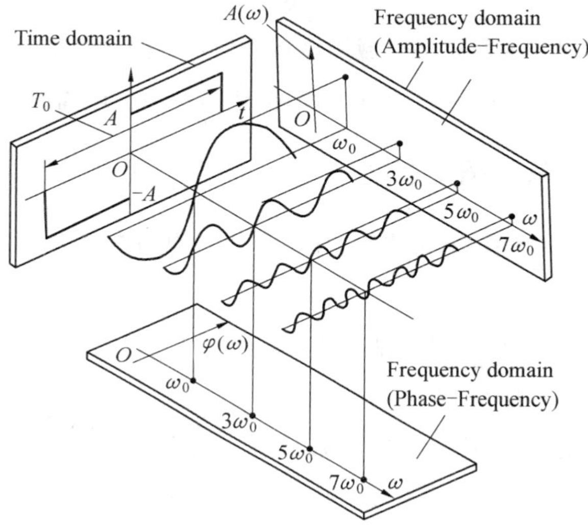

Figure 2.7　The relations between the time-domain graph, amplitude spectrum, and phase spectrum of the periodic rectangular

Two same period square waves and their amplitude spectrum and phase spectrum are listed in Table 2.1. It is not difficult to see that in the time domain, the two square waves are exactly the same except for their relative translation $T_0/4$, but both had the same amplitude spectrum, phase spectrum is different. Shifting causes each frequency component to produce $n\pi/2$ phase angle, and n is the harmonic order. In short, each signal has its own amplitude spectrum and phase spectrum, therefore each signal in the frequency domain needs to be described by them at the same time.

The time domain description of the signal intuitively reflects the change of the instantaneous value of the signal over time; and the frequency domain description reflects the frequency composition of the signal, its amplitude, and the size of the initial phase angle. The master of characteristics of different aspects of the signals are needed to solve different problems so that different description methods can be used. For example, it is necessary to use the root mean square value of the vibration intensity as a criterion to evaluate the vibration intensity of the machine. The root mean square value can be quickly obtained with the speed signal described in the time domain. When looking for the vibration source, it is necessary to use the frequency domain description due to the need for grasping the frequency components of the vibration signal. Actually, the two description methods can be interchanged with containing the

握振动信号的频率分量,这就需采用频域描述。实际上,两种描述方法能相互转换,而且包含同样的信息量。

信号的时域描述和频域描述是对同一客观物理量的两个角度描述方式。其中,时域描述更直观,是一种表象上的描述方式;频域描述更抽象,是一种本质上的描述方式。两种描述方式之间不存在因果、先后。由于时间变量更容易直接测量,因此实际应用中一般也可以直接获得以时间为自变量的时域描述信号。当需要进一步对信号进行分析、提取信号特征时,多数情况下就需要将时域描述转换为频域描述。基于这一要求,后面的章节将分别介绍将周期信号和非周期信号由时域描述转换为频域描述的方法。

2.3.3 周期信号的强度表述

周期信号的强度以峰值、绝对均值、有效值和平均功率来表述,周期信号的强度表示如图 2.8 所示。

峰值 x_p 是信号可能出现的最大瞬时值,即

$$x_p = |x(t)|_{max} \tag{2.3}$$

峰-峰值 x_{p-p} 是在一个周期中最大瞬时值与最小瞬时值之差。

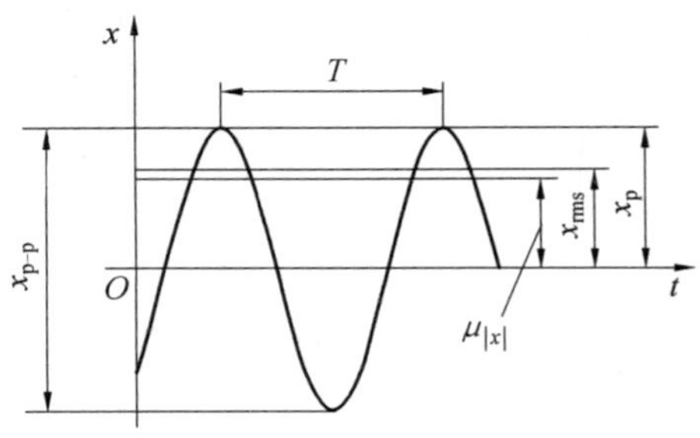

图 2.8 周期信号的强度表示

对信号的峰值和峰-峰值应有足够的估计,以便确定测试系统的动态范围。一般希望信号的峰-峰值在测试系统的线性区域内,使所观测(记录)到的信号正比于被测量的变化状态。如果进入非线性区域,则信号将发生畸变,结果不仅不能正比于被测信号的幅值,而且会增生大量谐波。

周期信号的均值 μ_x 为

$$\mu_x = \frac{1}{T_0}\int_0^{T_0} x(t)\,\mathrm{d}t \tag{2.4}$$

它是信号的常值分量。

周期信号全波整流后的均值就是信号的绝对均值 $\mu_{|x|}$,即

$$\mu_{|x|} = \frac{1}{T_0}\int_0^{T_0} |x(t)|\,\mathrm{d}t \tag{2.5}$$

有效值是信号的均方根值 x_{rms},即

$$x_{rms} = \sqrt{\frac{1}{T_0}\int_0^{T_0} x^2(t)\,\mathrm{d}t} \tag{2.6}$$

same amount of information.

Table 2.1　Two same period square waves and their amplitude spectrum and phase spectrum

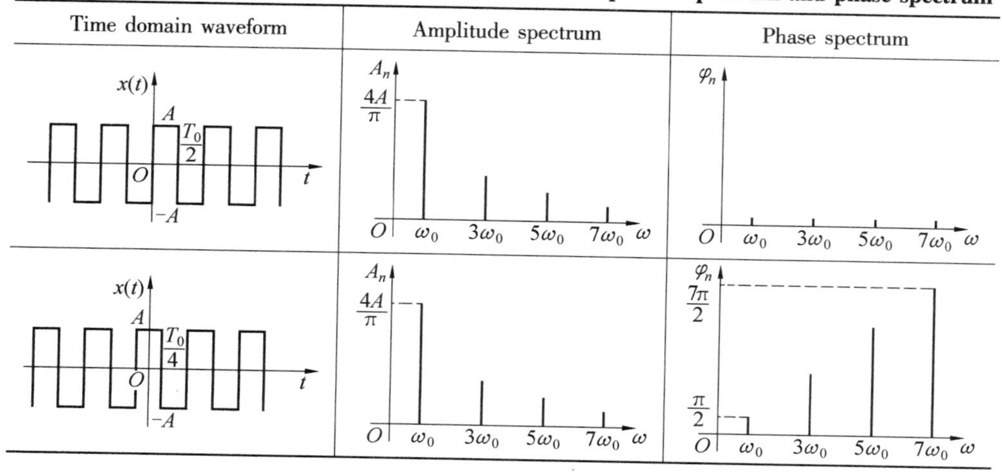

The time domain description and the frequency domain description of the signal are two ways of describing the same objective physical quantity. Compared with the time domain description that is more intuitive and is a representational description method; the frequency domain description is more abstract and is an essential way of description. There is no cause and effect or sequence between the two description methods. Easier to directly measure the time variable, it is generally possible to directly obtain the time domain description of signal with time as the independent variable in practical applications. With needing to further analyze the signal and extract signal features, the time domain description is needed to be converted to the frequency domain description in most cases. Based on this requirement, the following chapters will respectively introduce the methods of converting periodic signals and nonperiodic signals from time domain descriptions to frequency domain descriptions.

2.3.3　The Intensity Description of Periodic Signals

The intensity of periodic signals is expressed in terms of peak value, absolute mean value, effective value and average power, intensity representation of periodic signal is shown in Figure 2.8.

The peak value x_p is the maximum instantaneous value that the signal may appear, namely

$$x_p = |x(t)|_{max} \tag{2.3}$$

Peak-to-peak value x_{p-p} is the difference between the maximum instantaneous value and the minimum instantaneous value in a cycle.

The peak value and peak-to-peak value of the signal should be adequately estimated to determine the dynamic range of themeasurement system. It is generally expected that the peak-to-peak value of the signal is within the linear region of the measurement system so that the observed (recorded) signal is proportional to the change of measured quantity. Entering the non-linear region, the signal will be distorted, which the result cannot be proportional to the amplitude of the measured signal rather than that a large number of harmonics will be added.

有效值的平方——均方值就是信号的平均功率 p_{av}，即它反映信号的功率的大小，有

$$P_{av} = \frac{1}{T_0}\int_0^{T_0} x^2(t)\,dt \tag{2.7}$$

几种典型周期信号各值之间的数量关系见表 2.2。可见，信号的均值、绝对均值、有效值和峰值之间的关系与波形有关。

表 2.2　几种典型周期信号各值之间的数量关系

名称	波形图	傅里叶级数展开式	x_p	μ_x	$\mu_{\|x\|}$	x_{rms}
正弦波	x(t)、A、O、$\frac{T_0}{2}$、T_0、t	$x(t)=A\sin\omega_0 t$ $T_0=\frac{2\pi}{\omega_0}$	A	0	$\frac{2A}{\pi}$	$\frac{A}{\sqrt{2}}$
方波	x(t)、A、O、$\frac{T_0}{2}$、T_0、t	$x(t)=\frac{4A}{\pi}\left(\sin\omega_0 t+\frac{1}{3}\sin 3\omega_0 t+\frac{1}{5}\sin 5\omega_0 t+\cdots\right)$	A	0	A	A
三角波	x(t)、A、O、$\frac{T_0}{2}$、T_0、t	$x(t)=\frac{8A}{\pi^2}\left(\sin\omega_0 t-\frac{1}{9}\sin 3\omega_0 t+\frac{1}{25}\sin 5\omega_0 t+\cdots\right)$	A	0	$\frac{A}{2}$	$\frac{A}{\sqrt{3}}$
锯齿波	x(t)、A、O、T_0、t	$x(t)=\frac{A}{2}-\frac{A}{\pi}\left(\sin\omega_0 t+\frac{\sin 2\omega_0 t}{2}+\frac{\sin 3\omega_0 t}{3}+\cdots\right)$	A	$\frac{A}{2}$	$\frac{A}{2}$	$\frac{A}{\sqrt{3}}$
正弦整流	x(t)、A、O、$\frac{T_0}{2}$、T_0、t	$x(t)=\frac{2A}{\pi}\left(1-\frac{2}{3}\cos 2\omega_0 t-\frac{2}{15}\cos 4\omega_0 t-\frac{2}{35}\cos 6\omega_0 t-\cdots\right)$	A	$\frac{2A}{\pi}$	$\frac{2A}{\pi}$	$\frac{A}{\sqrt{2}}$

信号的峰值 x_p、绝对均值 $\mu_{|x|}$ 和有效值 x_{rms} 可用三值电压表来测量。峰值 x_p 可根据波形折算或用能记忆瞬峰示值的仪表测量，也可以用示波器来测量。均值可用直流电压表测量。信号是周期交变的，如果交流频率较高，则交流成分只影响表针的微小晃动，不

The mean value of the periodic signal μ_x is

$$\mu_x = \frac{1}{T_0}\int_0^{T_0} x(t)\,\mathrm{d}t \tag{2.4}$$

It is the constant component of the signal.

The mean value of the periodic signal after full-wave rectification is the absolute mean value of the signal $\mu_{|x|}$, namely

$$\mu_{|x|} = \frac{1}{T_0}\int_0^{T_0} |x(t)|\,\mathrm{d}t \tag{2.5}$$

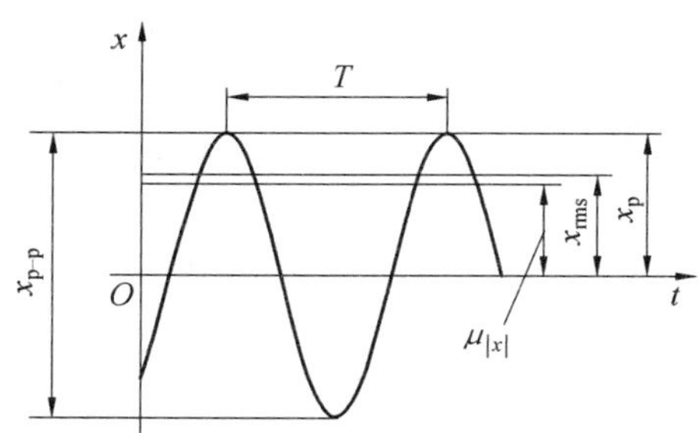

Figure 2. 8　Intensity representation of periodic signal

The effective value is the root mean square value of the signal x_{rms}, that is

$$x_{\mathrm{rms}} = \sqrt{\frac{1}{T_0}\int_0^{T_0} x^2(t)\,\mathrm{d}t} \tag{2.6}$$

The square of the effective value—the mean square value is the average power p_{av} of the signal, which reflects the power of the signal, that is

$$p_{\mathrm{av}} = \frac{1}{T_0}\int_0^{T_0} x^2(t)\,\mathrm{d}t \tag{2.7}$$

The quantitative relationships among the values of several typical periodic signals are listed in Table 2. 2. It can be seen from the table that the relationship between the mean value, absolute mean value, effective value and peak value of the signal is related to the waveform.

The peak value x_{p}, absolute mean value $\mu_{|x|}$ and effective value x_{rms} of the signal can be measured with a three-valued voltmeter. The peak value x_{p} can be converted according to the waveform or measured with a meter that can memorize the instantaneous peak value, or with an oscilloscope. The mean value can be measured with a DC voltmeter. Because of signals of periodic alternation, if the AC frequency is high, the AC component only affects the slight shaking of the hands without affecting the mean reading. When the frequency is low, the hands of the watch will oscillate to affect the reading. At this time, a capacitor can be connected in parallel with the AC voltmeter to bypass the AC component, but attention should be paid to the impact of this capacitor on the circuit under measurement.

It is worth pointing out that although the general AC voltmeter is scaled according to the effective value, its output (such as the deflection angle of the pointer) is not necessarily proportional to the effective value of the signal, while its output may be proportional to the effective value of the signal, or to the peak value or absolute mean value of the signal with the difference of the detection circuit of the voltmeter. The effective value scales on the voltmeters of different detection circuits are all based on a single harmonic signal, which ensures that various voltmeters can correctly measure the effective value of the signal to obtain consistent readings when measuring a single harmonic signal. However, the calibration process is actually equivalent to "solidifying" the relationship between the output of the detector circuit and the effective value of the harmonic signal in the voltmeter, which is not suitable for non-single harmonic signals due to that the relationship between the output of various detection circuits and the effective value of the signal has changed with the different waveforms, resulting in systematic errors when the voltmeter measures the effective value of complex signals. It's time

影响均值读数。当频率低时，表针将产生摆动，影响读数。这时，可用一个电容器与交流电压表并接将交流分量旁路，但应该注意这个电容器对被测电路的影响。

值得指出的是，虽然一般的交流电压表均按有效值刻度，但其输出量（如指针的偏转角）并不一定与信号的有效值成比例，而是随着电压表的检波电路的不同，其输出量可能与信号的有效值成正比例，也可能与信号的峰值或绝对均值成比例。不同检波电路的电压表上的有效值刻度都是依据单一简谐信号来刻度的，这就保证了用各种电压表在测量单一简谐信号时都能正确测得信号的有效值，获得一致的读数。然而，刻度过程实际上相当于把检波电路输出和简谐信号有效值的关系“固化”在电压表中，这种关系不适用于非单一简谐信号，因为随着波形的不同，各类检波电路输出和信号有效值的关系已经改变了，从而造成电压表在测量复杂信号有效值时的系统误差。这时，应根据检波电路和波形来修正有效值读数。

2.4 周期信号及其离散频谱

前文给出了信号时域描述和频率描述的方式。由于直接获取的信号一般是基于时域描述的，因此为了对信号进行分析和提取特征，从本节开始，将要分别研究如何将周期信号和非周期信号的时域描述转换为频域描述，即已知信号 $x(t)$ 的数学模型，怎样求出构成该信号的频率成分，其中包括各成分的幅值有多大、初相位是多少。

根据数学分析可知，凡满足狄利克雷（Dirichlet）条件的周期函数信号 $x(t)$ 都可以展开成傅里叶级数。工程实际中的周期函数一般均满足狄利克雷条件，所以可将它展开成收敛的傅里叶级数。傅里叶级数是描述周期信号的基本数学工具，通过它可以把任一周期信号展开成无穷多个正弦或余弦函数之和。把周期信号展开为傅里叶级数的主要目的是了解给定周期信号含有哪些频率分量，以及各分量幅值、相位的相对比例关系，这种关系就是信号的“频率特性”。其中，幅值与频率的关系称为幅频特性，相位与频率的关系称为相频特性。寻找信号频率特性的过程称为信号的频谱分析。傅里叶级数有两种表达形式：三角函数展开式和复指数函数展开式。

2.4.1 周期信号傅里叶级数的三角函数展开式

傅里叶级数的三角函数展开式为

$$x(t) = a_0 + \sum_{n=1}^{\infty} (a_n \cos n\omega_0 t + b_n \sin n\omega_0 t) \tag{2.8}$$

to correct the effective value reading according to the detection circuit and waveform.

Table 2.2　The quantitative relationships among the values of several typical periodic signals

Name	Oscillogram	Fourier series expansion	x_p	μ_x	$\mu_{\|x\|}$	x_{rms}
Sine wave	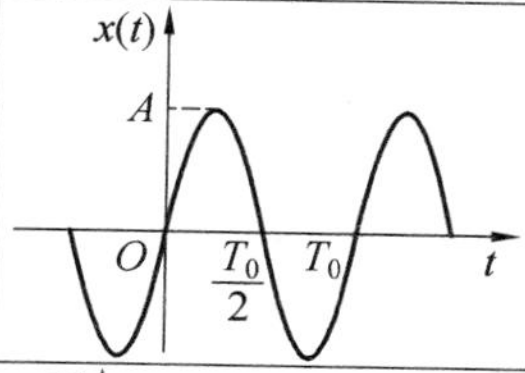	$x(t) = A\sin \omega_0 t$ $T_0 = \frac{2\pi}{\omega_0}$	A	0	$\frac{2A}{\pi}$	$\frac{A}{\sqrt{2}}$
Square wave	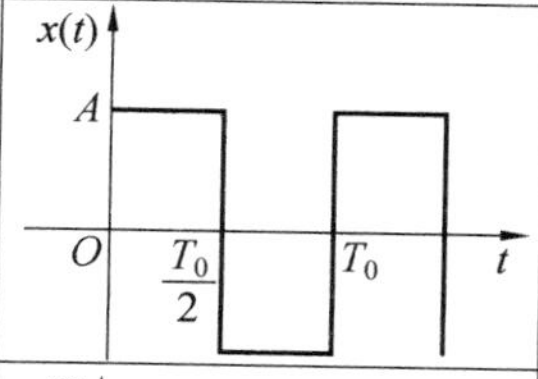	$x(t) = \frac{4A}{\pi}\left(\sin \omega_0 t + \frac{1}{3}\sin 3\omega_0 t + \frac{1}{5}\sin 5\omega_0 t + \cdots\right)$	A	0	A	A
Triangular wave	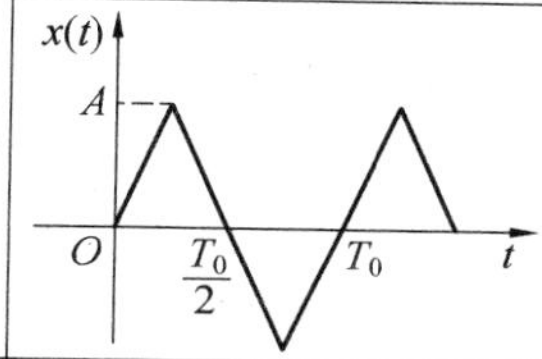	$x(t) = \frac{8A}{\pi^2}\left(\sin \omega_0 t - \frac{1}{9}\sin 3\omega_0 t + \frac{1}{25}\sin 5\omega_0 t + \cdots\right)$	A	0	$\frac{A}{2}$	$\frac{A}{\sqrt{3}}$
Sawtooth wave	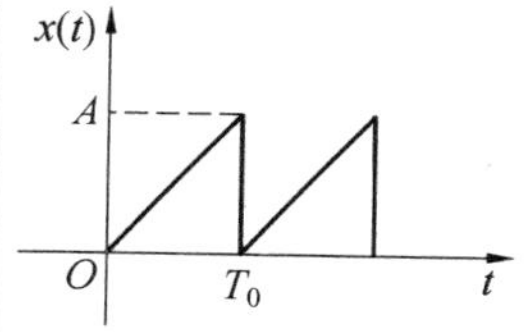	$x(t) = \frac{A}{2} - \frac{A}{\pi}\left(\sin \omega_0 t + \frac{\sin 2\omega_0 t}{2} + \frac{\sin 3\omega_0 t}{3} + \cdots\right)$	A	$\frac{A}{2}$	$\frac{A}{2}$	$\frac{A}{\sqrt{3}}$
Sinusoidal rectification	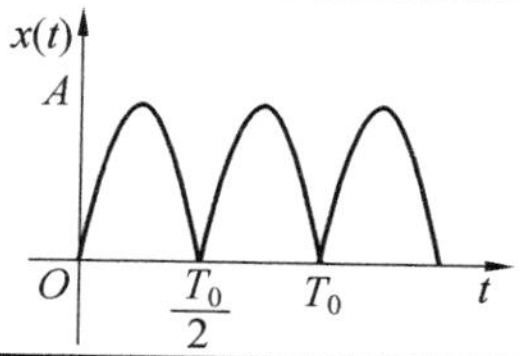	$x(t) = \frac{2A}{\pi}\left(1 - \frac{2}{3}\cos 2\omega_0 t - \frac{2}{15}\cos 4\omega_0 t - \frac{2}{35}\cos 6\omega_0 t - \cdots\right)$	A	$\frac{2A}{\pi}$	$\frac{2A}{\pi}$	$\frac{A}{\sqrt{2}}$

2.4　Periodic Signal and Its Discrete Spectrum

The signal time domain description and frequency description method have been described in the previous section. Due to the directly acquired signal generally based on the time domain description, starting from this section, we will separately study how to convert the time domain description of periodic signals and nonperiodic signals into frequency domain descriptions in order to analyze the signal and extract features, that is how to find the frequency components that make up the signal, including the amplitude of each component and the initial phase after knowing the mathematical model of the signal $x(t)$.

According to mathematical analysis, any periodic function signal $x(t)$ that satisfies the Dirichlet condition can be expanded into a Fourier series. The periodic function in engineering

其中，直流分量为

$$a_0 = \frac{1}{T_0}\int_{-\frac{T_0}{2}}^{\frac{T_0}{2}} x(t)\,\mathrm{d}t$$

余弦分量幅值为

$$a_n = \frac{2}{T_0}\int_{-\frac{T_0}{2}}^{\frac{T_0}{2}} x(t)\cos n\omega_0 t\mathrm{d}t,\quad n = 1, 2, 3, \cdots \tag{2.9}$$

正弦分量幅值为

$$b_n = \frac{2}{T_0}\int_{-\frac{T_0}{2}}^{\frac{T_0}{2}} x(t)\sin n\omega_0 t\mathrm{d}t$$

式中　T_0—— 基波周期；

ω_0—— 基波圆频率，$\omega_0 = \frac{2\pi}{T} = 2\pi f_0$，$f_0$ 为基波频率，简称基频；

$n\omega_0$—— 第 n 次谐波的圆频率。

根据以上计算公式可知，各系数的大小完全由信号 $x(t)$ 确定。利用三角函数的和差化积公式，周期信号傅里叶级数的三角函数展开式(2.8) 可以改写成

$$x(t) = a_0 + \sum_{n=1}^{\infty} A_n \sin(n\omega_0 t + \varphi_n) \tag{2.10}$$

式中

$$A_n = \sqrt{a_n^2 + b_n^2}$$

各频率分量的相位为

$$\varphi_n = \arctan\frac{a_n}{b_n}$$

式(2.10) 也可写成

$$x(t) = a_0 + \sum_{n=1}^{\infty} A_n \cos(n\omega_0 t + \varphi_n) \tag{2.11}$$

式中

$$A_n = \sqrt{a_n^2 + b_n^2}$$

$$\varphi_n = \arctan\frac{-b_n}{a_n}$$

将信号表示为不同频率正弦分量的线性组合，可以很方便地对不同信号进行比较。此外，可以将数学表示式较为复杂的周期信号分解开来，分别进行研究，再利用叠加原理获得综合分析结果。

practice generally satisfies the Dirichlet condition, thus it can be expanded into a convergent Fourier series, a basic mathematical tool for describing periodic signals, through which any periodic signals can be expanded into the sum of infinitely many sine or cosine functions. The main purpose of unfolding a periodic signal into a Fourier series is to understand what frequency components a given periodic signal contains, and the relative proportional relationship of the amplitude and phase of each component, which is the "frequency characteristic" of the signals. In it, the relationship between amplitude and frequency is called amplitude-frequency characteristics, and the relationship between phase and frequency is called phase-frequency characteristics. The process of finding the frequency characteristics of a signal is called signal spectrum analysis. A Fourier series has two expressions: trigonometric function expansion and complex exponential function expansion.

2.4.1 Trigonometric Function Expansion of the Fourier Series of the Periodic Signal

The trigonometric function expansion of Fourier series is

$$x(t) = a_0 + \sum_{n=1}^{\infty} (a_n \cos n\omega_0 t + b_n \sin n\omega_0 t) \tag{2.8}$$

where the DC component is

$$a_0 = \frac{1}{T_0} \int_{-\frac{T_0}{2}}^{\frac{T_0}{2}} x(t)\, dt$$

the cosine component amplitude is

$$a_n = \frac{2}{T_0} \int_{-\frac{T_0}{2}}^{\frac{T_0}{2}} x(t) \cos n\omega_0 t dt , \quad n = 1, 2, 3, \cdots \tag{2.9}$$

the sine component amplitude is

$$b_n = \frac{2}{T_0} \int_{-\frac{T_0}{2}}^{\frac{T_0}{2}} x(t) \sin n\omega_0 t dt , \quad n = 1, 2, 3, \cdots$$

where　T_0—fundamental wave period;

ω_0—fundamental circular frequency, $\omega_0 = \frac{2\pi}{T} = 2\pi f_0$, f_0 is the fundamental frequency, referred to as fundamental;

$n\omega_0$—circular frequency of the nth harmonic.

According to the above formula, the size of each coefficient is completely determined by the signal $x(t)$. Applying the sum-to-product formula of trigonometric functions, the trigonometric function expansion as formula (2.8) of the Fourier series of periodic signals can be rewritten as

$$x(t) = a_0 + \sum_{n=1}^{\infty} A_n \sin(n\omega_0 t + \varphi_n) \tag{2.10}$$

where

$$A_n = \sqrt{a_n^2 + b_n^2}$$

the phase of each frequency component $\varphi_n = \arctan \frac{a_n}{b_n}$.

根据式(2.8)，利用函数的奇偶性，可对周期信号的傅里叶三角函数展开式进一步简化。

(1) 如果周期信号 $x(t)$ 是奇函数，则傅里叶系数的直流分量 $a_0=0$，余弦分量幅值 $a_n=0$，傅里叶级数 $x(t)=\sum_{i=1}^{\infty} b_n \sin n\omega_0 t$。

(2) 如果周期信号 $x(t)$ 是偶函数，即 $x(t)=x(-t)$，此时傅里叶系数的正弦分量幅值 $b_n=0$，则傅里叶级数 $x(t)=a_0+\sum_{n=1}^{\infty} a_n \cos n\omega_0 t$。

例 2.1 求图 2.9 中周期性三角波的傅里叶级数。

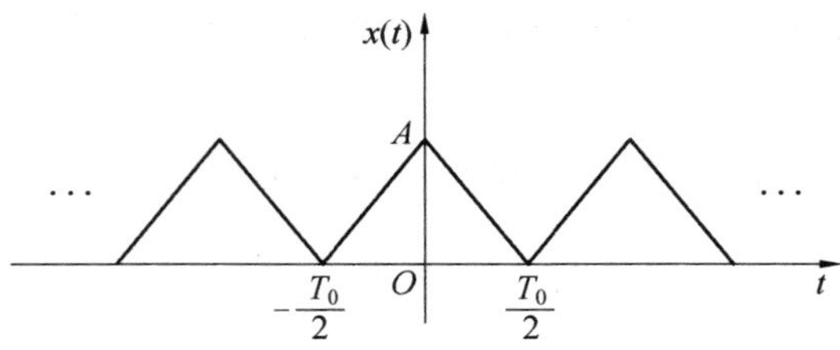

图 2.9 周期性三角波

解 (1) 该信号在一个周期内是连续的，满足狄利克雷条件。

(2) 写出该信号在一个周期内的数学表示式，即

$$x(t)=\begin{cases} A+\dfrac{2A}{T_0}t, & -\dfrac{T_0}{2}\leqslant t\leqslant 0 \\ A-\dfrac{2A}{T_0}t, & 0\leqslant t\leqslant \dfrac{T_0}{2} \end{cases}$$

(3) 计算直流分量，即

$$a_0=\frac{1}{T_0}\int_{-\frac{T_0}{2}}^{\frac{T_0}{2}} x(t)\,\mathrm{d}t=\frac{2}{T_0}\int_0^{\frac{T_0}{2}}\left(A-\frac{2A}{T_0}t\right)\mathrm{d}t=\frac{A}{2}$$

(4) 计算余弦分量幅值，即

$$a_n=\frac{2}{T_0}\int_{-\frac{T_0}{2}}^{\frac{T_0}{2}} x(t)\cos n\omega_0 t\,\mathrm{d}t=\frac{4}{T_0}\int_0^{\frac{T_0}{2}}\left(A-\frac{2A}{T_0}t\right)\cos n\omega_0 t\,\mathrm{d}t$$

$$=\frac{4A}{n^2\pi^2}\sin^2\frac{n\pi}{2}=\begin{cases} \dfrac{4A}{n^2\pi^2}, & n=1,3,5,\cdots \\ 0, & n=2,4,6,\cdots \end{cases}$$

(5) 计算正弦分量幅值，根据奇偶性原则可知为 0，即

$$b_n=\frac{2}{T_0}\int_{-\frac{T_0}{2}}^{\frac{T_0}{2}} x(t)\sin n\omega_0 t\,\mathrm{d}t=0$$

Formula (2.10) can be rewritten as

$$x(t) = a_0 + \sum_{n=1}^{\infty} A_n \cos(n\omega_0 t + \varphi_n) \tag{2.11}$$

where

$$A_n = \sqrt{a_n^2 + b_n^2}$$

$$\varphi_n = \arctan \frac{-b_n}{a_n}$$

The signal is expressed as a linear combination of sinusoidal components of different frequencies, easy to compare different signals. Besides, the periodic signals with more complex mathematical expressions can be decomposed and studied separately so as to obtain comprehensive analysis results by using superposition principle.

According to formula(2.8), the Fourier trigonometric function expansion of the periodic signal can be further simplified by using the parity of the function.

(1) If the periodic signal $x(t)$ is an odd function, the DC component of the Fourier coefficient is $a_0 = 0$, the amplitude of the cosine component is $a_n = 0$, and the Fourier series is $x(t) = \sum_{i=1}^{\infty} b_n \sin n\omega_0 t$.

(2) If the periodic signal $x(t)$ is an even function, that is $x(t) = x(-t)$, the amplitude of the sine component of the Fourier coefficient is $b_n = 0$ at this time, then the Fourier series is $x(t) = a_0 + \sum_{n=1}^{\infty} a_n \cos n\omega_0 t$.

Example 2.1 Find the Fourier series of the periodic triangular wave in Figure 2.9.

Solution (1) The signal is continuous within a period, and satisfies the Dirichlet condition.

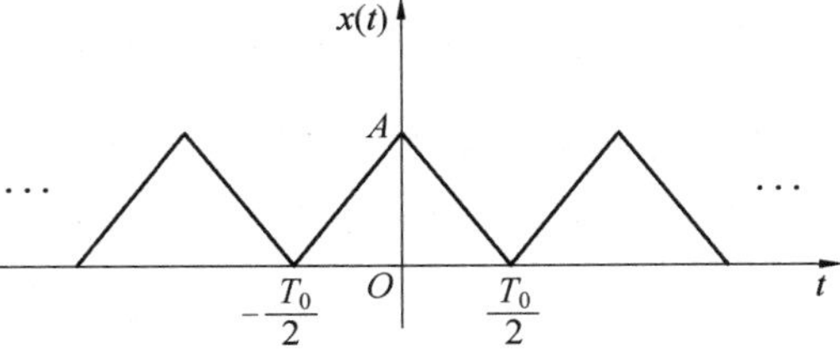

Figure 2.9　Periodic triangular wave

(2) Write the mathematical expression of the signal in one period, that is

$$x(t) = \begin{cases} A + \dfrac{2A}{T_0}t, & -\dfrac{T_0}{2} \leqslant t \leqslant 0 \\ A - \dfrac{2A}{T_0}t, & 0 \leqslant t \leqslant \dfrac{T_0}{2} \end{cases}$$

(3) Calculate the DC component, that is

$$a_0 = \frac{1}{T_0}\int_{-\frac{T_0}{2}}^{\frac{T_0}{2}} x(t)\,\mathrm{d}t = \frac{2}{T_0}\int_0^{\frac{T_0}{2}}\left(A - \frac{2A}{T_0}t\right)\mathrm{d}t = \frac{A}{2}$$

(4) Calculate the cosine component amplitude, that is

$$a_n = \frac{2}{T_0}\int_{-\frac{T_0}{2}}^{\frac{T_0}{2}} x(t)\cos n\omega_0 t\mathrm{d}t = \frac{4}{T_0}\int_0^{\frac{T_0}{2}}\left(A - \frac{2A}{T_0}t\right)\cos n\omega_0 t\mathrm{d}t$$

$$= \frac{4A}{n^2\pi^2}\sin^2\frac{n\pi}{2} = \begin{cases} \dfrac{4A}{n^2\pi^2}, & n = 1,3,5,\cdots \\ 0, & n = 2,4,6,\cdots \end{cases}$$

(6) 综上,该周期性三角波傅里叶级数的三角函数展开式为

$$\begin{aligned} x(t) &= \frac{A}{2} + \frac{4A}{\pi^2}\left(\cos \omega_0 t + \frac{1}{3^2}\cos 3\omega_0 t + \frac{1}{5^2}\cos 5\omega_0 t + \cdots\right) \\ &= \frac{A}{2} + \frac{4A}{\pi^2}\sum_{n=1}^{\infty} \frac{1}{n^2}\cos n\omega_0 t, \quad n = 1,3,5,\cdots \end{aligned}$$

(7) 绘制频谱图。

周期性三角波的频谱图如图 2.10 所示,其幅频谱只包含常值分量、基波和奇次谐波的频率分量,谐波的幅值以 $1/n^2$ 的规律收敛。在其相频谱中,基波和各次谐波的初相位 φ_n 均为 0。

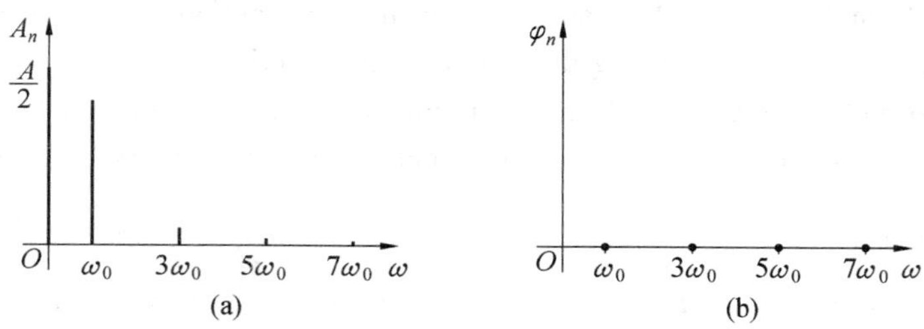

图 2.10　周期性三角波的频谱

2.4.2　周期信号傅里叶级数的复指数函数展开式

傅里叶级数按三角函数形式展开后进行各系数积分运算时涉及三角函数运算。因此,为方便于表达和运算,傅里叶级数常写成复指数函数形式。

根据复指数函数的定义,复变量 $z = x + \mathrm{j}y$ 的指数函数记作 e^z,通常写为

$$\mathrm{e}^z = \mathrm{e}^x(\cos y + \mathrm{j}\sin y)$$

这里,e^z 并无幂的意义,只是一种复数表示方式而已。

当 $y = 0$ 时,下式成立,其中 e^x 是 e^z 的模,即

$$\mathrm{e}^z = \mathrm{e}^x$$

当 $x = 0$ 时,下式成立,y 是 e^z 的辐角,此式又称欧拉公式,即

$$\mathrm{e}^z = \cos y + \mathrm{j}\sin y$$

根据欧拉公式,当 $x = 0, y = \omega t$ 时,有

$$\mathrm{e}^{\pm \mathrm{j}\omega t} = \cos \omega t \pm \mathrm{j}\sin \omega t \tag{2.12}$$

进一步推导可得

$$\cos \omega t = \frac{1}{2}(\mathrm{e}^{-\mathrm{j}\omega t} + \mathrm{e}^{\mathrm{j}\omega t}) \tag{2.13}$$

(5) Calculate the sine component amplitude, which can be known as 0 according to the parity principle, namely

$$b_n = \frac{2}{T_0}\int_{-\frac{T_0}{2}}^{\frac{T_0}{2}} x(t)\sin n\omega_0 t\mathrm{d}t = 0$$

(6) In summary, the trigonometric function expansion of the periodical triangular wave Fourier series is

$$\begin{aligned} x(t) &= \frac{A}{2} + \frac{4A}{\pi^2}\left(\cos \omega_0 t + \frac{1}{3^2}\cos 3\omega_0 t + \frac{1}{5^2}\cos 5\omega_0 t + \cdots\right) \\ &= \frac{A}{2} + \frac{4A}{\pi^2}\sum_{n=1}^{\infty}\frac{1}{n^2}\cos n\omega_0 t, \quad n = 1,3,5,\cdots \end{aligned}$$

(7) Draw a spectrum diagram.

The frequency spectrum of a periodic triangular wave is shown in Figure 2. 10, its amplitude spectrum only contains the constant value components, frequency components of fundamental waves and odd harmonics. The amplitude of the harmonics converges with the law of $1/n^2$. In its phase spectrum, the initial phase φ_n of the fundamental wave and each harmonic is 0.

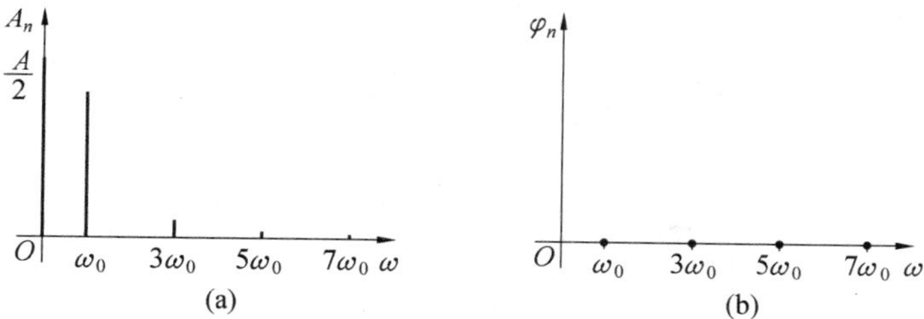

Figure 2. 10 The frequency spectrum of a periodic triangular wave

2.4.2 The Complex Exponential Function Expansion of the Fourier Series of the Periodic Signal

The Fourier series is expanded in the form of trigonometric function to perform the integral operation of each coefficient, involving the operation of trigonometric function. Therefore, the Fourier series is often written in the form of a complex exponential function in order to facilitate the expression and operation.

According to the definition of the complex exponential function, the exponential function of the complex variable $z = x + \mathrm{j}y$ denoted as e^z, often written as

$$\mathrm{e}^z = \mathrm{e}^x(\cos y + \mathrm{j}\sin y)$$

where e^z has no meaning of power, it is just a representation of a complex number.

When $y=0$, the following formula is established, where e^x is the modulus of e^z, that is

$$\mathrm{e}^z = \mathrm{e}^x$$

When $x=0$, the following formula is established, and y is the argument of e^z, this formula is also called Euler's formula, that is

$$\mathrm{e}^z = \cos y + \mathrm{j}\sin y$$

According to Euler's formula, when $x=0$ and $y = \omega t$, we get

$$\sin \omega t = \mathrm{j}\frac{1}{2}(\mathrm{e}^{-\mathrm{j}\omega t} - \mathrm{e}^{\mathrm{j}\omega t}) \tag{2.14}$$

根据式(2.13)，则有

$$\begin{aligned} A_n\cos(n\omega_0 t + \varphi_n) &= \frac{A_n}{2}[\mathrm{e}^{-\mathrm{j}(n\omega_0 t+\varphi_n)} + \mathrm{e}^{\mathrm{j}(n\omega_0 t+\varphi_n)}] \\ &= \frac{A_n}{2}\mathrm{e}^{-\mathrm{j}\varphi_n}\mathrm{e}^{-\mathrm{j}n\omega_0 t} + \frac{A_n}{2}\mathrm{e}^{\mathrm{j}\varphi_n}\mathrm{e}^{\mathrm{j}n\omega_0 t} \\ &= c_{-n}\mathrm{e}^{-\mathrm{j}n\omega_0 t} + c_n\mathrm{e}^{\mathrm{j}n\omega_0 t} \end{aligned} \tag{2.15}$$

式中

$$c_n = \frac{A_n}{2}\mathrm{e}^{\mathrm{j}\varphi_n} = \frac{1}{2}(a_n - \mathrm{j}b_n)$$

$$c_{-n} = \frac{A_n}{2}\mathrm{e}^{-\mathrm{j}\varphi_n} = \frac{1}{2}(a_n + \mathrm{j}b_n)$$

c_n、c_{-n} 称为余弦函数的复半幅，二者是互为共轭、模为 $A_n/2$、辐角分别为 φ_n 和 $-\varphi_n$ 的复数。

此时，傅里叶级数的三角函数展开式(2.8)可写为

$$x(t) = a_0 + \sum_{n=1}^{+\infty} c_{-n}\mathrm{e}^{-\mathrm{j}n\omega_0 t} + \sum_{n=1}^{+\infty} c_n\mathrm{e}^{\mathrm{j}n\omega_0 t}, \quad n = 1,2,\cdots \tag{2.16}$$

考虑到 $n = 0$ 时，$c_0 = \frac{A_0}{2} = a_0$，故式(2.16)又可进一步整理为

$$x(t) = \sum_{n=-\infty}^{+\infty} c_n\mathrm{e}^{\mathrm{j}n\omega_0 t}, \quad n = 0,\ \pm 1,\ \pm 2,\cdots \tag{2.17}$$

$$c_n = \frac{1}{T_0}\int_{-\frac{T_0}{2}}^{\frac{T_0}{2}} x(t)\mathrm{e}^{-\mathrm{j}n\omega_0 t}\mathrm{d}t \tag{2.18}$$

根据式(2.18)可知，系数 c_n 是一个复指数函数，完全由 $x(t)$ 决定，可表示为

$$c_n = c_{n\mathrm{R}} + \mathrm{j}c_{n\mathrm{I}} = |c_n|\mathrm{e}^{\mathrm{j}\varphi_n} \tag{2.19}$$

式中　$c_{n\mathrm{R}}$——复数 c_n 在实轴上的投影，称为复数 c_n 的实部；

$c_{n\mathrm{I}}$——复数 c_n 在虚轴上的投影，称为复数 c_n 的虚部。

c_n 的模和幅角分别是周期函数的幅频函数和相频函数，$|c_n|$—$n\omega_0$ 和 φ_n—$n\omega_0$ 的图形就是周期函数的幅频谱和相频谱，有

$$|c_n| = \sqrt{c_{n\mathrm{R}}^2 + c_{n\mathrm{I}}^2} = \frac{1}{2}\sqrt{a_n^2 + b_n^2} = \frac{1}{2}A_n \tag{2.20}$$

$$\varphi_n = \arctan\frac{c_{n\mathrm{I}}}{c_{n\mathrm{R}}} = -\arctan\frac{b_n}{a_n} \tag{2.21}$$

$$e^{\pm j\omega t} = \cos \omega t \pm j\sin \omega t \tag{2.12}$$

Further derivation can be obtained that

$$\cos \omega t = \frac{1}{2}(e^{-j\omega t} + e^{j\omega t}) \tag{2.13}$$

$$\sin \omega t = j\frac{1}{2}(e^{-j\omega t} - e^{j\omega t}) \tag{2.14}$$

According to formula (2.13), we have

$$\begin{aligned} A_n\cos(n\omega_0 t + \varphi_n) &= \frac{A_n}{2}[e^{-j(n\omega_0 t+\varphi_n)} + e^{j(n\omega_0 t+\varphi_n)}] \\ &= \frac{A_n}{2}e^{-j\varphi_n}e^{-jn\omega_0 t} + \frac{A_n}{2}e^{j\varphi_n}e^{jn\omega_0 t} \\ &= c_{-n}e^{-jn\omega_0 t} + c_n e^{jn\omega_0 t} \end{aligned} \tag{2.15}$$

where

$$c_n = \frac{A_n}{2}e^{j\varphi_n} = \frac{1}{2}(a_n - jb_n)$$

$$c_{-n} = \frac{A_n}{2}e^{-j\varphi_n} = \frac{1}{2}(a_n + jb_n)$$

c_n, c_{-n} is called the complex half-rang of the cosine function, the two are conjugated with each other, the modulus is $A_n/2$, and the argument is φ_n and $-\varphi_n$ respectively.

At this time, the trigonometric function expansion as formala (2.8) of the Fourier series can be written as

$$x(t) = a_0 + \sum_{n=1}^{+\infty} c_{-n}e^{-jn\omega_0 t} + \sum_{n=1}^{+\infty} c_n e^{jn\omega_0 t}, \quad n = 1,2,\cdots \tag{2.16}$$

Considering when $n = 0$, $c_0 = \frac{A_0}{2} = a_0$, so formula (2.16) can be further organized as

$$x(t) = \sum_{n=-\infty}^{+\infty} c_n e^{jn\omega_0 t}, \quad n = 0, \pm 1, \pm 2, \cdots \tag{2.17}$$

$$c_n = \frac{1}{T_0}\int_{-\frac{T_0}{2}}^{\frac{T_0}{2}} x(t)e^{-jn\omega_0 t}dt \tag{2.18}$$

According to formula (2.18), it can be seen that the coefficient c_n is a complex exponential function, completely determined by $x(t)$, and can be expressed as

$$c_n = c_{nR} + jc_{nI} = |c_n|e^{j\varphi_n} \tag{2.19}$$

where c_{nR} — the projection of the complex number c_n on the real axis, which is called the real part of the complex number c_n;

c_{nI} —the projection of the complex number c_n on the imaginary axis, which is called the imaginary part of the complex number c_n.

The modulus and argument of c_n are respectively the amplitude-frequency function and phase-frequency function of the periodic function, the graphs of $|c_n|$—$n\omega_0$ and φ_n—$n\omega_0$ are the amplitude frequency spectrum and phase frequency spectrum of the periodic function, we have

$$|c_n| = \sqrt{c_{nR}^2 + c_{nI}^2} = \frac{1}{2}\sqrt{a_n^2 + b_n^2} = \frac{1}{2}A_n \tag{2.20}$$

以上频谱函数是复指数形式的,其频谱呈“双边”形式。$|c_n|$是 ω 的偶函数,幅频谱偶对称。φ_n 是 ω 的奇函数,相频谱奇对称。

例 2.2　画出余弦、正弦函数的实、虚部频谱图。

解　余弦、正弦函数的复指数式为

$$\cos \omega_0 t = \frac{1}{2}(e^{-j\omega_0 t} + e^{j\omega_0 t})$$

$$\sin \omega_0 t = j\frac{1}{2}(e^{-j\omega_0 t} - e^{j\omega_0 t})$$

由上式可知,余弦函数只有实频谱图,与纵轴偶对称;正弦函数只有虚频谱图,与纵轴奇对称。图 2.11 所示为正、余弦函数的幅频谱图。方便起见,周期函数按傅里叶级数的复指数函数形式展开时,一般总是约定统一将信号的频率成分视为余弦谐波。这样,所绘制的幅频谱都是实频谱,其实频谱总是偶对称的。

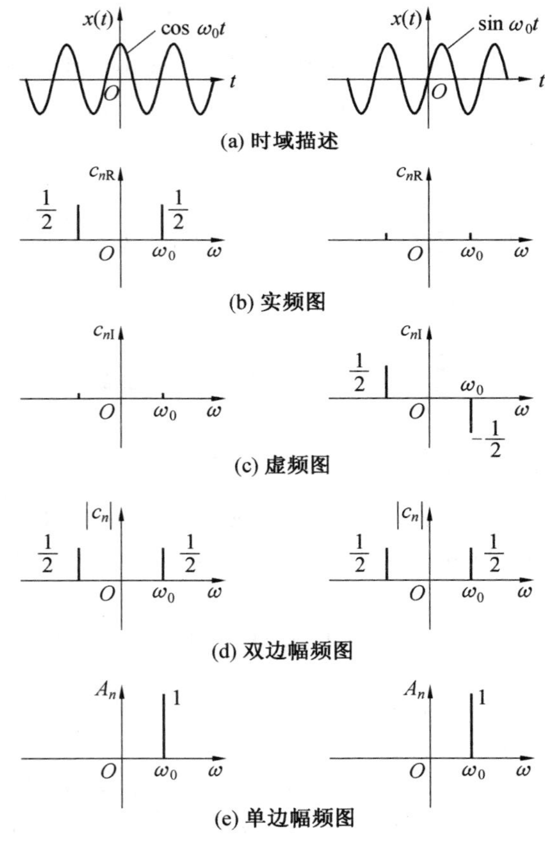

图 2.11　正、余弦函数的幅频谱图

$$\varphi_n = \arctan \frac{c_{nI}}{c_{nR}} = -\arctan \frac{b_n}{a_n} \tag{2.21}$$

The above spectrum function is in the form of a complex exponential, and its spectrum is in a "two-sided" form. $|c_n|$ is an even function of ω , and the amplitude frequency spectrum proves even symmetry. φ_n is an odd function of ω , and the phase spectrum proves odd symmetry.

Example 2.2　Draw the real and imaginary spectrograms of the cosine and sine functions.

Solution　The complex exponential formulas of cosine and sine functions are

$$\cos \omega_0 t = \frac{1}{2}(e^{-j\omega_0 t} + e^{j\omega_0 t})$$

$$\sin \omega_0 t = j\frac{1}{2}(e^{-j\omega_0 t} - e^{j\omega_0 t})$$

It can be seen from the above formula that the cosine function has only a real spectrogram, which is even symmetric with the longitudinal axis; while, the sine function has only the imaginary spectrogram, oddly symmetric with the longitudinal axis. Figure 2.11 shows the amplitude

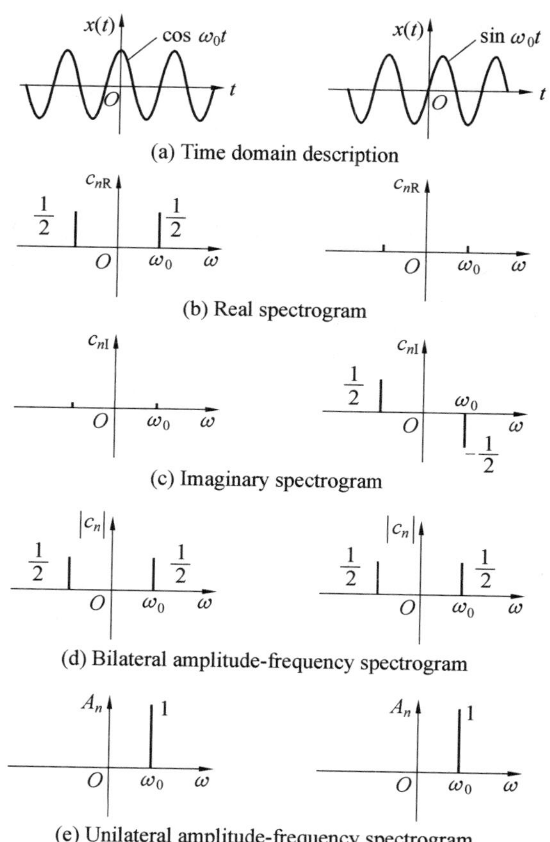

Figure 2.11　The amplitude spectrum of the sine and cosine functions

值得一提的是,负频率的出现是因为引入了复数描述方式。这里,频率的正负是基于共轭复数旋向相反而定义的数学描述方式,并不是实数空间中的正负。

周期信号的频谱具有以下三个特点。

(1) 周期信号的频谱是离散的,称为离散频谱。

(2) 每条谱线只出现在基波频率的整倍数上,基波频率是诸分量频率的公约数。

(3) 各频率分量的谱线高度表示该谐波的幅值或初始相位角。

工程中常见的周期信号,其谐波幅值总的趋势是随谐波次数的增高而减小。由此可见,信号的能量主要集中在低频分量,所以谐波次数过高的那些分量(高频分量)所占能量很少,在频谱分析中没有必要取那些次数过高的谐波分量。

2.4.3 三角函数展开式与复指数函数展开式的关系

三角函数形式的傅里叶级数和复指数函数形式的傅里叶级数并不是两种不同类型的级数,而只是同一级数的两种不同的表示方法。三角函数形式的傅里叶级数物理含义明确;复指数函数形式的傅里叶级数数学处理方便、简单,而且很容易与后面介绍的傅里叶变换统一起来。在实际应用中,特别是公式推导中,常将周期信号展开成复指数函数形式。傅里叶级数的三角函数展开与复指数展开的关系见表 2.3,周期信号展开为傅里叶级数有关公式汇总见表 2.4。

表 2.3 傅里叶级数的三角函数展开与复指数展开的关系

三角函数展开式	傅里叶系数	复指数展开式	傅里叶系数
常值分量	a_0	复指数常量	$c_0 = a_0$
余弦分量幅值	a_n	复数 c_n 的实部	$c_{nR} = \frac{a_n}{2}$
正弦分量幅值	b_n	复数 c_n 的虚部	$c_{nI} = -\frac{b_n}{2}$
幅值	A_n	复数 c_n 的模	$\lvert c_n \rvert = \frac{A_n}{2}$
相位角	$\theta_n = \arctan \frac{a_n}{b_n}$	辐角(相位角)	$\varphi_n = \arctan\left(-\frac{b_n}{a_n}\right)$

spectrum of these two functions. For the sake of convenience, when the periodic function is expanded in the form of the complex exponential function of the Fourier series, the frequency components of the signal are generally agreed to be uniformly regarded as cosine harmonics, so that the drawn amplitude spectrum is the real spectrum. In fact, the frequency spectrum is always even symmetry.

There are some details worth mentioning that the emergence of negative frequency is due to the introduction of a complex number description, where the positive and negative of the frequency is a mathematical description defined based on the opposite direction of the conjugate complex number, not the positive and negative in the real number space.

The frequency spectrum of a periodic signal has following three characteristics.

(1) The frequency spectrum is discrete, which is called the discrete spectrum.

(2) Each spectral line only appears on an integral multiple of the fundamental frequency, which is the common divisor of the component frequencies.

(3) The height of the spectral line of each frequency component represents the amplitude or initial phase angle of the harmonic.

For the common periodic signal seen in engineering, the general trend of its harmonic amplitude is to decrease with the increase of the harmonic order. Therefore, the energy of the signal is mainly concentrated in the low-frequency components, so those components with too high harmonic orders (high-frequency components) occupy little energy, not necessary to take those harmonic components with too high orders in the spectrum analysis.

2.4.3 The Relationship Between the Trigonometric Function Expansion and the Complex Exponential Function Expansion

The Fourier series in the form of trigonometric functions and in the form of complex exponential functions are not two different types of series, but only two different representations of the same series. The physical meaning of the Fourier series in the form of trigonometric functions is clear; while the mathematical processing of the Fourier series in the form of complex exponential functions is convenient and simple, easy to unify with the Fourier transform described later. In practical applications, especially in formula derivation, the periodic signal is often expanded into a complex exponential function. The relationship between the trigonometric function expansion and the complex exponential expansion of the Fourier series is shown in Table 2.3, and the summary of relevant formulas for the expansion of periodic signals into Fourier series is shown in Table 2.4.

Table 2.3 The relationship between the trigonometric function expansion and the complex exponential expansion of the Fourier series

Trigonometric function expansion	Fourier series	Complex exponential expansion	Fourier series expansion
Constant component	a_0	Complex exponential constant	$c_0 = a_0$

表 2.4　周期信号展开为傅里叶级数有关公式汇总

傅里叶级数形式	展开式	傅里叶系数	傅里叶系数之间关系
三角形式	$x(t)=a_0+\sum_{n=1}^{\infty}(a_n\cos n\omega_0 t+b_n\sin \omega_0 t)$ $=A_0+\sum_{n=1}^{\infty}A_n\sin(n\omega_0 t+\theta_n)$	$a_0=\frac{1}{T}\int_{-\frac{T}{2}}^{\frac{T}{2}}x(t)\mathrm{d}t$ $a_n=\frac{2}{T}\int_{-\frac{T}{2}}^{\frac{T}{2}}x(t)\cos n\omega_0 t\mathrm{d}t$ $b_n=\frac{2}{T}\int_{-\frac{T}{2}}^{\frac{T}{2}}x(t)\sin n\omega_0 t\mathrm{d}t$	$A_0=a_0$ $A_n=\sqrt{a_n^2+b_n^2}$ $\theta_n=\arctan\frac{a_n}{b_n}$
复指数形式	$x(t)=\sum_{n=-\infty}^{\infty}c_n\mathrm{e}^{\mathrm{j}n\omega_0 t}$	$c_n=\frac{1}{T}\int_{-\frac{T}{2}}^{\frac{T}{2}}x(t)\mathrm{e}^{-\mathrm{j}n\omega_0 t}\mathrm{d}t$	$c_0=a_0$ $c_{n\mathrm{R}}=a_n/2$ $c_{n\mathrm{I}}=-b_n/2$ $\varphi_n=\arctan\left(-\frac{b_n}{a_n}\right)$

通过傅里叶级数及其表达图形——频谱图,可以一目了然地知道周期信号是由哪些频率成分构成的、各频率成分的幅值和相位角是多大、各次谐波的幅值在周期信号中所占的比例等,这些统称为周期信号的频域描述。

任何一个周期信号 $x(t)$ 只要满足狄利克雷条件,就可以展开成三角函数形式或复指数函数形式的傅里叶级数,表明周期信号 $x(t)$ 必定是由有限个或无穷多个谐波分量叠加而成的,这一结论对于工程测试非常重要。

对周期信号进行频域描述不仅是为了研究信号本身的特性,其另一个重要用途是正确选择测试系统(装置)。因为在测试系统中,无论是转换电路还是放大电路,对可通过系统的信号的频率都有一定的选择性。而对于被测信号来说,任一谐波分量的丢失都会使原信号失真。例如,一个周期方波的基波频率为 1 000 Hz,为保证测试精度,所取的谐波分量应不小于 9 次,因此该方波信号所选用的测试系统的电路必须有通过 1 000 ~ 9 000 Hz 谐波信号的能力。又如,若某个周期信号的$a_0\neq 0$,即存在直流分量,则测试系统就必须能通过直流信号。

2.5　瞬变非周期信号及其连续频谱

本节所说的瞬变非周期信号仅指在时域内有收敛性的非周期信号(不包括无收敛性

Continued Table 2.3

Trigonometric function expansion	Fourier series	Complex exponential expansion	Fourier series expansion
Cosine component amplitude	a_n	The real part of the complex number c_n	$c_{nR} = \frac{a_n}{2}$
The sine component amplitude	b_n	The imaginary part of the complex number c_n	$c_{nI} = -\frac{b_n}{2}$
Amplitude	A_n	The modulus of the complex number c_n	$\|c_n\| = \frac{A_n}{2}$
Phase angle	$\theta_n = \arctan \frac{a_n}{b_n}$	Argument (phase angle)	$\varphi_n = \arctan\left(-\frac{a_n}{b_n}\right)$

Table 2.4　Summary of relevant formulas for the expansion of periodic signals into Fourier series

Fourier series form	Expansion	Fourier series	The relationship between Fourier coefficients
Triangular form	$x(t) = a_0 + \sum_{n=1}^{\infty}(a_n\cos n\omega_0 t + b_n \sin \omega_0 t)$ $= A_0 + \sum_{n=1}^{\infty} A_n \sin(n\omega_0 t + \theta_n)$	$a_0 = \frac{1}{T}\int_{-\frac{T}{2}}^{\frac{T}{2}} x(t)\,\mathrm{d}t$ $a_n = \frac{2}{T}\int_{-\frac{T}{2}}^{\frac{T}{2}} x(t)\cos n\omega_0 t\mathrm{d}t$ $b_n = \frac{2}{T}\int_{-\frac{T}{2}}^{\frac{T}{2}} x(t)\sin n\omega_0 t\mathrm{d}t$	$A_0 = a_0$ $A_n = \sqrt{a_n^2 + b_n^2}$ $\theta_n = \arctan \frac{a_n}{b_n}$
Complex exponential form	$x(t) = \sum_{n=-\infty}^{\infty} c_n \mathrm{e}^{\mathrm{j}n\omega_0 t}$	$c_n = \frac{1}{T}\int_{-\frac{T}{2}}^{\frac{T}{2}} x(t)\mathrm{e}^{-\mathrm{j}n\omega_0 t}\mathrm{d}t$	$c_0 = a_0$ $c_{nR} = a_n/2$ $c_{nI} = -b_n/2$ $\varphi_n = \arctan\left(-\frac{b_n}{a_n}\right)$

Through the Fourier series and its expression graph—spectrogram, it is possible to know at a glance what frequency components the periodic signal is composed of, how large the amplitude and phase angle of each frequency component are, and the proportion of the amplitude of each harmonic in the periodic signal etc., which are collectively called the frequency domain description of the periodic signal.

As long as any periodic signal $x(t)$ satisfies the Dirichlet condition, it can be expanded into a Fourier series in the form of a trigonometric function or a complex exponential function, indicating that the periodic signal $x(t)$ must be superimposed by a finite or infinite number of harmonic components, whose conclusions are very crucial for engineering measurement.

的非周期信号和准周期信号）。其特点是，当 $|t|\to\infty$ 时，信号 $x(t)$ 或趋于零，或趋于它的均值，此处用“瞬变”二字来形容它的特点。本节主要介绍当已知这类信号的数学模型之后，如何求它们的频谱（密度）函数。

前面通过傅里叶级数建立了周期信号时域与频域的对应关系，并得到了周期信号的离散频谱。本节再从傅里叶级数引出傅里叶变换，详细讨论非周期信号的分解问题，并建立非周期信号时域与频域的各种对应关系及连续频谱的概念。

2.5.1 傅里叶变换

如果信号 $x(t)$ 在 $|t|\to\infty$ 时收敛于均值 μ_x，则可以将该信号偏置 $-\mu_x$，这样它就与一般瞬态信号（图2.5(a)）一样，经过足够长的时间 T_0 以后信号消失。因此，对于具有收敛性的非周期信号，就可以每隔 T_0 时间或更长时间激发一次，使其成为周期信号，这就是所谓对瞬态非周期信号做周期化处理。只有能做周期化处理的信号原则上才可以用下式来计算它的频谱函数，即

$$x(t)=\sum_{n=-\infty}^{+\infty}c_n e^{jn\omega_0 t},\quad n=0,\ \pm1,\ \pm2,\cdots$$

$$c_n=\frac{1}{T_0}\int_{\frac{T_0}{2}}^{\frac{T_0}{2}}x(t)e^{-jn\omega_0 t}dt$$

对上式，按以下步骤处理。

（1）周期$T_0\to\infty$。这样的处理方法建立了周期信号和非周期信号之间的联系。

（2）基频$\omega_0=(2\pi/T_0)\to0$。由于谱线间距均为ω_0的整数倍，因此谱线最小间距ω_0演化为 $d\omega$，$n\omega_0$ 进而演化为连续变量 ω。离散频率谱线趋向于致密，频率由离散变量演化成为连续变量。

（3）$1/T_0=\omega_0/2\pi$ 演化为 $d\omega/2\pi$。

（4）求和运算 $\sum$ 演化为积分运算 $\int$，积分限从时间轴的局部 $(-T_0/2,T_0/2)$ 扩展到时间轴的全部 $(-\infty,\infty)$。谐波分量的频率范围从 $-\infty$ 到 ∞，占据整个频率域。

则傅里叶级数演化为

$$x(t)=\int_{-\infty}^{\infty}\frac{c_n}{d\omega}e^{j\omega t}d\omega \tag{2.22}$$

$$\frac{c_n}{d\omega}=\frac{1}{2\pi}\int_{-\infty}^{\infty}x(t)\ e^{-j\omega t}dt \tag{2.23}$$

此时，式(2.23)右侧称为函数 $x(t)$ 的傅里叶变换，左侧$\frac{c_n}{d\omega}$的物理意义是c_n 在

The frequency domain description of periodic signals is not only for studying the characteristics of the signal itself, but also for the correct selection ofmeasurement systems (devices). Because in the measurement systems, whether it is a conversion circuit or an amplifying circuit, there is a certain degree of selectivity for the frequency of the signal that can pass through the system. For the measured signals, the loss of any harmonic component will distort the original signal. For example, the fundamental frequency of a periodic square wave is 1 000 Hz, the taken harmonic component should not be less than 9th in order to ensure the accuracy of the measurement, thus the circuit of the measurement systems selected for the square wave signal must have the ability of passing 1 000–9 000 Hz harmonics. For another example, if a certain periodic signal $a_0 \neq 0$ has a DC component, the measurement systems must be able to pass the DC signal.

2.5 Transient Nonperiodic Signal and Its Continuous Spectrum

The transient nonperiodic signals mentioned in this section only refer to nonperiodic signals that have convergence in the time domain (not including non-convergence nonperiodic signals and quasi-periodic signals). The characteristic is that when $|t| \to \infty$, signal $x(t)$ may tend to zero, or tend to its mean value. The word "transient" is used here to describe its characteristics. This section mainly introduces how to find their frequency spectrum (density) function after the mathematical model of such signals is known.

The corresponding relationship between the time domain and the frequency domain of the periodic signal is established through the Fourier series, and the discrete frequency spectrum ofthe periodic signal is obtained. In this section, the Fourier transform is derived from the Fourier series, emphasis is placed on the decomposition of nonperiodic signals in detail, and various correspondences between the time domain and frequency domain of nonperiodic signals and the concept of continuous frequency spectrum are established.

2.5.1 Fourier Transform

If the signal $x(t)$ is converging to mean μ_x at time $|t| \to \infty$, then the signal can be offset by $-\mu_x$, so that it is the same as a general transient signal (see Figure 2.5(a)), and the signal disappears after a long enough time T_0. Therefore, the nonperiodic signal with convergence can be excited every T_0 time or longer to make it a periodic signal, which is the so-called periodic processing of transient nonperiodic signals. Only the signal processed periodically can be used to the following formula to calculate its spectral function, that is

$$x(t) = \sum_{n=-\infty}^{+\infty} c_n e^{jn\omega_0 t}, \quad n = 0, \pm 1, \pm 2, \cdots$$

$$c_n = \frac{1}{T_0} \int_{-\frac{T_0}{2}}^{\frac{T_0}{2}} x(t) e^{-jn\omega_0 t} dt$$

For the above formula, proceed as follows.

(1) Period $T_0 \to \infty$. This processing method can establish the connection between

$[\omega,\omega+\mathrm{d}\omega]$ 这一微小区间上的平均值，具有密度的含义，记为 $X(\omega)$，称为频谱密度函数，则以上两个公式可写成

$$x(t)=\int_{-\infty}^{\infty}X(\omega)\,\mathrm{e}^{\mathrm{j}\omega t}\mathrm{d}\omega \tag{2.24}$$

$$X(\omega)=\frac{1}{2\pi}\int_{-\infty}^{\infty}x(t)\,\mathrm{e}^{-\mathrm{j}\omega t}\mathrm{d}t \tag{2.25}$$

式(2.24)称为傅里叶逆变换(IFT)，实现信号由频域向时域的转换。式(2.25)称为傅里叶正变换(FT)，实现信号由时域向频域的转换。二者是一对互逆的运算。信号 $x(t)$ 和频谱密度函数 $X(\omega)$ 称为一个傅里叶变换对，记作

$$x(t)\underset{\mathrm{IFT}}{\overset{\mathrm{FT}}{\rightleftarrows}}X(\omega)$$

由于 $\omega=2\pi f$，因此式(2.24)和式(2.25)可变为

$$X(f)=\int_{-\infty}^{\infty}x(t)\,\mathrm{e}^{-\mathrm{j}2\pi ft}\mathrm{d}t \tag{2.26}$$

$$x(t)=\int_{-\infty}^{\infty}X(f)\,\mathrm{e}^{\mathrm{j}2\pi ft}\mathrm{d}f \tag{2.27}$$

这就避免了在傅里叶变换中出现常数因子 $1/2\pi$。

需要指出的是，以上是从形式上进行了推导，即从周期信号的周期 $T\to\infty$、“离散频谱”→“连续频谱”推导出傅里叶变换对，这在数学上是不严格的。严格来讲，非周期信号 $x(t)$ 的傅里叶变换存在的必要条件如下。

(1)$x(t)$ 在 $(-\infty,\infty)$ 范围内满足狄利克雷条件。

(2)$x(t)$ 的积分 $\int_{-\infty}^{\infty}|x(t)|\mathrm{d}t$ 收敛，即绝对可积。

(3)$x(t)$ 为能量有限信号。

由此可见，并不是所有信号都能进行傅里叶变换，但是这并不能说明不能进行傅里叶变换的信号就不能转化至频域。而且，借助奇异函数，一些不满足绝对可积条件的信号有时也可以进行傅里叶变换，如正弦函数、余弦函数、阶跃函数、符号函数等。

例 2.3　求单边指数脉冲的频谱：

$$x(t)=\begin{cases}E\mathrm{e}^{-at}, & a>0,t\geqslant 0\\ 0, & t<0\end{cases}$$

其时域波形如图 2.12(a) 所示。

解　该非周期信号的频谱函数为

$$X(\omega)=\frac{1}{2\pi}\int_{-\infty}^{\infty}x(t)\,\mathrm{e}^{-\mathrm{j}\omega t}\mathrm{d}t=\frac{1}{2\pi}\int_{0}^{\infty}E\,\mathrm{e}^{-at}\,\mathrm{e}^{-\mathrm{j}\omega t}\mathrm{d}t$$

periodic signals and nonperiodic signals.

(2) Fundamental frequency $\omega_0 = (2\pi/T_0) \to 0$. Since the spectral line spacing is an integer multiple of ω_0, the minimum spectral line spacing ω_0 evolves to $d\omega$, $n\omega_0$ then evolves into a continuous variable ω. The discrete frequency spectrum tends to be dense, and the frequency evolves from a discrete variable to a continuous variable.

(3) $1/T_0 = \omega_0/2\pi$ evolves into $d\omega/2\pi$.

(4) The summation operation $\sum$ evolves into the integration operation $\int$, integral limit extends from the local $(-T_0/2, T_0/2)$ of the time axis to the entire $(-\infty, \infty)$ of the time axis. The frequency range of the harmonic components is from $-\infty$ to ∞, occupying the entire frequency domain.

Then the Fourier series evolves into

$$x(t) = \int_{-\infty}^{\infty} \frac{c_n}{d\omega} e^{j\omega t} d\omega \tag{2.22}$$

$$\frac{c_n}{d\omega} = \frac{1}{2\pi}\int_{-\infty}^{\infty} x(t)\ e^{-j\omega t} dt \tag{2.23}$$

By now, the right side of formula (2.23) is called the Fourier transform of the function of $x(t)$, and the physical meaning of $\frac{c_n}{d\omega}$ on the left side of the formula is the average value of c_n in the small interval of $[\omega, \omega + d\omega]$, with the meaning of density, which is denoted as $X(\omega)$, called spectral density function, the above two formulas can be written as

$$x(t) = \int_{-\infty}^{\infty} X(\omega)\ e^{j\omega t} d\omega \tag{2.24}$$

$$X(\omega) = \frac{1}{2\pi}\int_{-\infty}^{\infty} x(t)\ e^{-j\omega t} dt \tag{2.25}$$

Formula (2.24) is called the inverse Fourier transform (IFT), which can realize the conversion of the signal from the frequency domain to the time domain; formula (2.25) is called the Fourier transform (FT), realizing the conversion from the time domain to the frequency domain, the two of which are a pair of reciprocal operations. The signal $x(t)$ and the spectral density function $X(\omega)$ are called a transform pair, denoted as

$$x(t) \underset{\text{IFT}}{\overset{\text{FT}}{\rightleftarrows}} X(\omega)$$

Because of $\omega = 2\pi f$, formula (2.24) and formula (2.25) can be changed to

$$X(f) = \int_{-\infty}^{\infty} x(t)\ e^{-j2\pi ft} dt \tag{2.26}$$

$$x(t) = \int_{-\infty}^{\infty} X(f)\ e^{j2\pi ft} df \tag{2.27}$$

This avoids the constant factor $1/2\pi$ in the Fourier transform.

It should be pointed out that the above is derived from the form, that is, the transform pair is derived from the period $T \to \infty$ of the periodic signal, "discrete spectrum" → "continuous spectrum", which is not mathematically strict. Strictly speaking, the necessary conditions for the existence of the Fourier transform of the nonperiodic signal $x(t)$ are as follows.

(1) $x(t)$ satisfies the Dirichlet condition within $(-\infty, \infty)$.

$$= \frac{1}{2\pi} \frac{E}{a + j\omega} = \frac{1}{2\pi} \frac{E}{a^2 + \omega^2}(a - j\omega)$$

其幅值频谱函数为

$$|X(\omega)| = \frac{1}{2\pi} \frac{E}{\sqrt{a^2 + \omega^2}}$$

如图 2.12(b) 所示。

其相位频谱函数为

$$\varphi(\omega) = \arctan\left(-\frac{\omega}{a}\right)$$

如图 2.12(c) 所示。

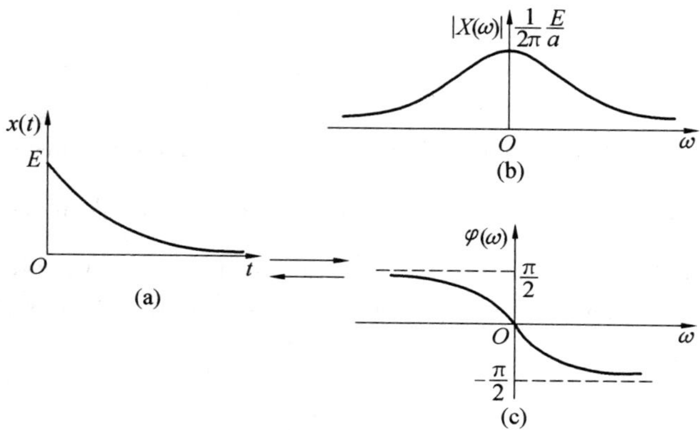

图 2.12　单边指数信号及其频谱

例 2.4　求矩形窗函数 $x(t)$ 的傅里叶变换(图 2.13)。

解　(1) 写出该信号的数学表示式,即

$$x(t) = \begin{cases} 1, & |t| < \frac{T}{2} \\ 0, & |t| > \frac{T}{2} \end{cases}$$

(2) 对 $x(t)$ 进行傅里叶变换,有

$$\begin{aligned} X(\omega) &= \frac{1}{2\pi}\int_{-\infty}^{\infty} x(t)\, e^{-j\omega t} dt \\ &= \frac{1}{2\pi}\int_{-\frac{T}{2}}^{\frac{T}{2}} e^{-j\omega t} dt \\ &= \frac{1}{2\pi} \cdot \frac{-1}{j\omega}(e^{-j\omega T/2} - e^{j\omega T/2}) \end{aligned}$$

(2) The integral $\int_{-\infty}^{\infty} |x(t)| dt$ of $x(t)$ converges, that is, absolutely integrable.

(3) $x(t)$ is a signal with limited energy.

Thus, not all signals can be Fourier transformed, however, this does not mean that signals that cannot be Fourier transformed cannot be converted to the frequency domain. Moreover, with the help of singular functions, some signals that do not satisfy the condition of absolute integrability can sometimes be Fourier transformed, such as sine function, cosine function, step function, sign function, etc.

Example 2.3　Find the frequency spectrum of a unilateral exponential pulse:

$$x(t) = \begin{cases} E\mathrm{e}^{-at}, & a > 0, t \geqslant 0 \\ 0, & t < 0 \end{cases}$$

The time domain waveform is shown in Figure 2.12(a).

Solution　The spectrum function of this nonperiodic signal is

$$\begin{aligned} X(\omega) &= \frac{1}{2\pi}\int_{-\infty}^{\infty} x(t)\,\mathrm{e}^{-\mathrm{j}\omega t}\mathrm{d}t = \frac{1}{2\pi}\int_{0}^{\infty} E\,\mathrm{e}^{-at}\,\mathrm{e}^{-\mathrm{j}\omega t}\mathrm{d}t \\ &= \frac{1}{2\pi}\frac{E}{a + \mathrm{j}\omega} = \frac{1}{2\pi}\frac{E}{a^2 + \omega^2}(a - \mathrm{j}\omega) \end{aligned}$$

Its amplitude spectrum function is

$$|X(\omega)| = \frac{1}{2\pi}\frac{E}{\sqrt{a^2 + \omega^2}}$$

as shown in Figure 2.12(b).

Its phase spectrum function is

$$\varphi(\omega) = \arctan\left(-\frac{\omega}{a}\right)$$

as shown in Figure 2.12(c).

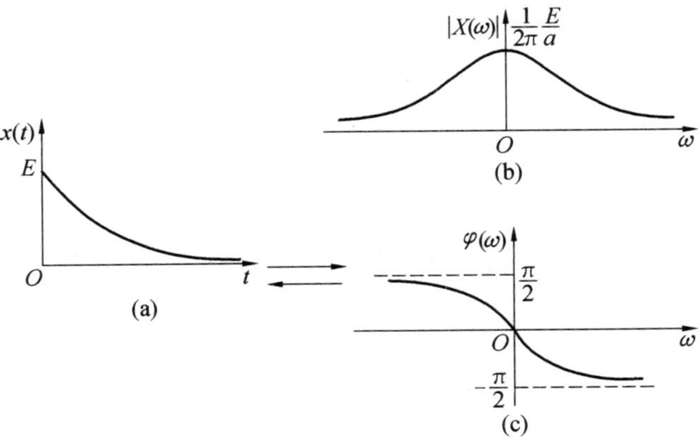

Figure 2.12　A unilateral exponential signal and its spectrum

Example 2.4　Find the Fourier transform of the rectangular window function $x(t)$ (Figure 2.13(a)).

Solution　(1) Write the mathematical expression of the signal, that is

$$= \frac{T}{2\pi} \frac{\sin \frac{\omega T}{2}}{\frac{\omega T}{2}} = \frac{T}{2\pi} \mathrm{sinc}\left(\frac{\omega T}{2}\right)$$

$\mathrm{sinc}(x) = \sin x/x$ 是一个特定表达的函数,该函数称为采样函数,又称滤波函数或内插函数,该函数在信号分析中经常用到。$\mathrm{sinc}(x)$ 函数的图像如图 2.14 所示,其函数值有专门的数学表可查,它以 2π 为周期并随 x 的增加而做衰减振荡。$\mathrm{sinc}(x)$ 函数为偶函数,在 $n\pi(n = 0,\ \pm 1,\ \pm 2,\cdots)$ 处,其值为 0。

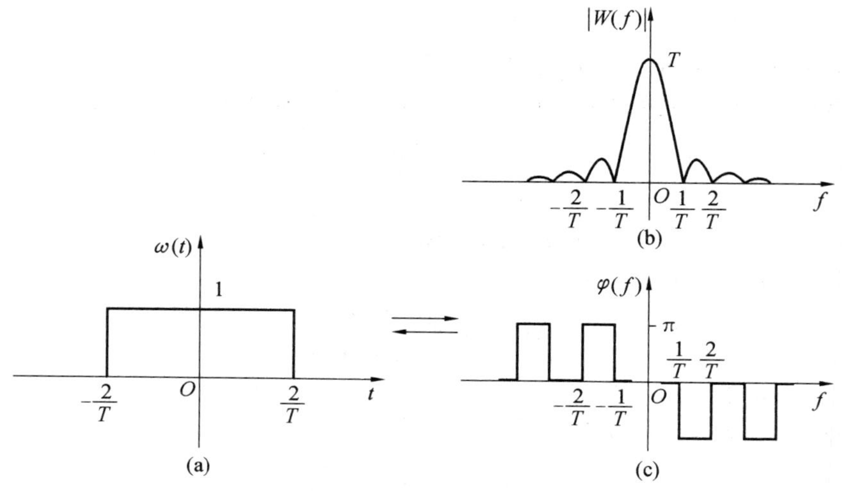

图 2.13　矩形窗函数及其频谱

(3) 图 2.13(b)、(c) 是窗函数的幅频谱和相频谱。

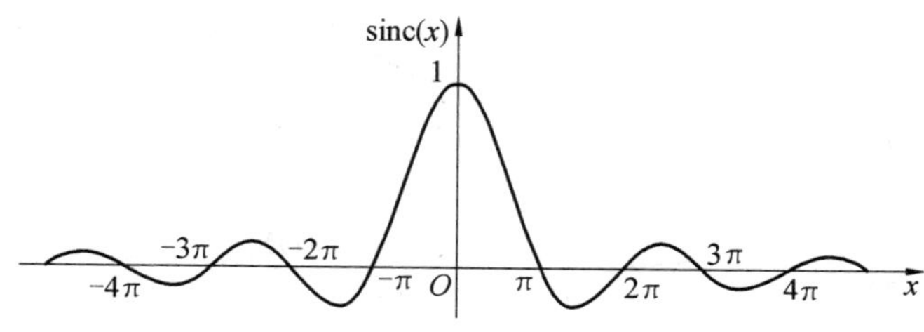

图 2.14　$\mathrm{sinc}(x)$ 的图像

根据矩形窗的频谱,运用非周期信号周期化的思路可以进一步估计出以下信号的频谱。

① 当脉冲宽度 T 很大时,信号的能量大部分集中在 $\omega = 0$ 附近(图 2.15(a))。

② 当脉冲宽度 $T \to \infty$ 时,脉冲信号变成直流信号,频谱函数 $X(\omega)$ 只在 $\omega = 0$ 处存在(图 2.15(b))。

③ 当脉冲宽度 T 减小时,频谱中的高频成分增加,信号频带宽度增大(图 2.15(c))。

$$x(t)=\begin{cases}1, & |t|<\dfrac{T}{2}\\ 0, & |t|>\dfrac{T}{2}\end{cases}$$

(2) Perform Fourier transform on $x(t)$, we have

$$\begin{aligned}X(\omega)&=\frac{1}{2\pi}\int_{-\infty}^{\infty}x(t)\,\mathrm{e}^{-\mathrm{j}\omega t}\mathrm{d}t\\&=\frac{1}{2\pi}\int_{-\frac{T}{2}}^{\frac{T}{2}}\mathrm{e}^{-\mathrm{j}\omega t}\mathrm{d}t\\&=\frac{1}{2\pi}\cdot\frac{-1}{\mathrm{j}\omega}(\mathrm{e}^{-\mathrm{j}\omega T/2}-\mathrm{e}^{\mathrm{j}\omega T/2})\\&=\frac{T}{2\pi}\frac{\sin\frac{\omega T}{2}}{\frac{\omega T}{2}}=\frac{T}{2\pi}\mathrm{sinc}\left(\frac{\omega T}{2}\right)\end{aligned}$$

$\mathrm{sinc}(x)=\sin x/x$ is a function of a specific expression, which is called a sampling function, also known as a filter function or an interpolate function. This function is often used in signal analysis. The image of $\mathrm{sinc}(x)$ function is shown in Figure 2.14. The function value has a special mathematical table to check, it takes 2π as the period and decays with the increase of x. $\mathrm{sinc}(x)$ function is an even function, and its value is 0 at $n\pi(n=0, \pm1, \pm2, \cdots)$.

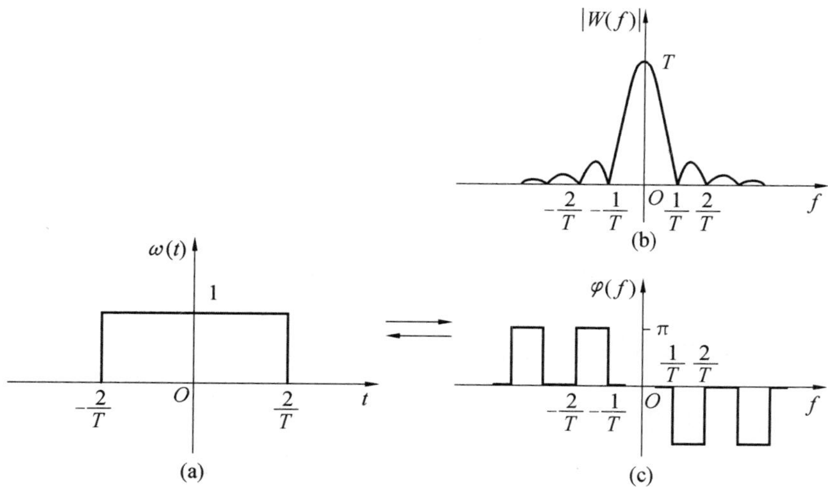

Figure 2.13　Rectangular window function and its frequency spectrum

(3) Figure 2.13 (b), (c) are the amplitude spectrum and phase spectrum of the window function.

According to the frequency spectrum of the rectangular window, with using the idea of nonperiodic signal periodicity, the frequency spectrum of the following signal can be further estimated.

①When the pulse width T is very large, the bulk of the signal energy is concentrated near

④ 当脉冲宽度 $T \to 0$ 时，矩形脉冲变成无穷窄的脉冲（相当于单位冲击信号），频谱函数 $X(\omega)$ 成为一条平行于 ω 轴的直线，并扩展到全部频率范围，信号的频带宽度趋于无穷大（图 2.15（d））。

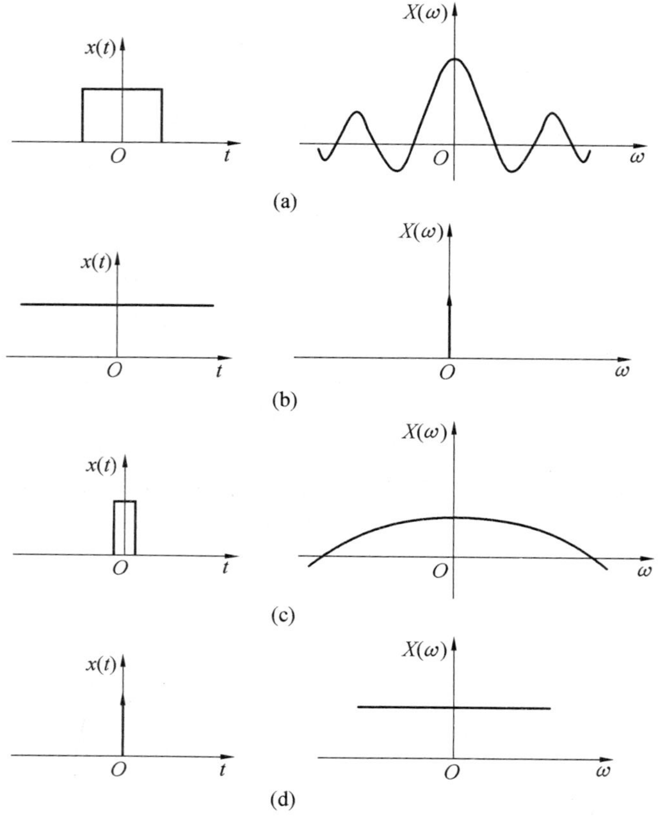

图 2.15　脉冲宽度与频谱的关系

2.5.2　傅里叶变换的主要性质

瞬态非周期信号的傅里叶变换对在时域与频域之间将 $x(t)$ 与 $X(f)$ 联系起来，是信号时域与频域之间的桥梁。熟悉傅里叶变换的主要性质会有助于对许多概念的理解，并简化某些运算。即使在依赖于计算机技术进行信号处理的今天，提前根据傅里叶变换性质对信号进行简化不仅会提高计算速度，而且对计算结果精度的提升也有重要意义。

实际上，通过数学运算求解一个信号的傅里叶变换不是最终的目的，重要的是在信号分析的理论研究与实际设计中能够了解信号在时域进行某种运算后在频域将发生何种变化，或反过来从频域的运算推测时域信号的变动。利用傅里叶变换的基本性质求解复杂信号变换不仅计算过程简单，而且物理概念清楚。傅里叶变换的主要性质见表 2.5。

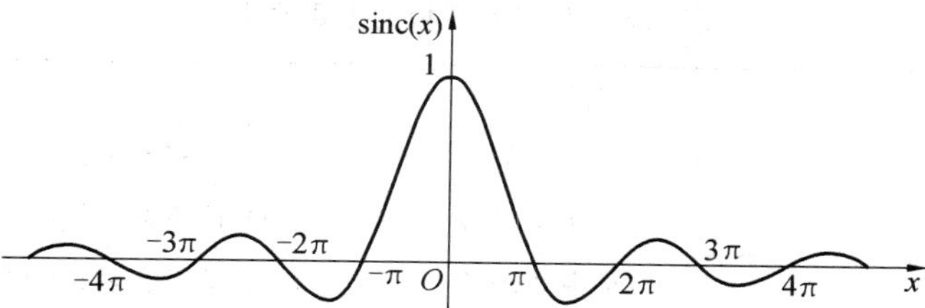

Figure 2.14　Image of sinc(x) function

$\omega = 0$ (Figure 2.15(a)).

②When the pulse width is $T \to \infty$, the pulse signal becomes a DC signal, and the spectrum function $X(\omega)$ only exists at $\omega = 0$ (Figure 2.15(b)).

③With the pulse width T decreasing, the high frequency components in the frequency spectrum will increase, and the frequency bandwidth of the signal will increase (Figure 2.15(c)).

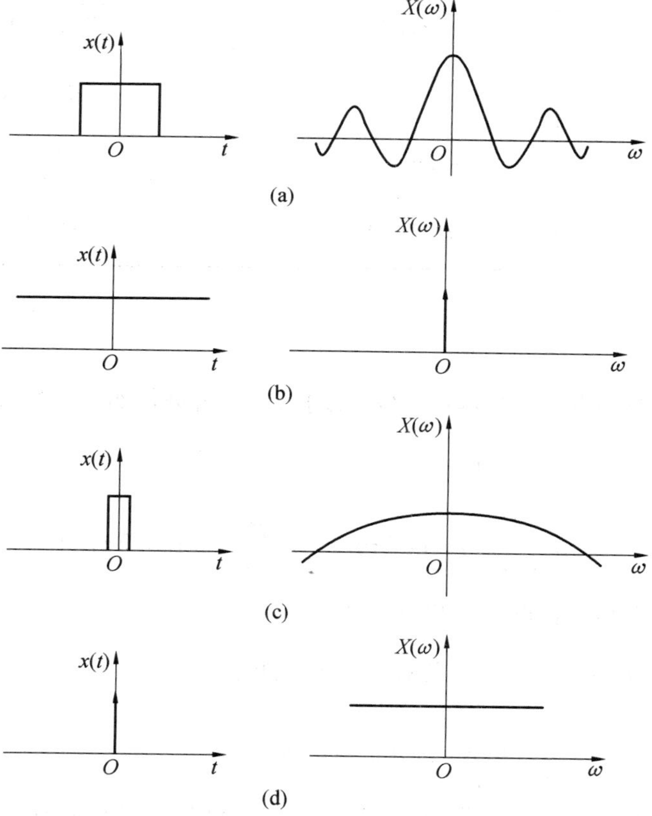

Figure 2.15　The relationship between pulse width and frequency spectrum

④When the pulse width is $T \to 0$, the rectangular pulse becomes an infinitely narrow pulse (equivalent to a unit shock signal), and the spectral function $X(\omega)$ becomes a straight line parallel to the ω axis and extends to the full frequency range, and the frequency bandwidth of

表 2.5　傅里叶变换的主要性质

性质	时域	频域
函数的奇偶虚实性	$x(t)$ 是实偶函数	$X(f)$ 是实偶函数（$X(f)$ 的虚部 = 0）
	$x(t)$ 是实奇函数	$X(f)$ 是虚奇函数（$X(f)$ 的实部 = 0）
	$x(t)$ 是虚偶函数	$X(f)$ 是虚偶函数（$X(f)$ 的实部 = 0）
	$x(t)$ 是虚奇函数	$X(f)$ 是实奇函数（$X(f)$ 的虚部 = 0）
线性叠加性	$ax(t)+by(t)$	$aX(f)+bY(f)$
对称性	$x(t)$	$x(-f)$
尺度改变特性	$x(kt)$	$X(f/k)/\lvert k\rvert$
时移特性	$x(t-t_0)$	$x(f)\mathrm{e}^{-\mathrm{j}2\pi f t_0}$
频移特性	$x(t)\mathrm{e}^{\pm \mathrm{j}2\pi f_0 t}$	$X(f\pm f_0)$
时域卷积	$x_1(t)*x_2(t)$	$X_1(f)X_2(f)$
频域卷积	$x_1(t)x_2(t)$	$X_1(f)*X_2(f)$
时域微分	$\dfrac{\mathrm{d}^n x(t)}{\mathrm{d}t^n}$	$(\mathrm{j}2\pi f)^n X(f)$
频域微分	$(-\mathrm{j}2\pi f)^n x(t)$	$\mathrm{d}^n X(f)/\mathrm{d}f^n$
积分	$\int_{-\infty}^{t} x(t)\mathrm{d}t$	$X(f)/(\mathrm{j}2\pi f)$

（1）奇偶虚实性。

一般 $X(f)$ 是实变量 f 的复函数。应用欧拉公式，可进行如下变换，即

$$\begin{aligned} X(f) &= \int_{-\infty}^{\infty} x(t)\mathrm{e}^{-\mathrm{j}2\pi f t}\mathrm{d}t \\ &= \int_{-\infty}^{\infty} x(t)\cos 2\pi f t\mathrm{d}t - \mathrm{j}\int_{-\infty}^{\infty} x(t)\sin 2\pi f t\mathrm{d}t \end{aligned}$$

只有 $x(t)$ 是非奇非偶的实函数时，才能断定：

① 实部 $\mathrm{Re}X(f)=\int_{-\infty}^{\infty} x(t)\cos 2\pi f t\mathrm{d}t$ 为偶函数；

② 虚部 $\mathrm{Im}X(f)=\int_{-\infty}^{\infty} x(t)\sin 2\pi f t\mathrm{d}t$ 为奇函数。

若 $x(t)$ 是或实或虚的偶函数，则式（2.33）为0，$X(f)$ 必是 f 的偶函数，且 $X(f)$ 与 $x(t)$ 保持相同的虚实性，有

$$X(f)=\int_{-\infty}^{\infty} x(t)\cos 2\pi f t\mathrm{d}t$$

若 $x(t)$ 是或实或虚的奇函数，则式（2.34）为0，$X(f)$ 必是 f 的奇函数，且 $X(f)$ 与 $x(t)$

the signal tends to infinity (Figure 2.15(d)).

2.5.2 The Main Properties of the Fourier Transform

The Fourier transform pair of transient nonperiodic signals, as a bridge, connects $x(t)$ and $X(f)$ between the time domain and the frequency domain. It will be helpful to understand many concepts and simplify some operations, being familiar with the main properties of Fourier transform. Even today relying on computer technology for signal processing, simplification of the signal according to the properties of Fourier transform in advance will not only increase the calculation speed, but also have important meanings for the improvement of the accuracy of the calculation results.

In fact, solving the Fourier transform of a signal through mathematical operations is not the ultimate goal. The important thing is that it is possible to understand what happens to the signal in the frequency domain after a certain operation in the time domain, or conversely infer the change in the time domain signal from the operation in the frequency domain in the theoretical research and actual design of signal analysis. The way to use the basic properties of Fourier transform to solve complex signal transformation proves a simple calculation process with a clear physical concept. The main properities of Fourier transform are listed in Table 2.5.

Table 2.5　The main properties of Fourier transform

Nature	Time domain	Frequency domain
The odd, even, virtual and real characteristics of function	$x(t)$ is a real even function	$X(f)$ is a real even function (imaginary part of $X(f) = 0$)
	$x(t)$ is a real odd function	$X(f)$ is a virtual odd function (imaginary part of $X(f) = 0$)
	$x(t)$ is a virtual even function	$X(f)$ is a virtual even function (imaginary part of $X(f) = 0$)
	$x(t)$ is a virtual odd function	$X(f)$ is a real odd function (imaginary part of $X(f) = 0$)
Linear superposition	$ax(t) + by(t)$	$aX(f) + bY(f)$
Symmetry	$x(t)$	$x(-f)$
Scale change characteristics	$x(kt)$	$X(f/k)/\|k\|$
Time shifting characteristics	$x(t - t_0)$	$x(f)\mathrm{e}^{-\mathrm{j}2\pi f t_0}$
Frequency shift characteristics	$x(t)\mathrm{e}^{\pm \mathrm{j}2\pi f_0 t}$	$X(f \pm f_0)$
Time domain convolution	$x_1(t) * x_2(t)$	$X_1(f)X_2(f)$
Frequency domain convolution	$x_1(t)x_2(t)$	$X_1(f) * X_2(f)$
Time domain differentiation	$\frac{\mathrm{d}^n x(t)}{\mathrm{d}t^n}$	$(\mathrm{j}2\pi f)^n X(f)$
Frequency domain differentiation	$(-\mathrm{j}2\pi f)^n x(t)$	$\mathrm{d}^n X(f)/\mathrm{d}f^n$
Integral	$\int_{-\infty}^{t} x(t)\mathrm{d}t$	$X(f)/(\mathrm{j}2\pi f)$

的虚实性刚好相反，有

$$X(f)=\mathrm{j}\int_{-\infty}^{\infty}x(t)\sin 2\pi ft\mathrm{d}t$$

(2) 线性性质。

若

$$x_1(t)\leftrightarrows X_1(f)$$
$$x_2(t)\leftrightarrows X_2(f)$$

a_1、a_2 为任意常数，则有

$$a_1x_1(t)+a_2x_2(t)\leftrightarrows a_1X_1(f)+a_2X_2(f)$$

即信号的线性组合的变换等于变换的线性组合。

(3) 对称性。

若

$$x(t)\leftrightarrows X(f)$$

则有

$$x(t)\leftrightarrows X(-f)$$

对称性如图 2.16 所示。

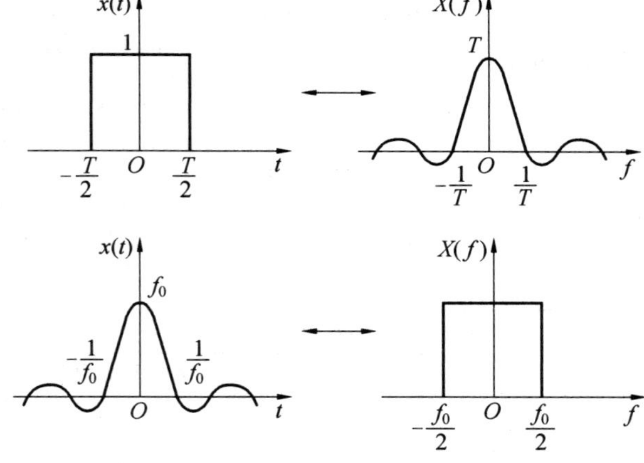

图 2.16 对称性

(4) 时间展缩性。

若

$$x(t)\leftrightarrows X(f)$$

则有

(1) Odd, even, virtual and real characteristics.

$X(f)$ is generally a complex function of the real variable f. Applying Euler's formula, the following transformations can be made, that is

$$X(f) = \int_{-\infty}^{\infty} x(t)\,\mathrm{e}^{-\mathrm{j}2\pi ft}\,\mathrm{d}t$$
$$= \int_{-\infty}^{\infty} x(t)\cos 2\pi ft\,\mathrm{d}t - \mathrm{j}\int_{-\infty}^{\infty} x(t)\sin 2\pi ft\,\mathrm{d}t$$

Only when $x(t)$ is a non-odd and non-even real function, can it be concluded:

①The real part $\mathrm{Re}X(f) = \int_{-\infty}^{\infty} x(t)\cos 2\pi ft\,\mathrm{d}t$ is an even function;

②The virtual part $\mathrm{Im}X(f) = \int_{-\infty}^{\infty} x(t)\sin 2\pi ft\,\mathrm{d}t$ is an odd function.

If $x(t)$ is a real or virtual even function, then the formula (2.33) is 0, $X(f)$ must be an even function of f, $X(f)$ and $x(t)$ maintain the same virtual and real properties, we have

$$X(f) = \int_{-\infty}^{\infty} x(t)\cos 2\pi ft\,\mathrm{d}t$$

If $x(t)$ is a real or virtual odd function, then the formula (2.34) is 0, $X(f)$ must be an odd function of f, the virtual and real function of $X(f)$ and $x(t)$ reveals the opposite, we have

$$X(f) = \mathrm{j}\int_{-\infty}^{\infty} x(t)\sin 2\pi ft\,\mathrm{d}t$$

(2) Linear property.

If

$$x_1(t) \leftrightarrows X_1(f)$$
$$x_2(t) \leftrightarrows X_2(f)$$

a_1, a_2 is arbitrary constant, then

$$a_1x_1(t) + a_2x_2(t) \leftrightarrows a_1X_1(f) + a_2X_2(f)$$

That is, the transformation of the linear combination of signals is equal to the linear combination of transformations.

(3) Symmetry.

If

$$x(t) \leftrightarrows X(f)$$

we have

$$x(t) \leftrightarrows X(-f)$$

Symmetry is shown in Figure 2.16.

(4) Time scaling characteristic.

If

$$x(t) \leftrightarrows X(f)$$

we have

$$x(kt) \Rightarrow \frac{1}{|k|}X\left(\frac{f}{k}\right)$$

If the tape is recorded slowly and played quickly, it means to compress the time scale, which can improve the signal processing efficiency, but the frequency band of the signal (playback signal) obtained will be widened. If the frequency passband of subsequent

$$x(kt) \Rightarrow \frac{1}{|k|}X\left(\frac{f}{k}\right)$$

把记录磁带慢录快放，即使时间尺度压缩，这样虽可以提高处理信号效率，但是所得到的信号（放演信号）频带就会加宽。倘若后续处理设备（放大器、滤波器等）的通频带不够宽，则会导致失真。反之，快录慢放，则放演信号的带宽变窄，对后续处理设备的通频带要求可以降低，但信号处理效率也随之降低。时间展缩性如图 2.17 所示。

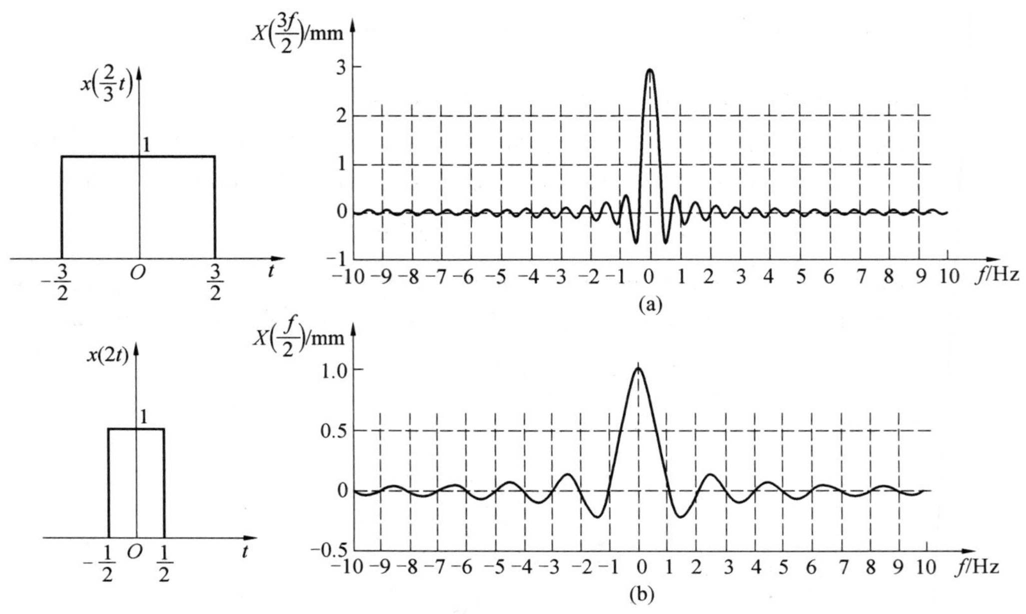

图 2.17 时间展缩性

（5）时移特性。

若

$$x(t) \leftrightarrows X(f)$$

则在时域中信号沿时间轴平移一常值 t_0 时，它要表示为 $x(t-t_0)$，那么将有

$$x(t-t_0) \leftrightarrows X(f)\mathrm{e}^{-\mathrm{j}2\pi f t_0}$$

时移特性举例如图 2.18 所示。

（6）频移特性。

若

$$x(t) \leftrightarrows X(f)$$

则在频域内信号频谱 $X(f)$ 沿频率轴平移一常值 $\pm f_0$ 时，它要表示为 $X(f \pm f_0)$，那么将有

$$x(t)\mathrm{e}^{\mp \mathrm{j}2\pi f_0 t} \leftrightarrows X(f \pm f_0)$$

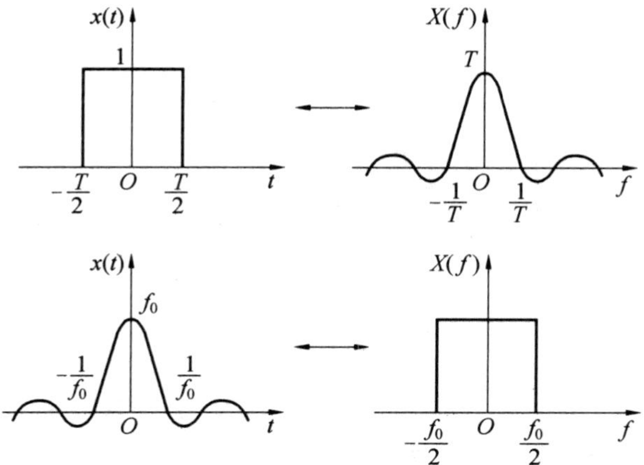

Figure 2.16　Symmetry

processing equipment (amplifiers, filters, etc.) is not wide enough, it will cause distortion. Conversely, if the tape is drecorded fastly and played slowly, the bandwidth of the playback signal is narrowed, and the frequency passband requirements for subsequent processing equipment can be reduced, but the signal processing efficiency is also reduced. Time scaling characteristic is shown in Figure 2.17.

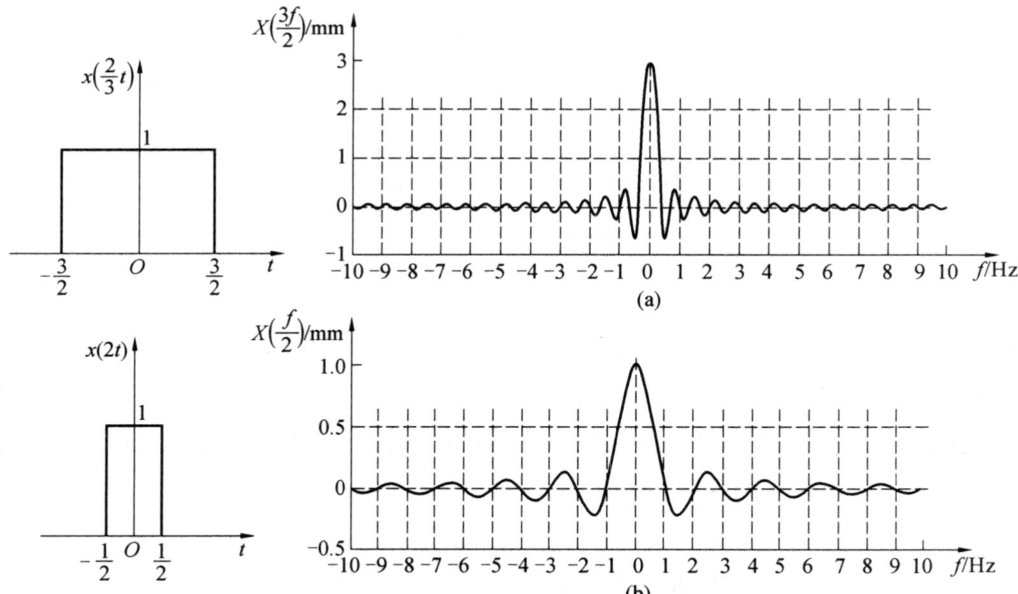

Figure 2.17　Time scaling characteristic

(5) Time shifting characteristics.

If

$$x(t) \leftrightarrows X(f)$$

then in the time domain, when the signal is translated along the time axis by a constant value

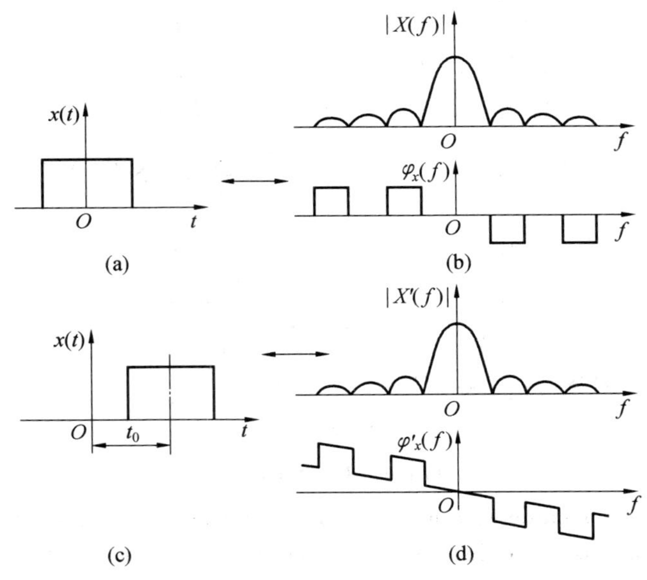

图 2.18　时移特性举例

（7）卷积特性。

若

$$x_1(t) \leftrightarrows X_1(f)$$

$$x_2(t) \leftrightarrows X_2(f)$$

则有

$$x_1(t) \cdot x_2(t) \leftrightarrows X_1(f) * X_2(f)$$

$$x_1(t) * x_2(t) \leftrightarrows X_1(f) \cdot X_2(f)$$

式中　$X_1(f) * X_2(f)$、$x_1(t) * x_2(t)$——$X_1(f)$ 与 $X_2(f)$ 的卷积及 $x_1(t)$ 与 $x_2(t)$ 的卷积。

上式说明信号之积的谱是谱的卷积，谱之积的像原是像原的卷积。

卷积积分是一种数学方法，它是沟通时域和频域的一个桥梁，在信号与系统的理论研究中占有重要的地位。在很多情况下，卷积积分的计算较困难，但根据卷积特性可以将卷积积分变为乘法运算，从而使信号分析工作大为简化。

2.6　典型信号的频谱

2.6.1　矩形窗函数及其频谱

在例 2.4 中已经讨论过矩形窗函数的频谱，即在时域有限区间内幅值为常数的一个窗信号，其频谱延伸至无限频率。矩形窗函数在信号处理中有着重要的应用。在时域中，

t_0, it should be expressed as $x(t-t_0)$, then there will be

$$x(t-t_0) \leftrightarrows X(f)\mathrm{e}^{-\mathrm{j}2\pi f t_0}$$

Examples of time shifting characteristics is shown in Figure 2.18.

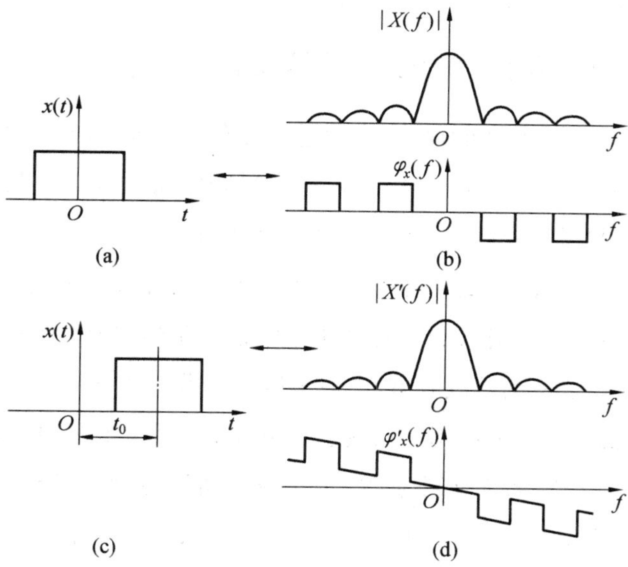

Figure 2.18　Examples of time shifting characteristics

(6) Frequency shift characteristics.

If

$$x(t) \leftrightarrows X(f)$$

then in the frequency domain, when the signal spectrum $X(f)$ is translated along the time axis by a constant value $\pm f_0$, it should be expressed as $X(f \pm f_0)$, then there will be

$$x(t)\mathrm{e}^{\mp \mathrm{j}2\pi f_0 t} \leftrightarrows X(f \pm f_0)$$

(7) Convolution characteristics.

If

$$x_1(t) \leftrightarrows X_1(f)$$
$$x_2(t) \leftrightarrows X_2(f)$$

then there are

$$x_1(t) \cdot x_2(t) \leftrightarrows X_1(f) * X_2(f)$$
$$x_1(t) * x_2(t) \leftrightarrows X_1(f) \cdot X_2(f)$$

where　$X_1(f) * X_2(f)$, $x_1(t) * x_2(t)$ —the convolution of $X_1(f)$ and $X_2(f)$ and the convolution of $x_1(t)$ and $x_2(t)$ respectively.

The above formula explains that the spectrum of the product of signals is the convolution of the spectrum, and the image of the product of the spectrum is the convolution of the original.

Convolution integral, as a bridge between the time domain and the frequency domain, is a mathematical method, which occupies an important position in the theoretical research of signals and systems. In many cases, the calculation of convolution integral is more difficult, but according to the convolution characteristics, the convolution integral can be converted into a

若截取某信号的一段记录长度,则相当于原信号和矩形窗函数的乘积,因此所得频谱将是原信号频域函数和 sinc(x) 函数的卷积。由于 sinc(x) 函数的频谱是连续的,频率是无限的,因此信号截取后频谱将是连续的,频率将无限延伸。

2.6.2 单位脉冲函数(δ 函数) 及其频谱

在研究某种量在时空中分布时,除遇到连续分布的量外,还会遇到集中于空间某一点,时域中某一时刻的量,如空间点电荷、发生在某一时刻的冲力或电脉冲、在频率轴上某一点上的幅值,它们都是集中量。习惯上用离散函数描述它们的分布,为方便研究,有必要将集中的量和连续分布的量统一处理,δ 函数便是把集中量也按连续分布的量来处理时要借助的数学工具。本书引入 δ 函数的目的就是借助 δ 函数将周期信号和非周期信号的频谱统一起来,统一表示为连续频谱,以便于运算。另外,δ 函数还具有采样等功能,在测试与数据处理方面有广泛的应用。在数学中,δ 函数又称单位脉冲函数,记为 $\delta(t)$。其定义为:在 ε 时间内激发一个矩形脉冲$S_{\varepsilon}(t)$ (或三角形脉冲、双边指数脉冲、钟形脉冲等),其面积为 1(图 2.19)。当 $\varepsilon \to 0$ 时,在保持其面积不变的前提下,$S_{\varepsilon}(t)$ 的极限就称为单位脉冲函数。它是一个特定条件下的线密度的概念,其特点如下。

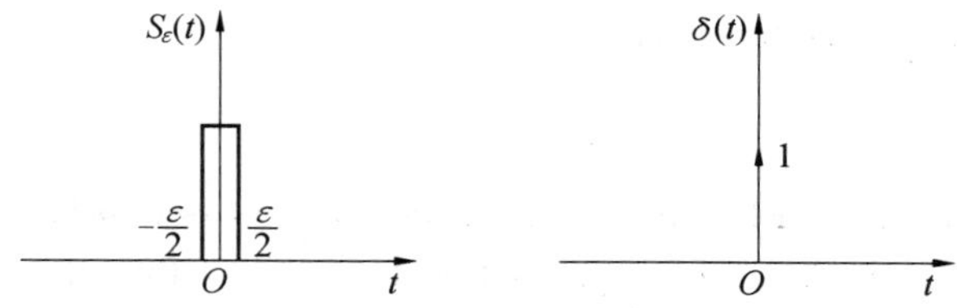

图 2.19 矩形脉冲与 δ 函数

(1)δ 函数的数学表示式。

$$\delta(t)=\begin{cases}\infty, & t=0\\ 0, & t\neq 0\end{cases} \tag{2.28}$$

(2)δ 函数的强度。

$$\int_{-\infty}^{\infty}\delta(t)\,\mathrm{d}t=\lim_{\varepsilon\to 0}\int_{-\infty}^{\infty}S_{\varepsilon}(t)\,\mathrm{d}t=1 \tag{2.29}$$

δ 函数的幅值在实际应用中是没有价值的,但依据其形象——幅值无穷大,可以断定它的存在。在实际应用中,真正有价值的是脉冲的强度在数值上等于 1。某些具有冲击性的物理现象,如电网线路中的短时冲击干扰、数字电路中的采样脉冲、力学中的瞬间作用力,以及材料的突然断裂、撞击及爆炸等,都可以通过函数来分析,只是函数面积(能量或强度) 不一定为 1,而是某一常数。

multiplication operation, greatly simplifying the signal analysis work.

2.6 Frequency Spectrum of a Typical Signal

2.6.1 Rectangular Window Function and Its Frequency Spectrum

The frequency spectrum of the rectangular window function has been discussed in Example 2.4, that is, a window signal with a constant amplitude in a finite interval in the time domain, and its frequency spectrum extends to an infinite frequency. The rectangular window function has an important application in signal processing. In the time domain, if a record length of a certain signal is intercepted, it is equivalent to the product of the original signal and the rectangular window function, so the resulting spectrum will be the convolution of frequency domain of original signal function and $\text{sinc}(x)$ function. Since the frequency spectrum of $\text{sinc}(x)$ function is continuous and the frequency is infinite, the frequency spectrum will be continuous and the frequency will extend infinitely after the signal is intercepted.

2.6.2 Unit Pulse Function (δ Function) and Its Frequency Spectrum

When studying the distribution of a certain quantity in space and time, there are also quantities concentrated at a certain point in space and at a certain moment in the time domain except for the quantities encountered in continuous distribution, all of which are concentrated quantities, such as space point charges, the impulse or electrical impulse that occurs at a certain moment, the amplitude at a point on the frequency axis. Customary to use discrete functions to describe their distribution, it is necessary to process the concentrated quantity and the continuously distributed quantity in a unified manner for the convenience of research. It is a mathematical tool that δ function is to be used when processing the concentrated quantity as a continuously distributed quantity. The purpose of introducing δ function in this book is to use δ function to unify the frequency spectrum of the periodic signal and the nonperiodic signal to express a continuous spectrum for easy calculation. In addition, δ function also has functions such as sampling, which has a wide range of applications in measurement and data processing. In mathematics, δ function is also called the unit impulse function, denoted as $\delta(t)$. It is defined as: at time ε , exciting a rectangular pulse $S_\varepsilon(t)$ (or triangular pulse, bilateral exponential pulse, bell-shaped pulse, etc.), whose area is 1 (Figure 2.19). When $\varepsilon \to 0$, the limit of $S_\varepsilon(t)$ is called the unit impulse function under the premise of keeping its area constant. It is a concept of linear density under specific conditions, and its characteristics are as follow.

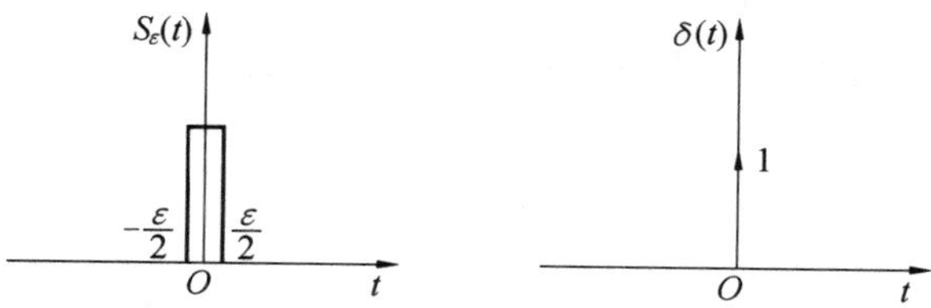

Figure 2.19　Rectangular pulse and δ function

(3)δ 函数的采样性质。

如果δ 函数与某一连续函数$f(t)$ 相乘,则显然其乘积仅在$t=0$处为$f(0)\delta(t)$,其余各点($t \neq 0$) 的乘积均为0。其中,$f(0)\delta(t)$ 是一个强度为$f(0)$ 的δ 函数。如果δ 函数与某一连续函数$f(t)$ 相乘,并在($-\infty,\infty$) 区间中积分,则有

$$\int_{-\infty}^{\infty}\delta(t)f(t)\,\mathrm{d}t=\int_{-\infty}^{\infty}\delta(t)f(0)\,\mathrm{d}t$$

$$=f(0)\int_{-\infty}^{\infty}\delta(t)\,\mathrm{d}t=f(0) \tag{2.30}$$

同理,对于有延时t_0的δ 函数$\delta(t-t_0)$,它与连续函数$f(t)$ 的乘积只有在$t=t_0$时刻不等于0,而等于强度为$f(t_0)$ 的δ 函数。在($-\infty,\infty$) 区间内,该乘积的积分为

$$\int_{-\infty}^{\infty}\delta(t-t_0)f(t)\,\mathrm{d}t=\int_{-\infty}^{\infty}\delta(t-t_0)f(t_0)\,\mathrm{d}t$$

$$=f(t_0) \tag{2.31}$$

式(2.37) 和式(2.38) 表示δ 函数的采样性质。此性质表明任何函数$f(t)$ 和$\delta(t-t_0)$的乘积是一个强度为$f(t_0)$ 的δ 函数$\delta(t-t_0)$,而该乘积在无限区间的积分则是$f(t)$ 在$t=t_0$ 时刻的函数值$f(t_0)$。

(4)δ 函数与其他函数的卷积。

任何函数和δ 函数$\delta(t)$ 卷积是一种最简单的卷积积分。例如,一个矩形函数$x(t)$ 与δ 函数$\delta(t)$ 的卷积为

$$x(t)*\delta(t)=\int_{-\infty}^{\infty}x(\tau)\delta(t-\tau)\,\mathrm{d}\tau$$

$$=\int_{-\infty}^{\infty}x(\tau)\delta(\tau-t)\,\mathrm{d}\tau=x(t)$$

同理,当δ 函数为$\delta(t\pm t_0)$ 时,有

$$x(t)*\delta(t\pm t_0)=\int_{-\infty}^{+\infty}x(\tau)\delta(t\pm t_0-\tau)\,\mathrm{d}\tau$$

$$=x(t\pm t_0) \tag{2.32}$$

δ 函数与其他函数的卷积示例如图2.20所示。从图形上看,δ 函数与任意函数的卷积就是将该函数搬运到δ 函数所在位置上重新构图,这个特点对连续信号的离散采样是十分重要的。

(5)δ 函数的频谱。

根据图2.15,利用非周期信号周期化处理的思路,可以得到δ 函数的频谱。δ 函数及其频谱如图2.21 所示。

(1) The mathematical expression of δ function.

$$\delta(t)=\begin{cases}\infty, & t=0\\ 0, & t\neq 0\end{cases} \tag{2.28}$$

(2) The strength of δ function.

$$\int_{-\infty}^{\infty}\delta(t)\,\mathrm{d}t=\lim_{\varepsilon\to 0}\int_{-\infty}^{\infty}S_{\varepsilon}(t)\,\mathrm{d}t=1 \tag{2.29}$$

The amplitude of δ function is of no value in practical applications, but its existence can be determined based on its image—infinite amplitude. What is really valuable in practical applications is that the intensity of the pulse is numerically equal to 1. Some shocking physical phenomena can be analyzed by functions, such as short-term impact interference in power grid lines, sampling pulses in digital circuits, instantaneous forces in mechanics, and sudden fracture, impacts, explosions of materials. However, the function area (energy or intensity) is not necessarily 1, but a constant.

(3) The sampling properties of δ function.

If δ function is multiplied by a continuous function $f(t)$, it is obvious that the product is only $f(0)\delta(t)$ at $t=0$, and the multiplication of the other points ($t\neq 0$) is 0, where $f(0)\delta(t)$ is a δ function with strength $f(0)$. If δ function is multiplied by a continuous function $f(t)$ and integrated in the interval $(-\infty,\infty)$, then there is

$$\begin{aligned}\int_{-\infty}^{\infty}\delta(t)f(t)\,\mathrm{d}t&=\int_{-\infty}^{\infty}\delta(t)f(0)\,\mathrm{d}t\\&=f(0)\int_{-\infty}^{\infty}\delta(t)\,\mathrm{d}t=f(0)\end{aligned} \tag{2.30}$$

In the same way, for δ function $\delta(t-t_0)$ with delay t_0, its product with the continuous function $f(t)$ is not equal to 0 at time $t=t_0$, but equal to the δ function with intensity $f(t_0)$. In the interval $(-\infty,\infty)$, the integral of the product is

$$\begin{aligned}\int_{-\infty}^{\infty}\delta(t-t_0)f(t)\,\mathrm{d}t&=\int_{-\infty}^{\infty}\delta(t-t_0)f(t_0)\,\mathrm{d}t\\&=f(t_0)\end{aligned} \tag{2.31}$$

Formulas (2.37) and (2.38) represent the sampling properties of δ function. This property indicates that the product of any function $f(t)$ and $\delta(t-t_0)$ is a δ function $\delta(t-t_0)$ with intensity $f(t_0)$, and the integral of the product in the infinite interval is the function value $f(t_0)$ of $f(t)$ at time $t=t_0$.

(4) Convolution of δ function and other functions.

The convolution of any function and δ function $\delta(t)$ is the simplest kind of convolution integral. For example, the convolution of a rectangular function $x(t)$ and δ function $\delta(t)$ is

$$\begin{aligned}x(t)*\delta(t)&=\int_{-\infty}^{\infty}x(\tau)\delta(t-\tau)\,\mathrm{d}\tau\\&=\int_{-\infty}^{\infty}x(\tau)\delta(\tau-t)\,\mathrm{d}\tau=x(t)\end{aligned}$$

And by the same logic, when δ function is $\delta(t\pm t_0)$, we have

$$\begin{aligned}x(t)*\delta(t\pm t_0)&=\int_{-\infty}^{\infty}x(\tau)\delta(t\pm t_0-\tau)\,\mathrm{d}\tau\\&=x(t\pm t_0)\end{aligned} \tag{2.32}$$

Examples of convolution of δ function and other functions are shown in Figure 2.20. From

本节将利用傅里叶变换的时移和频移特性,进一步可得

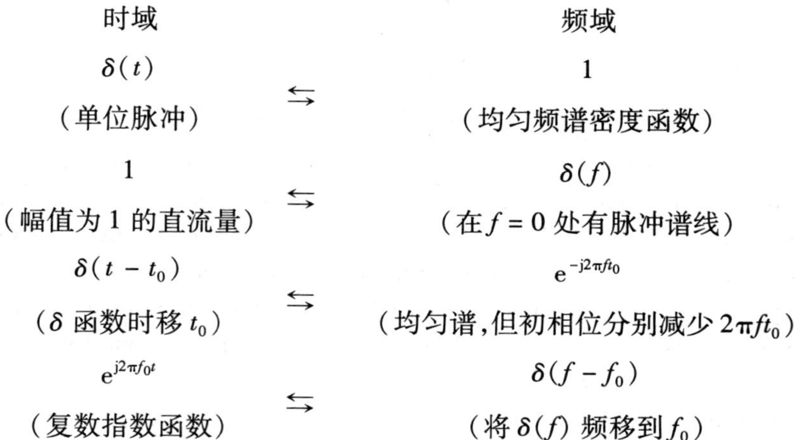

时域		频域
$\delta(t)$ (单位脉冲)	$\leftrightarrows$	1 (均匀频谱密度函数)
1 (幅值为 1 的直流量)	$\leftrightarrows$	$\delta(f)$ (在 $f=0$ 处有脉冲谱线)
$\delta(t-t_0)$ (δ 函数时移 t_0)	$\leftrightarrows$	$e^{-j2\pi f t_0}$ (均匀谱,但初相位分别减少 $2\pi f t_0$)
$e^{j2\pi f_0 t}$ (复数指数函数)	$\leftrightarrows$	$\delta(f-f_0)$ (将 $\delta(f)$ 频移到 f_0)

(2.33)

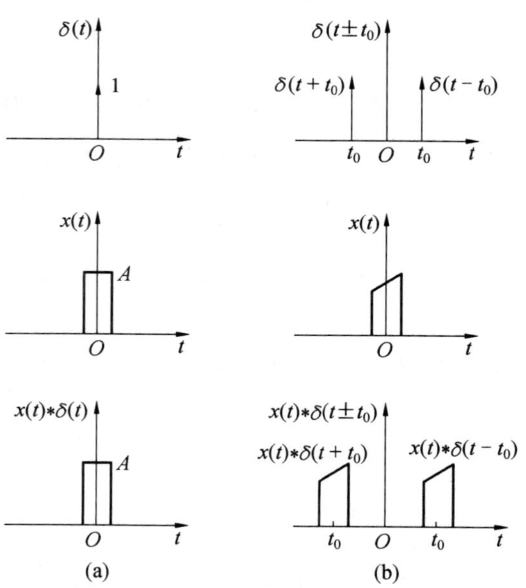

图 2.20　δ 函数与其他函数的卷积示例

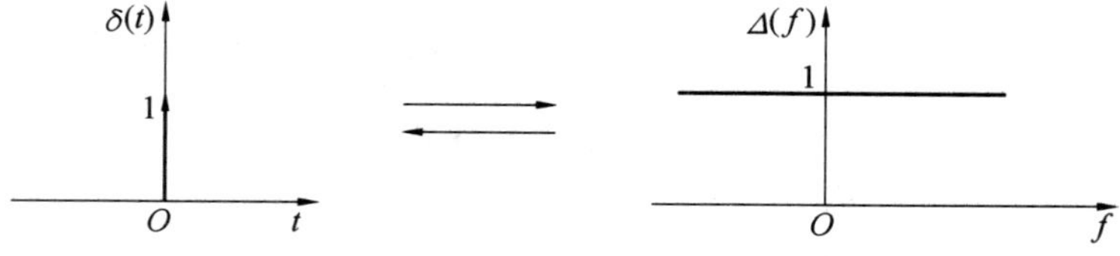

图 2.21　δ 函数及其频谱

a graphical point of view, the convolution of the δ function and any function is to transport the function to the position of δ function to recompose the picture. This feature is very important for the discrete sampling of continuous signals.

(5) Spectrum of δ function.

According to Figure 2.15, with using the idea of periodic processing of nonperiodic signals, the spectrum of δ function can be obtained. The δ function and its spectrum are shown in Figure 2.21.

In this section, the time-shift and frequency-shift characteristics of the Fourier transform can be used to further obtain that

Time domain		Frequency domain
$\delta(t)$ (Unit impulse)	$\leftrightarrows$	1 (Uniform spectrum density function)
1 (DC quantity with amplitude 1)	$\leftrightarrows$	$\delta(f)$ (There is a pulse spectrum at $f=0$)
$\delta(t-t_0)$ (δ Function time shift t_0)	$\leftrightarrows$	$e^{-j2\pi f t_0}$ (Uniform spectrum, but the initial phase is reduced by $2\pi f t_0$)
$e^{j2\pi f_0 t}$ (Complex exponential function)	$\leftrightarrows$	$\delta(f-f_0)$ ($\delta(f)$ shifts the frequency to f_0)

(2.33)

2.6.3 The Frequency Spectrum of Sine and Cosine Functions

Since the sine and cosine functions do not satisfy the absolute integrable condition, the Fourier transform cannot be performed directly, and δ function needs to be introduced in the Fourier transform.

According to Euler's formula, the sine and cosine functions can be written as

$$\sin 2\pi f_0 t = \mathrm{j}\frac{1}{2}(e^{-j2\pi f_0 t} - e^{j2\pi f_0 t})$$

$$\cos 2\pi f_0 t = \frac{1}{2}(e^{-j2\pi f_0 t} + e^{j2\pi f_0 t})$$

Using the time shifting and frequency shifting characteristics of δ function given in formula (2.33), it can be considered that the sine and cosine functions are the inverse Fourier transform of the difference or sum of two δ functions in the frequency domain after frequency shifting in different directions. Therefore, the Fourier transform of the sine and cosine functions is

$$\sin 2\pi f_0 t \leftrightarrows \mathrm{j}\frac{1}{2}(\delta(f+f_0) - \delta(f-f_0))$$

$$\cos 2\pi f_0 t \leftrightarrows \frac{1}{2}(\delta(f+f_0) + \delta(f-f_0))$$

2.6.3 正、余弦函数的频谱

由于正、余弦函数不满足绝对可积条件,因此不能直接进行傅里叶变换,需在傅里叶变换时引入 δ 函数。

根据欧拉公式,正、余弦函数可以写成

$$\sin 2\pi f_0 t = \mathrm{j}\,\frac{1}{2}\left(\mathrm{e}^{-\mathrm{j}2\pi f_0 t} - \mathrm{e}^{\mathrm{j}2\pi f_0 t}\right)$$

$$\cos 2\pi f_0 t = \frac{1}{2}\left(\mathrm{e}^{-\mathrm{j}2\pi f_0 t} + \mathrm{e}^{\mathrm{j}2\pi f_0 t}\right)$$

应用式(2.33)中给出的 δ 函数的时移和频移特性,可认为正、余弦函数是把频域中的两个 δ 函数向不同方向频移后之差或和的傅里叶逆变换。因此,求得正、余弦函数的傅里叶变换为

$$\sin 2\pi f_0 t \leftrightarrows \mathrm{j}\,\frac{1}{2}\left(\delta(f+f_0) - \delta(f-f_0)\right)$$

$$\cos 2\pi f_0 t \leftrightarrows \frac{1}{2}\left(\delta(f+f_0) + \delta(f-f_0)\right)$$

正、余弦函数及其频谱如图 2.22 所示。

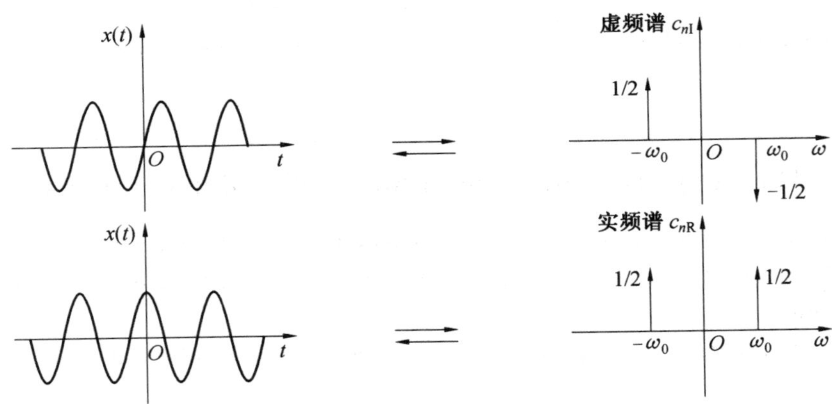

图 2.22　正、余弦函数及其频谱

2.6.4 周期单位脉冲序列的频谱

等间隙的周期单位脉冲序列常称为梳状函数,并用 comb (t, nT_s) 表示,即

$$\mathrm{comb}\,(t, T_s) \overset{\text{def}}{=\!=} \sum_{n=-\infty}^{\infty} \delta(t - nT_s)$$

式中　T_s—— 周期;

n—— 整数,$n = 0, \pm 1, \pm 2, \cdots$。

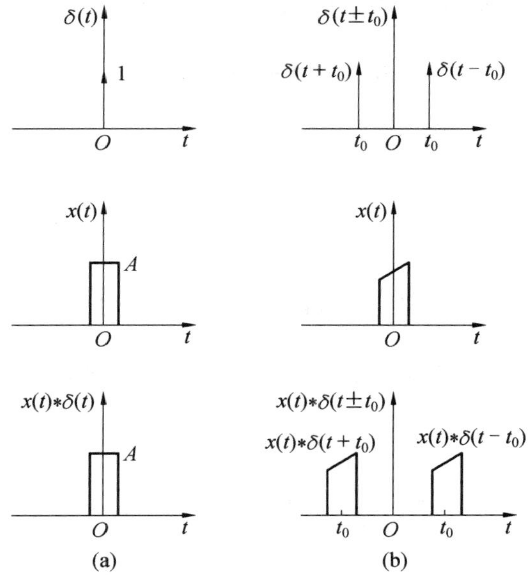

Figure 2.20 Examples of convolution of δ function and other functions

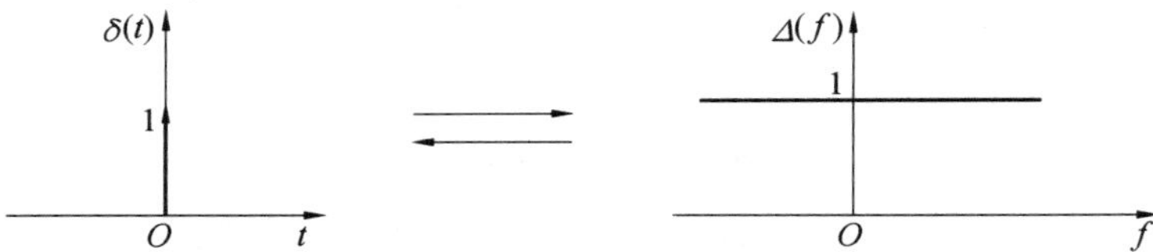

Figure 2.21 The δ function and its spectrum

Sine and cosine functions and their frequency spectrum are shown in Figure 2.22.

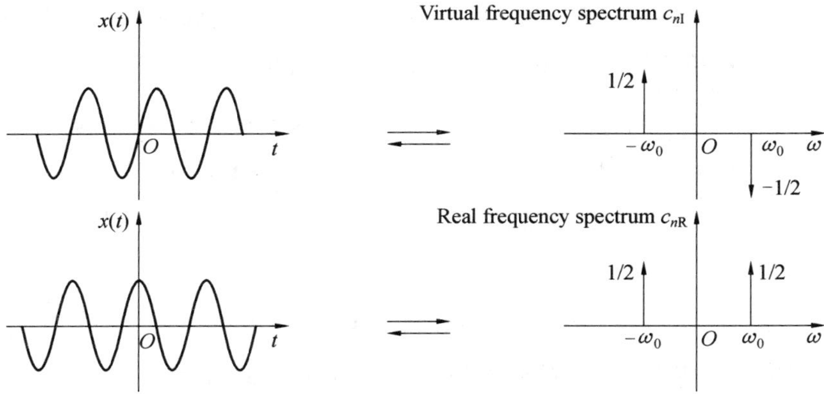

Figure 2.22 Sine and cosine functions and their frequency spectrum

2.6.4 The Frequency Spectrum of Periodically Unit Impulse-Train

A comb function refers to the periodically unit impulse-train of equal clearance, which is denoted by comb (t, nT_s) , that is

因为此函数是周期函数，所以可以把它表示为傅里叶级数的复指数函数形式，即

$$\text{comb}\ (t,T_s)=\sum_{k=-\infty}^{\infty}c_k e^{j2\pi kf_s t} \tag{2.34}$$

式中

$$f_s=1/T_s$$

系数 c_k 为

$$c_k=\frac{1}{T_s}\int_{-\frac{T_s}{2}}^{\frac{T_s}{2}}\text{comb}\ (t,T_s)\,e^{-j2\pi kf_s t}dt$$

因为在 $(-T_s/2,T_s/2)$ 区间内，式(2.34)只有一个 δ 函数 $\delta(t)$，而当 $t=0$ 时，$e^{-j2\pi kf_s t}=e^0=1$，所以

$$c_k=\frac{1}{T}\int_{-\frac{T_s}{2}}^{\frac{T_s}{2}}\delta(t)\ e^{-j2\pi kf_s t}dt=\frac{1}{T}$$

这样，式(2.34)可写成

$$\text{comb}\ (t,T_s)=\frac{1}{T_s}\sum_{k=-\infty}^{\infty}e^{j2\pi kf_s t}$$

根据式(2.34)，有

$$e^{j2\pi kf_s t}\rightleftarrows\delta(f-kf_s)$$

可得 $\text{comb}\ (t,nT_s)$ 的频谱 $\text{Comb}\ (f,f_s)$（图 2.23）也是梳状函数，即

$$\text{Comb}\ (f,T_s)=\frac{1}{T_s}\sum_{k=-\infty}^{\infty}\delta(f-kf_s)=\frac{1}{T_s}\sum_{k=-\infty}^{\infty}\delta\left(f-\frac{k}{T_s}\right) \tag{2.35}$$

由图 2.23 可见，时域周期单位脉冲序列的频谱也是周期脉冲序列。若时域周期为 T_s，则频域脉冲序列的周期为 $1/T_s$，时域脉冲强度为 1，频域中强度为 $1/T_s$。

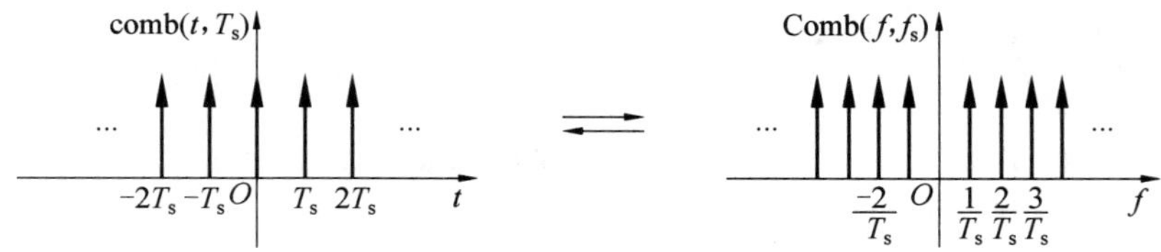

图 2.23　周期单位脉冲序列及其频谱

根据 δ 函数的采样特性可知，周期单位脉冲序列 $\text{comb}\ (t,nT_s)$ 和某一信号 $x(t)$ 的乘积就是周期单位脉冲序列在序列出现时刻 $nT_s(n=0,\pm1,\pm2,\pm3,\cdots)$ 对信号 $x(t)$ 在该时刻的值进行采样获取的过程，该过程称为信号的采样，如图 2.24 所示。

时域周期单位脉冲序列 $\text{comb}\ (t,nT_s)$ 的频谱也是周期脉冲序列。时域周期为 T_s 时，频域脉冲序列的周期为 $1/T_s$。时域脉冲强度为 1 时，频域中的脉冲强度为 $1/T_s$。周期单位脉冲序列 $\text{comb}\ (t,nT_s)$ 常用作模拟信号数字化时的采样信号。

$$\mathrm{comb}\ (t, T_s) \xlongequal{\mathrm{def}} \sum_{n=-\infty}^{\infty} \delta(t - nT_s)$$

where　T_s — the period;

n—an integer, $n = 0, \pm 1, \pm 2, \cdots$.

Because this function is a periodic function, it can be expressed as a complex exponential function of Fourier series, that is

$$\mathrm{comb}\ (t, T_s) = \sum_{k=-\infty}^{\infty} c_k \mathrm{e}^{\mathrm{j}2\pi k f_s t} \tag{2.34}$$

where

$$f_s = 1/T_s$$

coefficient c_k is

$$c_k = \frac{1}{T_s} \int_{-\frac{T_s}{2}}^{\frac{T_s}{2}} \mathrm{comb}\ (t, T_s) \mathrm{e}^{-\mathrm{j}2\pi k f_s t} \mathrm{d}t$$

Because in the interval of $(-T_s/2, T_s/2)$, formula (2.34) has only one δ function $\delta(t)$, and when $t = 0$, $\mathrm{e}^{-\mathrm{j}2\pi k f_s t} = \mathrm{e}^0 = 1$, so

$$c_k = \frac{1}{T} \int_{-\frac{T_s}{2}}^{\frac{T_s}{2}} \delta(t)\ \mathrm{e}^{-\mathrm{j}2\pi k f_s t} \mathrm{d}t = \frac{1}{T}$$

In this way, formula (2.34) can be expressed as

$$\mathrm{comb}\ (t, T_s) = \frac{1}{T_s} \sum_{k=-\infty}^{\infty} \mathrm{e}^{\mathrm{j}2\pi k f_s t}$$

According to formula (2.34), we have

$$\mathrm{e}^{\mathrm{j}2\pi k f_s t} \leftrightarrows \delta(f - kf_s)$$

The spectrum of comb (t, nT_s), Comb (f, f_s), is available (Figure 2.23), which is also a comb function, that is

$$\mathrm{Comb}\ (f, T_s) = \frac{1}{T_s} \sum_{k=-\infty}^{\infty} \delta(f - kf_s) = \frac{1}{T_s} \sum_{k=-\infty}^{\infty} \delta\left(f - \frac{k}{T_s}\right) \tag{2.35}$$

As seen from Figure 2.23, the frequency spectrum of the time-domain periodically unit impulse-train is also a periodical pulse sequence. If the the time domain period is T_s, the period of the frequency domain pulse sequence is $1/T_s$, the intensity of the time domain pulse is 1, and the intensity in the frequency domain is $1/T_s$.

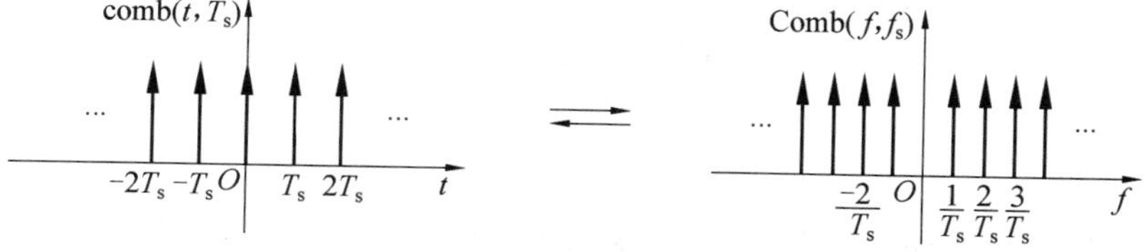

Figure 2.23　Periodically unit impulse-train and its frequency spectrum

According to the sampling characteristics of δ function, the product of the periodically unit impulse-train comb (t, nT_s) and a certain signal $x(t)$ is the process of sampling and acquiring the value of the signal $x(t)$ at the time $nT_s (n = 0, \pm 1, \pm 2, \pm 3, \cdots)$ of the sequence of the periodic unit pulse sequence, this process is called signal sampling, as shown in Figure 2.24.

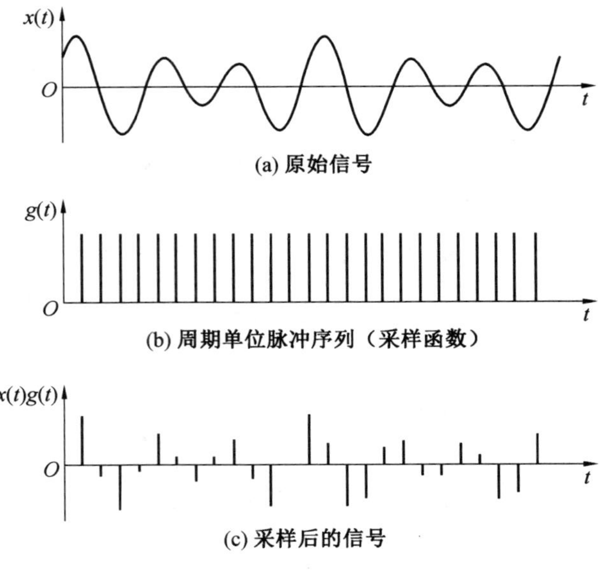

图 2.24　信号的采样

思考题与练习题

2.1　周期三角波如图 2.25 所示,其数学表达式为

$$x(t)=\begin{cases}A+\dfrac{4A}{T}t, & -\dfrac{T}{2}<t<0\\ A-\dfrac{4A}{T}t, & 0<t<\dfrac{T}{2}\end{cases}$$

求其傅里叶级数三角函数展开式并画出单边频谱图。

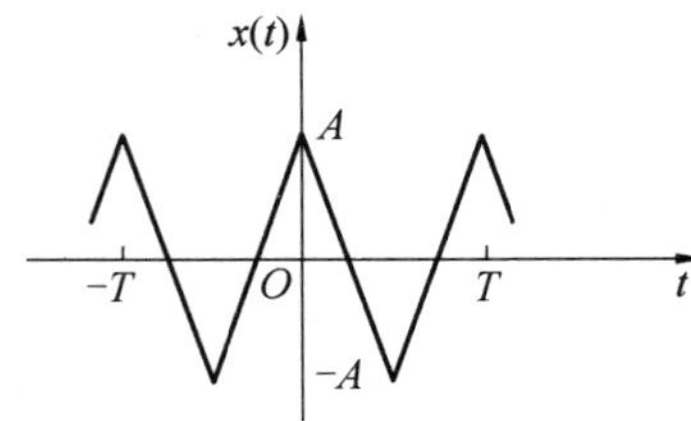

图 2.25　周期三角波

2.2　周期锯齿波如图 2.26 所示。求傅里叶级数三角函数展开式,并画出其单边频谱图。

2.3　已知方波的傅里叶级数展开式为

$$f(t)=\frac{4A_0}{\pi}\left(\cos\omega_0t-\frac{1}{3}\cos 3\omega_0t+\frac{1}{5}\cos 5\omega_0t-\cdots\right)$$

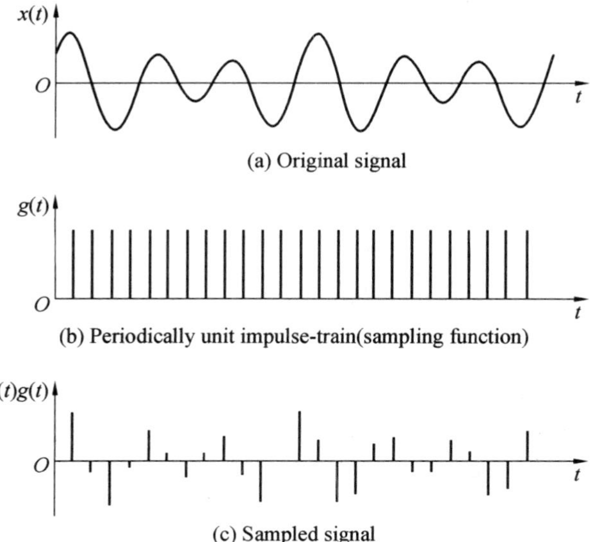

Figure 2.24　Signal sampling

The frequency spectrum of the time domain periodically unit impulse-train comb (t, nT_s) is also a periodic pulse train. If the time domain period is T_s, the period of the frequency domain pulse sequence is $1/T_s$. If the intensity of the time domain pulse is 1, the intensity in the frequency domain is $1/T_s$. The periodically unit impulse-train comb (t, nT_s) is often used as a sampling signal when the analog signal is digitized.

Questions and Exercises

2.1　The periodic triangular wave is shown in Figure 2.25, and its mathematical expression is

$$x(t) = \begin{cases} A + \dfrac{4A}{T}t, & -\dfrac{T}{2} < t < 0 \\ A - \dfrac{4A}{T}t, & 0 < t < \dfrac{T}{2} \end{cases}$$

Find the Fourier series trigonometric function expansion and draw a one-sided spectrogram.

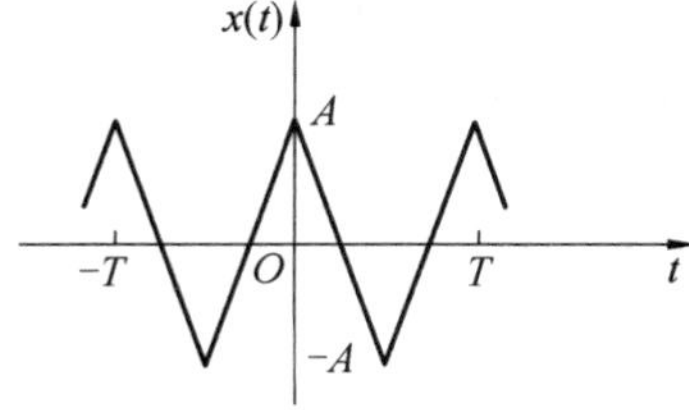

Figure 2.25　Periodic triangle wave

2.2　The periodic sawtooth signal is shown in Figure 2.26. Find the Fourier series trigonometric function expansion and draw its one-sided spectrum.

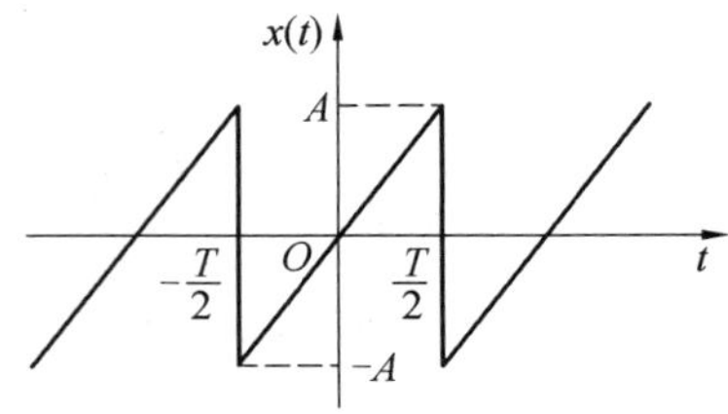

图 2.26　周期锯齿波

求该方波的均值、频率成分、各频率的幅值，并画出其频谱图。

2.4　$f(t)$ 的时域波形及其频谱如图 2.27 所示。已知函数 $x(t)=f(t)\cos\omega_0 t$，设 $\omega_0>\omega_m$（ω_m 为 $f(t)$ 中最高频率分量的角频率）。试画出 $x(t)$ 和 $X(j\omega)$ 的示意图形。当 $\omega_0<\omega_m$ 时，$X(j\omega)$ 的图形会出现什么样的情况？

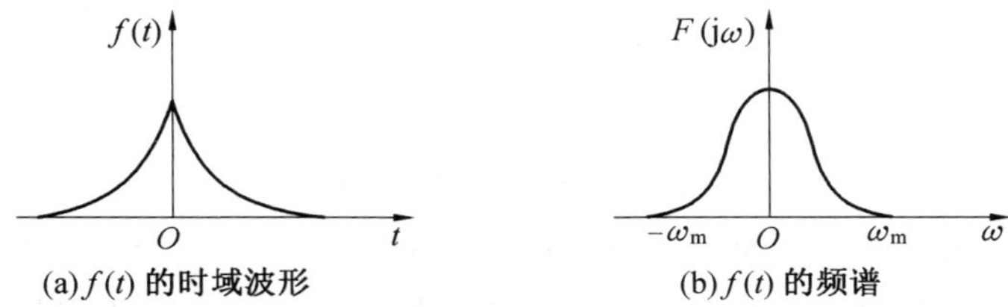

图 2.27　$f(t)$ 的时域波形及其频谱

2.5　求指数衰减振荡信号 $x(t)=e^{-at}\sin\omega_0 t(a>0,t\geqslant 0)$ 的频谱。

2.6　求正弦信号 $x(t)=x_0\sin\omega t$ 的绝对均值 $\mu_{|x|}$ 和方均根值 x_{rms}。

2.7　求指数函数 $x(t)=Ae^{-at}(a>0,t\geqslant 0)$ 的频谱。

2.8　求被截断的余弦函数

$$x(t)=\begin{cases}\cos\omega_0 t, & |t|<T\\ 0, & |t|>T\end{cases}$$

的傅里叶变换。$x(t)$ 的波形如图 2.28 所示。

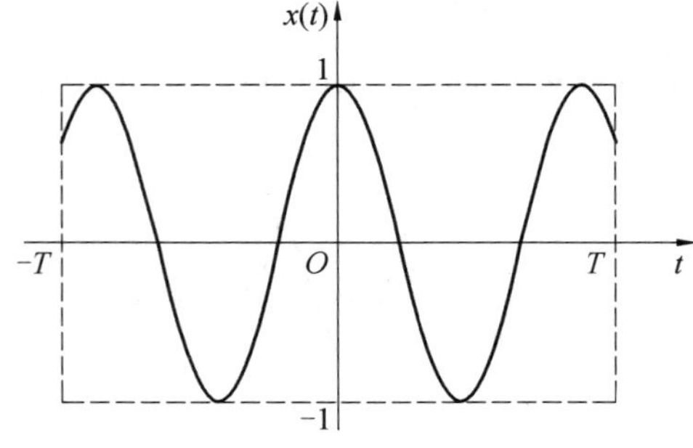

图 2.28　$x(t)$ 的波形

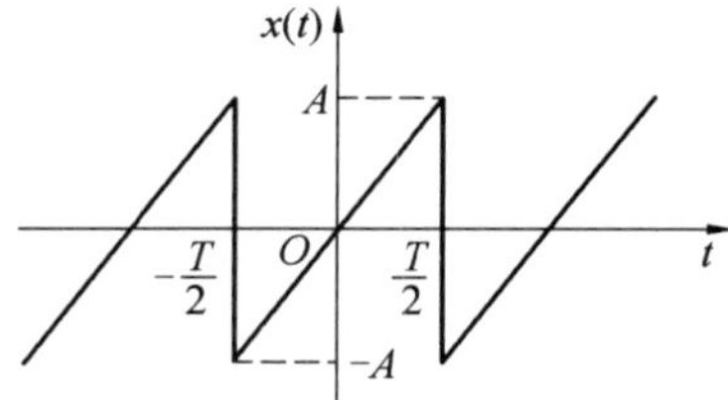

Figure 2.26　Periodic sawtooth wave

2.3　The Fourier series expansion of square wave is

$$f(t) = \frac{4A_0}{\pi}\left(\cos \omega_0 t - \frac{1}{3}\cos 3\omega_0 t + \frac{1}{5}\cos 5\omega_0 t - \cdots\right)$$

Calculate the average value, frequency components and amplitude of each frequency of the square wave, and draw its frequency spectrum.

2.4　The time domain waveform of $f(t)$ and its frequency spectrum are shown in Figure 2.27. Given function $x(t) = f(t)\cos \omega_0 t$, let $\omega_0 > \omega_m$ (ω_m is the angular frequency of the highest frequency component in $f(t)$). Try to draw the schematic diagrams of $x(t)$ and $X(j\omega)$. When $\omega_0 < \omega_m$, what will happen to the graph of $X(j\omega)$?

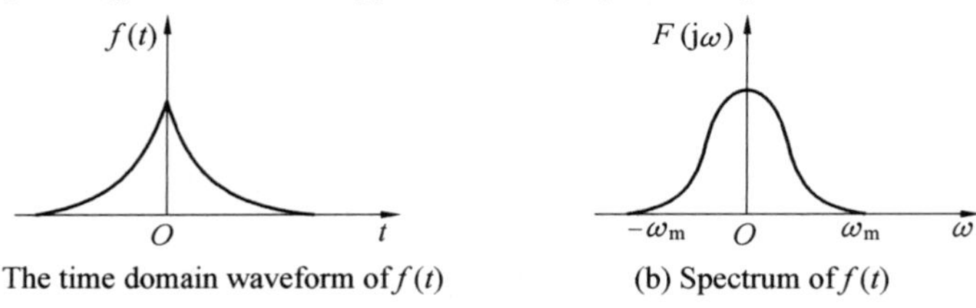

(a) The time domain waveform of $f(t)$　(b) Spectrum of $f(t)$

Figure 2.27　The time domain waveform of $f(t)$ and its frequency spectrum

2.5　Find the frequency spectrum of the exponentially decayed oscillation signal $x(t) = e^{-at}\sin \omega_0 t$ ($a > 0, t \geqslant 0$).

2.6　Find the absolute mean value $\mu_{|x|}$ and the root mean square value x_{rms} of the sine signal $x(t) = x_0 \sin \omega t$.

2.7　Find the frequency spectrum of the exponential function $x(t) = Ae^{-at}$ ($a > 0, t \geqslant 0$).

2.8　Find the Fourier transform of the truncated cosine function

$$x(t) = \begin{cases} \cos \omega_0 t, & |t| < T \\ 0, & |t| < T \end{cases}$$

The waveform of $x(t)$ is shown in Figure 2.28.

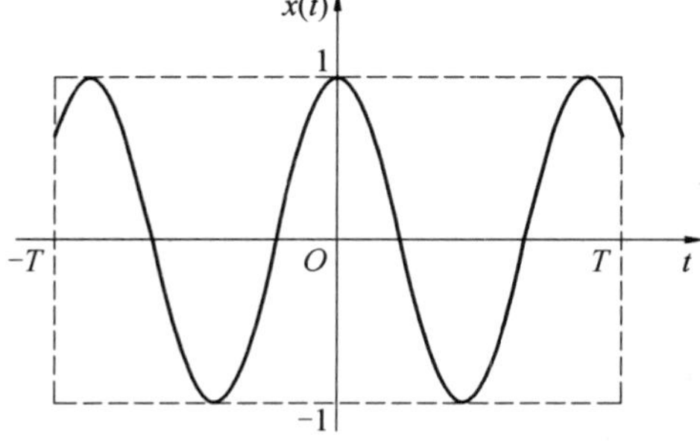

Figure 2.28　The waveform of $x(t)$

第3章　测试系统的基本特性

【本章学习目标】

1. 掌握线性系统、代数系统及其主要特性；

2. 掌握测试系统静态特性参数的定义；

3. 掌握测试系统动态特性的三种表述方法（线性微分方程、传递函数、频率响应函数）；

4. 掌握一、二阶测试系统动态特性参数（一阶系统的时间常数 τ、二阶系统的固有频率 ω_n 与阻尼比 ξ）及其获取方法；

5. 掌握并能应用测试系统的不失真测试条件。

Chapter 3　The Basic Features of Measurement Systems

【Learning Objectives】

1. To be able to grasp linear systems, algebra systems and their main characteristics;

2. To be able to know the definition of static characteristic parameters of the measurement systems;

3. To be able to master three expression methods of the dynamic characteristics of the measurement systems (linear differential equations, transfer function, frequency response function);

4. To be able to master the dynamic characteristic parameters of the first-order and second-order measurement systems (time constant τ of the first-order system, natural frequency ω_n and damping ratio ξ of the second-order system) and their obtaining methods;

5. To be able to apply the non-distortionmeasurement conditions of the measurement systems.

3.1 概 述

3.1.1 对测试系统的基本要求

首先介绍计量、测量和测试的基本概念，这三者关系密切。

(1)计量是指实现单位统一和量值准确可靠的测量。

(2)测量是指以确定被测对象量值为目的的全部操作。根据被测量是否随时间变化，可将测量分为动态测量和静态测量：静态测量是指测量期间被测量值可以认为恒定的测量；而动态测量是为确定被测量的瞬时值及（或）其随时间变化的历程所进行的测量。

(3)测试是具有试验性质的测量，也可理解为测量和试验的综合。

测试是从客观事物中提取有用信息的过程，是人类认识客观世界的手段，需要借助专门的设备——测试装置。最古老最简单的测试装置是“度、量、衡”，出于测量的目的，它比单凭人的感观估计来得精确。随着生产和科学技术的发展，测试所涵盖的内容越来越丰富，测试装置所肩负的任务越来越繁多，其科技含量及技术复杂程度也越来越高，但归根到底它还是人们感观的延伸和神经系统的扩大。在测试过程中，测试装置担负着信号的拾取、传递、处理、记录和显示等多重任务。要正确选择、使用测试装置，必须了解与考查它的基本特性，以求获得存在于客观世界中的真实信息。

在测量工作中，一般把测试装置作为一个系统来看待，系统是由若干相互作用、相互依赖的事物组合而成的具有特定功能的整体，系统遵从某些物理规律。系统的特性是指系统的输出和输入的关系。测试系统大体总可以分为以下三部分。

(1)感受外界信息的部分——传感器。

(2)显示、记录测试结果的部分或向下一系统传递本系统测试结果的输出部分，也可以笼统地称为系统终端。

(3)以上二者之间的部分，负责将传感器捕捉到的信号送往系统终端，但常常因为传递技术的实际需要而要将所传递的信号首先做某种转化（如将电路参数——电阻、电感等转化为电压、电流信号），或进一步转化为某种更易传输的形式，到达终端之前再将其恢复成原来的电信号，这一部分称为中间电路。

系统、输入和输出如图3.1所示，被处理的信号称为系统的激励或输入，处理后的信

3.1 Overview

3.1.1 Basic Requirements for Measurement Systems

Firstly, the basic concepts of metrology, measurement and test will be introduced, because the three are closely related.

(1) Metrology refers to achieve the unity of measurement units and accurate and reliable measurement of quantity.

(2) Measurement refers to all operations that determine the magnitude of the object to be measured. According to whether the measured quantity changes with time, the measurement can be divided into dynamic measurement and static measurement. Static measurement refers to the measurement in which the measured value can be regarded as constant during the measurement, while dynamic measurement is the measurement to determine the instantaneous value of the measured value and(or) its course of change over time.

(3) Test is a measurement of experimental nature, and can also be understood as a combination of measurement and experiment.

Measurement, as a means for human beings to understand the objective world, is a process of extracting useful information from objective things, which requires special equipment—measurement devices. The oldest and simplest measurement device is "measures, capacities and weights", which is more accurate than human perception estimation alone for measurement purposes. With the development of production and science and technology, the content covered by measurements is becoming more and more abundant, and measurement devices are taking on more and more tasks with its technological content and technical complexity becoming increasingly higher, ultimately, though, it is the extension of people's senses and the enlargement of the nervous system. During measuring, the measurement device is responsible for multiple tasks such as signal pickup, transmission, processing, recording, and display. To select and use measurement devices correctly, it is necessary to understand and investigate its basic characteristics in order to obtain real information that exists in the objective world.

In measurement work, the measurement device is generally regarded as a system that is a whole with specific function composed of a number of interacting and interdependent things, and obeys certain physical laws. System characteristics refer to the relationship between the output and input of the system. The measurement systems can be roughly divided into the following three parts.

(1) The part of feeling external information—sensor.

(2) The part that displays and records the measurement results or the output part that transmits the measurement results of the system to the next system can also be called the system terminal in general.

(3) The part between the above two is responsible for sending the signal captured by the

号为系统的响应或输出，任一系统的响应取决于系统本身及其输入。在测试信号传输过程中，连接输入、输出并有特定功能的部分均可视为测试系统。测试的内容、目的和要求不同，而测量对象又千变万化，因此测试系统的组成及其复杂程度也会有很大的差别。由测试装置自身的物理结构决定的测试系统对信号传输变换的影响称为测试系统的传输特性，简称系统的传输特性或系统的特性。任何测试系统都有自己的传输特性。为正确地描述或反映被测的物理量，或者根据测试系统的输出来识别其输入，必须研究测试系统输出、输入及测试系统传输特性之间的关系。测试系统与输入/输出量之间一般有以下几种关系。

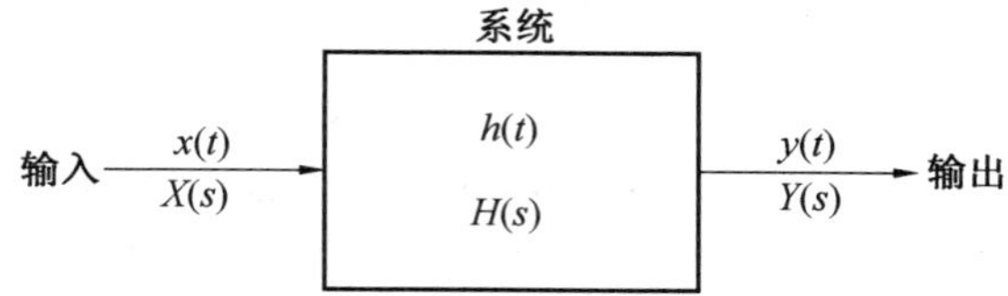

图 3.1　系统、输入和输出

（1）测量、测试过程。系统特性 $h(t)$ 已知，输出 $y(t)$ 可以观测，根据特性和测取的输出来推断输入 $x(t)$，这就是最常见的测量工作，其目的是获取测试对象的某种量值，包括它在时域内的变化历程。

（2）标定、校准过程。系统特性 $h(t)$ 已知，输入 $x(t)$ 是人为规定的已知量，观测输出 $y(t)$ 是否是由特性所规定的理论结果、相差多少、是否被规定误差所允许等。这是常见的对测试装置的标定、校准，其主要目的是对装置的质量、当前的技术状态、精确程度等做出判定。

（3）系统参数辨识过程。输入 $x(t)$ 已知，输出 $y(t)$ 可观测，通过输入和输出来研究系统的特性 $h(t)$ 或对系统运行状态、特性的变化等做出判断是更为复杂的测试工作。

对于测量过程来说，输入 $x(t)$ 是待测的未知量，测试系统的输出 $y(t)$ 能否正确地反映输入量 $x(t)$ 显然与测试系统本身的特性 $h(t)$ 有密切关系。由于测试系统传输特性的影响和外界各种干扰的入侵，因此输入量 $x(t)$ 难免会产生不同程度的失真，即输出量 $y(t)$ 是输入量 $x(t)$ 在经过测试系统传输、外界干扰双重影响后的一种结果。只有掌握了测试系统的特性，才能找出正确的使用方法，将失真控制在允许的范围之内，并对失真的大小做出定量分析。或者说，只有掌握了测试系统的特性，才能根据测试要求合理选用测试仪器。

测试系统的输出信号应该真实地反映被测物理量的变化过程，即实现不失真测试。从输入到输出，系统对输入信号进行传输和变换，系统的传输特性将对输入信号产生影

sensor to the system terminal, but often due to the actual needs of the transmission technology, it must first transform the transmitted signal (such as circuit parameters—resistance, inductance, etc. are transformed into voltage and current signals), or further transformed into a form that is easier to transmit, and then restored to the original electrical signal before reaching the terminal, the part of which is called the intermediate circuit.

System, input and output is shown in Figure 3. 1, the processed signal is called the excitation or input of the system, and the processed signal is the response or output of the system. The response of any system depends on the system itself and its input. A measurement system can be regarded as the part connecting the input and output with a specific function in the process of measurement signal transmission. The content, purpose and requirements of the measurement are different, and the test objects are ever-changing, so the composition and complexity of measurement systems will also be very different. The influence of the measurement systems on the signal transmission and transformation determined by the physical structure of the testing device is called the transmission characteristic of measurement systems, which is referred to as the transmission characteristic of systems or the characteristic of systems for short. Any measurement system has its own transmission characteristics. In order to correctly describe or reflect the physical quantity being measured, or to identify its input according to the output of measurement systems, it is necessary to study the relationship between the output and input of the measurement system and the transmission characteristics of the measurement system. There are generally the following relationships between the measurement system and the input/output volume.

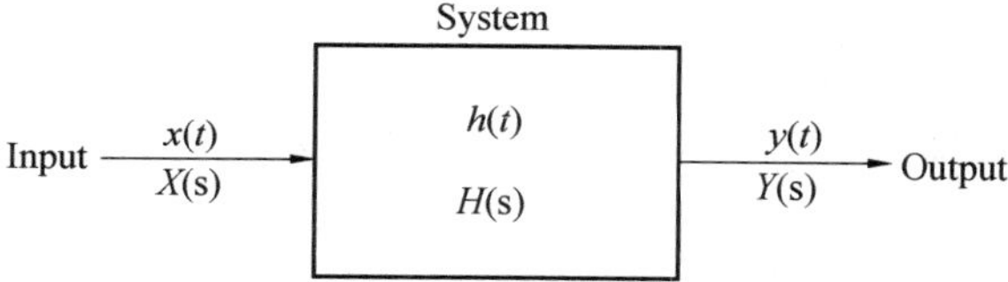

Figure 3. 1 System, input and output

(1) Measurement and test process. The most common measurement task is that the system characteristic $h(t)$ is known and the output $y(t)$ can be observed, and the input $x(t)$ is inferred based on the characteristic and the measured output, which its purpose is to obtain a certain value of the measurement subject, including its history of change in the time domain.

(2) Calibration process. The system characteristic $h(t)$ is known, and the input $x(t)$ is an artificially prescribed known quantity, observe whether the output $y(t)$ is the theoretical result specified by the characteristic? What is the difference? Is it within the specified error range? This is a common calibration of measuring device, and its main purpose is to make judgments on the quality of the equipment, the current technical state, and the degree of accuracy.

(3) System parameter identification process. Input $x(t)$ is known and output $y(t)$ is observable. It is a more complicated measurement work to study the characteristics of the system $h(t)$ through input and output, or to make judgments on the operating status of the system and changes in characteristics.

For the measurement process, input $x(t)$ is the unknown quantity to be measured, and whether the output $y(t)$ of the measurement system can correctly reflect the input quantity $x(t)$

响。因此，要使输出真实反映输入的状态，理想的测试系统应满足如下要求：其传输特性应该具有单值的、确定的输入 - 输出关系，即对应于每个确定的输入量，都应有唯一的输出量与之对应，并且以输出和输入呈线性关系为最佳。

实际测试系统不可能是理想的线性系统。在静态测试时，测试系统最好具有线性关系，但不是必须的（不是线性关系也可以）。一般只要求测试系统的静态特性是单值函数，因为在静态测量中可用曲线校正或输出补偿技术做非线性校正。对于动态测试，目前只能对线性系统做较完善的数学处理与分析，而且在动态测试中做非线性校正相当困难。因此，要求动态测试系统的传输特性必须是线性的，否则输出信号会产生畸变。然而，实际的测试系统只能在允许误差范围内和一定的工作范围内满足这一要求。

无论是动态测试还是静态测试，都是以系统的输出量去估计输入量的，测试的目的是准确地了解被测物理量。但人们通过测试只能得到被测量经过测试系统的各个环节后的输出量，无法测到被测量的真值。研究系统的特性就是为使系统尽可能在准确、真实地反映被测量方面做得更好，同时也是为了对现有的测试系统优劣提供客观评价。

综上所述，理想的测试系统，其输出信号与输入信号一一对应，且输出与输入之间最好是线性关系。根据系统输出和输入的关系，大体上可分为两类：第一类称为线性系统；第二类称为代数系统。

3.1.2 理想的测试系统——线性时不变系统

代数系统的输出函数常常取简单的代数形式，线性系统输出和输入之间的关系可以用常系数线性微分方程来描述。代数系统和线性系统在动态测试中多数负责信号的拾取捕捉或记录与显示，常作为传感器、显示和记录仪器来使用。一阶线性系统在一定条件下具有微、积分功能，同时对信号的频率成分具有很强的选择性，故一阶线性系统还可作为微分器、积分器或滤波器使用。

1. 线性时不变系统的概念

当测试系统的输入 $x(t)$ 和输出 $y(t)$ 之间可以用下列常系数线性微分方程

$$\begin{aligned} & a_n \frac{\mathrm{d}^n y(t)}{\mathrm{d}t^n} + a_{n-1} \frac{\mathrm{d}^{n-1} y(t)}{\mathrm{d}t^{n-1}} + \cdots + a_1 \frac{\mathrm{d}y(t)}{\mathrm{d}t} + a_0 y(t) \\ = & b_m \frac{\mathrm{d}^m x(t)}{\mathrm{d}t^m} + b_{m-1} \frac{\mathrm{d}^{m-1} x(t)}{\mathrm{d}t^{m-1}} + \cdots + b_1 \frac{\mathrm{d}x(t)}{\mathrm{d}t} + b_0 x(t) \end{aligned} \tag{3.1}$$

来描述时，则称该系统为线性时不变系统（Linear Time Invariant，LTI），又称定常线性系统。时不变系统包含以下两层含义。

is obviously closely related to the characteristic $h(t)$ of the measurement system itself. Due to the influence of transmission characteristics of the measurement system and the intrusion of various external interference, it is inevitable that the input quantity $x(t)$ will produce different degrees of distortion, that is, the output quantity $y(t)$ is a result of the input quantity $x(t)$ after the double influence of the transmission of the measurement system and the external interference. Only by mastering the characteristics of the measurement system can we find out the correct use method to control the distortion within the allowable range and make a quantitative analysis of the magnitude of the distortion. In other words, only when the characteristics of measurement systems are mastered, can the measurement equipment be reasonably selected according to measuring instruments.

The output signal of themeasurement system should truly reflect the change process of measured physical quantity, that is, to realize the non-distortion measurement. From input to output, the system transmits and transforms the input signal, of which the transmission characteristics will affect the input signal. Therefore, to make the output truly reflect the state of the input, an ideal measurement system should meet the following requirements: its transmission characteristics should have a single-valued, definite input-output relationship, that is, there should be a unique output quantity corresponding to each certain input quantity and the linear relationship between output and input is the best.

The actual measurement system cannot be an ideal linear system. In static measurement, the measurement system is best to have a linear relationship, but it is not necessary (it may not be linear). Generally, only the static characteristics of measurement system are required to be a single value function, because curve correction or output compensation technology can be used for nonlinear correction in static measurement. For dynamic measurement, we can only do more complete mathematical processing and analysis on linear systems at present, and it is quite difficult to make nonlinear corrections. Therefore, the transmission characteristics of dynamic measurement systems must be linear, otherwise the output signal will be distorted. However, the actual measurement system can only meet this requirement within the allowable error range and within a certain working range.

Whether it is a dynamic measurement or a static measurement, the input is estimated by the output of system. The purpose of the measurement is to accurately understand the measured physical quantity. However, people can only get the output of the measured quantity after each link of the measurement system through the measurement, and the true value of the measured quantity cannot be measured. The purpose of studying the system characteristics is not only to make the system as accurate and true as possible in reflecting the measured, but also to provide an objective evaluation of the pros and con of the existing measurement system.

As mentioned above, an ideal measurement system has a one-to-one correspondence between output signals and input signals, and the relationship between them proves preferably linear. According to the relationship between system output and input, it is roughly divided into two categories: the first category is called linear system; and the second category is called algebraic system.

3.1.2　The Ideal Measurement System—Linear Time Invariant System

The output function of algebraic systems often takes a simple algebraic form. The

(1) 式(3.1) 中 $a_n, a_{n-1}, \cdots, a_1, a_0$ 及 $b_m, b_{m-1}, \cdots, b_1, b_0$ 是与系统硬件参数有关的常数。不随时间变化而变化。输出信号 $y(t)$ 的波形与输入 $x(t)$ 加入的时间无关,时不变系统举例如图 3.2 所示。严格地说,很多物理系统是时变的,因为构成系统的材料、元件、部件的特性参数并非稳定。例如,材料的弹性模量,电子元件的电阻、电容,半导体器件的特性都易受温度的影响,而环境温度是时间的缓变量,因此导致微分方程各系数有时变性。但在工程上,常可以足够精确地认为多数常见的物理系统的参数是时不变的常数。

(2) 各阶微分项之间是线性叠加关系。在上述微分方程中,$n \geqslant m$。当 $n = 1$ 时,称系统为一阶线性系统;当 $n = 2$ 时,称系统为二阶线性系统,以此类推。

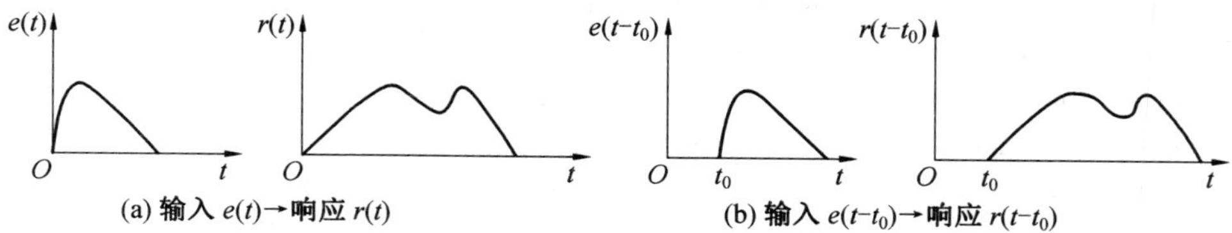

图 3.2　时不变系统举例

2. 线性时不变系统的主要性质

(1) 叠加性。几个输入所产生的总输出是各个输入所产生的输出叠加的结果,即若

$$x_1(t) \rightarrow y_1(t)$$

$$x_2(t) \rightarrow y_2(t)$$

则有

$$x_1(t) \pm x_2(t) \rightarrow y_1(t) \pm y_2(t) \tag{3.2}$$

叠加原理意味着线性系统的各个输入是互不影响的,一个输入的存在绝不影响另一个输入的输出。在分析众多输入总的效果时,可以先分别分析单个输入(假定其他输入不存在)的效果,然后将这些效果叠加起来以表示总的效果。当分析线性系统在复杂输入作用下的总输出时,可先将复杂输入分解成许多简单的输入分量,分别求出各简单分量输入时所对应的输出,然后求出这些输出之和,即得总的输出。这就给试验工作带来很大的方便,测试系统的正弦试验就是采用这种方法。

(2) 比例特性。对于任意常数 a,必有

$$ax(t) \rightarrow ay(t) \tag{3.3}$$

(3) 微分特性。系统对输入导数的响应等于原响应的导数,即

$$\frac{\mathrm{d}x(t)}{\mathrm{d}t} \rightarrow \frac{\mathrm{d}y(t)}{\mathrm{d}t} \tag{3.4}$$

(4) 积分特性。若系统的初始状态均为0,则系统对输入积分的响应等于原响应的积

relationship between the output and input of a linear system can be described by a linear differential equation with constant coefficients. In dynamic measurement, these two systems are mostly responsible for signal pickup and capture, or recording and display, often used as a sensor, display and recording instrument. The first-order linear system has functions of differentiation and integration under certain conditions, and has strong selectivity for the frequency components of the signal at the same time, thus the first-order linear system can also be used as a differentiator, integrator or filter.

1. The Concept of Linear Time Invariant System

When the input $x(t)$ and output $y(t)$ of the measurement system can be described by the following constant coefficient linear differential equation, namely

$$\begin{aligned} & a_n \frac{\mathrm{d}^n y(t)}{\mathrm{d}t^n} + a_{n-1} \frac{\mathrm{d}^{n-1} y(t)}{\mathrm{d}t^{n-1}} + \cdots + a_1 \frac{\mathrm{d}y(t)}{\mathrm{d}t} + a_0 y(t) \\ = {} & b_m \frac{\mathrm{d}^m x(t)}{\mathrm{d}t^m} + b_{m-1} \frac{\mathrm{d}^{m-1} x(t)}{\mathrm{d}t^{m-1}} + \cdots + b_1 \frac{\mathrm{d}x(t)}{\mathrm{d}t} + b_0 x(t) \end{aligned} \tag{3.1}$$

the system is called linear time invariant (LTI), also known as a time-invariant linear system. The system has the following two meanings.

(1) In formula (3.1), $a_n, a_{n-1}, \cdots, a_1, a_0$ and $b_m, b_{m-1}, \cdots, b_1, b_0$ are constants related to the system hardware parameters, and the internal parameters of the system do not change with time. The waveform of the output signal $y(t)$ has nothing to do with the time added by the input $x(t)$. Examples of time-invariant system are shown in Figure 3.2. Strictly speaking, many physical systems are time-varying because the characteristic parameters of the materials, elements, and parts that make up the system are not stable. For example, the elastic modulus of materials, the resistance and capacitance of electronic components, and the characteristics of semiconductor devices are all easily affected by temperature. The ambient temperature is a slow variable of time, which causes the coefficients of the differential equations changing sometimes. But it is often possible to believe that the parameters of most common physical systems are time-invariant constants with sufficient accuracy in engineering.

(2) There is a linear superposition relationship between the differential terms of each order. In the above differential equation, $n \geqslant m$. When $n = 1$, the system is called a first-order linear system; when $n = 2$, the system is called a second-order linear system, and so on.

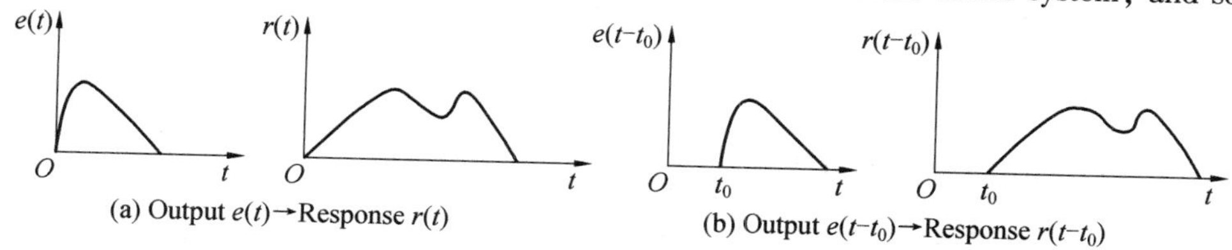

(a) Output $e(t)$→Response $r(t)$　　(b) Output $e(t-t_0)$→Response $r(t-t_0)$

Figure 3.2　Examples of time-invariant system

2. The Main Properties of Linear Time Invariant System

(1) Superposition. The total output produced by several inputs is the result of the superposition of the outputs produced by each input, that is, if

$$x_1(t) \rightarrow y_1(t)$$
$$x_2(t) \rightarrow y_2(t)$$

分，即

$$\int_0^t x(t)\,\mathrm{d}t \to \int_0^t y(t)\,\mathrm{d}t \tag{3.5}$$

(5) 频率保持性。若输入为某一频率的简谐(正弦或余弦) 信号，则系统的稳态输出必然也只能是同频率的简谐信号。

由于

$$x(t) \to y(t)$$

因此按线性系统的比例特性，对于某一已知频率 ω，有

$$\omega^2 x(t) \to \omega^2 y(t)$$

又根据线性系统的微分特性，有

$$\frac{\mathrm{d}^2 x(t)}{\mathrm{d}t^2} \to \frac{\mathrm{d}^2 y(t)}{\mathrm{d}t^2}$$

应用叠加原理，有

$$\frac{\mathrm{d}^2 x(t)}{\mathrm{d}t^2} + \omega^2 x(t) \to \frac{\mathrm{d}^2 y(t)}{\mathrm{d}t^2} + \omega^2 y(t)$$

现令输入为某一频率的简谐信号，即

$$x(t) = X_0 \mathrm{e}^{\mathrm{j}\omega t} = X_0(\cos \omega t + \mathrm{j}\sin \omega t)$$

那么其二阶导数应为

$$\frac{\mathrm{d}^2 x(t)}{\mathrm{d}t^2} = \frac{\mathrm{d}^2(X_0 \mathrm{e}^{\mathrm{j}\omega t})}{\mathrm{d}t^2} = (\mathrm{j}\omega)^2 X_0 \mathrm{e}^{\mathrm{j}\omega t} = -\omega^2 x(t)$$

由此可得

$$\frac{\mathrm{d}^2 x(t)}{\mathrm{d}t^2} + \omega^2 x(t) = 0$$

相应的输出也应为

$$\frac{\mathrm{d}^2 y(t)}{\mathrm{d}t^2} + \omega^2 y(t) = 0$$

于是输出 $y(t)$ 的唯一可能解只能是

$$y(t) = Y_0 \mathrm{e}^{\mathrm{j}(\omega t+\varphi_0)} = Y_0[\cos(\omega t + \varphi_0) + \mathrm{j}\sin(\omega t + \varphi_0)] \tag{3.6}$$

线性系统的这些主要特性，特别是符合叠加原理和频率保持性，在测试工作中具有重要的作用。假如已知系统是线性的，并且知道其输入的频率，那么根据频率保持性，可以认定测得的信号中只有与输入频率相同的成分才真正是由该输入引起的输出，而其他频率成分都是噪声(干扰)。进而可以根据这一特性，采用相应的滤波技术，在很强的噪声干扰下把有用的信息提取出来。

then we have

$$x_1(t) \pm x_2(t) \rightarrow y_1(t) \pm y_2(t) \tag{3.2}$$

The superposition principle means that the various inputs of a linear system do not affect each other. The existence of one input will never affect the output of another input. When analyzing the total effect of many inputs, we can analyze a single input separately (assuming that other inputs do not exist), and then superimpose these effects to represent the total effect. When analyzing the total output of a linear system under the action of a complex input, the complex input can be decomposed into many simple input components, and the corresponding output of each simple component input can be calculated, and then the sum of these outputs, all which means to get the total output. This method that brings great convenience to the measurement work is adopted in the sine measurement of the testing measurement.

(2) Proportional characteristic. For any constant a, there must be

$$ax(t) \rightarrow ay(t) \tag{3.3}$$

(3) Differential characteristic. The response of the system to the input derivative is equal to the derivative of the original response, namely

$$\frac{dx(t)}{dt} \rightarrow \frac{dy(t)}{dt} \tag{3.4}$$

(4) Integral characteristics. If the initial state of the system is 0, the response of the system to the input integral is equal to the integral of the original response, that is

$$\int_0^t x(t)\,dt \rightarrow \int_0^t y(t)\,dt \tag{3.5}$$

(5) Frequency retention. If the input is a simple harmonic (sine or cosine) signal of a certain frequency, the steady-state output of the system must be and can only be a simple harmonic signal of the same frequency.

Because of

$$x(t) \rightarrow y(t)$$

according to the proportional characteristic of the linear system, for a certain known frequency ω, we have

$$\omega^2 x(t) \rightarrow \omega^2 y(t)$$

And according to the differential characteristics of the linear system, we have

$$\frac{d^2x(t)}{dt^2} \rightarrow \frac{d^2y(t)}{dt^2}$$

Applying the superposition principle, we have

$$\frac{d^2x(t)}{dt^2} + \omega^2 x(t) \rightarrow \frac{d^2y(t)}{dt^2} + \omega^2 y(t)$$

Now let the input be a simple harmonic signal of a certain frequency, that is

$$x(t) = X_0 e^{j\omega t} = X_0(\cos \omega t + j\sin \omega t)$$

then its second derivative should be

$$\frac{d^2x(t)}{dt^2} = \frac{d^2(X_0 e^{j\omega t})}{dt^2} = (j\omega)^2 X_0 e^{j\omega t} = -\omega^2 x(t)$$

From this, we get

$$\frac{d^2x(t)}{dt^2} + \omega^2 x(t) = 0$$

或者说，若输入为正弦信号 $x(t) = A\sin(\omega t + \alpha)$，则输出信号必为 $y(t) = B\sin(\omega t + \beta)$。

该特性表明，当系统处于线性工作范围内时，若输入信号频率已知，则输出信号与输入信号具有相同的频率分量。如果输出信号中出现与输入信号频率不同的分量，则说明系统中存在非线性环节（噪声等干扰）或者超出了系统的线性工作范围，应采用滤波等方法进行处理。

线性系统的频率保持性在动态测试中具有重要作用。例如，在振动测试中，若已知输入的激励频率，则测得的输出信号中只有与激励频率相同的成分才可能是由该激励引起的振动，而其他频率信号都属于干扰，应予以剔除。利用这一特性，就可以采用相应的滤波技术，在有很强的噪声干扰情况下提取出有用的信息。

3.1.3 代数系统及其主要性质

当测试系统的输出 $y(t)$ 和输入 $x(t)$ 之间的函数可以用一般的代数形式来描述时，称该系统为代数系统，常见的代数系统其输出函数有以下两种形式，即

$$y(t) = b + Sx(t),\ y(t) = Sx(t) \tag{3.7}$$

式中 S —— 灵敏度。

$$y(t) = \frac{D}{x(t)} \tag{3.8}$$

式中 D —— 时不变常数。

代数系统的输出函数曲线如图 3.3 所示。此外，还有其他形式。在实际测试系统中，式(3.7) 用得最多，因为它的输出和输入保持定比例线性关系，是很理想的测试装置。式(3.8) 从整体上不满足线性关系，$y(t)$ 与 $x(t)$ 是反比例关系。它之所以在动态测试中获得应用，是因为在 $x(t)$ 的一个微小区间内可以将 $y(t)$ 曲线近似的视为一段斜率为负值的直线（图 3.3(c)）。若在实际使用中 $x(t)$ 的变化范围能保持在这一微小区间里，又为规定的误差所允许，那么 $y(t)$ 与 $x(t)$ 的关系就可以表示为式(3.7) 所示的形式，只是此时 S 为负值而已。当以过 $y(t)$ 曲线上的点 (x_0, y_0) 的切线来代替 $y(t)$ 曲线时，则在 x_0 左右的微小区间内，$y(t)$ 与 $x(t)$ 的关系可以近似表示为

$$y(t) = y_0 + S(x(t) - x_0) = y_0 + S\Delta x(t)$$

式中

$$y_0 = \frac{D}{x_0}$$

The corresponding output should also be

$$\frac{d^2y(t)}{dt^2} + \omega^2 y(t) = 0$$

So the only possible solution for the output $y(t)$ can only be

$$y(t) = Y_0 e^{j(\omega t+\varphi_0)} = Y_0[\cos(\omega t + \varphi_0) + j\sin(\omega t + \varphi_0)] \tag{3.6}$$

These main characteristics of the linear system, especially in accordance with the superposition principle and frequency retention, play an important role in the measurement work. If the linear system and other input frequencies are known, then based on the frequency retention, it can be concluded that only the components of the measured signal with the same frequency as the input are the output caused by the input, and other frequency components are all noise (interference). Furthermore, based on this characteristic, the corresponding filtering technology can be used to extract useful information under strong noise interference.

In other words, if the input is a sin signal $x(t) = A\sin(\omega t + \alpha)$, then the output signal must be $y(t) = B\sin(\omega t + \beta)$.

This characteristic shows that when the system is in the linear operating range, the output signal and the input signal have the same frequency components with the input signal frequency being known. If, in the output signal, there are components different with frequencies in the input signal, it indicates that there is a non-linear link (interference such as noise) in the system or it exceeds the linear working range of the system, therefore filtering and other methods should be used for processing.

The frequency retention of linear systems plays an important role in dynamic measurement. For example, in a vibration measurement, if the input excitation frequency is known, only the components of the measured output signal that are the same as the excitation frequency can be the vibration caused by the excitation, and other frequency signals belonging to interference should be considered eliminated. To take full advantage of this feature, corresponding filtering technology can be used to extract useful information in the presence of strong noise interference.

3.1.3 Algebraic Systems and Its Main Properties

When the function between the output $y(t)$ and the input $x(t)$ of the measurement system can be described in a general algebraic form, it is called an algebraic system, of which the output function usually has the following two forms, namely

$$y(t) = b + Sx(t), y(t) = Sx(t) \tag{3.7}$$

where S —— sensitivity.

$$y(t) = \frac{D}{x(t)} \tag{3.8}$$

where D —— time-invariant constant.

The output function curves of algebraic systems are shown in Figure 3.3. In addition, there are other forms. In the actual measurement system, the formula (3.7) is used the most, because its output and input maintain a proportional linear relationship, which proves an ideal measurement device. The formula (3.8) does not satisfy the linear relationship as a whole, and the relationship between $y(t)$ and $x(t)$ is inversely proportional. The reason why it is applied in the dynamic measurement is that in a small interval of $x(t)$, $y(t)$ curve can be ap-

x_0 是 $x(t)$ 的初始值，是一个常数。

$$S = \frac{\mathrm{d}y}{\mathrm{d}x}\Big|_{x=x_0}$$

S 是过(x_0, y_0) 点的 $y(t)$ 曲线的切线的斜率，即系统的灵敏度。

若将纵坐标平移至 $x(t) = x_0$ 处(图 3.3(c))，则上式就可以写为

$$y(t) = y_0 + Sx(t)$$

$x(t)$ 仅发生微小变化时，可以认为输出与输入保持(近似)线性关系。只要近似关系所引起的线性误差不超过规定值，且灵敏度合适，则该代数系统就具有使用价值，可以化为定比例代数线性系统，在 x_0 附近的微小区间内对 $x(t)$ 的微小变化进行测量。

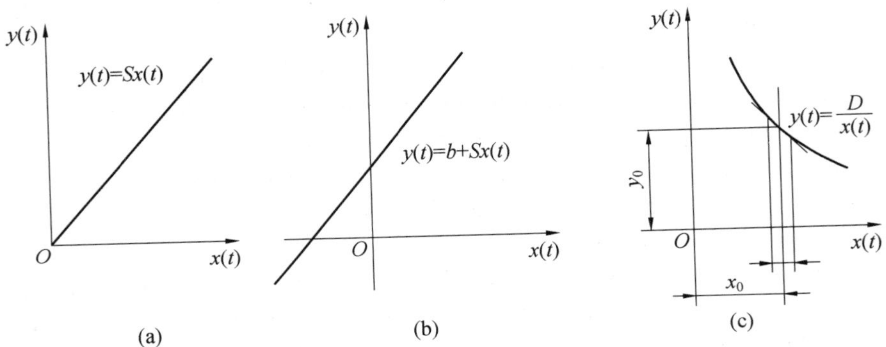

图 3.3　代数系统的输出函数曲线

例 3.1　求图 3.4 所示两个平行金属板组成电容器的灵敏度。

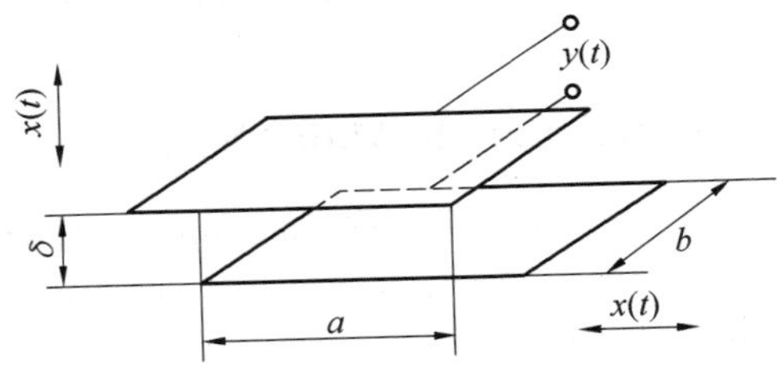

图 3.4　电容式位移传感器

电容量为

$$C = \frac{\varepsilon_0 \varepsilon ab}{\delta}$$

式中　ε_0—— 真空中的介电常数；

ε—— 极板间介质的相对介电常数；

δ—— 极板间距离；

b—— 极板的宽度；

proximated as a straight line with a negative slope(Figure 3.3(c)). If the variation range of $x(t)$ can be kept within this small interval in actual use, and within the specified error range, then the relationship between $y(t)$ and $x(t)$ can be expressed as the form shown in formula (3.7), but at this time S is just a negative value. When the tangent to point (x_0, y_0) on the $y(t)$ curve is used instead of the $y(t)$ curve, within a small interval of about x_0, the relationship between $y(t)$ and $x(t)$ can be approximately expressed as

$$y(t) = y_0 + S(x(t) - x_0) = y_0 + S\Delta x(t)$$

where

$$y_0 = \frac{D}{x_0}$$

x_0 is the initial values of $x(t)$, which is a constant.

$$S = \frac{\mathrm{d}y}{\mathrm{d}x}\Big|_{x=x_0}$$

S is the slope of the tangent to the curve $y(t)$ passing through point (x_0, y_0), and it is the sensitivity of the system.

If the ordinate is translated to $x(t) = x_0$(Figure 3.3(c)), then the above formula can be written as

$$y(t) = y_0 + Sx(t)$$

When $x(t)$ changes only slightly, it can be considered that the output and input maintain (approximately) linear relationship. As long as the linear error caused by the approximate relationship does not exceed the specified value and the sensitivity is appropriate, the algebraic system has use value and can be transformed into a proportional algebraic linear system to measure the small change of $x(t)$ in a small interval near x_0.

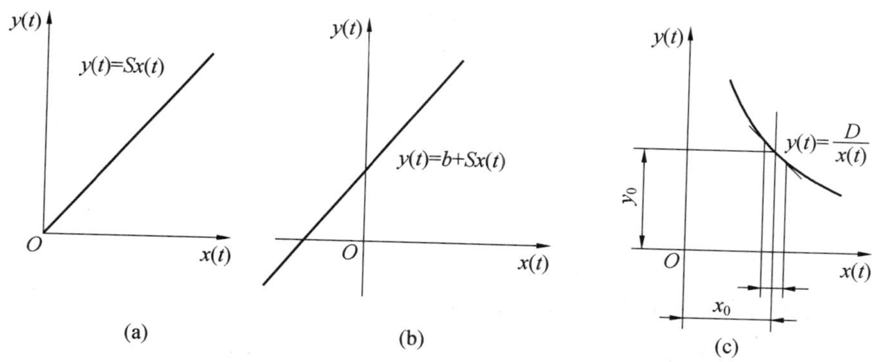

Figure 3.3　The output function curves of algebraic systems

Example 3.1 Find the sensitivity of the capacitor composed of two parallel metal plates shown in Figure 3.4.

Capacitance is

$$C = \frac{\varepsilon_0 \varepsilon ab}{\delta}$$

where　ε_0—the permittivity in vacuum;

ε—the relative permittivity of the medium between the polar plates;

δ—the distance between the polar plates;

a—— 上下极板的重合度。

解 在结构形式和有关参数不变的情况下，将两个平行金属板作为电容式传感器使用时，随输入形式的不同有以下两种模式。

(1) 面积变化型。

当被测对象的水平位移 $x(t)$ 引起重合度 a 变化时，以 C 为传感器的输出 $y(t)$，则有

$$y(t)=\frac{\varepsilon_0\varepsilon b}{\delta}a=C_0+Sx(t)$$

式中

$$S=\frac{\varepsilon_0\varepsilon b}{\delta}$$

S 是系统灵敏度，为常数。

$$C_0=Sa_0$$

其中 a_0 是原始重合度。

在这种测试装置中，$y(t)$ 与 $x(t)$ 保持定比例线性关系。

(2) 极距变化型。

当被测对象的垂直位移 $x(t)$ 引起极距 δ 变化时，系统的输出 $y(t)$（电容器电容 C）将与极距 δ 保持反比例关系，即

$$y(t)=\frac{\varepsilon_0\varepsilon ab}{\delta}=\frac{D}{x(t)}$$

式中

$$D=\varepsilon_0\varepsilon ab$$

D 为常数。

(3) 对比分析。

以上两种传感器都可以作为位移传感器使用。前者 $y(t)$ 与 $x(t)$ 保持定比例线性关系，后者 $y(t)$ 与 $x(t)$ 成反比例关系，它们都属于代数系统。前者十分理想，后者在使用中受到一定限制，仅适用于微小位移(0.01 μm ~ 数百微米）的测量。

当极距变化型传感器的初始极距定为 δ_0 时，$y_0=D/\delta_0=C_0$。此后，极距随被测对象位移产生一微小增量 $x(t)$ 时，输出为

$$y(t)=y_0+Sx(t)$$

式中

$$S=\left.\frac{\mathrm{d}y}{\mathrm{d}x}\right|_{x=\delta_0}=-\frac{\varepsilon_0\varepsilon ab}{\delta_0^2}$$

b —the width of the polar plates;

a —the coincidence degree of the upper and lower plates.

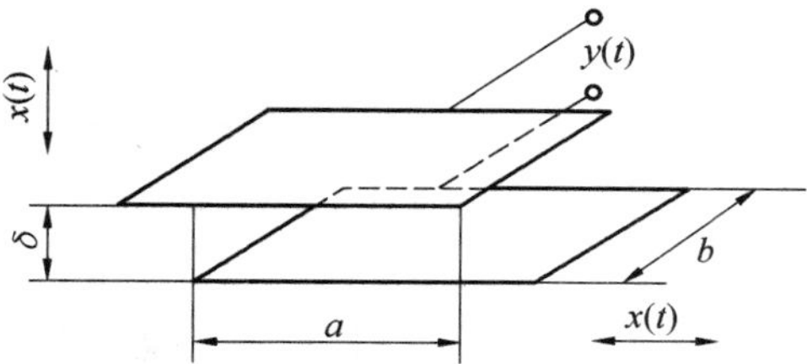

Figure 3.4　Capacitive displacement sensors

Solution　Under the condition that the structure and related parameters remain unchanged, when two parallel metal plates are used as a capacitive sensor, there are the following two modes depending on the input form.

(1) Area change type.

When the horizontal displacement $x(t)$ of the measured object causes the coincidence degree a to change, taking C as the output $y(t)$ of the sensor, then we have

$$y(t) = \frac{\varepsilon_0 \varepsilon b}{\delta} a = C_0 + Sx(t)$$

where

$$S = \frac{\varepsilon_0 \varepsilon b}{\delta}$$

S is the system sensitivity, a constant.

$$C_0 = Sa_0$$

a_0 is the original coincidence degree.

This measurement device $y(t)$ and $x(t)$ maintain a proportional linear relationship.

(2) Polar distance change type.

When the vertical displacement $x(t)$ of the measured object causes the pole distance δ to change, the output $y(t)$ (capacitor capacitance C) of the system will maintain an inverse proportional relationship with the pole distance δ, that is

$$y(t) = \frac{\varepsilon_0 \varepsilon ab}{\delta} = \frac{D}{x(t)}$$

where

$$D = \varepsilon_0 \varepsilon ab$$

D is a constant.

(3) Comparative analysis.

The above two kinds of sensors can be used as displacement sensors. The former $y(t)$ and $x(t)$ maintain a fixed proportional linear relationship, and the latter $y(t)$ and $x(t)$ are in an inverse proportional relationship, all which belong to an algebraic system. The former is ideal, and the latter, subject to certain restrictions in use, is only suitable for measurement of small displacements (0.01 μm to hundreds of microns).

When the initial pole distance of polar distance change type sensor is set as δ_0, $y_0 = D/\delta_0 = C_0$. After that, when the polar distance produces a small increment $x(t)$ with the displacement of the measured object, the output is

$$y(t) = y_0 + Sx(t)$$

where

$$S = \left.\frac{\mathrm{d}y}{\mathrm{d}x}\right|_{x=\delta_0} = -\frac{\varepsilon_0 \varepsilon ab}{\delta_0^2}$$

S 为传感器的灵敏度。

当 δ_0 很小时，S 会很大，系统非常灵敏，因此适于测量微位移。若 $x(t) < 0.1\delta_0$，则最大线性误差不会超过 1%。

3.2 测试系统的静态特性

如果被测试的信号是不随时间变化的常量，或者在测试阶段信号的变化极缓慢，以至于在所观察的时间段内几乎不变，则可以认为是常量，这样的信号称为静态信号。此外，还有一些情况下，被测试的量虽然在不停地变化，但却在一段时间内围绕一个波动中心忽上忽下。观察的目的不是寻找它随时间的变化历程，而在于寻找这段时间内的波动中心。这种不以测试被测量随时间的变化历程为目的的测试又称静态测试。静态测试所用的测试装置称为静态测试装置，这类装置只能用于静态测试而不能用于动态测试，但动态测试装置往往（但不是全部）也能在静态测试中使用。

测试系统的静态特性是指当被测信号 $x(t)$ = 常数时，测得信号与被测信号之间所呈现的关系，即在静态测量情况下，实际测试系统与理想系统的接近程度，即

$$y(t) = Sx(t) \tag{3.9}$$

式中 S——测试装置的静态灵敏度，又称标度因子，一般是有量纲的，并由输入输出的量纲来决定，理想的灵敏度应当是时不变的常数。

动态测试中所用的代数系统和线性系统当用于静态测试时多数都能满足式(3.9)的要求。所谓静态测量，实际上就是在已知静态灵敏度 S 的情况下，根据输出 $y(t)$ 的量值来推断输入 $x(t)$ 的量值的一切操作。所谓测试装置的静态特性，就是输出函数 $y(t)=Sx(t)$ 的特性。常用灵敏度、非线性度、回程误差、分辨率、重复性等指标来定量描述实际测试系统的静态特性。

3.2.1 线性度

线性度是指测量装置输出、输入之间保持常值比例关系的程度。

在静态测量的情况下，用实验来确定被测量的实际值和测量装置示值之间的函数关系的过程称为静态校准，所得到的关系曲线称为校准曲线或定度曲线。通常，校准曲线并非直线。为使用简便，总是以线性关系来代替实际关系。为此，需用直线来拟合校准曲线。校准曲线接近拟合直线的程度（图 3.5）就是线性度。相应地，校准曲线偏离拟合直线的程度就是非线性度。作为技术指标，则采用非线性误差来表示，即用在装置标称输出

S is the sensitivity of sensors.

When δ_0 is very small, S will be very large, and the system is very sensitive, so it is suitable for measuring micro-displacement. If $x(t) < 0.1\delta_0$, the maximum linearity error will not exceed 1%.

3.2 Static Characteristics of the Measurement Systems

If the signal being measured is a constant that does not change with time, or the signal changes so slowly during the measurement phase that it is almost unchanged during the observed time period, it can be considered as a constant. Such a signal is called a static signal. In addition, in some cases, although the measured quantity is constantly changing, it goes up and down around a fluctuation center within a period of time. The purpose of observation is not to find its course of change over time, but to find the fluctuation center during this period of time. This kind of measurement that does not test the course of the measured quantity change over time can also be called static measurement. A static measurement device refers to a device used for static measurement, which can only be used for static measurement, not for dynamic measurement, but dynamic measurement device can often (but not all) be used in static measurement.

The static characteristic of the measurement system refers to the relationship between the signal which has measured and the signal which is hoped to measured when the measured signal $x(t)$ = constant, that is, the closeness of the actual measurement system to the ideal system in the case of static measurement, namely

$$y(t) = Sx(t) \tag{3.9}$$

where S—the static sensitivity of the measurement device, also known as the scale factor, is generally dimensioned, and determined by the dimension of input and output, and the ideal sensitivity should be a time-invariant constant.

Most of the algebraic systems and linear systems used in dynamic measurement can meet the requirements of formula (3.9) when used in static measurement. The so-called static measurement is actually all operations to infer the input value $x(t)$ based on the output value $y(t)$ when the static sensitivity S is known. The so-called static characteristics of the measurement device are the characteristics of the output function $y(t) = Sx(t)$. Indexes such as sensitivity, nonlinearity, return error, resolution, repeatability, etc. are often used to quantitatively describe the static characteristics of the actual measurement system.

3.2.1 Linearity

Linearity refers to the degree to which the output and input of the measuring device maintain a constant proportional relationship.

In the case of static measurement, the process of using experiments to determine the functional relationship between the measured actual value and the value indicated by the measuring device is called static calibration, and the relationship curve obtained is called the calibration curve. Normally, the calibration curve is not a straight line. For ease of use, the actual relationship is always replaced by a linear relationship. For this reason, a straight line is

范围 A 内校准曲线与该拟合直线的最大偏差 B 来表示。非线性误差也可用相对误差来表示，如

$$非线性误差 = \frac{B}{A} \times 100\% \tag{3.10}$$

至于拟合直线应如何确定，目前国内外尚无统一的标准。较常用的有两种：端基直线和独立直线。端基直线是一条通过测量范围的上下限点的直线(图 3.6)。若拟合的直线与校准曲线间的偏差B_i 的平方和最小，即 $\sum_i B_i^2$ 最小，则该直线称为独立直线。拟合直线不同会使得线性度的值有明显不同。因此，在测量线性度指标时应明确规定拟合直线的确定方法。任何测试系统都有一定的线性范围，线性范围越宽，表明测试系统的有效量程越大。因此，设计测试系统时，应尽可能保证其在近似线性的区间内工作。必要时，也可以对特性曲线进行线性补偿(采用电路或软件补偿均可)。实际工作中，经常会遇到非线性较为严重的系统。此时，可以采取限制测量范围、采用非线性拟合或非线性放大器等技术措施来提高系统的线性。

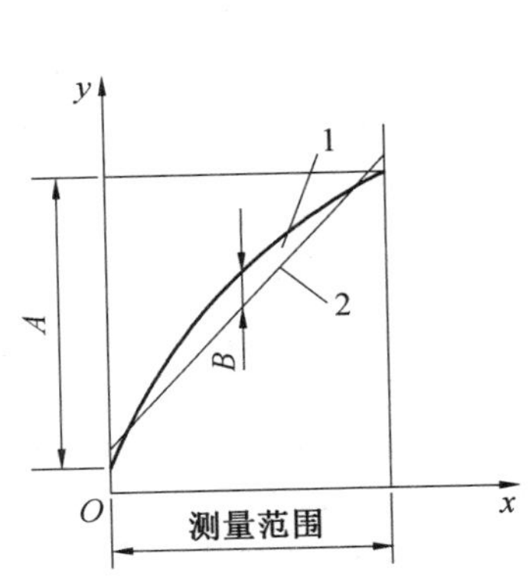

图 3.5　校准曲线与非线性度

1— 校准曲线；2— 拟合直线

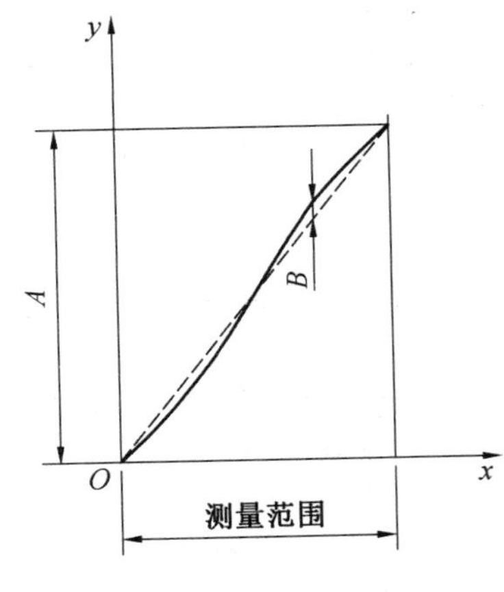

图 3.6　端基直线

3.2.2　灵敏度

灵敏度表征测试系统对输入信号 $x(t)$ 变化的一种反应能力。一般情况下，当系统的输入 x 有一个微小增量 Δx 时，将引起系统的输出 y 也发生相应的微量变化 Δy，则定义该系统的灵敏度为 $S = \Delta y/\Delta x$。对于静态测量，当系统的输入 / 输出特性为线性关系时，则有

$$S = \frac{\Delta y}{\Delta x} = \frac{y}{x} = \frac{b_0}{a_0} = 常数 \tag{3.11}$$

needed to fit the calibration curve. The degree to which the calibration curve is close to the fitting straight line (Figure 3.5) is the linearity. Correspondingly, the degree of deviation of the calibration curve from the fitting straight line is the non-linearity. As a technical indicator, non-linear error is used to indicate, that is, within the nominal output range A of the device, the maximum deviation B between the calibration curve and the fitted straight line is used to indicate. Non-linear error can also be expressed by relative error, such as

$$\text{Non-linear error} = \frac{B}{A} \times 100\% \tag{3.10}$$

As for how to determine the fitting straight line, there is currently no unified standard at home and abroad. There are two more commonly used: terminal-base straight line and independent straight line. The terminal-base straight line is a straight line that passes through the upper and lower limit points of the measurement range (Figure 3.6). If the sum of the squares of the deviation B_i between the fitting straight line and the calibration curve is the smallest, that is, $\sum_i B_i^2$ proves the smallest, then the straight line is called an independent straight line. The difference of the fitting straight line will make the linearity value significantly different. Therefore, the method of determining the fitting straight line should be clearly defined when measuring the linearity index. All measurement systems have a certain linear range. The wider the linear range, the larger the effective range of the measurement system. Therefore, when designing the measurement system, we should try to ensure it working in an approximate linear range. If necessary, the characteristic curve can also be linearly compensated (either circuit or software compensation can be used). In actual work, systems with more serious nonlinearities are often encountered. At this time, technical measures such as limiting the measurement range, using nonlinear fitting or nonlinear amplifiers can be taken to improve the linearity of the system.

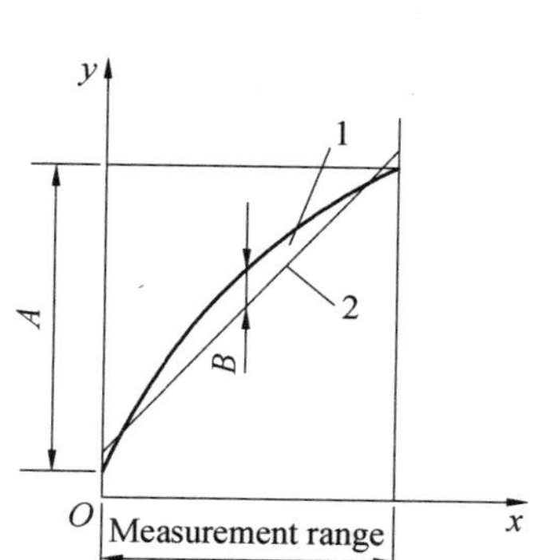

Figure 3.5　Calibration curve and nonlinearity
1— calibration curve;2—fitting straight line

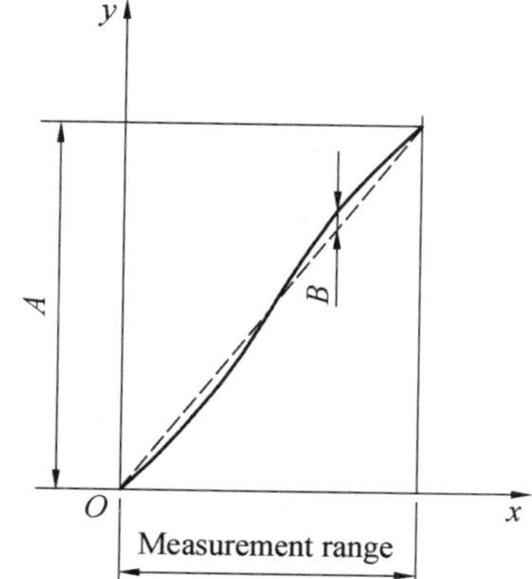

Figure 3.6　Terminal-base straight line

3.2.2　Sensitivity

Sensitivity characterizes the ability of the measurement system to respond to changes in the input signal $x(t)$. In general, when the input x of the system has a small increment Δx, it will

即测试系统的静态灵敏度(又称绝对灵敏度)等于拟合直线的斜率(图3.5)。线性测试系统的静态特性曲线为一条直线,直线的斜率即灵敏度,且是一个常数。实际测试系统并非理想的线性系统,其特性曲线不是直线,即灵敏度随输入量的变化而改变,说明不同的输入量对应的灵敏度大小是不相同的。通常用一条拟合直线代替实际特性曲线,该拟合直线的斜率作为测试系统的平均灵敏度。也可以用特性曲线的斜率 $S = \lim\limits_{\Delta x \to 0} \frac{\Delta y}{\Delta x} = \frac{dy}{dx}$ 来表示系统的瞬时灵敏度。

灵敏度的量纲取决于输入/输出的量纲。当测试系统的输出和输入量纲不同时,灵敏度是有单位的。例如,当某位移传感器在位移上变化1 mm时,输出电压变化300 mV,则该传感器的灵敏度 $S = 300$ mV/mm。

测试系统的输出、输入量纲相同时,灵敏度又称放大倍数或增益。此时,常用放大倍数来替代灵敏度。例如,一个最小刻度值为0.001 mm的千分表,若其刻度间隔为1 mm,则其放大倍数 = 1 mm/0.001 mm = 1 000倍。

灵敏度反映了测试系统对输入量变化的反应能力。灵敏度的高低可以由系统的测量范围、抗干扰能力等决定。通常,灵敏度越高,就越容易引入外界干扰,从而使稳定性变差,测量范围变窄。

值得说明的是,测试系统除对被测量敏感外,还可能对各种干扰量有反应,从而影响测量精度。这种对干扰量敏感的灵敏度称为有害灵敏度。在设计测试系统时,应尽可能使有害灵敏度降到最低限度。此外,灵敏度与系统的量程、固有频率等是相互制约的。在选择测试系统的灵敏度时,要综合考虑其合理性。一般来说,系统的灵敏度越高,其测量范围往往越窄,稳定性也会越差。

3.2.3 误差

装置的示值总是有误差的。测量装置的示值和被测量的真值之间的差值称为装置的示值误差。在不引起混淆的情况下,可简称为测量装置的误差。

一般来说,真值是未知的。只有按规定在特定条件下保存在国际计量局的基准可以被认为是某量的真值。例如,国际千克基准可认为是真值1 kg。在实际测量中,只能用所谓的约定真值来代替真值。在实际测量中,通常利用被测量的实际值、已修正过的算术平均值和计量标准器所复现的量值作为约定真值。其中,实际值是指满足规定准确度的可用来代替真值使用的量值。例如,在计量检定中,通常把高一等级计量标准所复现的量值称为实际值。因此,在使用时需注意区分以下概念。

cause the output y of the system to have a corresponding small change Δy. Then the sensitivity of the system is defined as $S = \Delta y / \Delta x$. For static measurement, if the input/output characteristics of the system are linear, then we have

$$S = \frac{\Delta y}{\Delta x} = \frac{y}{x} = \frac{b_0}{a_0} = \text{constant} \tag{3.11}$$

That is, the static sensitivity (also called absolute sensitivity) of the measurement system is equal to the slope of the fitting straight line (Figure 3.5). The static characteristic curve of the linear measurement system proves a straight line, and the slope of the straight line is the sensitivity, and it is a constant. The actual measurement system is not an ideal linear system, and its characteristic curve is not a straight line, that is, the sensitivity changes with the change of the input quantity, indicating that the sensitivity of different input quantities is not the same. Usually a fitting straight line is used instead of the actual characteristic curve, and the slope of the fitting straight line is used as the average sensitivity of the measurement system. The slope $S = \lim_{\Delta x \to 0} \frac{\Delta y}{\Delta x} = \frac{\mathrm{d}y}{\mathrm{d}x}$ of the characteristic curve can also be used to express the instantaneous sensitivity of the system.

The dimension of sensitivity depends on the dimension of input or output. When the output and input dimension of the measurement system are different, the sensitivity has a unit. For example, when a displacement sensor changes by 1 mm, the output voltage changes by 300 mV, then the sensor's sensitivity $S = 300$ mV/mm.

When the output and input dimensions of the measurement system are the same, the sensitivity is also called the amplification factor or gain. At this time, the amplification factor is often used to replace the sensitivity. For example, if a dial gauge with a minimum scale of 0.001 mm with 1 mm scale interval, its amplification factor = 1 mm/0.001 mm = 1 000 times.

Sensitivity reflects the ability of the measurement system to respond to changes in input. The level of sensitivity can be determined according to the measurement range and anti-interference ability of the system. Generally, the higher the sensitivity, the easier it is to introduce external interference, which will make the stability worse and the measurement range narrower.

It is worth noting that, in addition to being sensitive to the measured quantity, the measurement system may also respond to various interferences, thereby affecting the measurement accuracy. This sensitivity to the amount of interference is called harmful sensitivity. When designing the measurement system, the harmful sensitivity should be minimized as much as possible. In addition, the sensitivity and the measuring range and natural frequency of the system are mutually restricted. When choosing the sensitivity of the measurement system, it is necessary to consider its rationality comprehensively. Generally speaking, the higher the sensitivity of the system, the narrower the measurement range and the worse the stability.

3.2.3 Error

The indicating value of devices always has an error. The difference between the indication value of the measuring device and the true value of the measured quantity is called indication error. Without causing confusion, it can be referred to simply as the error of the measuring de-

在实际工作中，往往使用到测量装置的引用误差一词，有时又称满量程误差。它是指测量装置的示值绝对误差与引用值之比，并以百分数表示。引用值往往是指测量装置的量程或示值范围的最高值。例如，示值范围为 0 ～ 150 V 的电压表，当其示值为 100.0 V 时，测得电压实际值为 99.4 V，则该电压表的引用误差为

$$\frac{100.0 - 99.4}{150} \times 100\% = 0.4\%$$

引用误差实际上可能包含以下两类误差来源。

(1) 随机误差。

当在相同的测量条件下多次测量同一物理量时，误差的绝对值与符号以不可预知的方式发生变化，这样的误差就称为随机误差。从单次测量结果来看，随机误差没有规律性，但就其总体来说，随机误差服从一定的统计规律。

(2) 系统误差。

当在相同的测量条件下多次测量同一物理量时，误差不变或按一定规律变化，这样的误差称为系统误差。系统误差等于引用误差减去随机误差，是具有确定性规律的误差。

随机误差(主要以重复测量误差体现) 可以通过以下方法测得。

在测试条件不变的情况下，测试系统按同一方向做全量程多次(3 次以上) 测量时，对于同一个输入量，其测量结果的不一致程度即随机误差，可表示为

$$\delta_{\mathrm{n}} = \frac{\Delta R}{Y_{\mathrm{FS}}} \times 100\% \tag{3.12}$$

式中 ΔR—— 同一输入量对应多次循环的同向行程响应量的绝对误差；

Y_{FS}—— 测试系统的量程。

重复性表征了系统的随机误差的大小，可以根据标准偏差来计算 ΔR，即

$$\Delta R = K\sigma / \sqrt{n} \tag{3.13}$$

式中 σ—— 测量值的标准偏差；

K—— 置信因子，$K = 2$ 时置信度为 95%，$K = 3$ 时置信度为 99.73%。

$K\sigma$ 为置信区间或随机不确定度，其物理意义是：在整个测量范围内，测试系统相对于满量程输出的随机误差不超过 ΔR 的置信度为 95%($K = 2$ 时) 或 99.73%($K = 3$ 时)。

重复测量误差又有单次装夹和多次装夹的情况。单次装夹可以考查测试装置的电气误差；多次装夹则在单次装夹基础上增加了夹具装夹误差。

vice.

Generally speaking, the true value is unknown. Only the benchmarks kept in the International Bureau of Weights and Measures under certain conditions can be regarded as the true value of a certain quantity. For example, the international kilogram benchmark can be considered as the true value of 1 kg. In actual measurement, that's the only way to use the so-called conventional true value to replace true value. In actual measurement, the actual value being measured, the corrected arithmetic average, and the value reproduced by the measurement standard are usually used as the conventional true value. In it, the actual value refers to the value that can be used instead of the true value that meets the specified accuracy. For example, in the metrological verification, the value reproduced by the higher-level measurement standard is usually called the actual value. Thus, it is necessary to distinguish the following concepts in use.

In actual work, the termfiducial error of the measuring device is often used, sometimes also called full-scale error. It refers to the ratio of the absolute indication error of the measuring device to the quoted value, and is expressed as a percentage. The fiducial value often refers to the measuring range of the measuring device or the highest value of indication range. For example, for a voltmeter with an indication range of 0 ~ 150 V, when its indication value is 100.0 V, the actual value of the measured voltage is 99.4 V, then the fiducial error of the voltmeter is

$$\frac{100.0-99.4}{150}\times 100\% = 0.4\%$$

Fiducial errors may actually include the following two types of error sources.

(1) Random error.

When the same physical quantity is measured multiple times under the same measurement conditions, the absolute value and sign of the error change in an unpredictable way, this kind of which is called a random error. From a single measurement result, random errors have no regularity, but as a whole, they obey certain statistical laws.

(2) Systematic error.

When the same physical quantity is measured multiple times under the same measurement conditions, the error does not change or changes according to a certain rule, such an error is called systematic error. The systematic error is equal to a kind of error that the reference error minus the random error, with a deterministic law.

Random error (mainly reflected by repeated measurement error) can be measured by the following methods.

Under the unchanged measurement conditions, when the measurement system makes multiple (more than 3 times) measurements of the full range in the same direction, the inconsistency degree of the measurement results for the same input quantity is called the random error, which can be expressed as

$$\delta_{n} = \frac{\Delta R}{Y_{FS}} \times 100\% \tag{3.12}$$

where ΔR —the absolute error of the same input corresponding to multiple cycles in the same direction stroke response;

Y_{FS} —the measuring range of the measurement system.

3.2.4 精密度、准确度和精确度

测量的精密度、准确度和精确度是人们常常容易混淆的三个名词。虽然它们都是评价测量结果好坏的，但其含义有较大的差别。

1. 精密度(Precision)

仪表指示值之间的不一致程度称为精密度，即在相同条件下，多次重复测量所得的测量结果彼此间相互接近、相互密集的程度，又称重复精度。

精密度与随机误差及仪表的有效位数有关。精密度高是指重复误差较小。这时，测量数据比较集中，但系统误差的大小并不明确。

2. 准确度(Correctness)

准确度是仪表的指示值对于真值的接近程度，即测量结果与被测真值偏离的程度。它由系统误差引起，是重复误差和线性度等的综合。

测量装置的准确度是表示测量装置给出接近于被测量真值的示值的能力，反映测量装置的总误差，包括系统误差和随机误差两部分。测量装置的随机误差分量可用对同一被测量在同一行程方向连续进行多次测量后其示值的分散性来表述，通常又称测量装置的重复测量误差。准确度不高可能是仪器本身或计量基准的不完善两方面原因造成的。

测量装置的准确度等级是用来表示该装置在符合一定的计量要求情况下保持其误差在规定的极限范围内能力的参数。多数的电工、热工仪表和部分无线电测量仪器采用引用误差的形式来表示其准确度等级，以允许引用误差值来作为准确度级别的代号。例如，“0.2 级”电压表表示该电压表允许的示值误差不超过电压表引用值的 0.2%。

3. 精确度

精确度是精密度与准确度的综合，有时简称精度，但容易与精密度混淆。精确度可综合反映系统误差和随机误差。精密度高，但准确度差，其精确度就不会高；反之，准确度好，精密度差，其精确度也不会好。只有当精密度及准确度都高时，精确度才会高。精确度高，是指重复误差与系统误差都比较小，这时测量数据集中在真值附近。

以打靶为例，可形象理解上述三个度之间的关系，测试系统三个度的解释如图 3.7 所示。假定靶心为真值，射击点为测试结果，有以下三种情况。

(1)打得很集中，但都偏离靶心(图 3.7(a))，表示射击的精密度高但准确度较差，即系统误差较大。

(2)打得很离散，但对称地分布在靶心周围(图 3.7(b))，表示射击的准确度高，但精

Repeatability characterizes the size of the random error of the system, and ΔR can be calculated according to the standard deviation, namely

$$\Delta R = K\sigma / \sqrt{n} \tag{3.13}$$

where σ —the standard deviation of the measured value;

K—confidence factor, when $K=2$, the confidence is 95%, when $K=3$, the confidence is 99.73%.

$K\sigma$ is the confidence interval or random uncertainty, and its physical meaning is: in the entire measurement range, the confidence that the random error of the measurement system relative to the full-scale output does not exceed 95% (when $K=2$) or 99.73% (when $K=3$) of the confidence level of ΔR.

Repeated measurement errors include single clamping and multiple clamping. A single clamping can examine the electrical error of themeasurement device, and multiple clamping increases the clamping error of the fixture on the basis of a single clamping.

3.2.4 Precision, Correctness and Accuracy

Precision, correctness and accuracy of measurement are three terms easy to confuse. Although they are used to evaluate the quality of measurement results, their meanings are quite different.

1. Precision

The inconsistency degree between the indicating values of the instrument is called precision, that is, the degree to which the measurement results obtained from repeated measurements are close to each other and dense with each other under the same conditions, also calledrepeat precision.

Precision is related to random error and effective number of bits of the instrument. High precision means that the repetitive error is small. At this time, the measurement data is relatively concentrated, but the size of the systematic error is not clear.

2. Correctness

Correctness is the closeness of the indicated value of the instrument to the true value, that is, the degree of deviation between the measured result and the measured true value. Caused by system error, correctness is a combination of repeat error and linearity.

The correctness of a measuring device is the ability of the measuring device to give an indication close to the true value of the measured value. It reflects the total error of the measuring device, including systematic error and random error two parts. The random error component of the measuring device can be used to measure the same measured multiple times in the same stroke direction continuously, and the dispersion of the displayed value can be expressed. It is also commonly referred to as the repeated measurement error of the measuring device. The two reasons of the instrument itself or the imperfect measurement standard may cause the low correctness.

The corectness level of a measuring device is used to indicate the device's ability to keep its error within the specified limit under the condition that the device meets certain measurement resquirements. Most electrical, thermal instruments and some radio measuring instruments use the form of fiducial error to indicate their accuracy level, and allow the fiducial

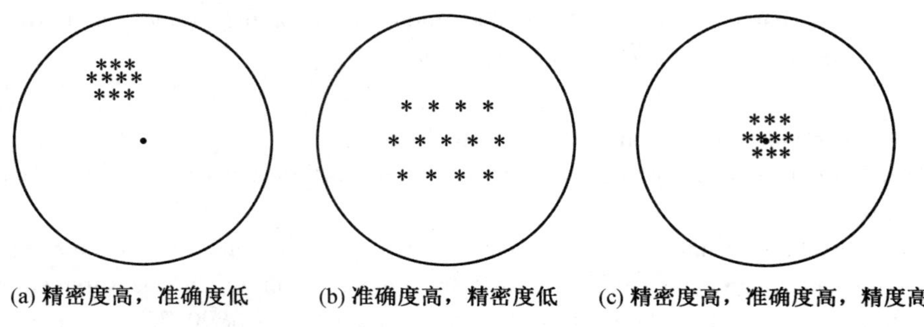

(a) 精密度高，准确度低
(b) 准确度高，精密度低
(c) 精密度高，准确度高，精度高

图 3.7　测试系统三个度的解释

密度较差,即随机误差较大。

(3)打得很集中,又都分布在靶心附近(图 3.7(c)),表示射击的精密度、准确度都较好,即精确度高,这时随机误差和系统误差都较小。

上述三种情况中,图 3.7(a)、(b)所示的结果都不理想。

3.2.5　量程和测量范围

测量装置示值范围上、下限之差的模称为量程。其测量范围则是指使该装置的误差处于允许极限内它所能测量的被测量值的范围。对于用于动态测量的装置,还标明在允许误差极限内所能测量的频率范围。

3.2.6　回程误差

回程误差又称迟滞或滞后量,表征测试系统在全量程范围内,输入量由小到大(正行程)和由大到小(反行程)的静态特性不一致的程度。回程误差如图 3.8 所示,对于理想的测试系统,某一个输入量只对应一个输出;然而,对于实际的测试系统,当输入信号由小变大,再由大变小时,对应于同一个输入量有时会出现数值不同的输出量。在测试系统的全量程范围内,将这种不同输出量中差值最大者($h_{max}=y_{2i}-y_{1i}$)定义为系统的回程误差,即

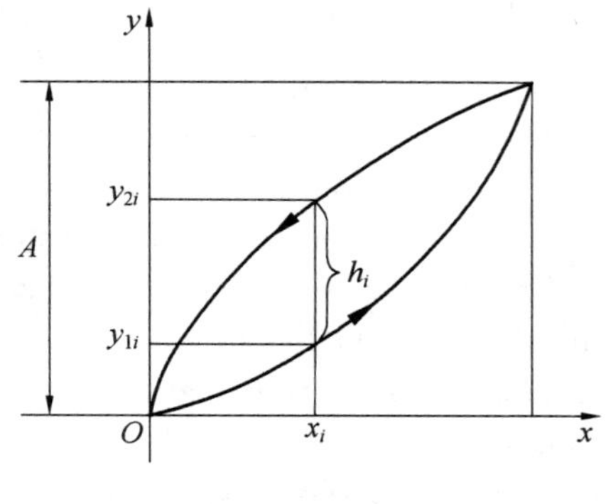

图 3.8　回程误差

$$\text{回程误差}=\frac{h_{max}}{A}\times 100\% \tag{3.14}$$

产生回程误差的原因主要有两个:一是测试系统中有吸收能量的元件,如磁性元件(磁滞)和弹性元件(弹性滞后、材料的受力变形);二是在机械结构中存在摩擦和间隙等

error value as the code for the accuracy level. For example, a “0.2-level” voltmeter means that the allowable indication error of the voltmeter does not exceed 0.2% of the quoted value of the voltmeter.

3. Accuracy

Accuracy is a combination of precision and correctness, but it is easily confused with precision. Accuracy can comprehensively reflect system error and random error. If the precision is high, but the correctness is poor, the accuracy will not be high; on the contrary, if the correctness is good and the precision is poor, the accuracy will not be good. Only when the precision and correctness are high, will the accuracy be high. High accuracy means that the repetitive error and system error are relatively small, and the measured data are concentrated near the true value.

Take shooting as an example to visually understand the relationship between the above three degrees, interpretation for above three terms in measurement systems is shown in Figure 3.7. Assuming that the bullseye is true value and the shooting point is the measurement result, there are three situations as follows.

(1) The shooting is very concentrated, but all deviate from the bullseye (Figure 3.7(a)), which means high shooting precision but poor correctness, that is, the system error is large.

(2) The shooting is very discrete, but symmetrically distributed around the bullseye (Figure 3.7(b)), indicating high shooting accuracy, but poor precision, that is, the random error is large.

(3) The shooting is very concentrated, and they are all distributed near the bullseye (Figure 3.7(c)), showing that the precision and correctness of the shooting are good, that is, the accuracy is high, which means that the random error and the system error are both small this moment.

In the above three cases, the results shown in Figures 3.7(a) and (b) are not ideal.

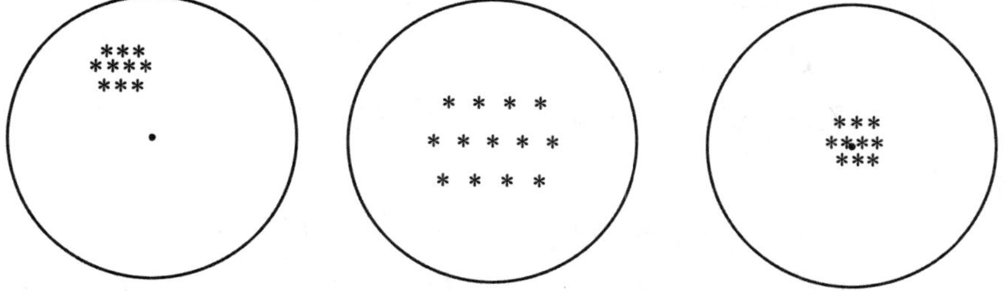

(a)High precision,low correctness (b) High correctness,low precision (c)High precision,high accuracy

Figure 3.7 Interpretation for above three terms in measurement systems

3.2.5 Range and Measurement Range

The modulus of the difference between the upper and lower limits of the display range of the measuring device is called range. Measurement range refers to the range of the measured value that the device can measure within the allowable limit. For devices used for dynamic measurement, the frequency range that can be measured within the allowable error limit is also

缺陷,如仪表传动机构的间隙和运动部件的摩擦也可能反映仪器死区的存在。

3.2.7 稳定度和漂移

稳定度是指测量装置在规定条件下保持其测量特性恒定不变的能力。通常在不指明影响量时,稳定度是指装置不受时间变化影响的能力。如果是对其他影响量来考查稳定度,则需特别说明。

测量装置的测量特性随时间的慢变化称为漂移。在规定条件下,对一个恒定的输入在规定时间内的输出变化称为点漂;标称范围内低值处的点漂称为零点漂移,简称零漂。产生漂移的原因有两方面:一是仪器自身结构参数的变化;二是外界工作环境参数的变化对响应的影响。最常见的漂移问题是温漂,即因外界工作温度的变化而引起输出的变化。对多数测试系统而言,不仅存在零点漂移,而且还存在灵敏度漂移,即测试系统的输入/输出特性曲线的斜率产生变化。因此,在工程测试中,必须对漂移进行观测和度量,减小漂移对测试系统的影响,从而有效提高稳定性。发生漂移后,需要对测试系统进行重新进行校准或标定。

3.3 测试系统的动态特性

3.3.1 动态特性测试的目的

前文已述系统的基本特性就是当信号的性态确定后,系统输出函数的特性。动态特性指的就是动态测试装置在动态信号 $x(t)$ 的激励下,其输出函数 $f[x(t)]$ 的特性,包括输出与输入到底呈什么关系,是线性关系、微分关系还是积分关系,它们之间的关系到底与什么因素有关等。

上述问题对于代数系统是非常简单的,因为代数系统在规定的使用条件下,在测量静态信号时,代数系统的输出 $y(t)$ 与输入 $x(t)$ 一律保持定比例线性关系,无论输入是简谐波还是由谐波叠加而成的复杂信号,输出 $y(t)$ 时时刻刻都与 $x(t)$ 保持确定比例,或者说 $y(t)$ 以确定比例在时域内再现了 $x(t)$ 的变化历程。因此,测量和记录过程不受时间限制,这正是测试工作所希望的。所以说代数系统在动态测试中作为传感器或显示仪器使用时是非常理想的,应用也很广泛。

对于一般形式的线性时不变系统来说,回答上述问题要复杂得多。线性时不变系统输出 $y(t)$ 和输入 $x(t)$ 的关系不是用函数形式来描述的,而是用微分方程式来描述的,于

indicated.

3.2.6 Return Error

Return error is also called hysteresis or lag, which represents the inconsistency of the static characteristics of the measurement system from small to large (positive stroke) and from large to small (reverse stroke) within the full range of the measurement system. Return error is shown in Figure 3.8, that is, for an ideal measurement system, a certain input corresponds to only one output; however, for an actual measurement system, when the input signal changes from small to large, and then from large to small, output quantities with different values sometimes appear corresponding to the same input quantity. Within the full range of the measurement system, the largest difference ($h_{max} = y_{2i} - y_{1i}$) among these different output quantities is defined as the return error of the system, namely

$$\text{return error} = \frac{h_{max}}{A} \times 100\% \tag{3.14}$$

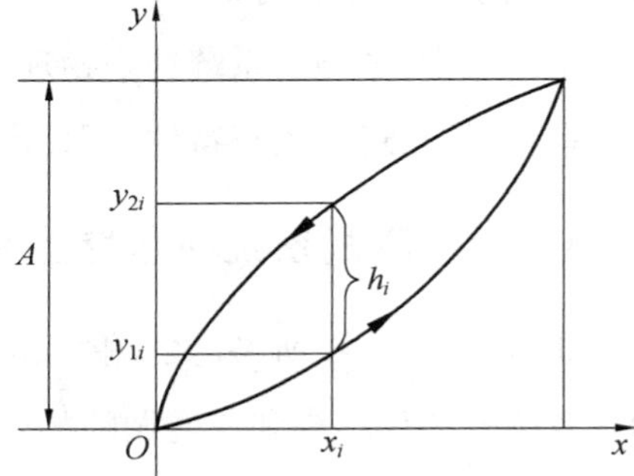

Figure 3.8　Return error

There are two main reasons for the return error: one is that there are energy-absorbing components in the measurement system, such as magnetic components (magnetic hysteresis) and elastic components (elastic hysteresis, material deformation); the other is that there are defects such as friction and clearance in the mechanical structure. For example, the gap of the instrument transmission mechanism and the friction of the moving parts may also reflect the existence of the dead zone of the instrument.

3.2.7 Stability and Drift

Stability refers to the ability of a measuring device to keep its measurement characteristics constant under specified conditions. Usually when the amount of influence is not specified, it refers to the ability of devices not to be affected by time changes. If it is to examine the stability of other influence quantities, special explanation is required.

Drift refers to the slow change of the measurement characteristics of the measuring device over time. Under specified conditions, the output change of a constant input within a specified time is called point drift; the point drift at the low value within the nominal range is called zero point drift, or zero drift for short. There are two reasons for the drift: one is the change of the structural parameters of the instrument itself; the other is the influence of changes in external working environment parameters to response. The most common drift problem is temperature drift, that is, changes in output due to changes in outside working temperature. For most measurement systems, there is not only a zero drift, but also a sensitivity drift, that is, the slope of the input/output characteristic curve of the measurement system changes. Therefore, the drift must be observed and measured to reduce the influence of drift on the measurement system in the engineering measurement, thereby effectively improving the stability. After drifting, the measurement system needs to be recalibrated or demarcated.

是人们首先想到的是求解这个微分方程，以求了解输出与输入在时域内的函数关系。如果对于每一个输入都用解微分方程的方法来得到输出，再通过考查它是否是输入的线性函数来判断 $y(t)$ 是否真实地反映了 $x(t)$ 的变化历程，那将是十分烦琐的。并且，由于时域信号有的相当复杂，有的是随机信号，根本没有描述它的数学模型，因此求解微分方程不仅很困难，而且有时根本办不到。本节要讨论的内容就是寻找一种切实可行的方法来回答上述的各个问题。

测试系统的动态特性不仅取决于测试系统的结构参数，而且与输入信号有关。因此，研究测试系统的动态特性实质上就是建立输入信号、输出信号和测试系统结构参数之间的关系。通常，把测试系统这一物理系统抽象成数学模型，分析输入信号与输出信号之间的关系，以便描述其动态特性、研究产生动态误差的原因并加以改进。

3.3.2 测试系统动态特性的时域求解——线性微分方程

通常情况下，在所考虑的测量范围内，将实际的测试系统视为线性时不变系统。根据测试系统的物理结构和所遵循的物理定律，总可以建立起如式(3.1)所示的常系数线性微分方程，并用来描述系统的输出量 $y(t)$ 与输入量 $x(t)$ 的关系。求解该微分方程，就可以得到系统的动态特性。这种方程的通式为

$$a_n \frac{\mathrm{d}^n y(t)}{\mathrm{d}t^n}+a_{n-1}\frac{\mathrm{d}^{n-1} y(t)}{\mathrm{d}t^{n-1}}+\cdots+a_1 \frac{\mathrm{d}y(t)}{\mathrm{d}t}+a_0 y(t)$$

$$=b_m \frac{\mathrm{d}^m x(t)}{\mathrm{d}t^m}+b_{m-1}\frac{\mathrm{d}^{m-1} x(t)}{\mathrm{d}t^{m-1}}+\cdots+b_1 \frac{\mathrm{d}x(t)}{\mathrm{d}t}+b_0 x(t)$$

线性时不变系统有两个十分重要的性质，即叠加性和频率保持性。根据叠加性，当一个系统有 n 个激励同时作用时，那么它的响应就等于这 n 个激励单独作用的响应之和，即

$$\sum_{i=1}^{n} x_i(t) \rightarrow \sum_{i=1}^{n} y_i(t)$$

即各个输入所引起的输出是互不影响的。这样，在分析常系数线性系统时，可将一个复杂的激励信号分解成若干个简单的激励。例如，利用傅里叶变换，将复杂信号分解成一系列谐波或若干个小的脉冲激励，然后求出这些分量激励的响应之和。频率不变性(频率保持性)表明，当线性系统的输入为某一频率时，则系统的稳态响应也为同一频率的信号。下面以一阶及二阶线性系统为例来分析当它们受到单位正弦信号 $x(t)=\sin \omega t$ 的激励时会产生什么样的响应。

例 3.2　分析基于一、二阶线性微分方程搭建的测试系统对正弦信号激励 $x(t)=\sin \omega t$的响应。

3.3 Dynamic Characteristics of the Measurement System

3.3.1 The Purpose of Dynamic Characteristics Measurement

The basic properties of the system mentioned above are the characteristics of the output function of the system when the signal ' s behavior is determined. Then the dynamic characteristics refer to the characteristics of the output function $f[x(t)]$ of the dynamic measurement device under the excitation of the dynamic signal $x(t)$, including what the relationship between output and input is, whether it is a linear relationship, a differential relationship or an integral relationship, and what factors are related to the relationship between them.

The above problems are very simple, because the algebraic system under the specified conditions of use, when measuring static signals, the output $y(t)$ and input $x(t)$ of the algebraic system will always maintain a proportional linear relationship, regardless of whether the input is a simple harmonic or a complex signal formed by the superposition of harmonics, output $y(t)$ always maintains a certain ratio with $x(t)$ at all times, or the change history of $x(t)$ is reproduced in the time domain with the proportion of $y(t)$. Therefore, the measurement and recording process is not limited by time, this is exactly what the measurement work hopes for. So we can say that, the algebraic system used as a sensor or display instrument in a dynamic measurement is very ideal and is widely used.

For a general form of time-invariant linear system, the answer to the above questions is much more complicated. The relationship between output $y(t)$ and input $x(t)$ of the time-invariant linear system is described by a differential equation instead of a function form, so the first thing that people think of is to solve this differential equation in order to understand the specific functional relationship between the output and the input in the time domain. If we use the method of solving differential equations for each input to get the output, and then use the linear function of the input to determine whether $y(t)$ truly reflects the change process of $x(t)$, it will be very cumbersome. Also, because some time domain signals are quite complex, and some are random signals, and there is no mathematical model describing them, solving differential equations is not only difficult, but sometimes impossible. The content to be discussed in this section is to find a practical way to answer the above questions.

The dynamic characteristics of the measurement system not only depend on the structural parameters, but also related to the input signals. Therefore, the study of the dynamic characteristics of the measurement system is essentially to establish the relationship between the input signals, the output signals and the structural parameters of the measurement system. Usually, the physical system of the measurement system will be abstracted into a mathematical model to analyze the relationship between the input signal and the output signal in order to describe its dynamic characteristics, study the causes of dynamic errors and make improvements.

解 (1)对于一阶系统,有

$$\tau y'(t)+y(t)=x(t)$$

当输入为 $x(t)=\sin\omega t$ 时,解微分方程,其解为

$$y(t)=A[\sin(\omega t+\varphi)-e^{-t/\tau}\sin\varphi]$$

式中

$$A=\frac{1}{\sqrt{1+(\tau\omega)^2}}$$

$$\varphi=-\arctan(\tau\omega)$$

$$\sin\varphi=-\frac{\tau\omega}{\sqrt{1+(\tau\omega)^2}}$$

解中的第一部分为

$$A\sin(\omega t+\varphi)=A\sin\omega(t+t_0)=y_1(t)$$

式中

$$t_0=\varphi/\omega$$

$y_1(t)$是非齐次方程

$$\tau y'(t)+y(t)=\sin\omega t$$

的特解,是系统响应中的强制部分,即输入 $x(t)$强制作用的结果。说得更通俗一点,输入信号 $x(t)$从输入端进入系统,从输出端输出来,$y_1(t)$就是输出来的 $x(t)$。因为系统的固有特性,所以输出来的已与进入之前的有了差别,$y_1(t)$是系统对正弦激励的稳态响应。

解中的第二部分$-Ae^{-t/\tau}\sin\varphi=y_2(t)$称为暂态解,是齐次方程

$$\tau y'(t)+y(t)=0$$

的解,即系统在 $t=0$ 时刻突然收到输入信号,使系统的原始平衡状态遭受破坏。$y_2(t)$就是失平后,扰动消失,系统重新回到平衡状态的过程——受扰位移按指数规律做衰减振荡,最终消失。经过不太长的时间,如 $t=7\tau$,$e^{-t/\tau}\approx0.001$,仅凭此项就可近似地认为 $y_2(t)$已消失,从此系统进入稳态。

(2)对于二阶系统,当二阶线性系统输出和输入的关系用定常二阶线性微分方程

$$y''(t)+2\zeta\omega_n y(t)+\omega_n^2 y(t)=x(t)$$

来描述,且输入为 $x(t)=\sin\omega t$ 时,解微分方程,其解便是系统对正弦激励的全部响应,即

$$y(t)=A\sin(\omega t+\varphi)-\frac{\omega}{\omega_d}e^{-\zeta\omega_n t}\sin(\omega_d t+\varphi_3)$$

式中

3.3.2　Time Domain Solvement of the Dynamic Characteristics of the Measurement System—Linear Differential Equations

Under normal circumstances, the actual measurement system is regarded as atime-invariant linear system within the considered measurement range. According to the physical structure of the measurement system and the laws of physics followed, the linear differential equation with constant coefficients as shown in formula(3.1) can always be established, and used to describe the relationship between the output $y(t)$ and the input $x(t)$ of the system. The dynamic characteristics of the system can be obtained by solving the differential equation. The general formula is

$$a_n \frac{d^n y(t)}{dt^n} + a_{n-1} \frac{d^{n-1} y(t)}{dt^{n-1}} + \cdots + a_1 \frac{dy(t)}{dt} + a_0 y(t) = b_m \frac{d^m x(t)}{dt^m} + b_{m-1} \frac{d^{m-1} x(t)}{dt^{m-1}} + \cdots + b_1 \frac{dx(t)}{dt} + b_0 x(t)$$

Linear time-invariant systems have two very important properties, namely superposition and frequency retention. According to superposition, when a system has n stimuli acting simultaneously, its response is equal to the sum of the responses of the n stimuli acting separately, namely

$$\sum_{i=1}^{n} x_i(t) \rightarrow \sum_{i=1}^{n} y_i(t)$$

That is, the output caused by each input does not affect each other. In this way, when analyzing constant coefficient linear system, a complex excitation signal can be decomposed into several simple excitations. For example, using Fourier transform to decompose a complex signal into a series of harmonics or several small pulse excitations, and then find the sum of the responses of these component excitations. Frequency invariance (frequency retention) indicates that when the input of a linear system is a certain frequency, the steady-state response of the system is also a signal of the same frequency. Let's take first-order and second-order linear systems as examples to analyze what kind of response they will produce when they are excited by a unit sine signal $x(t) = \sin \omega t$.

Example 3.2　Analyzing of the response of the measurement system based on the first and second order linear differential equations to the sine signal excitation $x(t) = \sin \omega t$.

Solution　(1) For the first-order system, we have

$$\tau y'(t) + y(t) = x(t)$$

When input is $x(t) = \sin \omega t$, solve the differential equation, its solution is

$$y(t) = A[\sin(\omega t + \varphi) - e^{-t/\tau} \sin \varphi]$$

where

$$A = \frac{1}{\sqrt{1 + (\tau\omega)^2}}$$

$$\varphi = -\arctan(\tau\omega)$$

$$\sin \varphi = -\frac{\tau\omega}{\sqrt{1 + (\tau\omega)^2}}$$

The first part of the solution is

$$A=\frac{1}{\sqrt{\left[1-\left(\frac{\omega}{\omega_n}\right)^2\right]^2+4\zeta\left(\frac{\omega}{\omega_n}\right)^2}}$$

$$\varphi=-\arctan\frac{2\zeta\left(\frac{\omega}{\omega_n}\right)}{1-\left(\frac{\omega}{\omega_n}\right)^2}$$

$$\varphi_3=-\arctan\frac{2\zeta\sqrt{1-\zeta^2}}{1-\left(\frac{\omega}{\omega_n}\right)^2-2\zeta^2}$$

ω_d 是系统在阻尼比为 $\zeta(\zeta<1)$ 时做有阻尼振荡时的圆频率，$\omega_d=\omega_n\sqrt{1-\zeta^2}$。

同样，解中 $y_1(t)=A\sin(\omega t+\varphi)=A\sin\omega(t+t_0)$ 是系统对正弦激励的稳态响应，其余部分是暂态响应。其中，$t_0=\varphi/\omega$。

通过分析系统对正弦激励的响应会发现以下问题。

(1)线性系统确实具有频率保持性。在正弦信号 $\sin\omega t$ 激励下系统稳定后输出确实是与输入同频的正弦波

$$y_1(t)=A\sin(\omega t+\varphi)=A\sin\omega(t+t_0)$$

或

$$y(t)=A\sin\omega\lambda$$

式中

$$\lambda=t+t_0$$

这样就可以将 $y(t)=A\sin\omega\lambda$ 理解为在单位正弦信号 $x(t)$ 激励下，系统的输出函数 $f[x(t)]$。

(2)稳态响应 $A\sin(\omega t+\varphi)$ 与激励 $\sin\omega_0 t$ 相比。

①幅值改变了，不再为 1，被放大为原来的 A 倍($A>1$ 时)，或被缩小为原来的 $\frac{1}{A}$ ($A<1$ 时)。其中，缩放倍率 A 是频率 ω 的函数，且随系统而异。

一阶系统为

$$A(\omega)=\frac{1}{\sqrt{1+(\tau\omega)^2}} \tag{3.15}$$

二阶系统为

$$A(\omega)=\frac{1}{\sqrt{\left[1-\left(\frac{\omega}{\omega_n}\right)^2\right]^2+4\zeta^2\left(\frac{\omega}{\omega_n}\right)^2}} \tag{3.16}$$

$$A\sin(\omega t+\varphi)=A\sin\omega(t+t_0)=y_1(t)$$

where

$$t_0=\varphi/\omega$$

$y_1(t)$ is the special solution of the non-homogeneous equation

$$\tau y'(t)+y(t)=\sin\omega t$$

and it is also the mandatory part of the system response, that is, the result of the forced action of input $x(t)$. To put it more simply, the input signal $x(t)$ enters the system from the input end and outputs from the output end, $y_1(t)$ is the output $x(t)$. Due to the inherent characteristics of the system, the output has been different from the one before entering. $y_1(t)$ is the steady-state response of the system to the sin excitation.

The second part of the solution $-Ae^{-t/\tau}\sin\varphi=y_2(t)$ is called the transient solution which is the solution of the homogeneous equation

$$\tau y'(t)+y(t)=0$$

that is, the system suddenly receives the input signal at $t=0$, which destroys the original equilibrium state of the system. $y_2(t)$ refers to the process in which the disturbance disappears and the system returns to the equilibrium state after the loss of level—the disturbed displacement decays and oscillates exponentially, and finally disappears. After a short time, such as $t=7\tau$, $e^{-t/\tau}\approx 0.001$, $y_2(t)$ can approximately be considered to have disappeared based on this item alone, and the system enters a steady state from then on.

(2) For the second-order system, when the relationship between the output and input of a second-order linear system is described by the steady second-order linear differential equation

$$y''(t)+2\zeta\omega_n y(t)+\omega_n^2 y(t)=x(t)$$

and the input is $x(t)=\sin\omega t$, solve the differential equation, and the solution is the total response of the system to sine excitation, that is

$$y(t)=A\sin(\omega t+\varphi)-\frac{\omega}{\omega_d}e^{-\zeta\omega_n t}\sin(\omega_d t+\varphi_3)$$

where

$$A=\frac{1}{\sqrt{\left[1-\left(\frac{\omega}{\omega_n}\right)^2\right]^2+4\zeta\left(\frac{\omega}{\omega_n}\right)^2}}$$

$$\varphi=-\arctan\frac{2\zeta\left(\frac{\omega}{\omega_n}\right)}{1-\left(\frac{\omega}{\omega_n}\right)^2}$$

$$\varphi_3=-\arctan\frac{2\zeta\sqrt{1-\zeta^2}}{1-\left(\frac{\omega}{\omega_n}\right)^2-2\zeta^2}$$

and ω_d is the circular frequency of the system when oscillates damped at a damping ratio of $\zeta(\zeta<1)$, $\omega_d=\omega_n\sqrt{1-\zeta^2}$.

In the same solution, $y_1(t)=A\sin(\omega t+\varphi)=A\sin\omega(t+t_0)$ is the steady-state response of the system to sinusoidal excitation, and the rest is the transient response. In it, $t_0=\varphi/\omega$.

The following problems can be found by analyzing the system' s response to sine excitation.

②初相改变了，不再为 0，而是增加了 φ 角。此增量也是频率 ω 的函数，并随系统而异。

一阶系统为

$$\varphi(\omega)=-\arctan(\tau\omega) \tag{3.17}$$

二阶系统为

$$\varphi(\omega)=-\arctan\frac{2\zeta\left(\frac{\omega}{\omega_{\mathrm{n}}}\right)}{1-\left(\frac{\omega}{\omega_{\mathrm{n}}}\right)^2} \tag{3.18}$$

综上所述，一阶、二阶线性系统对简谐信号的响应是同频同型的简谐信号。它以确定的比例 A 和确定的时差t_0在时域内再现了简谐激励的变化历程。

3.3.3　测试系统动态特性的频域描述——频率响应函数

对式(3.1)两边分别进行傅里叶变换，可得

$$(\mathrm{j}\omega)^n a_n Y(\omega)+(\mathrm{j}\omega)^{(n-1)}a_{n-1}Y(\omega)+\cdots+\mathrm{j}\omega a_1 Y(\omega)+a_0 Y(\omega)$$
$$=(\mathrm{j}\omega)^m b_m X(\omega)+(\mathrm{j}\omega)^{(m-1)}b_{m-1}X(\omega)+\cdots+\mathrm{j}\omega b_1 X(\omega)+b_0 X(\omega)$$

经整理后，得到系统输出 $y(t)$ 与输入 $x(t)$ 的频谱密度函数之比，即

$$\frac{Y(\omega)}{X(\omega)}=\frac{(\mathrm{j}\omega)^m b_m+(\mathrm{j}\omega)^{(m-1)}b_{m-1}+\cdots+\mathrm{j}\omega b_1+b_0}{(\mathrm{j}\omega)^n a_n+(\mathrm{j}\omega)^{(n-1)}a_{n-1}+\cdots+\mathrm{j}\omega a_1+a_0}$$

令

$$H(\omega)=\frac{Y(\omega)}{X(\omega)}=\frac{(\mathrm{j}\omega)^m b_m+(\mathrm{j}\omega)^{m-1}b_{m-1}+\cdots+\mathrm{j}\omega b_1+b_0}{(\mathrm{j}\omega)^n a_n+(\mathrm{j}\omega)^{n-1}a_{n-1}+\cdots+\mathrm{j}\omega a_1+a_0} \tag{3.19}$$

式中　$X(\omega)$——输入信号 $x(t)$ 的频谱密度函数；

$Y(\omega)$——当系统稳定后，稳态输出 $y(t)$ 的频谱密度函数；

$H(\omega)$——系统稳态输出的频谱密度函数与输入的频谱密度函数之比，称为线性系统的频率响应函数。

一般情况下，它是一个以实变量 ω 为自变量的复函数。因此，定义其模

$$|H(\omega)|=A(\omega)=\frac{|Y(\omega)|}{|X(\omega)|} \tag{3.20}$$

为线性系统的幅频特性函数，简称幅频特性，它是系统稳态输出与输入的幅值谱密度函数之比，也是上文提到的系统对简谐信号 $\sin\omega t$ 或 $\cos\omega t$ 的幅值响应函数。

定义其辐角

(1) The linear system does have frequency retention. After the system is stabilized under the excitation of the sine signal sin ωt , the output is indeed a sine wave with the same frequency as the input

$$y_1(t) = A\sin(\omega t + \varphi) = A\sin \omega(t + t_0)$$

or

$$y(t) = A\sin \omega\lambda$$

where

$$\lambda = t + t_0$$

So that $y(t) = A\sin \omega\lambda$ can be understood as under the excitation of the unit sine signal $x(t)$, the system's output function $f[x(t)]$.

(2) The comparison of steady-state response $A\sin(\omega t + \varphi)$ and excitation sin $\omega_0 t$.

① The amplitude has changed and is no longer 1, and it is enlarged by A times that of the origin value(when $A>1$), or reduced by $\frac{1}{A}$ that of the origin value (when $A<1$). In it, the zoom magnification A is a function of frequency ω and varies with the system.

First order system is

$$A(\omega) = \frac{1}{\sqrt{1 + (\tau\omega)^2}} \tag{3.15}$$

Second order system is

$$A(\omega) = \frac{1}{\sqrt{\left[1 - \left(\frac{\omega}{\omega_n}\right)^2\right]^2 + 4\zeta^2\left(\frac{\omega}{\omega_n}\right)^2}} \tag{3.16}$$

② The initial phase has changed, no longer being 0, but angle φ has been increased. This increment is also a function of frequency ω and varies with the system.

First order system is

$$\varphi(\omega) = -\arctan(\tau\omega) \tag{3.17}$$

Second order system is

$$\varphi(\omega) = -\arctan\frac{2\zeta\left(\frac{\omega}{\omega_n}\right)}{1 - \left(\frac{\omega}{\omega_n}\right)^2} \tag{3.18}$$

In summary, the response of the first-order and second-order linear systems to harmonic signals is the same frequency and the same type of harmonic signal. The determined ratio A and the determined time difference t_0 reproduce the change history of the harmonic excitation in the time domain.

3.3.3 The Frequency Domain Description of the Dynamic Characteristics of the Measurement System—Frequency Response Function

Carry out Fourier transform on both sides of formula (3.1), and we get

$$\begin{aligned}&(\mathrm{j}\omega)^n a_n Y(\omega) + (\mathrm{j}\omega)^{(n-1)} a_{n-1} Y(\omega) + \cdots + \mathrm{j}\omega a_1 Y(\omega) + a_0 Y(\omega)\\ &= (\mathrm{j}\omega)^m b_m X(\omega) + (\mathrm{j}\omega)^{(m-1)} b_{m-1} X(\omega) + \cdots + \mathrm{j}\omega b_1 X(\omega) + b_0 X(\omega)\end{aligned}$$

$$\mathrm{Arg}H(\omega)=\varphi(\omega)=\mathrm{Arg}Y(\omega)-\mathrm{Arg}X(\omega)=\varphi_y(\omega)-\varphi_x(\omega) \tag{3.21}$$

为线性系统的相频特性函数，简称相频特性，它是系统稳态输出与输入的频率成分的初相之差，即输出中的频率为 ω 的成分的相位滞后于输入的同频成分的相位角，也是上文提到的系统对简谐信号 $\sin\,\omega t$ 或 $\cos\,\omega t$ 的初始相位的响应函数。

根据以上概念，可以说，如果对系统输入一个正弦信号 $x(t)=\sin\,\omega_0 t$，则系统的稳态输出为

$$y(t)=A(\omega_0)\sin(\omega_0 t+\varphi(\omega_0))$$

这与用解微分方程所求得的稳态解在形式上是一致的。因此，又称 $H(\omega)$ 为正弦传递函数。

用傅里叶变换的方法求频率响应函数 $H(\omega)$ 不仅是一种方法的改变，其意义还在于借助于频率响应函数把在时域中用微分方程来描述的系统输出与输入的关系转化到了频域之中，成为稳态输出的频谱与输入频谱的简单代数关系，即

$$Y(\omega)=H(\omega)X(\omega) \tag{3.22}$$

而且此式不是方程式，不必求解，只要知道输入频谱 $X(\omega)$ 就可以直接求出输出的频谱 $Y(\omega)$，十分简捷，意义相当明确：线性系统对输入信号传递过程实际上就是对其频谱按式(3.22)进行改造的过程。系统不同，输出输入所构成的微分方程就不同，频率响应函数随之就不同，系统的动态特性就不一样。

如果将用式(3.22)计算出来的 $Y(\omega)$ 做傅里叶逆变换，那么其结果就是时域内系统对输入的稳态响应，即

$$\begin{aligned} y(t) &= F^{-1}[H(\omega)\cdot X(\omega)] \\ &= h(t)*x(t) \end{aligned} \tag{3.23}$$

式中　$h(t)$——$H(\omega)$ 的傅里叶逆变换，即 $F^{-1}[H(\omega)]$，称为系统的脉冲响应函数。

它也是微分方程的稳态解。

脉冲响应函数 $h(t)$ 是线性系统的另一个重要概念。设想对上述的线性系统输入一单位脉冲函数 $\delta(t)$，那么在频域内的响应为

$$Y(\omega)=H(\omega)\cdot\Delta(\omega)$$

式中　$\Delta(\omega)$——单位脉冲 $\delta(t)$ 的频谱，前已论证 $\Delta(\omega)=1$。

因此，$Y(\omega)=H(\omega)$。对上式做傅里叶逆变换，有

$$y(t)=h(t)$$

可见，$h(t)$ 就是系统对单位脉冲函数 $\delta(t)$ 的响应，故称 $h(t)$ 为脉冲响应函数，它是在时域内联系输出和输入的重要函数。$y(t)=h(t)*x(t)$，可见线性系统对任意信号 $x(t)$

After sorting, the ratio of the spectral density function of the system's output $y(t)$ to input $x(t)$ is obtained, that is

$$\frac{Y(\omega)}{X(\omega)}=\frac{(j\omega)^m b_m+(j\omega)^{(m-1)}b_{m-1}+\cdots+j\omega b_1+b_0}{(j\omega)^n a_n+(j\omega)^{(n-1)}a_{n-1}+\cdots+j\omega a_1+a_0}$$

Order

$$H(\omega)=\frac{Y(\omega)}{X(\omega)}=\frac{(j\omega)^m b_m+(j\omega)^{m-1}b_{m-1}+\cdots+j\omega b_1+b_0}{(j\omega)^n a_n+(j\omega)^{n-1}a_{n-1}+\cdots+j\omega a_1+a_0} \tag{3.19}$$

where $X(\omega)$ — the spectral density function of the input signal $x(t)$;

$Y(\omega)$ — the steady-state output spectral density function of $y(t)$ when the system is stable;

$H(\omega)$ —the ratio of the spectral density function of the steady-state output of the system to the spectral density function of the input, which is called frequency response function of the linear system.

In general, it is a complex function with a real variable ω as its independent variable. So, its modulus is

$$|H(\omega)|=A(\omega)=\frac{|Y(\omega)|}{|X(\omega)|} \tag{3.20}$$

It is the amplitude-frequency characteristic function of the linear system, referred to as the amplitude-frequency characteristic, which is the ratio of the amplitude spectral density function of the steady-state output of the system to the input. It is also the amplitude response function of the above-mentioned system to the harmonic signal sin ωt or cos ωt .

Define its argument is

$$\mathrm{Arg}H(\omega)=\varphi(\omega)=\mathrm{Arg}Y(\omega)-\mathrm{Arg}X(\omega)=\varphi_y(\omega)-\varphi_x(\omega) \tag{3.21}$$

It is the phase-frequency characteristic function of the linear system, referred to as the phase-frequency characteristic, which is the initial phase difference between the steady-state output of the system and the input frequency component, that is, the phase of ω component in the output lags the phase angle of the input co-frequency component. It is also the response function of initial phase of the above-mentioned system to the harmonic signal sin ωt or cos ωt .

According to the above concept, it can be said that if sine signal $x(t)=\sin\omega_0 t$ is input to the system, the steady-state output of the system is

$$y(t)=A(\omega_0)\sin(\omega_0 t+\varphi(\omega_0))$$

This is formally consistent with the steady-state solution obtained by solving differential equations. Therefore, $H(\omega)$ is also called the sine transfer function.

Using the Fourier transform method to find the frequency response function $H(\omega)$ is not only a change of the method, but its significance is that the relationship between the output and input of the system described by the differential equation in the time domain is transformed into the frequency domain by means of the frequency response function, it becomes the simple algebraic relationship between the steady-state output frequency spectrum and the input frequency spectrum, that is

$$Y(\omega)=H(\omega)X(\omega) \tag{3.22}$$

Moreover, this formula is not an equation, so we don't need to solve it. As long as we know the input spectrum $X(\omega)$, we can directly find the output spectrum $Y(\omega)$. It is very simple and has a clear meaning: the linear system's transfer process of the input signal is actually a

的稳态响应 $y(t)$ 是脉冲响应函数 $h(t)$ 与输入信号 $x(t)$ 的卷积。这样看一切都简单化了，所以对于线性系统，可以不考虑它的内部结构，只注意它的数学模型，用图 3.9 所示的框图来表示它，框内的 $H(\omega)$ 和 $h(t)$ 既用来联系输出与输入，又用来表示线性系统的动态特性。

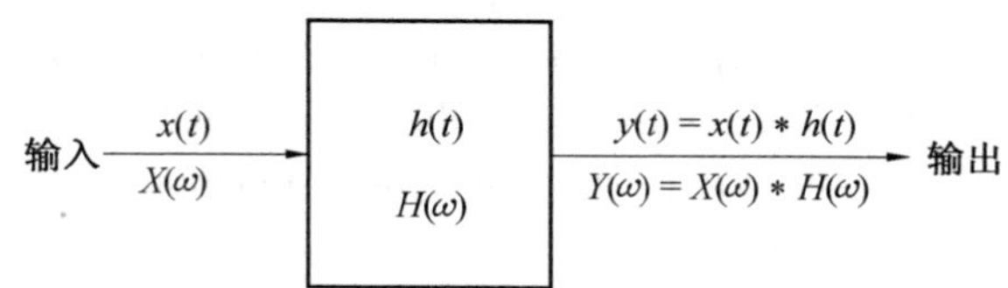

图 3.9　线性系统及其输入与输出

在时域内，线性系统输出与输入之间的关系还可以用如下方法得到。假设线性系统的输入为任意信号 $x(t)$，现将其分割成众多个并列的矩形脉冲，每一脉冲的宽度都为 $\Delta\tau$，任意输入下系统的响应如图 3.10 所示，若 $\Delta\tau$ 足够小，在 $t=\tau$ 时刻，$x(t)$ 和 $\Delta\tau$ 的乘积可看作发生在 $t=\tau$ 时刻的脉冲 $\delta(t-\tau)$ 的强度，系统对此脉冲的响应为 $(x(\tau)\Delta\tau)h(t-\tau)$，它是 $t=\tau$ 时刻输入对输出的一份贡献。在 t 时刻观察时，瞬时输出 $y(t)$ 应该是 t 时刻之前所发生的所有脉冲的贡献之和，即

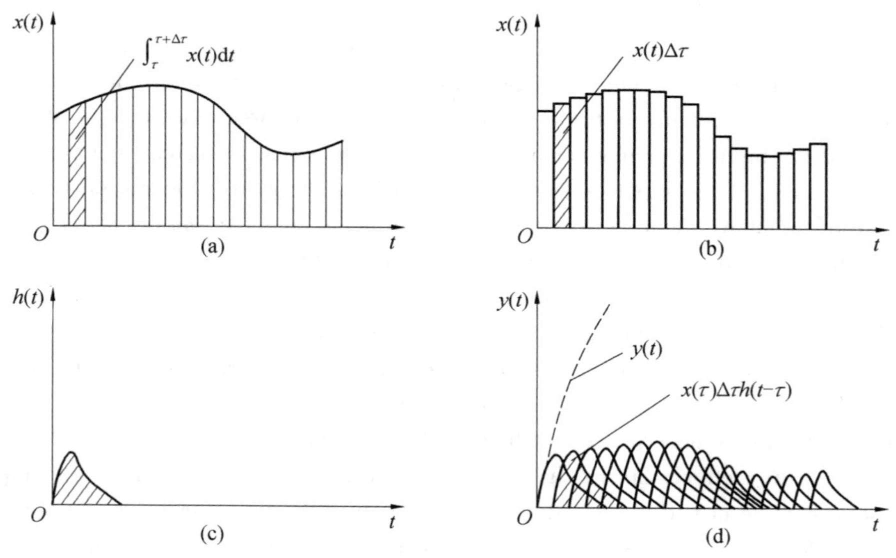

图 3.10　任意输入下系统的响应

$$y(t) = \sum_{\tau=0}^{t} [x(\tau)\Delta\tau] h(t - \tau)$$

当 $\Delta\tau\to 0$ 时，上式的右方演变为积分

process of transforming its frequency spectrum according to formula (3.22). Different systems have different differential equations composed of input and output, and the frequency response function will be different accordingly, it works the same way with the dynamic characteristics of the system.

If $Y(\omega)$ calculated by formula(3.22) is used as the inverse Fourier transform, then the result is the steady-state response of the system to the input in the time domain, that is

$$\begin{aligned} y(t) &= F^{-1}[H(\omega) \cdot X(\omega)] \\ &= h(t) * x(t) \end{aligned} \tag{3.23}$$

where $h(t)$ —the inverse Fourier transform of $H(\omega)$, which is $F^{-1}[H(\omega)]$, and it is called impulse response function of the system.

It is also the steady-state solution of the differential equation.

Impulse response function $h(t)$ is another important concept of linear systems. Imagine that a unit impulse function $\delta(t)$ is input to the above linear system, then the response in the frequency domain is

$$Y(\omega) = H(\omega) \cdot \Delta(\omega)$$

where $\Delta(\omega)$ — the frequency spectrum of unit pulse $\delta(t)$, and $\Delta(\omega) = 1$ has been demonstrated before.

So $Y(\omega) = H(\omega)$. Take the inverse Fourier transform of the above formula, we have

$$y(t) = h(t)$$

It can be seen that $h(t)$ is the response of the system to the unit impulse function $\delta(t)$, thus $h(t)$ is called the impulse response function. It is an important function linking output and input in the time domain. $y(t) = h(t) * x(t)$. It can be seen that the steady-state response $y(t)$ of the linear system to any signal $x(t)$ is the convolution of the impulse response function $h(t)$ and the input signal $x(t)$. In this way, everything is simplified, so for a linear system, you can ignore its internal structure and only pay attention to its mathematical model, which was represented by using the block diagram shown in Figure 3.9. $H(\omega)$ and $h(t)$ in the box are used to connect output and input, in turn, represent the dynamic characteristics of a linear system.

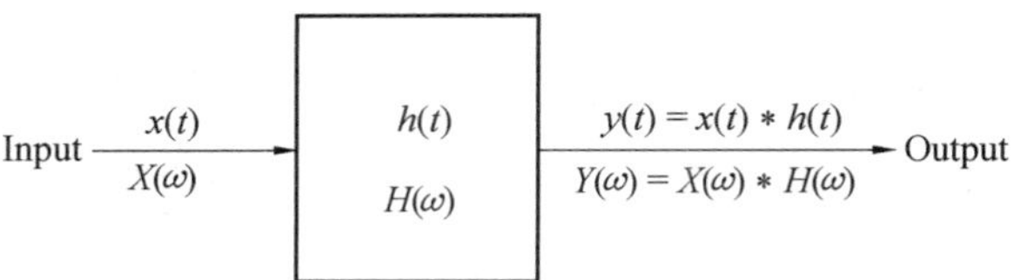

Figure 3.9 Linear system and its input and output

The relationship between the output and input of a linear system in the time domain can also be obtained by the following method. Assuming that the input of the linear system is any signal $x(t)$, it is now divided into a large number of parallel rectangular pulses, of which each pulse width is ΔT. System response under arbitrary input is shown in Figure 3.10. If $\Delta\tau$ is small enough, at time $t = \tau$, the product of $x(t)$ and $\Delta\tau$ can be regarded as the intensity of the pulse $t = \tau$ occurring at time $\delta(t-\tau)$, the system's response to this pulse is $(x(\tau)\Delta\tau)h(t-\tau)$, which is a contribution of input to output at time $t = \tau$. When observed at time t, the instantaneous output $y(t)$ should be the sum of the contributions of all pulses that occurred before

$$y(t)=\int_{0}^{t}x(\tau)h(t-\tau)\mathrm{d}\tau$$

即 $y(t)$ 是 $x(t)$ 与 $h(t)$ 的卷积。因为 $t<0$ 时，$x(t)=0$，故上式可写成

$$y(t)=\int_{-\infty}^{t}x(t)h(t-\tau)\mathrm{d}\tau=x(t)*h(t) \tag{3.24}$$

此式又称线性系统对任意输入信号的响应。

3.3.4 测试系统动态特性的复频域描述——传递函数

对式(3.1)进行拉普拉斯变换(简称拉氏变换)，并假定 $x(t)$ 和 $y(t)$ 及它们的各阶时间导数的初值($t=0$ 时)为 0，定义输出 $y(t)$ 的拉氏变换 $Y(s)$ 和输入的拉氏变换 $X(s)$ 之比为传递函数，并记为 $H(s)=Y(s)/X(s)$，则得

$$\frac{Y(s)}{X(s)}=\frac{b_m s^m+b_{m-1}s^{m-1}+\cdots+b_1 s+b_0}{a_n s^n+a_{n-1}s^{n-1}+\cdots+a_1 s+a_0} \tag{3.25}$$

$$Y(s)=H(s)X(s)=\frac{b_m s^m+b_{m-1}s^{(m-1)}+\cdots+b_1 s+b_0}{a_n s^n+a_{n-1}s^{(n-1)}+\cdots+a_1 s+a_0}X(s)$$

可见，传递函数 $H(s)$ 是用代数方程的形式来表示测试系统的动态特性，便于分析与计算，其作用等同于时域内的定常线性微分方程。而且在传递函数的表达式中，s 只是一种算符。尽管 s 未知，但并不影响已知 $X(s)$ 时对 $Y(s)$ 的求解。参数 $a_n,a_{n-1},\cdots,a_1,a_0$ 和 $b_m,b_{m-1},\cdots,b_1,b_0$ 是由系统的固有属性唯一确定的，与输入无关。因此，传递函数描述了系统的动态特性，与输入量无关。只要动态特性相似的系统，无论是电路系统或机械系统等，都可用同一类型的传递函数描述其特性。

传递函数有以下几个特点。

(1) $H(s)$ 与输入 $x(t)$ 及系统的初始状态无关，因为它是用式(3.25)来定义的，只不过当初始条件为 0 时它才是 $Y(s)$ 与 $X(s)$ 的比。如果今后未加特殊说明，便是指系统的初始条件全为 0，它只表现系统对输入信号的传输特性，对任意输入 $x(t)$ 都确定地给出唯一的输出 $y(t)$。

(2) $H(s)$ 是把线性系统的定常线性微分方程(即式(3.1))取拉普拉斯变换求得的，只反映系统的传输特性，而不拘泥于物理系统的实际结构。同一形式的传递函数可以表示具有相同传输特性的不同物理系统，如液柱式温度计、RC 电路及忽略质量的弹性系统，它们具有相同形式的传递函数，都属于一阶线性系统，但其中有热力学、电学和机械系统，它们的数学性质完全相同，或者说它们是在数学上相似的系统。

(3) 实际的物理系统，输入 $x(t)$ 和输出 $y(t)$ 都具有各自的量纲。用传递函数描述输

time t , that is

$$y(t)=\sum_{\tau=0}^{t}[x(\tau)\Delta\tau]h(t-\tau)$$

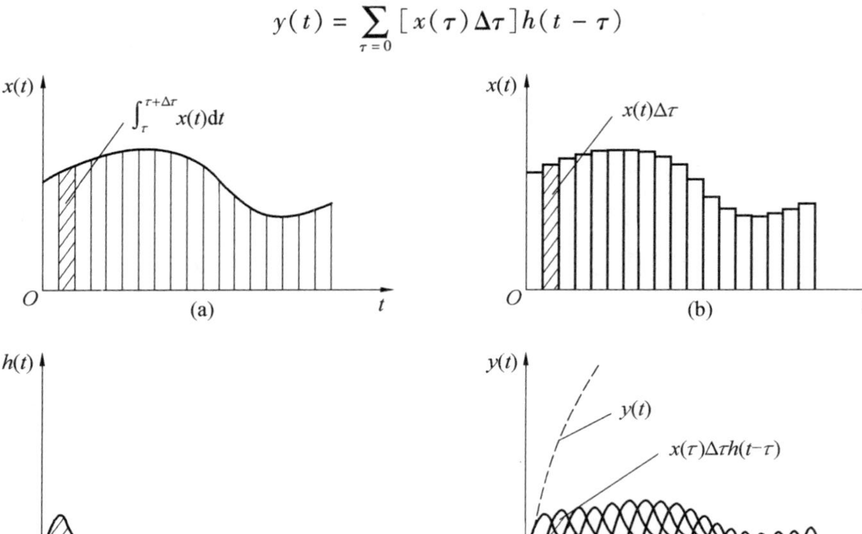

Figure 3. 10　System response under arbitrary input

When $\Delta\tau\to 0$, the right side of the above formula evolves into integral

$$y(t)=\int_{0}^{t}x(\tau)h(t-\tau)\mathrm{d}\tau$$

That is, $y(t)$ is the convolution of $x(t)$ and $h(t)$. Since $t<0$, $x(t)=0$, the above formula can be written as

$$y(t)=\int_{-\infty}^{t}x(t)h(t-\tau)\mathrm{d}\tau=x(t)*h(t) \tag{3.24}$$

This formula is also called the response of a linear system to any input signal.

3.3.4 The Complex Frequency Domain Description of the Dynamic Characteristics of the Measurement System—Transfer Function

We should use Laplace transform to formula (3.1), and assume that the initial values of $x(t)$ and $y(t)$ and their respective time derivatives (at $t=0$) are 0, define the ratio of Laplace transform $Y(s)$ of output $y(t)$ to the Laplace transform $X(s)$ of the input as the transfer function, and record it as $H(s)=Y(s)/X(s)$, then

$$\frac{Y(s)}{X(s)}=\frac{b_m s^m+b_{m-1}s^{m-1}+\cdots+b_1 s+b_0}{a_n s^n+a_{n-1}s^{n-1}+\cdots+a_1 s+a_0} \tag{3.25}$$

$$Y(s)=H(s)X(s)=\frac{b_m s^m+b_{m-1}s^{(m-1)}+\cdots+b_1 s+b_0}{a_n s^n+a_{n-1}s^{(n-1)}+\cdots+a_1 s+a_0}X(s)$$

It can be seen that the transfer function $H(s)$ expresses the dynamic characteristics of the measurement system in the form of algebraic equations, which is convenient for analysis and calculation, and its function is equivalent to the steady linear differential equation in the time domain. Moreover, in the expression of the transfer function, s is just an operator. Although s

出输入的关系及转换特性时，理应真实地反映量纲间的转化关系，这种转化关系是通过 $H(s)$ 中有纲量的各系数 $a_n \sim a_0$ 及 $b_m \sim b_0$ 参与运算来实现的。因此，$H(s)$ 也是有量纲的。

(4) $H(s)$ 是一个分式，分母是 s 的多项式，由 s 的各次幂与系统硬件参数组成，一些文献中称其为原微方程的特征多项式，该多项式完全由系统来决定，s 的最高幂次等于系统微分方程的阶次；分子则与系统与外界的联系有关，所谓与外界的联系，是指取什么量作为输入，作用在系统的什么部位，以什么量为输出，在系统的什么部位来测取它，实际上就是怎么样使用系统的硬件。使用方法不同，系统传递函数的分子就不同，传递函数就随之不同。

传递函数可用理论计算求取，也可用实验方法获得，这对于不便列出微分方程式的系统具有实际意义。例如，对于一个复杂的线性时不变测试系统，无须了解其具体内容，只要给系统一个激励（输入）$x(t)$，得到系统对 $x(t)$ 的响应（输出）$y(t)$，则系统的传递函数就可由 $H(s)=Y(s)/X(s)=L[y(t)]/L[x(t)]$ 确定。

传递函数 $g(t)\,(t\geqslant 0)$ 的拉普拉斯变换即组合函数 $g(t)u(t)\mathrm{e}^{-\beta t}\,(\beta>0)$ 的傅里叶变换。则当 $\beta\to 0$ 时，若下列极限存在，则它就是 $g(t)$ 的傅里叶变换，即

$$F[g(t)]=\lim_{\beta\to 0}F(s)$$

由于传递函数中的 $s=\beta+\mathrm{j}\omega$，因此当 $\beta\to 0$ 时，s 演化为 $\mathrm{j}\omega$。由此可知，傅里叶变换实际上是拉普拉斯变换的一个特例，复频域中的传递函数在实频为 0 的平面上的投影即频域中的频率响应函数。这也是拉普拉斯变换对信号需要满足的前提条件比傅里叶变换减弱的原因。拉普拉斯变换的条件比傅里叶变换宽松得多，不需要满足绝对可积条件，只需满足以下条件。

(1) 在 $t\geqslant 0$ 的任一有限区间上分段连续。

(2) 在 $t\to\infty$ 时，$g(t)$ 的增长速度不超过某一指数函数，即存在常数 $M>0$ 及 $C\geqslant 0$，使得

$$|g(t)|\leqslant M\mathrm{e}^{ct},\quad 0\leqslant t<\infty$$

这为求取那些不满足绝对可积条件的动态信号的频谱开辟了一条新路，也为前文所述的“动态信号都可以分解为许多简谐波”这一结论提供了更为充分的理论依据，如求单位阶跃函数 $u(t)$ 的傅里叶变换。

单位阶跃函数不满足绝对可积条件，不能按定义计算它的傅里叶变换。但是它的拉普拉斯变换存在，即

$$L[u(t)]=\frac{1}{s}=\frac{1}{\beta+\mathrm{j}\omega}$$

is unknown, it does not affect the solution of $Y(s)$ when $X(s)$ is known. Parameter a_n, a_{n-1}, ⋯, a_1, a_0 and b_m, b_{m-1}, ⋯, b_1, b_0 are uniquely determined by the inherent properties of the system and has nothing to do with the input. Therefore, the transfer function describes the dynamic characteristics of the system and has nothing to do with the input. As long as the dynamic characteristics of the system are similar, whether it is a circuit system or a mechanical system, the same type of transfer function can be used to describe its characteristics.

The transfer function has the following characteristics.

(1) The reason why $H(s)$ has nothing to do with input $x(t)$ and the initial state of the system is that it is defined by formula (2.68), this is only the case when it is the ratio of $Y(s)$ to $X(s)$ when the initial condition is 0. If no special instructions are added in the future, it means that the initial conditions of the system are all 0, only showing the transmission characteristics of the system to the input signal, and it definitely gives a unique output $y(t)$ for any input $x(t)$.

(2) $H(s)$ is obtained by taking the Laplace transform of the steady linear differential equation of the linear system (that is, formula (3.1)), which only reflects the transmission characteristics of the system and is not restricted to the actual structure of the physical system. The same form of transfer function can represent different physical systems with the same transmission characteristics, such as liquid column thermometers, RC circuits, and elastic systems that ignore mass, all which have the same form of transfer function, all belonging to first-order linear systems, but there are thermodynamics, electrical and mechanical systems, and their mathematical properties are exactly the same, or they are mathematically similar systems.

(3) In the actual physical system, both input $x(t)$ and output $y(t)$ have their own dimensions. When the transfer function is used to describe the relationship between output and input and the conversion characteristics, it should truly reflect the conversion relationship between the dimensions, which is realized through the participation of the dimensional coefficients $a_n - a_0$ and $b_m - b_0$ in $H(s)$ in the calculation. Therefore, $H(s)$ is also dimensional.

(4) $H(s)$ is a fraction, and the denominator is a polynomial of s. It is composed of the powers of s and the system hardware parameters. In some literatures, it is called the characteristic polynomial of the original differential equation. The polynomial is completely determined by the system. The highest power of s is equal to the order of the differential equation of the system. The molecule is related to the connection between the system and the outside world. The so-called connection with the outside world refers to what quantity is taken as input, what part of the system is used, what quantity is used as output, and what part of the system is used to measure it, that is, how to use the hardware of the system. With different usage methods, the numerator of the system transfer function will be different, and the transfer function will be different accordingly.

The transfer function can be obtained by theoretical calculations or experimental methods, which has practical significance for systems that are inconvenient to list differential equations. For example, for a complex linear time-invariant measurement system, it's no need to know its specific content. As long as you give the system a stimulus (input) $x(t)$ and get the system's response (output) $y(t)$ to $x(t)$, then the transfer function of the system can be determined by $H(s) = Y(s)/X(s) = L[y(t)]/L[x(t)]$.

由于 $s=0$ 时存在极点,因此进行如下处理:

$$\frac{1}{\beta+\mathrm{j}\omega}=\frac{\beta}{\beta^2+\omega^2}-\mathrm{j}\frac{\omega}{\beta^2+\omega^2}$$

对于虚部,$\beta\to 0$ 时,有

$$\mathrm{j}\frac{1}{\beta+\mathrm{j}\omega}=\frac{1}{\mathrm{j}\omega}$$

对于实部,$\beta\to 0$ 时,有

$$\lim_{\beta\to 0}\frac{\beta}{\beta^2+\omega^2}=\begin{cases}\infty, & \omega=0\\ 0, & \omega\neq 0\end{cases}$$

此时,可以看出实部应为一个脉冲信号,其强度为

$$\int_{-\infty}^{+\infty}\frac{\beta}{\beta^2+\omega^2}\mathrm{d}\omega=\pi$$

所以,当 $\beta\to 0$ 时,有

$$F[u(t)]=\lim_{\beta\to 0}\frac{1}{\beta+\mathrm{j}\omega}=\pi\delta(\omega)-\mathrm{j}\frac{1}{\omega}=\pi\delta(\omega)+\frac{1}{\mathrm{j}\omega}$$

拉普拉斯变换的应用很广,如用来解微分方程、求系统的暂态响应。本书中应用它的主要目的是用拉普拉斯变换建立线性系统的传递函数,变时域内输出输入间的微分方程为复频域内的代数方程,以简化计算,便于分析。

3.3.5 线性系统的串联与并联

两个传递函数各为 $H_1(s)$ 和 $H_2(s)$ 的环节,串联时(图 3.11),如果它们之间没有能量交换,则串联后所组成的系统的传递函数 $H(s)$ 在初始条件为 0 时有

$$H(s)=\frac{Y(s)}{X(s)}=\frac{Z(s)Y(s)}{X(s)Z(s)}=H_1(s)H_2(s) \tag{3.26}$$

类似地,对几个环节串联组成的系统,有

$$H(s)=\prod_{i=1}^{n}H_i(s) \tag{3.27}$$

若两个环节并联(图 3.12),则因

$$Y(s)=Y_1(s)+Y_2(s)$$

而有

$$H(s)=\frac{Y(s)}{X(s)}=\frac{Y_1(s)}{X(s)}+\frac{Y_2(s)}{X(s)}=H_1(s)+H_2(s) \tag{3.28}$$

因此,由 n 个环节并联组成的系统也类似地有

The Laplace transform of the transfer function $g(t)$ $(t\geqslant 0)$ proves the Fourier transform of the combined function $(g(t)u(t)\mathrm{e}^{-\beta t})$ $(\beta > 0)$. Then when $\beta \to 0$, if the following limit exists, it is the Fourier transform of $g(t)$, that is

$$F[g(t)] = \lim_{\beta\to 0} F(s)$$

Because of $s = \beta + \mathrm{j}\omega$ in the transfer function, when $\beta \to 0$, s evolves to $\mathrm{j}\omega$. It can be seen that the Fourier transform is actually a special case of the Laplace transform, and the projection of the transfer function in the complex frequency domain on the plane where the real frequency is 0 is the frequency response function in the frequency domain. This is also the reason why the preconditions that the Laplace transform needs to meet for the signal are weaker than those of the Fourier transform. The conditions of the Laplace transform are much more relaxed than those of the Fourier transform, which does not need to satisfy the absolute integrability condition, but only needs to satisfy the following conditions.

(1) Segmental continuous on any finite interval where $t \geqslant 0$.

(2) When $t \to \infty$, the growth rate of $g(t)$ does not exceed a certain exponential function, that is, there are constants $M>0$ and $C\geqslant 0$, so that

$$|g(t)| \leqslant M\mathrm{e}^{ct}, \quad 0 \leqslant t < \infty$$

This opens up a new way to obtain the spectrum of dynamic signals that do not satisfy the absolute integrability condition. It also provides a more sufficient theoretical basis for the conclusion that "dynamic signals can be decomposed into many simple harmonics" mentioned above. For example, find the Fourier transform of the unit step function $u(t)$.

The unit step function does not satisfy the absolute integrability condition, and its Fourier transform cannot be calculated by definition. But there exists its Laplace transform, that is

$$L[u(t)] = \frac{1}{s} = \frac{1}{\beta + \mathrm{j}\omega}$$

Due to the existence of poles in $s=0$, the following processing is carried out:

$$\frac{1}{\beta + \mathrm{j}\omega} = \frac{\beta}{\beta^2 + \omega^2} - \mathrm{j}\frac{\omega}{\beta^2 + \omega^2}$$

For the imaginary part, when $\beta \to 0$, we have

$$\mathrm{j}\frac{1}{\beta + \mathrm{j}\omega} = \frac{1}{\mathrm{j}\omega}$$

For the real part, when $\beta \to 0$, we have

$$\lim_{\beta\to 0}\frac{\beta}{\beta^2 + \omega^2} = \begin{cases} \infty, & \omega = 0 \\ 0, & \omega \neq 0 \end{cases}$$

At this time, we can see that the real part should be a pulse signal, its strength is

$$\int_{-\infty}^{+\infty} \frac{\beta}{\beta^2 + \omega^2}\mathrm{d}\omega = \pi$$

So when $\beta \to 0$, we have

$$F[u(t)] = \lim_{\beta\to 0} \frac{1}{\beta + \mathrm{j}\omega} = \pi\delta(\omega) - \mathrm{j}\frac{1}{\omega} = \pi\delta(\omega) + \frac{1}{\mathrm{j}\omega}$$

Therefore, for a one-sided function defined only on the interval of $t \geqslant 0$, if the s in its Laplace transform is replaced by $\mathrm{j}\omega$ (that is, let $\beta = 0$), what is obtained is the Fourier transform of the one-sided function (assuming if it exists).

The Laplace transform has a wide range of applications, such as solving differential

$$H(s)=\sum_{i=1}^{n}H_i(s) \tag{3.29}$$

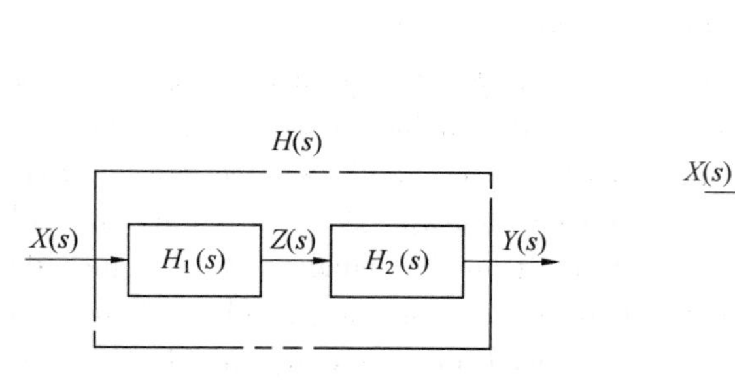

图 3.11　两个环节的串联

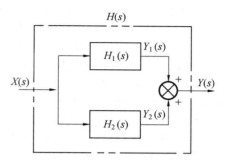

图 3.12　两个环节的并联

同样，将 $s=\mathrm{j}\omega$ 代入式(3.27)和式(3.29)，得到 n 个环节串联系统频率响应函数为

$$H(\omega)=\prod_{i=1}^{n}H_i(\omega) \tag{3.30}$$

其幅频、相频特性分别为

$$\begin{cases}A(\omega)=\prod_{i=1}^{n}A_i(\omega)\\ \varphi(\omega)=\sum_{i=1}^{n}\varphi_i(\omega)\end{cases} \tag{3.31}$$

而 n 个环节并联系统的频率响应函数为

$$H(\omega)=\sum_{i=1}^{n}H_i(\omega) \tag{3.32}$$

在对复杂线性系统动态特性进行描述时，可以先将复杂系统拆解为相对较为简单的基本环节，如一阶、二阶系统等，再利用线性系统的串、并联性质获得复杂线性系统动态特性的频域或复频域描述。

3.3.6　负载效应

在实际测试工作中，测试系统和被测对象之间、测试系统内部各环节之间相互连接并因此而产生相互作用是处处可见的。接入的测试装置就成为被测对象的负载。后接环节总是成为前面环节的负载。

现以图 3.13 所示测试装置和测试对象的串联为例来说明环节之间连接的复杂性。设被测对象是安装在电动机中的滚珠轴承，测试目的是判断轴承是否良好。此时，源信号之一 $x(t)$ 可以认为是轴承内滚道上的一个缺损所引起的滚珠在滚动中的周期性冲击。

equations and finding the transient response of the system. The main purpose of applying it in this book is to use Laplace transform to establish the transfer function of a linear system. The differential equation between output and input in the time domain is changed to an algebraic equation in the complex frequency domain to simplify calculations and facilitate analysis.

3.3.5 Series and Parallel Connection of Linear Systems

The two transfer functions are each link of $H_1(s)$ and $H_2(s)$. When they are connected in series (Figure 3.11), if there is no energy exchange between them, when the initial condition of $H(s)$ is 0, the transfer function of the system after the series connection will have

$$H(s) = \frac{Y(s)}{X(s)} = \frac{Z(s)}{X(s)} \frac{Y(s)}{Z(s)} = H_1(s) H_2(s) \tag{3.26}$$

Similarly, for a system composed of several links in series, there are

$$H(s) = \prod_{i=1}^{n} H_i(s) \tag{3.27}$$

If two links are connected in parallel (Figure 3.12), because of

$$Y(s) = Y_1(s) + Y_2(s)$$

then we have

$$H(s) = \frac{Y(s)}{X(s)} = \frac{Y_1(s)}{X(s)} + \frac{Y_2(s)}{X(s)} = H_1(s) + H_2(s) \tag{3.28}$$

Therefore, a system composed of n links in parallel similarly has

$$H(s) = \sum_{i=1}^{n} H_i(s) \tag{3.29}$$

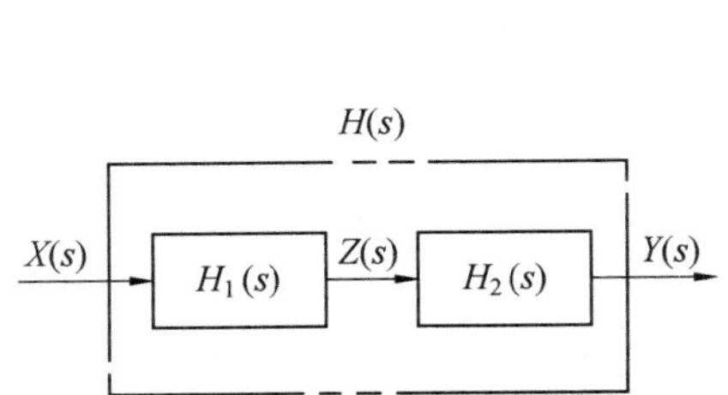

Figure 3.11　Series of two links

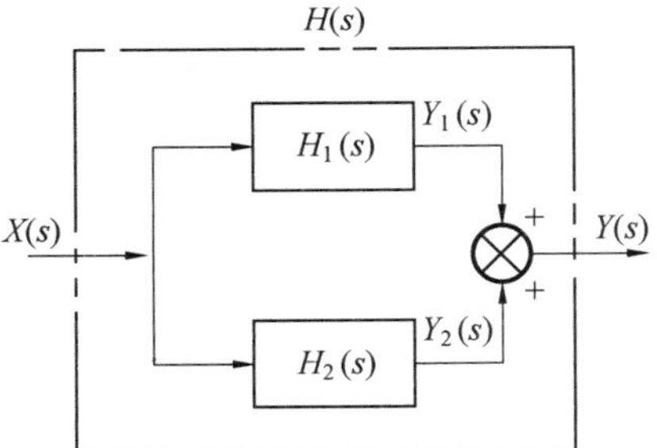

Figure 3.12　Parallel of two links

Similarly, substituting $s = j\omega$ into formula (3.27) and formula (3.29), the frequency response function of the n-link series system is obtained as

$$H(\omega) = \prod_{i=1}^{n} H_i(\omega) \tag{3.30}$$

Its amplitude-frequency and phase-frequency characteristics are respectively

$$\begin{cases} A(\omega) = \prod_{i=1}^{n} A_i(\omega) \\ \varphi(\omega) = \sum_{i=1}^{n} \varphi_i(\omega) \end{cases} \tag{3.31}$$

由于测振传感器只能安装在电动机外壳靠近轴承座的某处,因此对传感器来说,无法直接输入 $x(t)$,而只能间接地输入某一信号 $y(t)$。显然,$y(t)$ 是 $x(t)$ 经过电动机转轴—轴承—轴承座等机械结构传输后的输出,并且是测试系统的输入,而最后的测试结果却是 $y(t)$ 经过测试系统传输后的响应 $z(t)$。因此,$z(t)$ 和 $x(t)$ 的关系实际上取决于从电动机转轴—轴承—轴承座—测试装置整个系统的特性,取决于测试装置与被测对象两部分的串接。整个系统是由众多环节和装置连接而成的。后接环节总是成为前环节的负载,二者总是存在能量交换和相互影响的。因此,系统的传递函数不再是各组成环节传递函数的叠加(如并联时)或连乘(如串联时)。

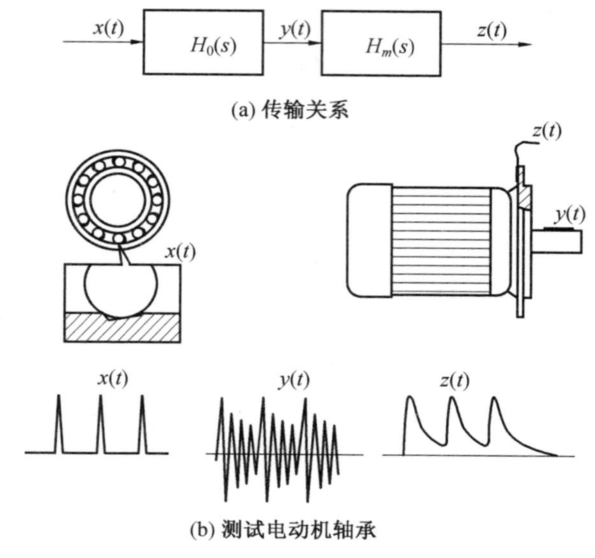

图 3.13　测试装置和测试对象的串联

前面曾假设在相连环节之间没有能量交换,因此在环节互联前后环节仍保持原有的传递函数。在此基础上,导出了环节串、并联后所形成的系统的传递函数表达式即式(3.27)和式(3.29)。然而,这种只有信息传递而没有能量交换的连接在实际系统中很少遇到。只有用不接触的辐射源信息探测器,如可见光和红外探测器或其他射线探测器,才可算是这类互联。

当一个装置连接到另一装置上并发生能量交换时,就会发生以下两种现象。

(1)前装置的连接处甚至整个装置的状态和输出都将发生变化。

(2)两个装置共同形成一个新的整体。该整体虽然保留其两组成装置的某些主要特征,但其传递函数已不能用式(3.27)和式(3.29)来表达。某装置因后接另一装置而产生的种种现象称为负载效应。

And the frequency response function of the n-link parallel system is

$$H(\omega) = \sum_{i=1}^{n} H_i(\omega) \tag{3.32}$$

When describing the dynamic characteristics of a complex linear system, we can first disassemble the complex system into relatively simple basic links, such as first-order, second-order systems, and use the series and parallel properties of the linear system to obtain the frequency domain or complex frequency domain description of the dynamic characteristics of complex linear systems.

3.3.6 Loading Effect

It can be seen everywhere that the links between the measurement system and the measured object, and the internal parts of the measurement system are connected and interact with each other in the actual measurement work. The access to the measurement device becomes the load of the measured object. The subsequent link always becomes the load of the previous link.

Now take series connection of measurement device and test object shown in Figure 3.13 as an example to illustrate the complexity of connections between links. Assuming that the measured object is a ball bearing installed in a motor, the purpose of the measurement is to determine whether the bearing is good. At this time, one of the source signals $x(t)$ can be considered as the periodical impact of the balls in the rolling caused by a defect on the inner raceway of the bearing. Because the vibration sensor can only be installed somewhere near the bearing seat of the motor housing. Therefore, for the sensor, it cannot directly input $x(t)$, but can only input a certain signal $y(t)$ indirectly. Obviously, $y(t)$ is the output of $x(t)$ after transmission through the motor shaft–bearing–bearing seat and other mechanical structures, and it is the input of the measurement system. And the final measurement result is tested by $y(t)$ response to the transmission system $z(t)$. Therefore, the relationship between $z(t)$ and $x(t)$ actually depends on the characteristics of the entire system from the motor shaft–bearing–the bearing seat – the measurement device, and it depends on the series connection of the measurement device and the measured object. The entire system is made up of many links and devices. The subsequent link always becomes the load of the previous link, and there is always energy exchange and mutual influence between the two. As a result, the transfer function of the system is no longer the superposition of the transfer functions of the components (such as in parallel) or multiplication (such as in series).

Although it was previously assumed that there is no energy exchange between the interconnected links, the original transfer function is maintained before and after the links are interconnected. On this basis, the transfer function expressions formula (3.27) and formula (3.29) of the system formed by the series and parallel links are derived. However, this kind of connection with only information transmission without energy exchange is rarely encountered in actual systems. Only non-contact radiation source information detectors, such as visible light and infrared detectors or other ray detectors, can be considered as this type of interconnection.

When one device is connected to another device and energy exchange occurs, the following two phenomena occur.

(1) The connection of the front device and even the state and output of the entire device

负载效应产生的后果有的可以忽略,有的却是很严重的,不能对其掉以轻心。下面举一些例子来说明负载效应的严重后果。

集成电路芯片温度虽高,但功耗很小,几十毫瓦,相当于一种功率小的热源。若用一个带探针的温度计去测试结点温度,显然温度计会从芯片吸收可观的热量而成为芯片的散热元件。这样不仅不能测出正确的结点工作温度,而且整个电路的工作温度都会下降。若在一个单自由度振动系统的质量块 m 上联结一个质量为 m_f 的传感器,致使参与振动的质量成为 $m+m_f$,则会导致系统固有频率的下降。

图 3.14 所示为简单的直流电路。不难算出电阻器 R_2 的电压降为 $U_0=\frac{R_2}{R_1+R_2}E$。为测量 U_0,可在 R_2 两端并联一个内阻为 R_m 的电压表。这时,由于 R_m 接入 R_2 和 R_m 两端,因此电压 U 为

$$U=\frac{R_L}{R_1+R_L}E=\frac{R_mR_2}{R_1(R_m+R_2)+R_mR_2}E$$

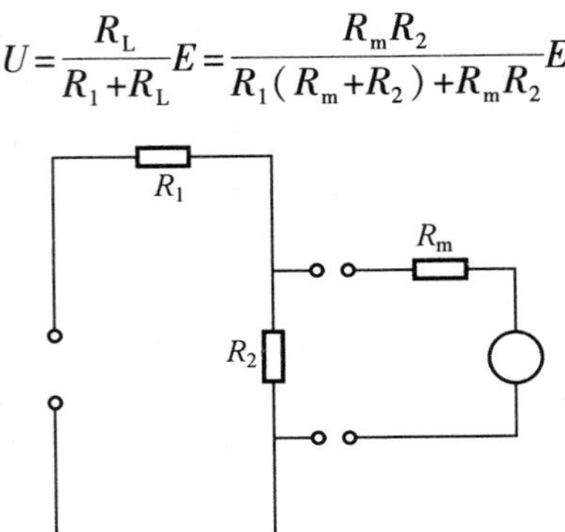

图 3.14　简单的直流电路

显然,由于接入测试装置(电表),因此被测系统(原电路)状态及被测量(R_2 的电压降)都发生了变化。原来电压降为 U_0,接入电表后,变为 $U\neq U_0$,二者的差值随 R_m 的增大而减小。为定量说明这种负载效应的影响程度,令 $R_1=100\ \text{k}\Omega$,$R_2=R_m=150\ \text{k}\Omega$,$E=150\ \text{V}$,代入上式,可以得到 $U_0=90\ \text{V}$,而 $U=64.3\ \text{V}$,误差达到 28.6%。若 R_m 改为 1 MΩ,其余不变,则 $U=84.9\ \text{V}$,误差为 5.7%。此例充分说明了负载效应对测量结果的影响有时是很大的。

图 3.15 中两个 RC 电路(一阶环节)的传递函数分别是

$$H_1(s)=\frac{1}{1+\tau_1 s},\tau_1=R_1C_1$$

$$H_2(s)=\frac{1}{1+\tau_2 s},\tau_2=R_2C_2$$

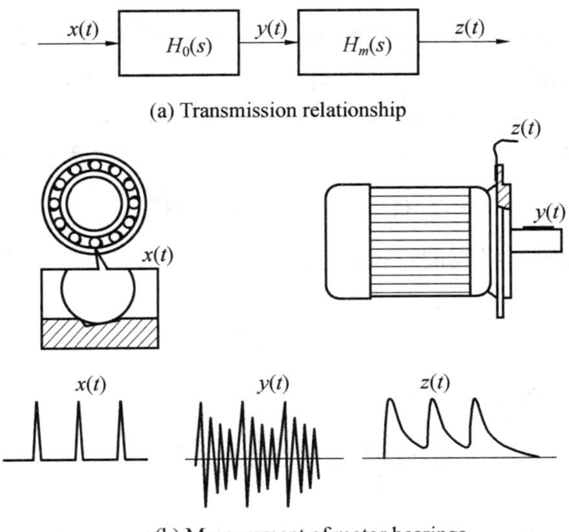

(a) Transmission relationship

(b) Measurement of motor bearings

Figure 3.13 Series connection of measurement device and test object

will change.

(2) The two devices together form a new whole. Although the whole retains some of the main features of the two components, its transfer function can no longer be expressed by formula(3.27) and formula(3.29). The various phenomena caused by a device connected to another device are called loading effect.

Some of the consequences of the loading effect can be ignored, while others are very serious and should not be taken lightly. Here are some examples to illustrate the serious consequences of the loading effect.

Although the temperature of the integrated circuit chip is high, the power consumption is very small, only tens of milliwatts, which is quite a heat source with low power. If a thermometer with a probe is used to test the junction temperature, it is obvious that the thermometer will absorb considerable heat from the chip and become the heat dissipation element of the chip. In this way, not only the correct junction operating temperature cannot be measured, but also the operating temperature of the entire circuit will drop. If a sensor with mass m_f is connected to the mass m of a single-degree-of-freedom vibration system, the mass participating in the vibration becomes $m + m_f$, which leads to a decrease in the natural frequency of the system.

Figure 3.14 shows a simple DC circuit. It is not difficult to calculate the voltage drop of resistor R_2 to $U_0 = \frac{R_2}{R_1 + R_2}E$. To measure U_0 , a voltmeter with internal resistance R_m can be connected in parallel to both ends of R_2. At this time, because R_m is connected to R_2 and R_m , the voltage U is

$$U = \frac{R_L}{R_1 + R_L}E = \frac{R_m R_2}{R_1(R_m + R_2) + R_m R_2}E$$

Obviously, due to the access to the measurement device (electric meter), the state of the

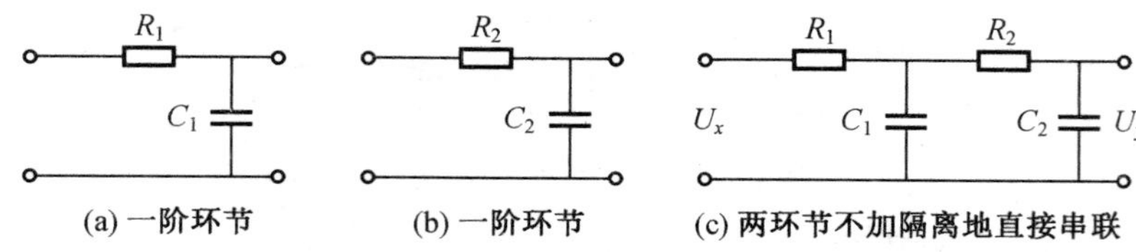

图 3.15　两个一阶环节的连接

若未加任何隔离措施而将它们直接串接，则令 $U_2(t)$ 为连接点的电压，可写成

$$\frac{U_y(s)}{U_2(s)}=\frac{1}{1+\tau_2 s}$$

两环节相连后，连接点右侧的阻抗为

$$Z_2=R_2+\frac{1}{C_2 s}=\frac{1+R_2C_2 s}{C_2 s}=\frac{1+\tau_2 s}{C_2 s}$$

令 Z 表示自 R_1 后的右侧电路的阻抗，它由电容 C_1 的阻抗和 Z_2 并联而成，有

$$Z=\frac{1}{C_1 s}//Z_2=\frac{1+\tau_2 s}{(C_1+C_2)s+\tau_2 C_1 s^2}$$

由于

$$\frac{U_x}{Z+R_1}=\frac{U_2}{Z}$$

因此

$$\frac{U_2(s)}{U_x(s)}=\frac{Z}{R_1+Z}=\frac{1+\tau_2 s}{1+(\tau_1+\tau_2+R_1C_2)s+\tau_1\tau_2 s^2}$$

连接后的传递函数为

$$H(s)=\frac{U_y(s)}{U_x(s)}=\frac{U_2(s)U_y(s)}{U_x(s)U_2(s)}$$

$$=\frac{1}{1+(\tau_1+\tau_2+R_1C_2)s+\tau_1\tau_2 s^2}$$

注意到

$$H_1(s)H_2(s)=\frac{1}{1+\tau_1 s}\frac{1}{1+\tau_2 s}=\frac{1}{1+(\tau_1+\tau_2)s+\tau_1\tau_2 s^2}$$

可见

$$H(s)\neq H_1(s)H_2(s)$$

此外，从上面推导过程可以看到，利用电阻抗的概念来求电路的传递函数是比较简便的，可免去求电路微分方程式的步骤。

负载效应造成的影响应根据具体环节和装置来具体分析。如果将电阻抗的概念推广

system under measurement (original circuit) and the measured (voltage drop of R_2) have changed. The original voltage drop is U_0. After the meter is connected, it becomes $U \neq U_0$, and the difference between the two decreases as R_m increases. In order to quantify the degree of influence of this loading effect, let $R_1 = 100\ \text{k}\Omega, R_2 = R_m = 150\ \text{k}\Omega, E = 150\ \text{V}$, substituting into the above formula, we can get $U_0 = 90\ \text{V}$, and $U = 64.3\ \text{V}$, the error reaches 28.6%. If R_m is changed to 1 MΩ and the rest remain unchanged, then $U = 84.9\ \text{V}$, with an error of 5.7%. This example fully illustrates that the influence of load effects on the measurement results is sometimes very large.

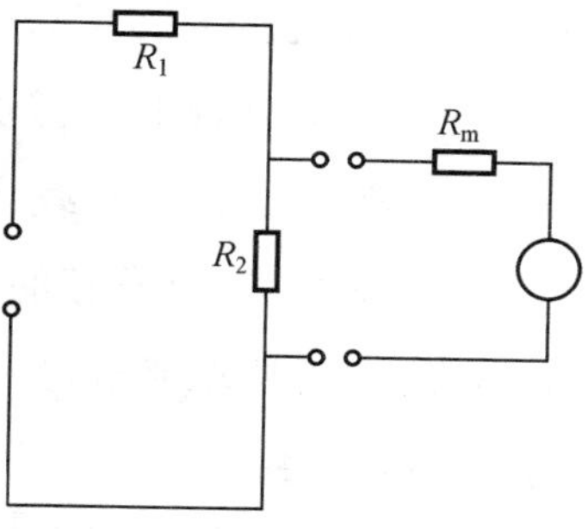

Figure 3.14　Simple DC circuit

The transfer functions of the two RC circuits (first-order links) in Figure 3.15 are

$$H_1(s) = \frac{1}{1 + \tau_1 s}, \tau_1 = R_1 C_1$$

$$H_2(s) = \frac{1}{1 + \tau_2 s}, \tau_2 = R_2 C_2$$

(a) First-order link　(b) First-order link　(c) Two links directly connected in series without isolation

Figure 3.15　The connection of two first-order links

If they are directly connected in series without any isolation measures, let $U_2(t)$ be the voltage of the connection point, which can be written as

$$\frac{U_y(s)}{U_2(s)} = \frac{1}{1 + \tau_2 s}$$

After the two links are connected, the impedance on the right side of the connection point is

$$Z_2 = R_2 + \frac{1}{C_2 s} = \frac{1 + R_2 C_2 s}{C_2 s} = \frac{1 + \tau_2 s}{C_2 s}$$

Let Z denote the impedance of the right circuit after R_1, which is formed by the impedance of capacitor C_1 and Z_2 in parallel, we have

$$Z = \frac{1}{C_1 s} // Z_2 = \frac{1 + \tau_2 s}{(C_1 + C_2)s + \tau_2 C_1 s^2}$$

Due to

$$\frac{U_x}{Z + R_1} = \frac{U_2}{Z}$$

So

$$\frac{U_2(s)}{U_x(s)} = \frac{Z}{R_1 + Z} = \frac{1 + \tau_2 s}{1 + (\tau_1 + \tau_2 + R_1 C_2)s + \tau_1 \tau_2 s^2}$$

The transfer function after connection is

为广义阻抗，那么就可以更简便地研究各种物理环节之间的负载效应。

对于电压输出的环节，减少负载效应的办法如下。

(1)提高后续环节(负载)的输入阻抗。

(2)在原来两个相连接的环节之中插入高输入阻抗、低输出阻抗的放大器，以便一方面减小从前环节吸取能量，另一方面在承受后一环节(负载)后又能减小电压输出的变化，从而减轻负载效应。

(3)使用反馈或零点测量原理，使后面环节几乎不从前环节吸取能量，如用电位差计测量电压等。

总之，在测试工作中，应当建立系统整体概念，充分考虑各种装置和环节相连接后可能产生的影响。接入的测量装置就成为被测对象的负载，将会产生测量误差。两环节连接时，后环节将成为前环节的负载，并产生相应的负载效应。进行测试工作时，在选择成品传感器时必须仔细考虑传感器对被测对象的负载效应。在组成测试系统时，要考虑各组成环节之间连接时的负载效应，尽可能减小负载效应的影响。对于成套仪器系统来说，仪器生产厂家应充分考虑各组成部分间的相互影响，使用者只需考虑传感器对被测对象所产生的负载效应。

3.4 不失真测试条件

当信号通过测试系统时，系统输出响应波形与激励波形通常是不同的，即产生了失真。

线性系统在动态测试中的失真问题一直为人们所关注，适合建立理论上不失真的测试系统模型。虽然线性系统在动态测试中的失真问题是不可避免的，但有了不失真模型，就可以给人们一个使用线性系统时应该注意的准则——尽量使试验条件下的线性系统能近似于不失真的理想模型，以减少失真。从时域上看，所谓不失真测试，就是指系统输出信号的波形与输入信号的波形完全相似的测试。

当将信号 $x(t)$ 接入测试系统时，其输出 $y(t)$ 可能出现以下三种情况。

(1)输出波形与输入波形相似，输出无滞后，只是幅值放大到原来的A_0倍(图3.16，这是最理想的情况)，即输出与输入之间满足

$$y(t)=A_0x(t) \tag{3.33}$$

(2)输出波形与输入波形相似，输出的幅值放大到原来的A_0倍，而且还相对于输入滞后了时间t_0(图3.16)，即满足

$$H(s)=\frac{U_y(s)}{U_x(s)}=\frac{U_2(s)}{U_x(s)}\frac{U_y(s)}{U_2(s)}$$
$$=\frac{1}{1+(\tau_1+\tau_2+R_1C_2)s+\tau_1\tau_2 s^2}$$

Note that

$$H_1(s)H_2(s)=\frac{1}{1+\tau_1 s}\frac{1}{1+\tau_2 s}=\frac{1}{1+(\tau_1+\tau_2)s+\tau_1\tau_2 s^2}$$

we have

$$H(s)\neq H_1(s)H_2(s)$$

In addition, from the above derivation process, it can be seen that using the concept of electrical impedance to find the transfer function of a circuit is relatively simple and can avoid the step of finding the circuit differential equation.

The impact caused by the loading effect should be analyzed in detail according to the specific links and devices. If the concept of electrical impedance is extended to generalized impedance, then the loading effect between various physical links can be studied more simply.

For the voltage output link, the ways to reduce the loading effect are as follow.

(1) Increase the input impedance of the subsequent link (load).

(2) Insert a high input impedance and low output impedance amplifier between the original two links, so as to reduce the energy absorbed from the previous link on the one hand, on the other hand, it can reduce the change in voltage output after receiving the latter link (load), thereby reducing the loading effect.

(3) Use feedback or zero point measurement principle to make sure the following links hardly absorbing energy from the previous links, such as measuring voltage with a potentiometer.

In short, in the measurement work, the overall concept of the system should be established with fully considering the possible effects of various devices and links. The connection of the measuring device becomes the load of the measured object, which will cause measurement errors. When the two links are connected, the back link will become the load of the front link and produce a corresponding loading effect. When selecting a finished sensor in measurement work, the loading effect of the sensor on the object under measurement must be carefully considered. When composing the measurement system, it is necessary to consider the loading effect of the connection between the components, and minimize the influence of the loading effect. For a complete set of instrument systems, the instrument manufacturer should fully consider the mutual influence between the various components, and the users only need to consider the loading effect of the sensor on the measured object.

3.4 Non-Distortion Measurement Conditions

When the signal passes through the measurement system, the system output waveform of response and the excitation waveform are usually different, that is, distortion occurs.

The distortion problem of linear system in dynamic measurement has always been concerned by people. It is suitable to establish a theoretically non-distortion measurement system model. Although the distortion problem of linear system in dynamic measurement is in-

$$y(t)=A_0x(t-t_0) \tag{3.34}$$

(3)输出与输入完全不一样,产生了波形畸变,即输出失真了。显然,这是测试系统不希望的情况。

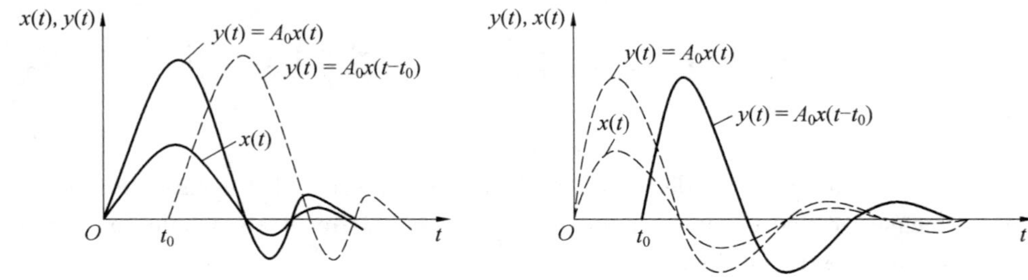

图 3.16　测试系统不失真条件

式(3.34)表示了测试系统不失真测试的时域条件,输出 $y(t)$ 是激励信号 $x(t)$ 的精确再现,因为输出响应波形与激励波形一样,只不过响应的幅度是原信号的A_0倍,并延迟了t_0时间。式(3.33)是式(3.34)在$t_0=0$ 时的特例。

对式(3.34)做傅里叶变换,可得

$$Y(\omega)=A_0X(\omega)\mathrm{e}^{-\mathrm{j}\omega t_0} \tag{3.35}$$

这样,不失真系统的频率响应函数应该是

$$H(\omega)=\frac{Y(\omega)}{X(\omega)}=A_0\mathrm{e}^{-\mathrm{j}\omega t_0} \tag{3.36}$$

即其幅频特性应该等于常数,有

$$A(\omega)=A_0 \tag{3.37}$$

其相频特性应该是 ω 的线性函数,即

$$\varphi(\omega)=-t_0\omega \tag{3.38}$$

式(3.37)表明,测试系统实现不失真测试的幅频特性曲线应是一条平行于 ω 轴的直线。式(3.38)则表明,系统实现不失真测试的相频特性曲线应是与水平坐标重合的真线(理想条件)或是一条通过坐标原点的斜直线,线性系统的不失真条件如图 3.17 所示。

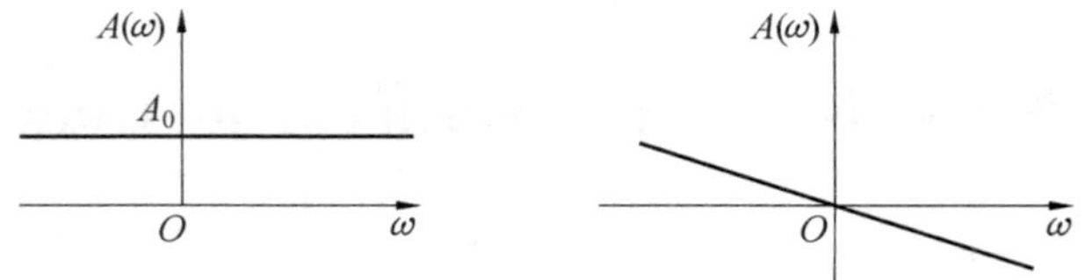

图 3.17　线性系统的不失真条件

$A(\omega)$不为常数所引起的失真称为幅值失真,因 $\varphi(\omega)$ 与 ω 之间呈非线性关系而引起

evitable, there is an non-distortion model which can give people a guideline that people should pay attention to when using a linear system—try to make the linear system under the measurement conditions approximate the ideal model without distortion to reduce distortion. From the time domain perspective, the so-called non-distortion measurement refers to a measurement in which the waveform of the output signal of the system is completely similar to the input signal.

When signal $x(t)$ is connected to the measurement system, its output $y(t)$ may appear in the following three situations.

(1) The output waveform is similar to the input waveform, when the output has no lag, but the amplitude is enlarged by A_0 times that of the original value (Figure 3.16, this is the most ideal situation), that is, the following relationship is satisfied between the output and the input:

$$y(t) = A_0 x(t) \tag{3.33}$$

(2) The output waveform is similar to the input waveform, the amplitude of the output is amplified by A_0 times that of the original value, and the input is delayed by time t_0 (Figure 3.16), that is, the following relationship is satisfied:

$$y(t) = A_0 x(t - t_0) \tag{3.34}$$

(3) The output is completely different from the input, and the waveform is distorted, that is, the output is distorted. Obviously, this is an undesirable situation for the measurement system.

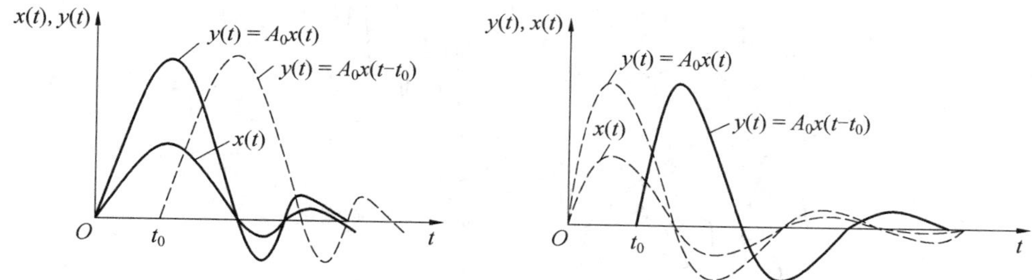

Figure 3.16　Distortion conditions of themeasurement system

Formula (3.34) expresses the time-domain conditions for the non-distortion measurement of the measurement system. Output $y(t)$ is an accurate reproduction of the excitation signal $x(t)$, because the output response waveform is the same as the excitation waveform, except that the amplitude of the response is A_0 times the original signal, and it is delayed by the t_0 time. Formula (3.33) is a special case of formula (3.34) in $t_0 = 0$.

Take the Fourier transform of formula (3.34) to get

$$Y(\omega) = A_0 X(\omega) e^{-j\omega t_0} \tag{3.35}$$

In this way, the frequency response function of the non-distortion system should be

$$H(\omega) = \frac{Y(\omega)}{X(\omega)} = A_0 e^{-j\omega t_0} \tag{3.36}$$

that is, its amplitude-frequency characteristics should be equal to a constant, we have

$$A(\omega) = A_0 \tag{3.37}$$

Its phase frequency characteristic should be a linear function of ω, that is

$$\varphi(\omega) = -t_0 \omega \tag{3.38}$$

Formula (3.37) shows that the amplitude-frequency characteristic curve of the

的失真称为相位失真。

由于任何测试系统不可能在很宽的频带范围内满足不失真测试的条件,因此在测试过程中要根据不同的测试目的,合理把握不失真的条件。

测试系统通常是由若干测试环节组成的,只有保证每一环节都满足不失真测试的条件,才能使最终的输出波形不失真,信号中不同频率成分通过测试装置后的输出如图3.18所示。四个不同频率的信号为通过一个具有图中 $A(\omega)$ 和 $\varphi(\omega)$ 特性的装置后的输出信号。四个输入信号都是正弦信号(包括直流信号),在某参考时刻 $t=0$,初始相角均为 0。图中形象地显示各输出信号相对输入信号有不同的幅值增益和相角滞后。对于单一频率成分的信号,因为定常线性系统具有频率保持性,所以只要其幅值未进入非线性区,输出信号的频率也是单一的,也就无所谓失真问题。对于含有多种频率成分的,显然既引起幅值失真,又引起相位失真,特别是频率成分跨越 ω_n 前、后的信号失真尤为严重。

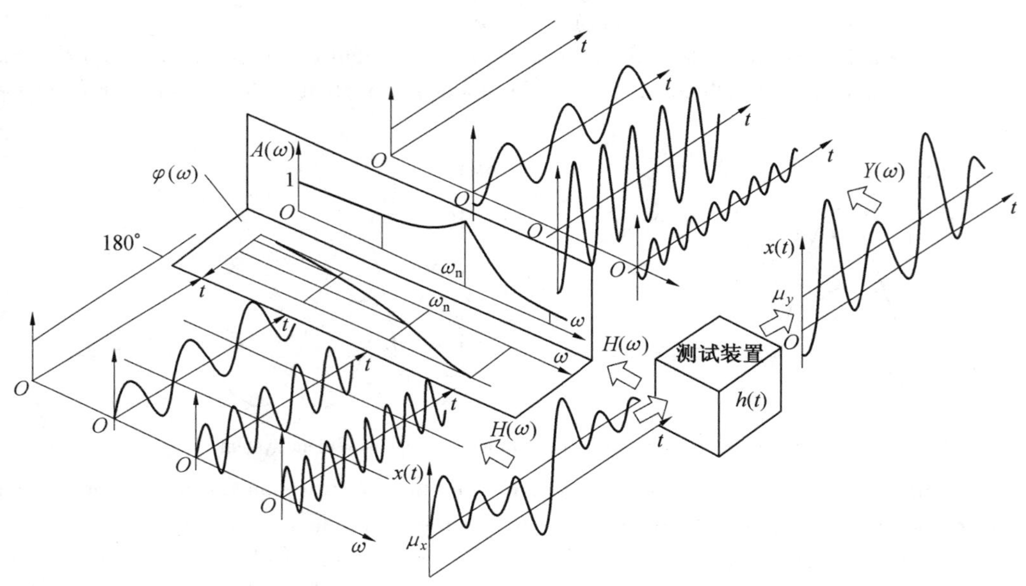

图 3.18　信号中不同频率成分通过测试装置后的输出

线性系统与代数系统在失真问题上前者有后者无,究其原因,表面看缘于二者输出与输入间的数学模型不同。实际上,二者在测试机理上不一样。线性系统的测试是依赖于它的硬件系统内所发生的真实动态物理过程,不免要牵扯到构件的质量、惯性力、阻尼等,有些量与输入的变化率有关。当输入变化极快时,系统来不及响应,或对信号的低频成分的响应迅速强烈,对高频成分的响应迟钝微弱,所以改变了输入信号的频谱,引起动态测试的失真。

measurement system to realize the undeformed measurement should be a straight line parallel to axis ω. *Formula* (3.38) indicates that the phase-frequency characteristic curve for the system to achieve undeformed measurement proves a true line (ideal condition) that coincides with the horizontal coordinate or an oblique line that passes through the origin of the coordinate, non-distortional conditions for linear systems are as shown in Figure 3.17.

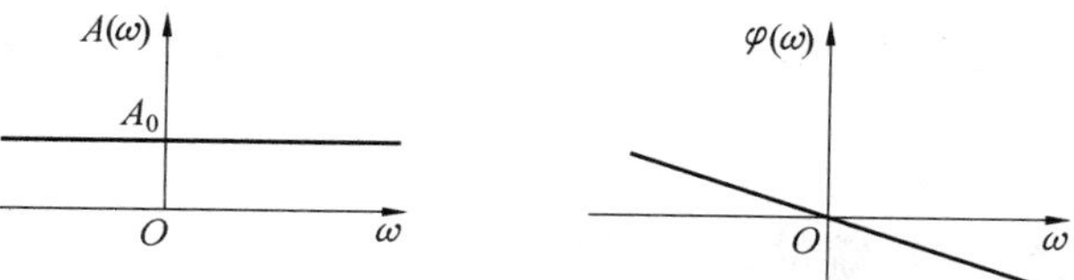

Figure 3.17　Non-distortional conditions for linear systems

The distortion caused by $A(\omega)$ not being a constant is called amplitude distortion, and the distortion caused by the nonlinear relationship between $\varphi(\omega)$ and ω is called phase distortion.

Since it is impossible for any measurement system to meet the conditions of non-distortion measurement in a wide frequency band, nondistortional conditions should be reasonably controlled according to different measurement purposes during the measurement.

The measurement system usually consists of several measurement links. Only by ensuring that each link meets the conditions of the non-distortion measurement, can the final output waveform be undistorted. The output of different frequency components in the signal after passing the measurement device is shown in Figure 3.18. Four signals with different frequencies are output signals after passing through a device with characteristics $A(\omega)$ and $\varphi(\omega)$ in the figure. The four input signals are all sine signals (including DC signals), and at a reference time $t=0$, the initial phase angles are all 0. The figure vividly shows that each output signal has different amplitude gain and phase angle lag relative to the input signal. For a signal with a single frequency component, because the steady linear system has frequency retention, as long as its amplitude does not enter the nonlinear region, the frequency of the output signal

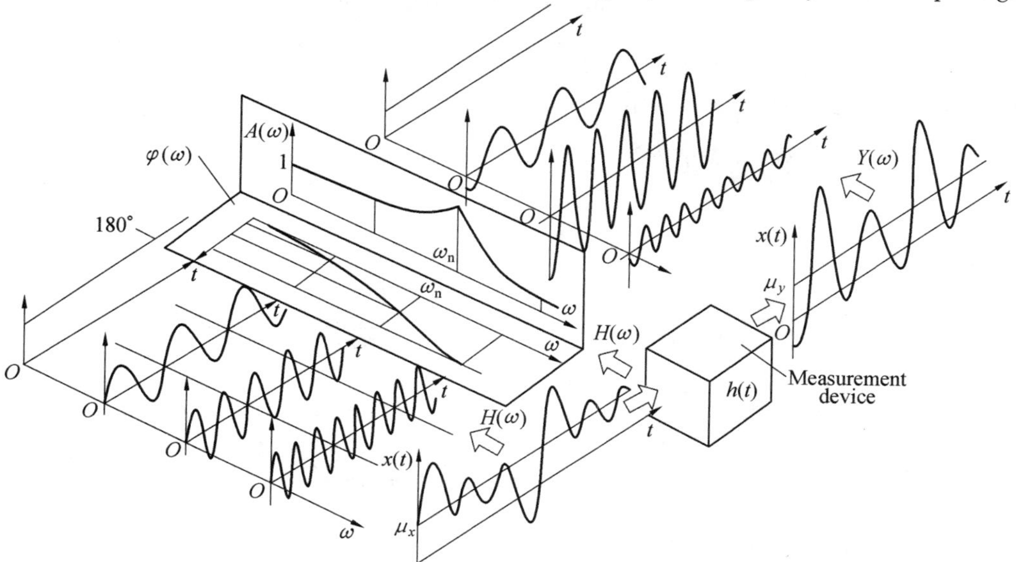

Figure 3.18　The output of different frequency components in the signal after passing the measurement device

3.5 一、二阶线性系统的动态特性分析

本节以典型的一阶、二阶线性系统为例来分析系统的动态特性，即分析它们的幅频特性和相频特性。弄清这些特性，就可以深入了解系统的各种功能，并知道怎样正确使用它。

3.5.1 一阶线性系统的动态特性

典型的一阶线性系统输出与输入的关系用下列一阶定常线性微分方程，即

$$\tau y'(t)+y(t)=x(t) \tag{3.39}$$

来描述。若对上式做傅里叶变换，得

$$\mathrm{j}\omega\tau Y(\omega)+Y(\omega)=X(\omega)$$

经整理后，得到系统的频率响应函数，即

$$H(\omega)=\frac{Y(\omega)}{X(\omega)}=\frac{1}{1+\mathrm{j}\tau\omega} \tag{3.40}$$

其幅频特性函数为

$$A(\omega)=|H(\omega)|=\frac{1}{\sqrt{1+(\tau\omega)^2}} \tag{3.41}$$

其相频特性函数为

$$\varphi(\omega)=\arg H(\omega)=-\arctan(\tau\omega) \tag{3.42}$$

图 3.19 所示为一阶系统的动态特性曲线（伯德图），其纵坐标即缩放尺度 $A(\omega)$ 用分贝表示，横坐标 ω 用对数 $\lg\omega$ 表示。

$H(\omega)$ 是一个复数，可表示为

$$H(\omega)=P+\mathrm{j}Q$$

式中，实部为

$$P=\mathrm{Re}H(\omega)=\frac{1}{1+(\tau\omega)^2}=\frac{1}{\sqrt{1+(\tau\omega)^2}}\cdot\frac{1}{\sqrt{1+(\tau\omega)^2}}=A(\omega)\cos\varphi(\omega)$$

虚部为

$$Q=\mathrm{Im}H(\omega)=-\frac{\tau\omega}{1+(\tau\omega)^2}=\frac{1}{\sqrt{1+(\tau\omega)^2}}\cdot\frac{-\tau\omega}{\sqrt{1+(\tau\omega)^2}}=A(\omega)\sin\varphi(\omega)$$

其中

$$\cos\varphi(\omega)=\frac{1}{\sqrt{1+(\tau\omega)^2}},\sin\varphi(\omega)=-\frac{\tau\omega}{\sqrt{1+(\tau\omega)^2}}$$

is also single, and there is no distortion problem. For those containing multiple frequency components, it is obvious that both amplitude distortion and phase distortion are caused. In particular, the signal distortion before and after the frequency components cross ω_n is particularly serious.

The reason why linear systems have distortion problems while algebraic systems do not is that the mathematical models between the output and the input of the two are different. In fact, the measurement mechanism of the two is different. The linear system measurement relies on the real dynamic physical process that occurs in its hardware system, which inevitably involves the mass of the component, inertial force, damping, etc., some of which are related to the rate of change of the input. When the input changes extremely fast, the system is too late to respond, or the response to the low-frequency components of the signal is fast and strong, and the response to the high-frequency components is slow and weak, so the frequency spectrum of the input signal is changed, causing the distortion of the dynamic measurement.

3.5 Analysis of the Dynamic Characteristics of the First- and Second-Order Linear Systems

This section takes typical first-order and second-order linear systems as an example to analyze the dynamic characteristics of the system, that is, analyze their amplitude frequency characteristics and phase frequency characteristics. To understand these characteristics can provide in-depth understanding of the various functions of the system and know how to use them correctly.

3.5.1 The Dynamic Characteristics of the First-Order Linear System

The relationship between output and input of a typical first-order linear system uses the following first-order stationary linear differential equation

$$\tau y'(t) + y(t) = x(t) \tag{3.39}$$

to describe. If we do the inverse Fourier transform of the above formula, we have

$$\mathrm{j}\omega\tau Y(\omega) + Y(\omega) = X(\omega)$$

After sorting, the frequency response function of the system can be obtained, that is

$$H(\omega) = \frac{Y(\omega)}{X(\omega)} = \frac{1}{1 + \mathrm{j}\tau\omega} \tag{3.40}$$

Its amplitude-frequency characteristic function is

$$A(\omega) = |H(\omega)| = \frac{1}{\sqrt{1 + (\tau\omega)^2}} \tag{3.41}$$

Its phase-frequency characteristic function is

$$\varphi(\omega) = \arg H(\omega) = -\arctan(\tau\omega) \tag{3.42}$$

Figure 3.19 shows the dynamic characteristic curve of the first-order system (Bode diagram), its ordinate, also called scaling scale $A(\omega)$, is expressed in decibels, and the abscissa ω is expressed in logarithm $\lg \omega$.

$H(\omega)$ is a complex number, which can be expressed as

$$H(\omega) = P + \mathrm{j}Q$$

where the real part is

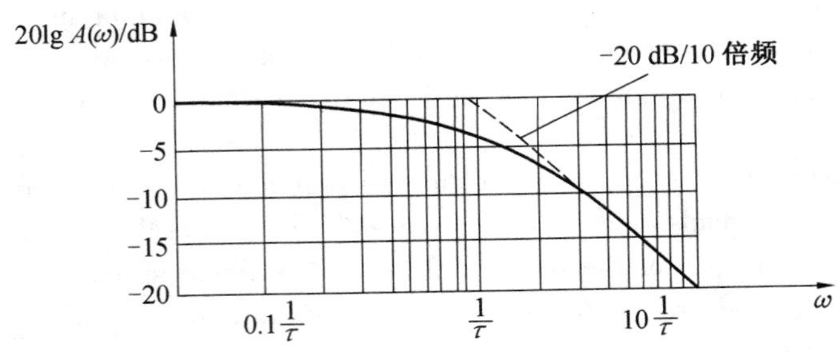

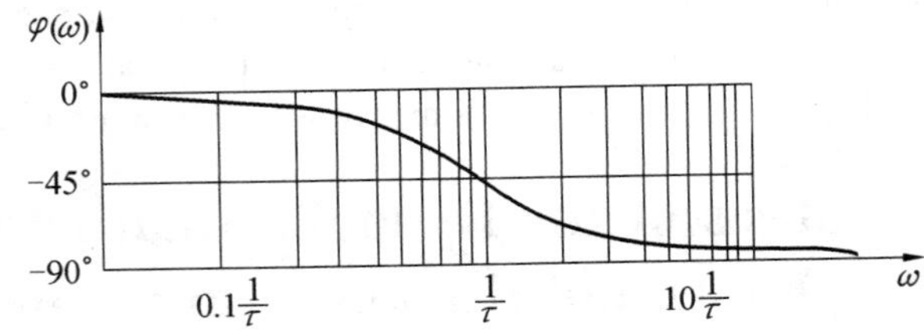

图 3.19　一阶系统的动态特性曲线(伯德图)

现借助伯德图对一阶系统的动态特性进行如下分析。

(1)$A(\omega)$曲线的总体趋势为随 ω 的增加而单调下降。曲线可以用两段直线来近似，$\omega<1/\tau$区段用 $20\lg A(\omega)=0$ dB 的一段横轴来近似,$\omega>1/\tau$ 区段用斜率为-20 dB/10 倍频的斜线来近似。两段直线相交于 $\omega=\omega_0=1/\tau$ 处,ω_0 称为转折频率,$A(\omega_0)=1/\sqrt{2}$,以两段直线近似的替代曲线 $A(\omega)$所产生的最大误差发生在 $\omega=\omega_0$ 处,数值为

$$0-20\lg\frac{1}{\sqrt{2}}=-3\ \text{dB}$$

或

$$1-\frac{1}{\sqrt{2}}=1-0.707=0.293$$

(2)$\varphi(\omega)$曲线变化于 0°～-90°,可用三段直线近似: $\omega\leqslant0.1\omega_0$,$\varphi(\omega)\approx0°$;$\omega\geqslant10\omega_0$,$\varphi\approx-90°$;$0.1\omega_0\leqslant\omega\leqslant10\omega_0$,用斜率为-45°/10 倍频的斜线近似。$\varphi(\omega)$曲线用三段折线近似替代的最大误差为±5.71°,分别发生在 $\omega=0.1\omega_0$ 和 $\omega=10\omega_0$ 处。

(3)当激励频率 $\omega\ll1/\tau$ 时,$\tau\omega$ 很小,可认为 $\tan(\tau\omega)\approx\tau\omega$,故 $\varphi(\omega)=-\arctan(\tau\omega)\approx-\tau\omega$。若略去 $\tau\omega$,则 $A(\omega)\approx1$,可见 $\omega\ll1/\tau$ 时可以认为系统基本上是无失真的。$\omega_m=1/5\tau$时的幅值及相位误差见表 3.1。

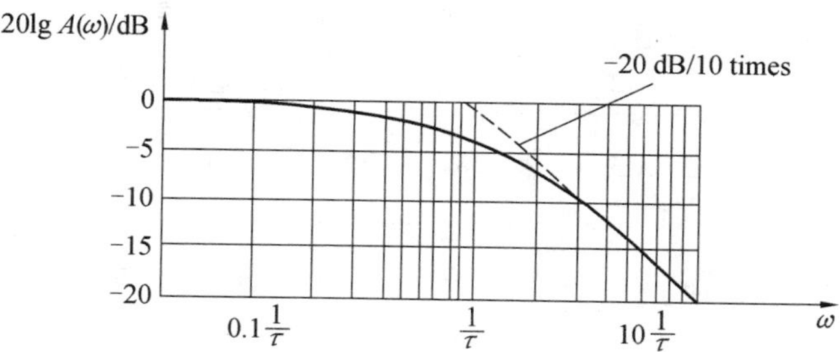

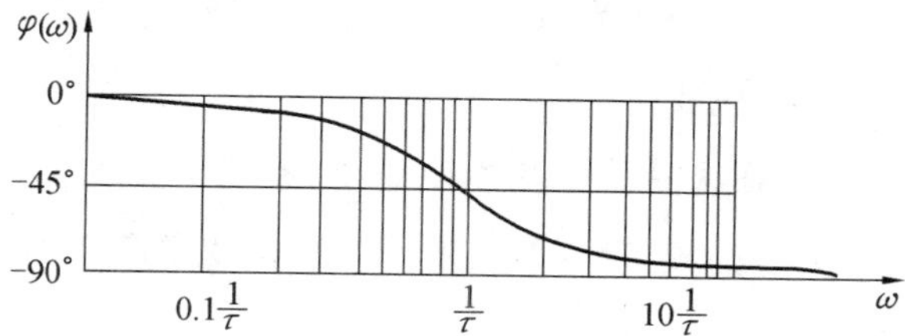

Figure 3. 19　The dynamic characteristic curve of the first-order system (Bode diagram)

$$P = \mathrm{Re}H(\omega) = \frac{1}{1 + (\tau\omega)^2} = \frac{1}{\sqrt{1 + (\tau\omega)^2}} \cdot \frac{1}{\sqrt{1 + (\tau\omega)^2}} = A(\omega)\cos\varphi(\omega)$$

The imaginary part is

$$Q = \mathrm{Im}H(\omega) = -\frac{\tau\omega}{1 + (\tau\omega)^2} = \frac{1}{\sqrt{1 + (\tau\omega)^2}} \cdot \frac{-\tau\omega}{\sqrt{1 + (\tau\omega)^2}} = A(\omega)\sin\varphi(\omega)$$

In it

$$\cos\varphi(\omega) = \frac{1}{\sqrt{1 + (\tau\omega)^2}} \qquad \sin\varphi(\omega) = -\frac{\tau\omega}{\sqrt{1 + (\tau\omega)^2}}$$

Now with the help of Bode diagram, the dynamic characteristics of the first-order system are analyzed as follows.

(1) The overall trend of $A(\omega)$ curve is monotonously decreasing with the increase of ω. The curve can be approximated by two straight lines, section of $\omega < 1/\tau$ is approximated by a horizontal axis of $20\lg A(\omega) = 0$ dB, and section of $\omega > 1/\tau$ is approximated by a diagonal line whose slope is − 20 dB/10 times. The two straight lines intersect at $\omega = \omega_0 = 1/\tau$, ω_0 is called the turning frequency, $A(\omega_0) = 1/\sqrt{2}$, the maximum error produced by replacing curve $A(\omega)$ with two straight lines approximating occurs at $\omega = \omega_0$, and the value is

$$0 - 20\lg\frac{1}{\sqrt{2}} = -3\ \mathrm{dB}$$

or

$$1 - \frac{1}{\sqrt{2}} = 1 - 0.707 = 0.293$$

(2) $\varphi(\omega)$ curve changes between 0° ~ −90°, which can be approximated by three straight lines: $\omega \leqslant 0.1\omega_0$, $\varphi(\omega) \approx 0°$; $\omega \geqslant 10\omega_0$, $\varphi \approx -90°$; $0.1\omega_0 \leqslant \omega \leqslant 10\omega_0$, approximated by a sloping line with a slope of − 45°/10 frequeney multiplier. The maximum error of approximate replacement of the curve $\varphi(\omega)$ with a three-segment polyline is ±5.71°, which

表 3.1　$\omega_m = 1/5\tau$ 时的幅值及相位误差

	实际值	理想(无失真)值	误差
幅值响应 $A(\omega)$	$\frac{1}{\sqrt{1+(\tau\omega)^2}}=0.98058$	1	<2%
初相响应 $\varphi(\omega)$	$-\arctan(\tau\omega)=-11.3099°$	-0.2 rad$=-11.4592°$	$-0.1492°$ $=-0.0026$ rad

可见,误差不大,而其余低频成分的误差会更小。

因此,一阶系统在测试中作为指示仪表或传感器使用时,适合测试缓变信号(此类信号所含的高频成分较少,主要是低频成分)。

时间常数 τ 是一阶系统的重要参数,τ 越小,系统反应速度越快,转折频率 $\omega_0 = 1/\tau$ 就越高,保持基本上不失真的频率区间(工作频宽)就越宽广。因此,一阶系统作为传感器或测量仪表使用时,希望 τ 越小越好。

(4)当选择系统元件参数使它的时间常数 τ 较大,以至于输入信号的最低频率成分的频率 $\omega>(2\sim3)/\tau$ 时,则可以认为 $\tau\omega\gg1$,因此

$$H(\omega)=\frac{1}{\mathrm{j}\tau\omega+1}\approx\frac{1}{\mathrm{j}\tau\omega}$$

于是有

$$Y(\omega)=H(\omega)X(\omega)\approx\frac{1}{\mathrm{j}\tau\omega}X(\omega)$$

因此,根据傅里叶变换的积分性质可知

$$y(t)=\frac{1}{\tau}\int_0^t x(t)\,\mathrm{d}t \tag{3.43}$$

可见,当输入信号各成分的频率都大于$(2\sim3)/\tau$,一阶系统将以幅频特性的斜率为-20 dB/10倍频的斜线区段(在此区段内 $A(\omega)\approx1/\tau\omega$,近似双曲线,$\varphi(\omega)\approx-90°$)来工作时,相当于一个积分器,即系统的输出与输入信号在时域内的积分成正比。

(5)当 ω 很低时,$A(\omega)\approx1$;当 ω 很高时,$A(\omega)\approx0$。可见,一阶系统对信号的频率成分具有明显的选择性,该系统几乎无衰减地使输入信号的低频成分通过,而对高频成分却给予很大的衰减,所以一阶系统又可以作为低通滤波器来使用。

例 3.3　用一个一阶系统测量 100 Hz 的正弦信号。

(1)如果要求限制振幅误差在 5% 以内,则时间常数 τ 应取多少?

(2)若用具有该时间常数的同一系统测试 50 Hz 的正弦信号,此时的振幅误差和相角差各是多少?

occurs at $\omega = 0.1\omega_0$ and $\omega = 10\omega_0$ respectively.

(3) When excitation frequency $\omega \ll 1/\tau$, $\tau\omega$ is very small, and it can be regarded as $\tan(\tau\omega) \approx \tau\omega$,so $\varphi(\omega) = -\arctan(\tau\omega) \approx -\tau\omega$. If $\tau\omega$ is omitted, then $A(\omega) \approx 1$,and it can be regarded as the system is basically without distortion when $\omega \ll 1/\tau$. Amplitude and phase error at $\omega_m = 1/5\tau$ is shown in Table 3.1.

Table 3.1 Amplitude and phase error at $\omega_m = 1/5\tau$

Response	Actual value	Ideal (non distortion) value	Error
Amplitude response $A(\omega)$	$\frac{1}{\sqrt{1+(\tau\omega)^2}} = 0.980\,58$	1	<2%
Initial phase response $\varphi(\omega)$	$-\arctan(\tau\omega) = -11.309\,9°$	-0.2 rad $= -11.459\,2°$	$-0.149\,2°$ $= -0.0026$ rad

It can be seen that the error is not large, and the error of the remaining low-frequency components will be smaller.

Therefore, when the first-order system is used as an indicator or sensor in the measurement, it is suitable to test the slowly changing signals (this kind of signal contain less high-frequency components, mainly low-frequency components).

Time constant τ is an important parameter of the first-order system. The smaller the τ is, the faster the system speed responses, the higher the corner frequency $\omega_0 = 1/\tau$ is, and the frequency range (operating bandwidth) that remains basically undistorted is wider, so when a first-order system is used as a sensor or a measuring instrument, it is hoped that τ is as small as possible.

(4) When the system component parameter is selected to make its time constant τ relatively large, so that the frequency of the lowest frequency component of the input signal $\omega > (2 \sim 3)/\tau$, it can be considered as $\tau\omega \gg 1$, so

$$H(\omega) = \frac{1}{j\tau\omega + 1} \approx \frac{1}{j\tau\omega}$$

So we have

$$Y(\omega) = H(\omega)X(\omega) \approx \frac{1}{j\tau\omega}X(\omega)$$

Therefore, according to the integral nature of the Fourier transform, we know that

$$y(t) = \frac{1}{\tau}\int_0^t x(t)\,dt \tag{3.43}$$

It can be seen that when each component frequency of the input signal is greater than $(2 \sim 3)/\tau$, when the first-order system will work with the slope of the amplitude-frequency characteristic of -20 dB/10 times the frequency of the oblique section (in this section, $A(\omega) \approx 1/\tau\omega$, approximate hyperbola, $\varphi(\omega) \approx -90°$), it is equivalent to an integrator, that is, the output of the system is proportional to the integral of the input signal in the time domain.

(5) When ω is very low, $A(\omega) \approx 1$;when ω is very high, $A(\omega) \approx 0$. It can be seen that the first-order system has obvious selectivity to the frequency components of the signal. The system allows the low-frequency components of the input signal to pass almost without attenuation, but gives great attenuation to the high-frequency components, so the first-order system can be used as a low-pass filter.

解 测试系统对某一信号测量后的幅值误差应为

$$\delta=\left|\frac{A_1-A_0}{A_1}\right|=|1-A(\omega)|$$

其相角差即相位移为 φ。对一阶系统，若设 $k=1$，则其幅频特性和相频特性分别为

$$A(\omega)=\frac{1}{\sqrt{(\omega\tau)^2+1}}$$

$$\varphi=\arctan(-\omega\tau)$$

(1)因为 $\delta=|1-A(\omega)|$，故当 $|\delta|=5\%=0.05$ 时，即要求 $1-A(\omega)=0.05$，所以

$$1-\frac{1}{\sqrt{(\omega\tau)^2+1}}=0.05$$

化简得

$$(\omega\tau)^2\leqslant\frac{1}{0.95^2}-1=0.108$$

则有

$$\tau\leqslant\sqrt{1.08}\times\frac{1}{2\pi f}=\sqrt{1.08}\times\frac{1}{2\pi\times100}=5.23\times10^{-4}$$

(2)当对 50 Hz 信号进行测试时，有

$$\delta=1-\frac{1}{\sqrt{(\omega\tau)^2+1}}=1-\frac{1}{\sqrt{(2\pi f\tau)^2+1}}$$

$$=1-\frac{1}{\sqrt{(2\pi\times50\times5.23\times10^{-4})^2+1}}$$

$$=1-0.986\ 8=1.32\%$$

$$\varphi=\arctan(-\omega\tau)$$

$$=\arctan(-2\pi f\tau)$$

$$=\arctan(-2\pi\times50\times5.23\times10^{-4})$$

$$=-9°18'50''$$

3.5.2 二阶线性系统的动态特性

分析以下二阶系统的动态特性，该系统的微分方程为

$$y''(t)+2\zeta\omega_n y'(t)+\omega_n^2 y(t)=\omega_n^2 x(t)$$

对上式两边做傅里叶变换，有

$$(j\omega)^2 Y(\omega)+(j\omega)2\zeta\omega_n Y(\omega)+\omega_n^2 Y(\omega)=\omega_n^2 X(\omega)$$

Example 3.3　Measure a 100 Hz sine signal by using a first-order system.

(1) If it is required to limit the amplitude error within 5%, how much time constant τ should be taken?

(2) If the same system with this time constant is used to test a 50 Hz sine signal, what is the amplitude error and phase angle difference at this time?

Solution　The amplitude error of a certain signal measured by the measurement system should be

$$\delta = \left| \frac{A_1 - A_0}{A_1} \right| = |1 - A(\omega)|$$

The phase angle difference, that is, the phase shift is φ. For a first-order system, if $k = 1$ is set, the amplitude-frequency characteristics and phase-frequency characteristics are respectively

$$A(\omega) = \frac{1}{\sqrt{(\omega\tau)^2 + 1}}$$

$$\varphi = \arctan(-\omega\tau)$$

(1) Because of $\delta = |1 - A(\omega)|$, when $|\delta| = 5\% = 0.05$, $1. A(\omega) = 0.05$ is required, so

$$1 - \frac{1}{\sqrt{(\omega\tau)^2 + 1}} = 0.05$$

it can be simplified to

$$(\omega\tau)^2 \leqslant \frac{1}{0.95^2} - 1 = 0.108$$

We get

$$\tau \leqslant \sqrt{1.08} \times \frac{1}{2\pi f} = \sqrt{1.08} \times \frac{1}{2\pi \times 100} = 5.23 \times 10^{-4}$$

(2) When testing the 50 Hz signal, we have

$$\begin{aligned}\delta &= 1 - \frac{1}{\sqrt{(\omega\tau)^2 + 1}} \\ &= 1 - \frac{1}{\sqrt{(2\pi f\tau)^2 + 1}} \\ &= 1 - \frac{1}{\sqrt{(2\pi \times 50 \times 5.23 \times 10^{-4})^2 + 1}} \\ &= 1 - 0.9868 = 1.32\%\end{aligned}$$

$$\begin{aligned}\varphi &= \arctan(-\omega\tau) \\ &= \arctan(-2\pi f\tau) \\ &= \arctan(-2\pi \times 50 \times 5.23 \times 10^{-4}) \\ &= -9°18'50''\end{aligned}$$

3.5.2 Dynamic Characteristics of the Second-Order Linear System

Analyze the dynamic characteristics of the following second-order system, the differential equation of the system is

$$y''(t) + 2\zeta\omega_n y'(t) + \omega_n^2 y(t) = \omega_n^2 x(t)$$

Do Fourier transform on both sides of the above formula, we have

$$(j\omega)^2 Y(\omega) + (j\omega)2\zeta\omega_n Y(\omega) + \omega_n^2 Y(\omega) = \omega_n^2 X(\omega)$$

Then the frequency response function of the system is

于是系统的频率响应函数为

$$H(\omega)=\frac{Y(\omega)}{X(\omega)}=\frac{\omega_{n}^{2}}{(j\omega)^{2}+2\zeta\omega_{n}(j\omega)+\omega_{n}^{2}}$$

$$=\frac{1}{\sqrt{\left[1-\left(\frac{\omega}{\omega_{n}}\right)^{2}\right]+j2\zeta\frac{\omega}{\omega_{n}}}} \tag{3.44}$$

幅频特性函数、相频特性函数分别为

$$A(\omega)=\frac{1}{\sqrt{\left[1-\left(\frac{\omega}{\omega_{n}}\right)^{2}\right]^{2}+4\zeta^{2}\left(\frac{\omega}{\omega_{n}}\right)^{2}}} \tag{3.45}$$

$$\varphi(\omega)=-\arctan\frac{2\zeta\left(\frac{\omega}{\omega_{n}}\right)}{1-\left(\frac{\omega}{\omega_{n}}\right)^{2}} \tag{3.46}$$

图3.20所示为二阶系统的幅频特性及相频特性曲线图。借助该图，对系统的动态特性分析如下。

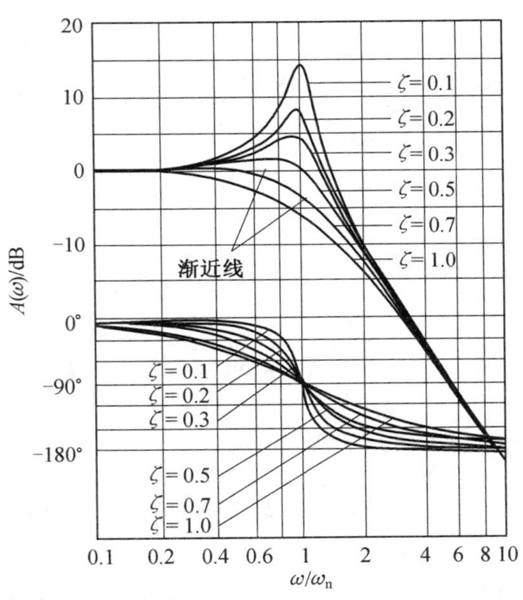

图3.20　二阶系统的幅频特性及相频特性曲线图

影响系统动态特性的参数有系统固有频率 ω_n 和阻尼比 ζ。实际测试中所使用的二阶系统 $\zeta<1$，$\zeta>1$ 属于过阻尼情况，不属于本书讨论范围。ζ 的大小影响 $A(\omega)$ 和 $\varphi(\omega)$ 曲线的走势。图3.20中分别绘出 ζ 不同的数条曲线。

(1)当 $\omega=0$ 时，$A(\omega)=1$。ζ 不同的数条曲线皆从该点出发，ζ 不同，各曲线走势不

$$H(\omega)=\frac{Y(\omega)}{X(\omega)}=\frac{\omega_n^2}{(j\omega)^2+2\zeta\omega_n(j\omega)+\omega_n^2}$$
$$=\frac{1}{\sqrt{\left[1-\left(\frac{\omega}{\omega_n}\right)^2\right]+j2\zeta\frac{\omega}{\omega_n}}} \tag{3.44}$$

The amplitude frequency characteristic function and the phase frequency characteristic function are respectively

$$A(\omega)=\frac{1}{\sqrt{\left[1-\left(\frac{\omega}{\omega_n}\right)^2\right]^2+4\zeta^2\left(\frac{\omega}{\omega_n}\right)^2}} \tag{3.45}$$

$$\varphi(\omega)=-\arctan\frac{2\zeta\left(\frac{\omega}{\omega_n}\right)}{1-\left(\frac{\omega}{\omega_n}\right)^2} \tag{3.46}$$

Figure 3.20 shows the amplitude frequency characteristic and phase frequency characteristic curve diagram of the second-order system. With the help of this figure, the dynamic characteristics of the system are analyzed as follows.

The parameters affecting the dynamic characteristics of the system are the natural frequency ω_n and the damping ratio ζ. That the second-order system $\zeta<1$, $\zeta>1$ used in the actual measurement belongs to the overcritical damping situation does not belong to the scope of this book. The size of ζ will affect the trend of $A(\omega)$ and $\varphi(\omega)$ curves. In Figure 3.20, several curves with different ζ are drawn.

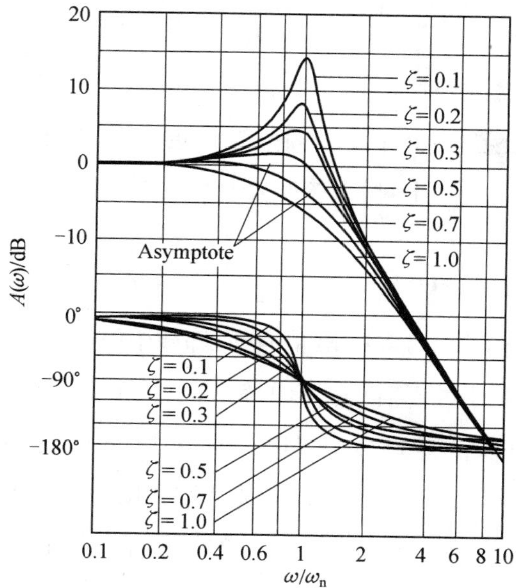

Figure 3.20　The amplitude frequency characteristic and phase frequency characteristic curve diagram of the second-order system

同。有的曲线出现峰值，与峰点相对应的频率称为谐振频率。根据导数为零，有

$$\mathrm{d}A\left(\frac{\omega}{\omega_n}\right)/\mathrm{d}\left(\frac{\omega}{\omega_n}\right)=0$$

求出的极值条件为

$$\frac{\omega}{\omega_n}=\sqrt{1-2\zeta^2}$$

或

$$\omega=\omega_n\sqrt{1-2\zeta^2}=\omega_r$$

式中　ω_r——谐振频率。

当 $\zeta=1/\sqrt{2}=0.707$ 时，$\omega_r=0$，说明系统不会发生谐振。这种情况下，$A(\omega)$ 曲线在 $\omega=0$ 处与 $A(\omega)=1$ 的横线相切。

当 $\zeta>1/\sqrt{2}$ 时，ω_r 将为虚数，说明 $A(\omega)$ 曲线无峰值，整个曲线在 $A(\omega)=1$ 的横线以下，也不会发生谐振。

当 $\zeta<1/\sqrt{2}$ 时，谐振频率为 $\omega_r=\omega_n\sqrt{1-2\zeta^2}$，将 $\omega/\omega_n=\sqrt{1-2\zeta^2}$ 代入式(2.57)会得到曲线的峰值，即

$$A(\omega)_{max}=\frac{1}{2\zeta\sqrt{1-\zeta^2}}$$

ζ 越少，$A(\omega)_{max}$ 越大，当 $\zeta\to0$ 时，$\omega_r\to\omega_n$，$A(\omega)_{max}\to\infty$。

也就是说，系统在无阻尼状态下受到频率为 ω_n 时，简谐性激励时，系统会发生共振，其输出幅值趋于无穷大。在使用二阶系统时要避免共振，即输入信号的任何频率成分都应尽量远离 ω_n。

(2)在系统的伯德图中(图3.20)，$A(\omega)$ 曲线有两条渐近线。当 ω 从 ω_n 趋于0时，ζ 不同的所有 $A(\omega)$ 曲线趋于 $A(\omega)=0$ dB 的直线，即趋于伯德图的横轴；当 ω 从 ω_n 趋于无穷大时，所有 $A(\omega)$ 曲线趋于过横轴上 ω_n 点的斜率为 −40 dB/10 倍频的斜线。

(3)各条相频特性曲线汇交于(ω_n，−90°)(图3.19)，该点为 $\varphi(\omega)$ 曲线的拐点。当 ω 跨越 ω_n 时，$\varphi''(\omega)$ 要变号，$\varphi(\omega)$ 的走势要发生不同程度的剧烈变化。

(4)当这类二阶系统在动态测试中当作传感器、测量仪表或记录仪器使用时，应尽量减少失真。通常的做法如下。

阻尼比 ζ 宜选为 0.65～0.7，输入信号中不可忽视的最高频率成分的频率应小于 $(0.6\sim0.8)\omega_n$，以使 $A(\omega)$ 曲线在低于 ω_n 区段尽量平坦，接近 $A(\omega)=1$ 的理想条件，$\varphi(\omega)$ 曲线尽量与 ω 成近似线性关系。

(1) When $\omega = 0$, $A(\omega) = 1$. Several different curves of ζ all start from this point, and different curves of ζ have different trends. Some curves have peaks, and the frequency corresponding to the peak point is called the resonant frequency. According to the derivative is zero, we have

$$\mathrm{d}A\left(\frac{\omega}{\omega_n}\right) / \mathrm{d}\left(\frac{\omega}{\omega_n}\right) = 0$$

The extreme condition is

$$\frac{\omega}{\omega_n} = \sqrt{1 - 2\zeta^2}$$

or

$$\omega = \omega_n \sqrt{1 - 2\zeta^2} = \omega_r$$

where　ω_r — the resonance frequency.

When $\zeta = 1/\sqrt{2} = 0.707$, $\omega_r = 0$ indicates that the system will not resonate. In this case, $A(\omega)$ curve is tangent to the horizontal line of $A(\omega) = 1$ at $\omega = 0$.

When $\zeta > 1/\sqrt{2}$, ω_r will be imaginary numbers, indicating that $A(\omega)$ curve has no peak, and the entire curve is below the horizontal line of $A(\omega) = 1$, and resonance will not occur.

When $\zeta < 1/\sqrt{2}$, the resonance frequency is $\omega_r = \omega_n\sqrt{1 - 2\zeta^2}$, and substituting $\omega/\omega_n = \sqrt{1 - 2\zeta^2}$ into formula (2.57) will get the peak value of the curve, that is

$$A(\omega)_{\max} = \frac{1}{2\zeta\sqrt{1 - \zeta^2}}$$

The less ζ is, the larger $A(\omega)_{\max}$ is, and when $\zeta \to 0$, $\omega_r \to \omega_n$, $A(\omega)_{\max} \to \infty$.

That is to say, the system will resonate when it is excited by the harmonic frequency of ω_n in the undamped state, and its output amplitude tends to infinity. Resonance should be avoided when using a second-order system, that is, any frequency components of the input signal should be as far away from ω_n as possible.

(2) In the Bode diagram of the system (Figure 3.20), curve $A(\omega)$ has two asymptotes. When ω tends to 0 from ω_n, all $A(\omega)$ curves of different ζ tend to a straight line with $A(\omega) = 0$ dB, that is, on the horizontal axis of the Bode diagram; when ω tends to infinity from ω_n, the slope of all $A(\omega)$ curves tends to cross the point ω_n on the horizontal axis with a slope of −40 dB/10 frequeney multiplier.

(3) Each phase-frequency characteristic curve converges at the point (ω_n, −90°) (Figure 3.19), which is the inflection point of $\varphi(\omega)$ curve. When ω crosses ω_n, $\varphi''(\omega)$ will change the sign, and the trend of $\varphi(\omega)$ will change drastically to varying degrees.

(4) When this type of second-order system is used as a sensor, measuring instrument, or recording instrument in dynamic measurement, distortion should be minimized. The usual practice is as follow.

The damping ratio ζ should be selected the range of 0.65–0.7, and the frequency of the highest frequency component that cannot be ignored in the input signal should be less than (0.6–0.8) ω_n. In order to make $A(\omega)$ curve as flat as possible in the section lower than ω_n section, close to the ideal condition of $A(\omega) = 1$, $\varphi(\omega)$ curve should have an approximately linear relationship with ω as much as possible.

The $A(\omega)$ value of the second-order system under different conditions listed in Table 3.2 are for reference.

二阶系统不同条件下的 $A(\omega)$ 值见表 3.2，可供参考。

表 3.2 二阶系统不同条件下的 $A(\omega)$ 值

阻尼比 ζ	谐振频率 ω_r	$A(\omega)$ 曲线峰值 $A(\omega)_{max}$	幅频函数 $A(\omega)$	
			$\omega/\omega_n=0.6$	$\omega/\omega_n=0.8$
0.600 0	$0.529\ 1\omega_n$	1.041 6	1.038 1	0.975 3
0.640 3	$0.424\ 3\omega_n$	1.016 6	1.000 0	0.920 9
0.650 0	$0.393\ 7\omega_n$	1.012 2	0.991 1	0.908 6
0.700 0	不产生谐振，$A(\omega)$ 曲线无峰值		0.946 9	0.850 0
0.800 0			0.866 7	0.752 1

(5) 当 $\omega \gg \omega_n$ 时，$A(\omega)\to 0$，$\varphi(\omega)\to -180°$

3.6 一、二阶线性系统对典型激励的瞬态响应

3.6.1 一、二阶线性系统的脉冲响应函数

脉冲响应函数 $h(t)$ 是线性系统对单位脉冲输入的响应，即

$$h(t)=F^{-1}[H(\omega)]$$

当引入系统传递函数的概念后，有

$$Y(s)=H(s)X(s)$$

因此

$$y(t)=h(t)*x(t)$$

式中

$$h(t)=L^{-1}[H(\omega)]=F^{-1}[H(\omega)]$$

可见，脉冲响应函数 $h(t)$ 还是系统传递函数 $H(s)$ 的像原函数。

(1) 一阶系统的脉冲响应函数。

低通形式的传递函数为

$$H(s)=\frac{1}{\tau s+1}$$

$$h(t)=L^{-1}[H(s)]=L^{-1}\left(\frac{1}{\tau s+1}\right)$$

根据拉普拉斯变换可知

$$e^{at} \leftrightarrows \frac{1}{s-a}$$

因此

$$h(t)=L^{-1}\left(\frac{1}{\tau s+1}\right)=\frac{1}{\tau}e^{-t/\tau} \tag{3.47}$$

(5) when $\omega \gg \omega_n$, $A(\omega) \to 0$, $\varphi(\omega) \to -180°$

Table 3.2　The $A(\omega)$ value of the second-order system under different conditions

Damping ratio ζ	Resonance frequency ω_r	$A(\omega)$ Curve peak $A(\omega)_{max}$	Amplitude spectrum function $A(\omega)$	
			$\omega/\omega_n = 0.6$	$\omega/\omega_n = 0.8$
0.600 0	$0.529\ 1\omega_n$	1.041 6	1.038 1	0.975 3
0.640 3	$0.424\ 3\omega_n$	1.016 6	1.000 0	0.920 9
0.650 0	$0.393\ 7\omega_n$	1.012 2	0.991 1	0.908 6
0.700 0	No resonance, $A(\omega)$ curve has no peak		0.946 9	0.850 0
0.800 0			0.866 7	0.752 1

3.6　Transient Response of First- and Second-Order Linear Systems to Typical Excitation

3.6.1　The Impulse Response Function of First-and Second-Order Linear Systems

The impulse response function $h(t)$ is the response of a linear system to a unit impulse input, that is

$$h(t) = F^{-1}[H(\omega)]$$

When the concept of system transfer function is introduced, there is

$$Y(s) = H(s)X(s)$$

Therefore

$$y(t) = h(t) * x(t)$$

where

$$h(t) = L^{-1}[H(\omega)] = F^{-1}[H(\omega)]$$

It can be seen that the impulse response function $h(t)$ is still the image source function of the system transfer function $H(s)$.

(1) Impulse response function of first-order systems.

The transfer function of the low-pass form is

$$H(s) = \frac{1}{\tau s + 1}$$

$$h(t) = L^{-1}[H(s)] = L^{-1}\left(\frac{1}{\tau s + 1}\right)$$

According to the Laplace transform, we can know that

$$e^{at} \rightleftarrows \frac{1}{s - a}$$

Therefore

$$h(t) = L^{-1}\left(\frac{1}{\tau s + 1}\right) = \frac{1}{\tau} e^{-t/\tau} \tag{3.47}$$

一阶系统的脉冲响应函数如图 3. 21 所示。

(2)二阶系统的脉冲响应函数。

型如$y''(t)+2\zeta\omega_n y'(t)+\omega_n^2 y(t)=\omega_n^2 x(t)$的二阶线性系统的传递函数为

$$H(s)=\frac{\omega_n^2}{s^2+2\zeta\omega_n s+\omega_n^2}$$

则

$$\begin{aligned}h(t)&=L^{-1}[H(s)]=L^{-1}\left(\frac{\omega_n^2}{s^2+2\zeta\omega_n s+\omega_n^2}\right)\\&=L^{-1}\left[\frac{\omega_n\sqrt{1-\zeta^2}}{s^2+2\zeta\omega_n s+\zeta^2\omega_n^2+\omega_n^2(1-\zeta^2)}\cdot\frac{\omega_n}{\sqrt{1-\zeta^2}}\right]\\&=\frac{\omega_n}{\sqrt{1-\zeta^2}}L^{-1}\left[\frac{\omega_d}{(s+\zeta\omega_n)^2+\omega_d^2}\right]\end{aligned}$$

式中

$$\omega_d=\omega_n\sqrt{1-\zeta^2}$$

根据拉普拉斯变换简表,有

$$e^{-bt}\sin at\leftrightarrows\frac{a}{(s+b)^2+a^2}$$

上式与之对比,可知

$$h(t)=\frac{\omega_n}{\sqrt{1-\zeta^2}}e^{-\zeta\omega_n t}\sin\omega_d t \tag{3.48}$$

二阶系统的脉冲响应函数如图 3. 22 所示。

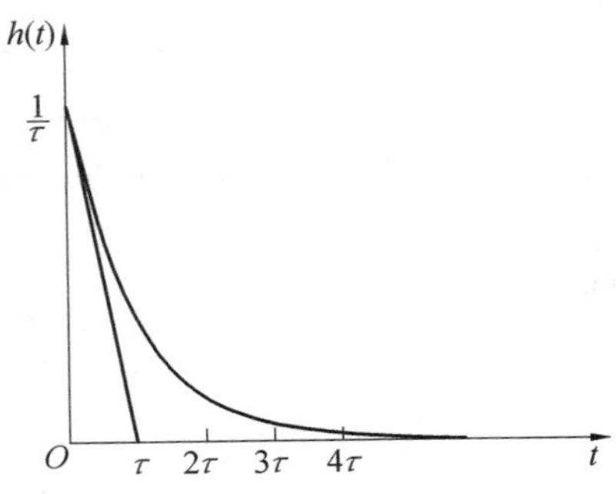

图 3. 21　一阶系统的脉冲响应函数

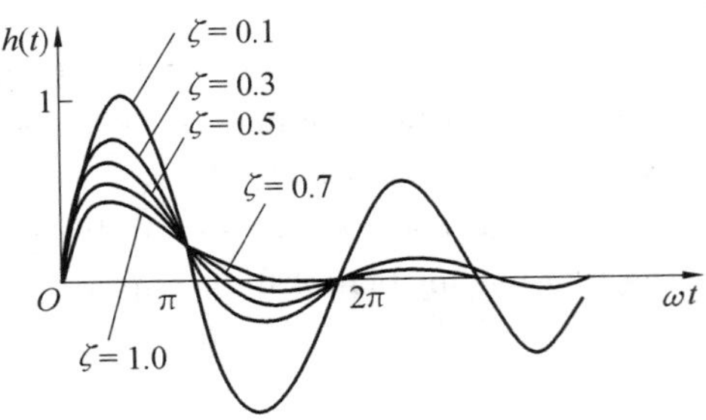

图 3. 22　二阶系统的脉冲响应函数

3.6.2　一、二阶线性系统对单位阶跃信号的响应

单位阶跃信号为

$$u(t)=\begin{cases}0, & t<0\\1, & t>0\end{cases}$$

The impulse response function of a first-order system is shown in Figure 3.21.

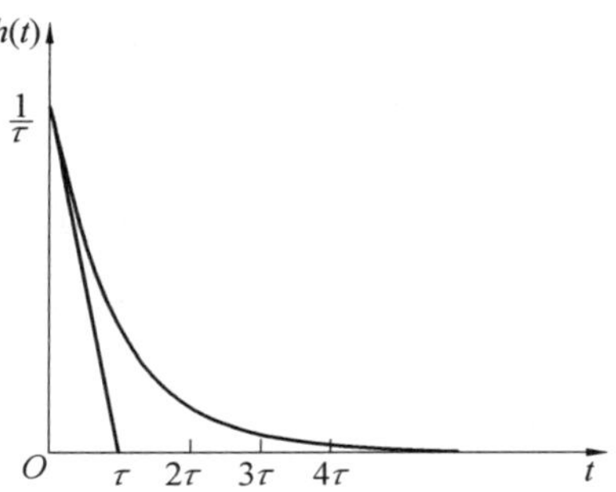

Figure 3.21　The impulse response function of a first-order system

(2) Impulse response function of second-order systems.

The transfer function of a second-order linear system of type $y''(t) + 2\zeta\omega_n y'(t) + \omega_n^2 y(t) = \omega_n^2 x(t)$ is

$$H(s) = \frac{\omega_n^2}{s^2 + 2\zeta\omega_n s + \omega_n^2}$$

There are

$$h(t) = L^{-1}[H(s)] = L^{-1}\left(\frac{\omega_n^2}{s^2 + 2\zeta\omega_n s + \omega_n^2}\right)$$

$$= L^{-1}\left[\frac{\omega_n\sqrt{1-\zeta^2}}{s^2 + 2\zeta\omega_n s + \zeta^2\omega_n^2 + \omega_n^2(1-\zeta^2)} \cdot \frac{\omega_n}{\sqrt{1-\zeta^2}}\right]$$

$$= \frac{\omega_n}{\sqrt{1-\zeta^2}} L^{-1}\left[\frac{\omega_d}{(s+\zeta\omega_n)^2 + \omega_d^2}\right]$$

where

$$\omega_d = \omega_n\sqrt{1-\zeta^2}$$

According to Laplace summary table, we have

$$e^{-bt}\sin at \leftrightarrows \frac{a}{(s+b)^2 + a^2}$$

In contrast with the above formula, we know that

$$h(t) = \frac{\omega_n}{\sqrt{1-\zeta^2}} e^{-\zeta\omega_n t}\sin\omega_d t \tag{3.48}$$

The impulse response function of the second-order system is shown in Figure 3.22.

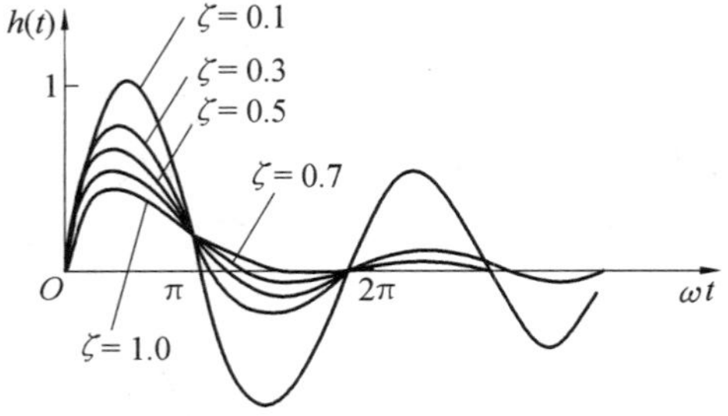

Figure 3.22　The impulse response function of the second-order system

3.6.2　The Response of First- and Second-Order Linear Systems to Unit Step Signals

Unit step signal is

$$u(t) = \begin{cases} 0, & t < 0 \\ 1, & t > 0 \end{cases}$$

它又是单位脉冲函数对时间的积分，即

$$u(t)=\int_0^t \delta(t)\,\mathrm{d}t$$

根据线性系统积分性质可知，系统对 $u(t)$ 响应等于对脉冲响应函数 $h(t)$ 的积分。

(1)一阶系统对单位阶跃信号的响应。

$$H(s)=\frac{1}{\tau s+1}$$

$$h(t)=\frac{1}{\tau}\mathrm{e}^{-\frac{t}{\tau}}$$

$$y(t)=\int_0^t h(t)\,\mathrm{d}t=1-\mathrm{e}^{-\frac{t}{\tau}} \tag{3.49}$$

一阶系统的单位阶跃响应如图 3.23 所示。

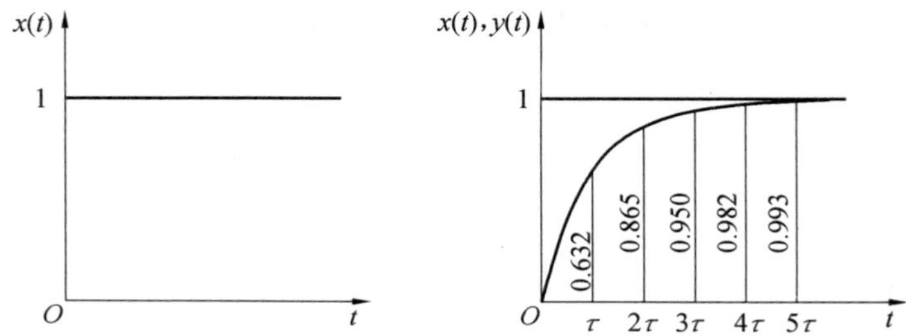

图 3.23　一阶系统的单位阶跃响应

(2)二阶系统对单位阶跃信号的响应。

$$H(s)=\frac{{\omega_n}^2}{s^2+2\zeta\omega_n s+{\omega_n}^2}$$

$$h(t)=\frac{\omega_n}{\sqrt{1-\zeta^2}}\mathrm{e}^{-\zeta\omega_n t}\sin\,\omega_d t$$

$$y(t)=\int_0^t h(t)\,\mathrm{d}t=1-\frac{\mathrm{e}^{-\zeta\omega_n t}}{\sqrt{1-\zeta^2}}\sin(\omega_d t+\varphi_2) \tag{3.50}$$

式中

$$\varphi_2=\arctan\frac{\sqrt{1-\zeta^2}}{\zeta}$$

二阶系统($\zeta<1$)的单位阶跃响应如图 3.24 所示。

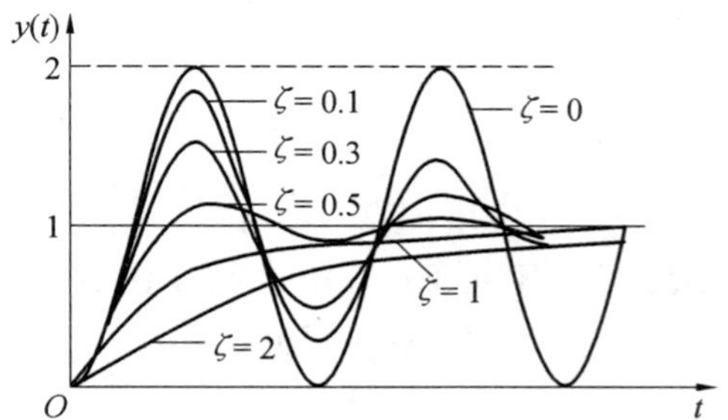

图 3.24　二阶系统($\zeta<1$)的单位阶跃响应

一阶和二阶系统对各种典型输入信号的响应见表 3.3。理想的单位脉冲输入实际上是不存在的，但若给系统以非常短暂的脉冲输入，其作用时间小于 0.1τ(τ

It is also the integral of the unit impulse function to time, that is

$$u(t) = \int_0^t \delta(t)\,\mathrm{d}t$$

According to the integral nature of the linear system, the response of the system to $u(t)$ is equal to the integral of the impulse response function $h(t)$.

(1) The response of first-order systems to a unit step signal.

$$H(s) = \frac{1}{\tau s + 1}$$

$$h(t) = \frac{1}{\tau}\mathrm{e}^{-\frac{t}{\tau}}$$

$$y(t) = \int_0^t h(t)\,\mathrm{d}t = 1 - \mathrm{e}^{-\frac{t}{\tau}} \tag{3.49}$$

Unit step response of a first-order system is shown in Figure 3.23.

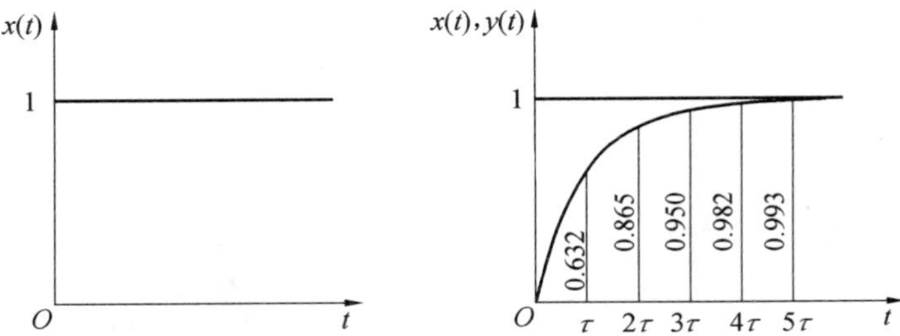

Figure 3.23　Unit step response of a first-order system

(2) The response of second-order system to a unit step signal.

$$H(s) = \frac{\omega_n^2}{s^2 + 2\zeta\omega_n s + \omega_n^2}$$

$$h(t) = \frac{\omega_n}{\sqrt{1-\zeta^2}}\mathrm{e}^{-\zeta\omega_n t}\sin \omega_d t$$

$$y(t) = \int_0^t h(t)\,\mathrm{d}t = 1 - \frac{\mathrm{e}^{-\zeta\omega_n t}}{\sqrt{1-\zeta^2}}\sin(\omega_d t + \varphi_2) \tag{3.50}$$

where

$$\varphi_2 = \arctan\frac{\sqrt{1-\zeta^2}}{\zeta}$$

Unit step response of second-order system ($\zeta < 1$) is shown in Figure 3.24.

Responses of the first-order and second-order systems to various typical input signals are listed in Table 3.3. The ideal unit pulse input does not actually exist, but if a very short pulse input is given to the system, and its action time is less than 0.1τ (τ is the time constant of the first-order system or the oscillation period of the second-order system), it can be approximately regarded as unit pulse input. The frequency domain function output by the system under unit impulse excitation is the frequency response function of the system, and the time domain response is the impulse response.

为一阶系统的时间常数或二阶系统的振荡周期)，则可近似地认为是单位脉冲输入。在单位脉冲激励下，系统输出的频域函数即该系统的频率响应函数，时域响应即脉冲响应。

表 3.3　一阶和二阶系统对各种典型输入信号的响应

输入		输出：一阶系统	输出：二阶系统
		$H(s)=\frac{1}{\tau s+1}$	$H(s)=\frac{\omega_n^2}{s^2+2\xi\omega_n s+\omega_n^2}$
脉冲响应	$X(s)=1$	$Y(s)=\frac{1}{\tau s+1}$	$Y(s)=\frac{\omega_n^2}{s^2+2\xi\omega_n s+\omega_n^2}$
	$x(t)=\delta(t)$	$y(t)=h(t)=\frac{1}{\tau}e^{-t/\tau}$	$y(t)=h(t)=\frac{\omega_n}{\sqrt{1-\xi_n^2}}e^{-\xi\omega_n t}\cdot \sin\sqrt{1-\xi^2}\,\omega_n t$
		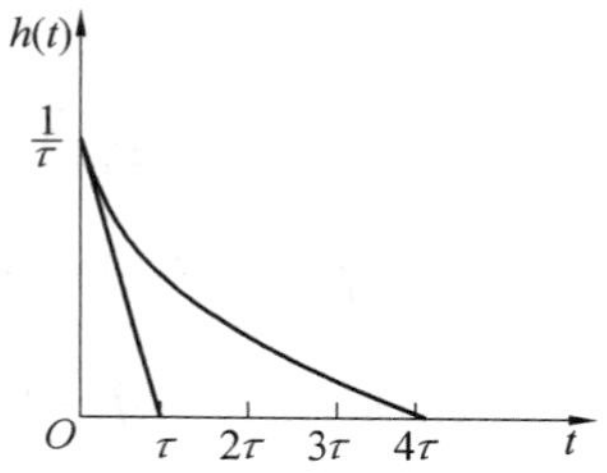	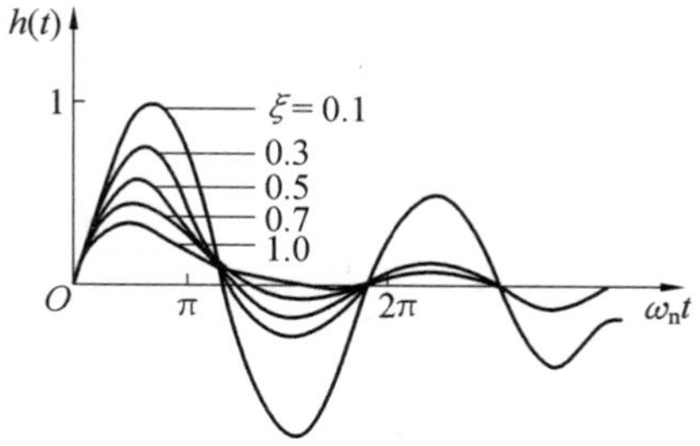
单位阶跃	$X(s)=\frac{1}{s}$	$Y(s)=\frac{1}{s(\tau s+1)}$	$Y(s)=\frac{\omega_n^2}{s(s^2+2\xi\omega_n s+\omega_n^2)}$
	$x(t)=\begin{cases}0, & t<0\\ 1, & t\geqslant 0\end{cases}$	$y(t)=1-e^{-t/\tau}$	$y(t)=1-\left(\frac{1}{\sqrt{1-\xi^2}}e^{-\xi\omega_n t}\right)\sin(\omega_d t+\varphi_2)$
		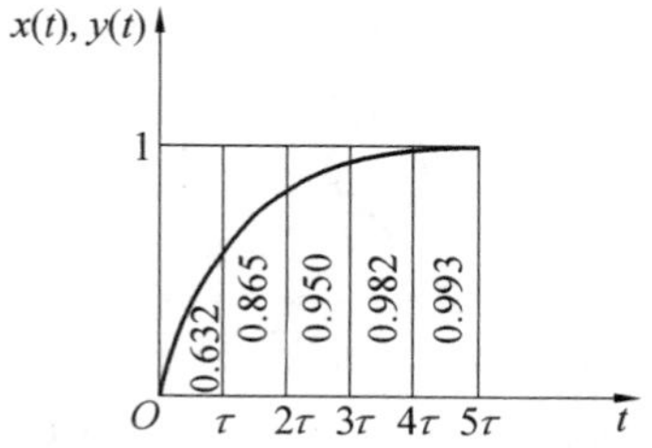	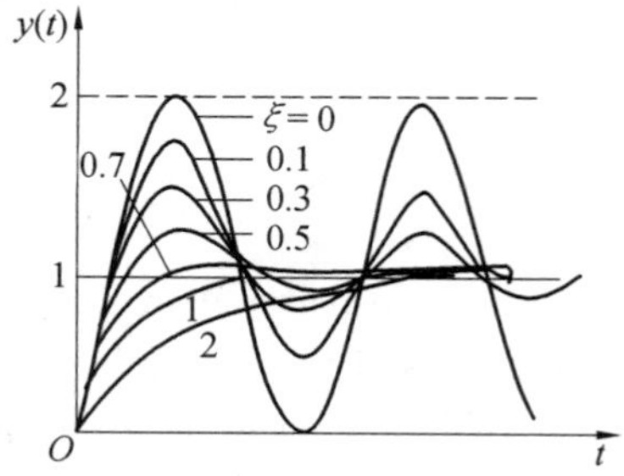

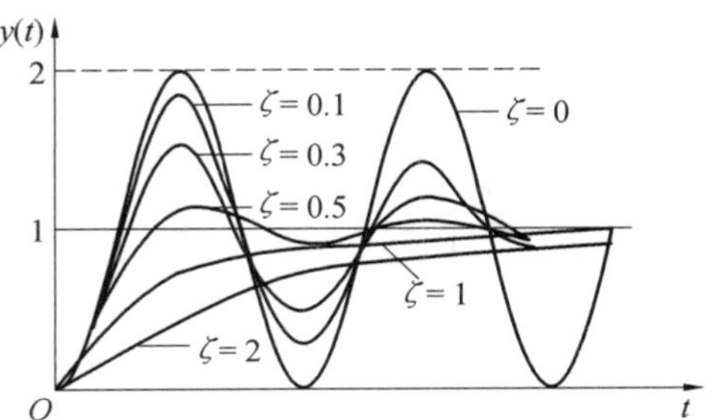

Figure 3.24　Unit step response of second-order system ($\zeta<1$)

Table 3.3　Responses of the first-order and second-order systems to various typical input signals

Input		Output: First-order system	Output: Second-order system
		$H(s)=\dfrac{1}{\tau s+1}$	$H(s)=\dfrac{\omega_n^2}{s^2+2\xi\omega_n s+\omega_n^2}$
Impulse response	$X(s)=1$	$Y(s)=\dfrac{1}{\tau s+1}$	$Y(s)=\dfrac{\omega_n^2}{s^2+2\xi\omega_n s+\omega_n^2}$
	$x(t)=\delta(t)$	$y(t)=h(t)=\dfrac{1}{\tau}e^{-t/\tau}$	$y(t)=h(t)=\dfrac{\omega_n}{\sqrt{1-\xi_n^2}}e^{-\xi\omega_n t}\cdot\sin\sqrt{1-\xi^2}\,\omega_n t$
		h(t); $\frac{1}{\tau}$; O; τ; 2τ; 3τ; 4τ; t	h(t); 1; ξ= 0.1; 0.3; 0.5; 0.7; 1.0; O; π; 2π; $\omega_n t$
Unit step	$X(s)=\dfrac{1}{s}$	$Y(s)=\dfrac{1}{s(\tau s+1)}$	$Y(s)=\dfrac{\omega_n^2}{s(s^2+2\xi\omega_n s+\omega_n^2)}$
	$x(t)=\begin{cases}0, & t<0\\1, & t\geqslant 0\end{cases}$	$y(t)=1-e^{-t/\tau}$	$y(t)=1-\left(\dfrac{1}{\sqrt{1-\xi^2}}e^{-\xi\omega_n t}\right)\sin(\omega_d t+\varphi_2)$
		x(t), y(t); 1; 0.632; 0.865; 0.950; 0.982; 0.993; O; τ; 2τ; 3τ; 4τ; 5τ; t	y(t); 2; 1; ξ= 0; 0.1; 0.3; 0.5; 0.7; 1; 2; O; t

续表 3.3

输入		输出	
		一阶系统	二阶系统
		$H(s)=\dfrac{1}{\tau s+1}$	$H(s)=\dfrac{\omega_{\mathrm{n}}^2}{s^2+2\xi\omega_{\mathrm{n}}s+\omega_{\mathrm{n}}^2}$
单位斜坡	$X(s)=\dfrac{1}{s^2}$	$Y(s)=\dfrac{1}{s^2(\tau s+1)}$	$Y(s)=\dfrac{\omega_{\mathrm{n}}^2}{s^2(s^2+2\xi\omega_{\mathrm{n}}s+\omega_{\mathrm{n}}^2)}$
	$x(t)=\begin{cases}0, & t<0\\ 1, & t\geqslant 0\end{cases}$	$y(t)=t-\tau(1-\mathrm{e}^{-t/\tau})$	$y(t)=t-\dfrac{2\xi}{\omega_{\mathrm{n}}}+\mathrm{e}^{-\xi\omega_{\mathrm{n}}t/\omega}\mathrm{d}\sin\left(\omega_{\mathrm{d}}t+\arctan\dfrac{2\xi\sqrt{1-\xi^2}}{2\xi^2-1}\right)$
单位正弦	$X(s)=\dfrac{\omega}{s^2+\omega^2}$	$Y(s)=\dfrac{\omega}{(s^2+\omega^2)(\tau s+1)}$	$Y(s)=\dfrac{\omega\omega_{\mathrm{n}}^2}{(s^2+\omega^2)(s^2+2\xi\omega_{\mathrm{n}}s+\omega_{\mathrm{n}}^2)}$
		$y(t)=\dfrac{1}{\sqrt{1+(\omega\tau)^2}}[\sin(\omega t+\varphi_1)-\mathrm{e}^{-t/\tau}\cos\varphi_1]$	$y(t)=A(\omega)\sin[\omega t+\varphi_2(\omega)]-\mathrm{e}^{-\xi\omega_{\mathrm{n}}t}[K_1\cos\omega_{\mathrm{d}}t+K_2\sin\omega_{\mathrm{d}}t]$
	$x(t)=\sin\omega t$, $t>0$	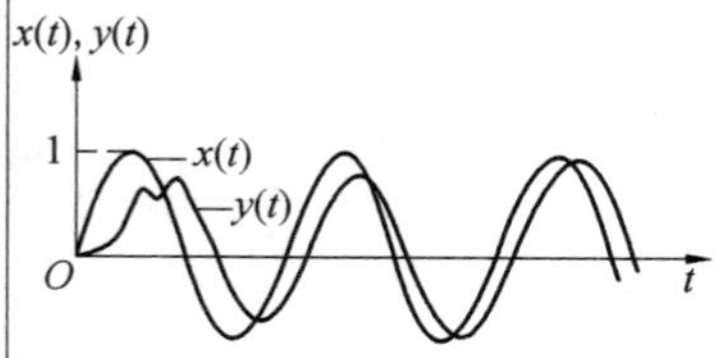	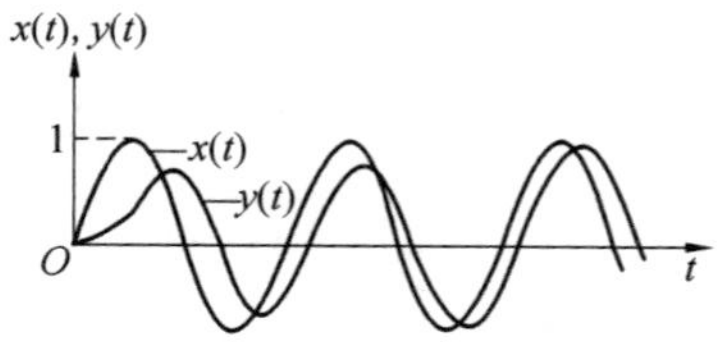

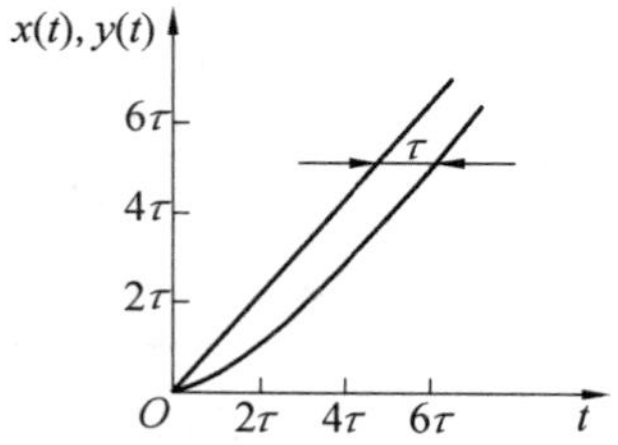

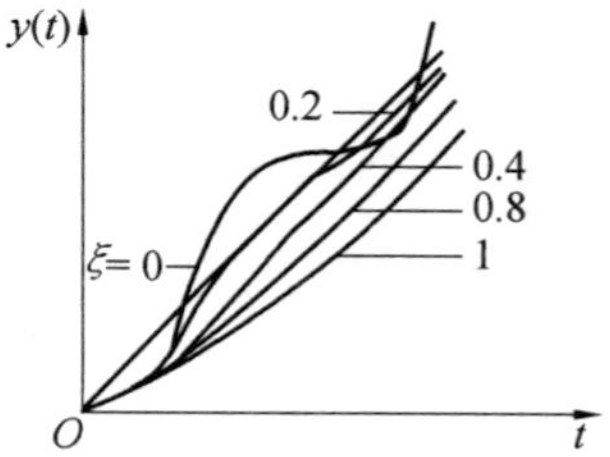

注：表中 $A(\omega)$ 和 $\varphi(\omega)$ 见式(3.31)，$\omega_{\mathrm{d}}=\omega_{\mathrm{n}}\sqrt{1-\xi^2}$，$\varphi_1=\arctan(\omega\tau)$，$k_1$ 和 k_2 均是取决于 ω_{n} 和 ξ 的系数，$\varphi_2=\arctan(\sqrt{1-\xi^2}/\xi)$。

3.7　一、二阶线性系统动态特性参数的获取

要使一阶、二阶线性系统精确可靠，不仅其定度应当精确，而且要定期校准。定度和校准的含义就是用试验方法对系统本身的特性参数进行测试，包括静态和动态参数两种。

Table 3.3

Input		Output: First-order system	Output: Second-order system
		$H(s) = \dfrac{1}{\tau s + 1}$	$H(s) = \dfrac{\omega_n^2}{s^2 + 2\xi\omega_n s + \omega_n^2}$
Unit slope	$X(s) = \dfrac{1}{s^2}$	$Y(s) = \dfrac{1}{s^2(\tau s + 1)}$	$Y(s) = \dfrac{\omega_n^2}{s^2(s^2 + 2\xi\omega_n s + \omega_n^2)}$
	$x(t) = \begin{cases} 0, & t < 0 \\ 1, & t \geqslant 0 \end{cases}$	$y(t) = t - \tau(1 - e^{-t/\tau})$ 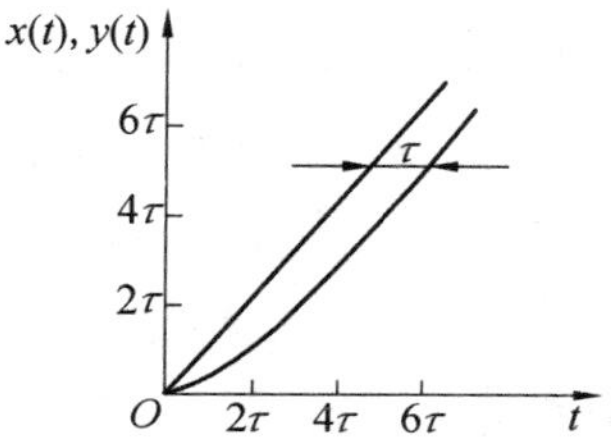	$y(t) = t - \dfrac{2\xi}{\omega_n} + e^{-\xi\omega_n t/\omega} d\sin\left(\omega_d t + \arctan\dfrac{2\xi\sqrt{1-\xi^2}}{2\xi^2 - 1}\right)$ 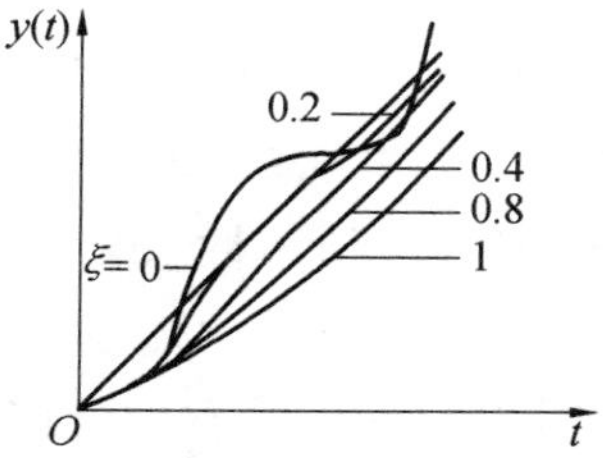
Unit sine	$X(s) = \dfrac{\omega}{s^2 + \omega^2}$	$Y(s) = \dfrac{\omega}{(s^2 + \omega^2)(\tau s + 1)}$	$Y(s) = \dfrac{\omega\omega_n^2}{(s^2 + \omega^2)(s^2 + 2\xi\omega_n s + \omega_n^2)}$
	$x(t) = \sin \omega t,\ t > 0$	$y(t) = \dfrac{1}{\sqrt{1 + (\omega\tau)^2}}[\sin(\omega t + \varphi_1) - e^{-t/\tau}\cos\varphi_1]$ 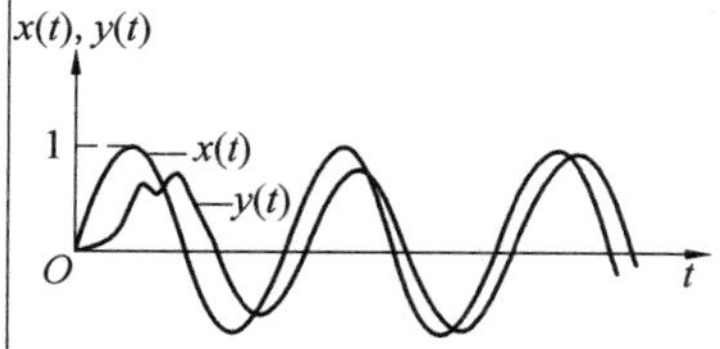	$y(t) = A(\omega)\sin[\omega t + \varphi_2(\omega)] - e^{-\xi\omega_n t}[K_1\cos\omega_d t + K_2\sin\omega_d t]$ 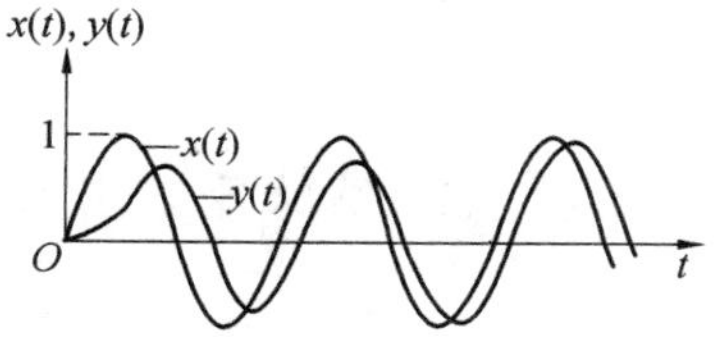

Note: In the table, $A(\omega)$ and $\varphi(\omega)$ are shown in formula (3.31), $\omega_d = \omega_n\sqrt{1-\xi^2}$, $\varphi_1 = \arctan\omega\tau$, k_1 and k_2 are all coefficients dependent on ω_n and ξ, $\varphi_2 = \arctan(\sqrt{1-\xi^2}/\xi)$.

3.7 Acquisition of Dynamic Characteristic Parameters of Second-Order Linear System

In order to make the first-order and second-order linear systems accurate and reliable, it is necessary not only to make it accurate, but also to calibrate it regularly. The meaning of calibration is to use measurement methods to test the characteristic parameters of the system itself, including static and dynamic parameters.

对装置的静态参数进行测试时，一般以经过校准的“标准”静态量作为输入，求出其输入—输出曲线，根据这条曲线确定其回程误差，整理和确定其校准曲线、线性误差和灵敏度。所用的输入量误差应当是所要求测试结果误差的1/3～1/5或更小。

本节主要叙述测试中常用的一阶、二阶测试装置的动态参数的测试方法。一阶系统的动态参数是时间常数τ，二阶系统的动态参数是阻尼比ζ和固有频率ω_n。当这些参数确定后，系统的传递函数就相应确定了。因此，常将以上动态参数的测试问题称为系统动态特性的测试。对于一、二阶线性系统，常用以下两种方法。

3.7.1 频率响应法——扫频

频率响应法的主导思想是：对系统输入正弦激励$x(t)=x_0\sin \omega t$，然后测取其稳态响应。若由低到高逐步改变激励的频率，则能获得该系统的频率响应函数的大量数据，然后根据这些数据计算系统的动态参数。扫频通常是属于校核性质的，即被测试的装置是某已知的一阶或二阶线性系统。测试目的只是校核其动态参数是否有变化，这时可参考该装置技术文件中所附的幅频特性和相频特性曲线。测试中输入的正弦信号的幅值通常是该装置量程的20%左右，其频率由接近0的足够低的频率开始，然后逐渐增加，直至输出的幅值已减少到输入幅值的一半为止。

获得有关测试装置的幅频特性$A(\omega)$和相频特性$\varphi(\omega)$的大量数据之后，计算动态参数的方法如下。

(1)一阶系统时间常数τ。

利用公式

$$A(\omega)=\frac{1}{\sqrt{1+(\tau\omega)^2}}$$

$$\varphi(\omega)=-\arctan \tau\omega$$

在已知ω和$A(\omega)$情况下，求出系统的时间常数为

$$\tau=\frac{1}{\omega}\sqrt{\frac{1}{A^2(\omega)}-1} \tag{3.51}$$

或

$$\tau=-\frac{1}{\omega}\cdot\tan\varphi(\omega) \tag{3.52}$$

(2)二阶系统阻尼比ζ和固有频率ω_n。

当根据测试数据绘制出幅频特性和相频特性曲线后，可用以下方法求系统的阻尼比ζ和固有频率ω_n。

①利用相频特性曲线。在$\omega=\omega_n$处，有

When testing the static parameters of the device, generally the calibrated "standard" static quantity will be used as input to find its input—output curve, determining its return error based on this curve, sorting out and determining its calibration curve, linearity error, and sensitivity. The input error used should be 1/3 ~ 1/5 or less of the required measurement result error.

This section mainly describes the measurement methods of the dynamic parameters of the first-order and second-order measurement devices commonly used in the measurement. The dynamic parameters of the first-order system are the time constant τ , and the dynamic parameters of the second-order system are the damping ratio ζ and the natural frequency ω_n . When these parameters are determined, the transfer function of the system is determined accordingly. Therefore, the measurement problem of the above dynamic parameters is often referred to as the measurement of the dynamic characteristics of the system. The following two methods are commonly used for first- and second-order linear systems.

3.7.1 Frequency Response Method—Sweep Frequency

The main idea of the frequency response method is: to input a sine excitation $x(t) = x_0 \sin \omega t$ to the system, and then measure its steady-state response. If the excitation frequency is gradually changed from low to high, a large amount of data of the frequency response function of the system can be obtained, and then calculate the dynamic parameters of the system based on these data. Usually sweep frequeney is of the nature of verification, that is, the device being measured is a known first-order or second-order linear system. The purpose of the measurement is to verify whether its dynamic parameters have changed. At this time, we can refer to the amplitude frequency characteristic and phase frequency characteristic curve attached to the technical file of the device. The amplitude of the sin signal input in the measurement is usually about 20% of the device's range whose frequency starts from a sufficiently low frequency close to 0, and then gradually increases until the output amplitude has been reduced to half of the input amplitude.

After obtaining a large amount of data on the amplitude frequency characteristic $A(\omega)$ and phase frequency characteristic $\varphi(\omega)$ of the measurement device, the method of calculating the dynamic parameters is as follows.

(1) First-order system time constant τ .

According to

$$A(\omega) = \frac{1}{\sqrt{1 + (\tau\omega)^2}}$$

$$\varphi(\omega) = -\arctan \tau\omega$$

given ω and $A(\omega)$, find the time constant of the system, that is

$$\tau = \frac{1}{\omega}\sqrt{\frac{1}{A^2(\omega)} - 1} \tag{3.51}$$

or

$$\tau = -\frac{1}{\omega} \cdot \tan \varphi(\omega) \tag{3.52}$$

(2) Second-order system damping ratio ζ and natural frequency ω_n .

After drawing the amplitude-frequency characteristic and phase-frequency characteristic curve according to themeasurement data, the following methods can be used to obtain the damping ratio ζ and natural frequency ω_n of the system.

① Utilizing phase frequency characteristic curve. At $\omega = \omega_n$,we have

$$\varphi(\omega_n)=-\arctan\frac{2\zeta\left(\frac{\omega}{\omega_n}\right)}{1-\left(\frac{\omega}{\omega_n}\right)^2}=-90°$$

相频特性曲线在该点的斜率$\frac{d\varphi(\omega)}{d\omega}=\frac{1}{\zeta}$。利用这一特点，可以求得阻尼比$\zeta$，即

$$\zeta=\frac{1}{\frac{d\varphi(\omega)}{d\omega}}=\left|\frac{1}{\tan\alpha}\right|$$

在$\varphi(\omega)$曲线上找到$\varphi(\omega)=\pi/2$的点，过此点作曲线的切线，切线的斜率为$\tan\alpha$，阻尼比ζ即该斜率的倒数的绝对值，切点对应的圆频率即该系统的固有频率ω_n。

②利用幅频特性曲线。一般相角的测量比较困难，所以可以通过幅频特性曲线来估计ζ和ω_n。

对于欠阻尼系统$\zeta<1$，当$1/\sqrt{2}>\zeta>0$时，幅频特性曲线$A(\omega)$在$\omega=\omega_r=\omega_n\sqrt{1-\zeta^2}$处出现峰值$A(\omega)_{max}$，即

$$A(\omega)_{max}=\frac{1}{2\zeta\sqrt{1-\zeta^2}}$$

当ζ较小，且只做粗略估计时，可以认为$\omega_r\approx\omega_n$，$A(\omega)_{max}\approx1/(2\zeta)$，即认为试验后所得到的$A(\omega)$曲线的峰值点的圆频率即$\omega_n$。二阶系统阻尼比的估计如图3.25所示。

又考虑到当$\omega_1=(1-\zeta)\omega_n$，$\omega_2=(1+\zeta)\omega_n$时，由理论幅频特性

$$A(\omega)=\frac{1}{\sqrt{\left[1-\left(\frac{\omega}{\omega_n}\right)^2\right]^2+4\zeta^2\left(\frac{\omega}{\omega_n}\right)^2}}$$

求得

$$A(\omega_1)=\frac{1}{\sqrt{8\zeta^2-12\zeta^3+5\zeta^4}}$$

$$A(\omega_2)=\frac{1}{\sqrt{8\zeta^2+12\zeta^3+5\zeta^4}}$$

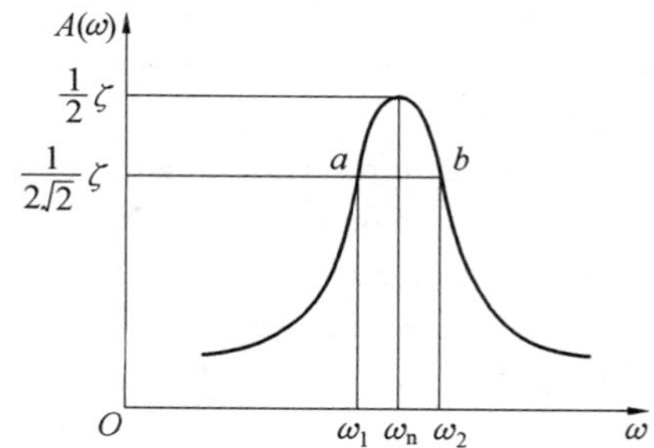

图3.25　二阶系统阻尼比的估计

若略去式中的ζ^3和ζ^4，则

$$A(\omega_1)=A(\omega_2)\approx\frac{1}{2\sqrt{2}\zeta}\approx\frac{1}{\sqrt{2}}A(\omega)_{max}$$

恰是上述近似峰值$1/(2\zeta)$的$1/\sqrt{2}$（即0.707）。

因此，在找到试验曲线的峰值之后，再在图中画一条高度为峰值的0.707的水平线交试验曲线于两点（图3.25），便可在横轴上得到两个对应点ω_1和ω_2，对应的幅频特性为

$$A(\omega_1)=\frac{1}{\sqrt{2}}\cdot\frac{1}{2\zeta}$$

$$\varphi(\omega_n) = -\arctan\frac{2\zeta\left(\frac{\omega}{\omega_n}\right)}{1-\left(\frac{\omega}{\omega_n}\right)^2} = -90°$$

The slope of the phase frequency characteristic curve at this point is $\frac{d\varphi(\omega)}{d\omega} = \frac{1}{\zeta}$. Using this feature, damping ratio ζ can be obtained, that is

$$\zeta = \frac{1}{\frac{d\varphi(\omega)}{d(\omega)}} = \left|\frac{1}{\tan\alpha}\right|$$

On $\varphi(\omega)$ curve, finding the point $\varphi(\omega) = \pi/2$, drawing the tangent to the curve through this point, the slope of the tangent is $\tan\alpha$, damping ratio ζ is the absolute value of the reciprocal of the slope, and the circular frequency corresponding to the tangent point is the natural frequency ω_n of the system.

② Utilization of phase frequency characteristic curve. Generally, it is difficult to measure the phase angle, so ζ and ω_n can be estimated through the amplitude frequency characteristic curve.

For the under-damped system $\zeta < 1$, if $1/\sqrt{2} > \zeta > 0$, the amplitude-frequency characteristic curve $A(\omega)$ has a peak $A(\omega)_{max}$ at $\omega = \omega_r = \omega_n\sqrt{1-\zeta^2}$, that is

$$A(\omega)_{max} = \frac{1}{2\zeta\sqrt{1-\zeta^2}}$$

If ζ is small and only a rough estimate is made, we get $\omega_r \approx \omega_n$, $A(\omega)_{max} \approx 1/(2\zeta)$, that is, the circular frequency of the peak point of $A(\omega)$ curve obtained after the measurement is ω_n. The damping ratio estimation of the second-order system is shown in Figure 3.25.

Taking into account that when $\omega_1 = (1-\zeta)\omega_n$, $\omega_2 = (1+\zeta)\omega_n$, from the theoretical amplitude frequency characteristic

$$A(\omega) = \frac{1}{\sqrt{\left[1-\left(\frac{\omega}{\omega_n}\right)^2\right]^2 + 4\zeta^2\left(\frac{\omega}{\omega_n}\right)^2}}$$

The result is obtained that

$$A(\omega_1) = \frac{1}{\sqrt{8\zeta^2 - 12\zeta^3 + 5\zeta^4}}$$

$$A(\omega_2) = \frac{1}{\sqrt{8\zeta^2 + 12\zeta^3 + 5\zeta^4}}$$

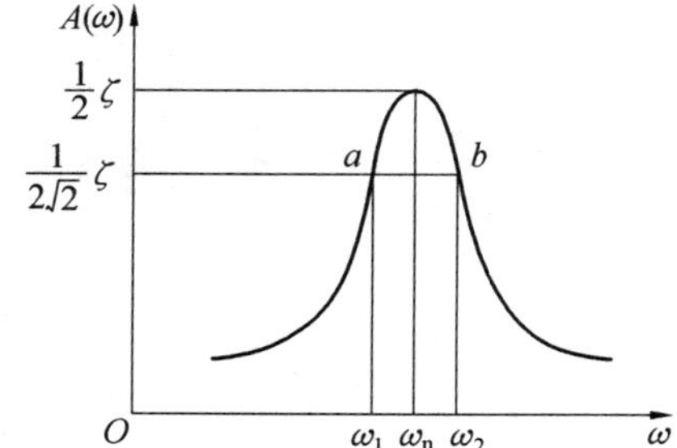

Figure 3.25　The damping ratio estimation of the second order system

If ζ^3 and ζ^4 in the formula are omitted, then

$$A(\omega_1) = A(\omega_2) \approx \frac{1}{2\sqrt{2}\zeta} \approx \frac{1}{\sqrt{2}}A(\omega)_{max}$$

It is exactly $1/\sqrt{2}$ (that is, 0.707) times the above approximate peak $1/(2\zeta)$.

Therefore, after finding the peak value of themeasurement curve, draw a horizontal line with a height of 0.707 times the peak value and cross the measurement curve at two points (Figure 3.25) to obtain two corresponding points ω_1 and ω_2 on the horizontal axis. The corresponding amplitude frequency characteristic is

$$A(\omega_2)=\frac{1}{\sqrt{2}}\cdot\frac{1}{2\zeta}$$

它们与峰值保持0.707的倍数关系，故可认为 $\omega_1\approx(1-\zeta)\omega_n$，$\omega_2\approx(1+\zeta)\omega_n$，则有

$$\omega_2-\omega_1=2\zeta\omega_n$$

故

$$\zeta=\frac{\omega_2-\omega_1}{2\omega_n}$$

因此，在 ζ 较小的情况下，粗略估计 ζ 时，只要在实验曲线 $A(\omega)$ 上查找出峰值点对应的圆频率，并记为 ω_n^*，同时作图求得 ω_1 和 ω_2，则估算的阻尼比约为

$$\zeta\approx\frac{\omega_2-\omega_1}{2\omega_n^*} \tag{3.53}$$

理论上来说，$A(\omega)$ 曲线的峰值出现在 $\omega=\omega_r=\omega_n\sqrt{1-2\zeta^2}$ 处，其峰值为

$$A(\omega)_{max}=\frac{1}{2\zeta\sqrt{1-\zeta^2}}$$

而当 ω 为试验中的最低频率 ω_{min} 时，$A(\omega_{min})$ 应近似为 $A(0)=1$，故有

$$\frac{A(\omega)_{max}}{A(\omega_{min})}=\frac{1}{2\zeta\sqrt{1-\zeta^2}}$$

当从实验曲线上或从试验数据中获得 $A(\omega)_{max}$ 和 $A(\omega_{min})$ 实际比值，并记为 C 时，则有

$$C=\frac{1}{2\zeta\sqrt{1-\zeta^2}}$$

解上面的方程并去掉 $\zeta>1/\sqrt{2}$ 的增根，求得阻尼比 ζ 为

$$\zeta=\sqrt{\frac{1}{2}-\frac{1}{2}\sqrt{1-\frac{1}{C^2}}} \tag{3.54}$$

再根据试验数据挑选出 $A(\omega)_{max}$ 所对应的 ω，并记为 ω_r，则系统的固有频率 ω_n 为

$$\omega_n=\frac{\omega_r}{\sqrt{1-2\zeta^2}} \tag{3.55}$$

3.7.2 阶跃响应法——激励

阶跃响法的主导思想是对被测试的一、二阶线性系统输入一个单位阶跃信号。测试中，要注意观察暂态部分消失之前的过渡过程并记录有关数据，根据它们来求取系统的动态参数，方法如下。

(1)一阶系统时间常数 τ。

①单点法。一阶系统对单位阶跃的响应为

$$y(t)=1-e^{-\frac{t}{\tau}}$$

$$A(\omega_1)=\frac{1}{\sqrt{2}}\cdot\frac{1}{2\zeta}$$

$$A(\omega_2)=\frac{1}{\sqrt{2}}\cdot\frac{1}{2\zeta}$$

Now that they maintain a 0. 707 times relationship with the peak value, it can be considered that $\omega_1 \approx (1-\zeta)\omega_n$, $\omega_2 \approx (1+\zeta)\omega_n$, so

$$\omega_2-\omega_1=2\zeta\omega_n$$

Thus

$$\zeta=\frac{\omega_2-\omega_1}{2\omega_n}$$

Therefore, when ζ is relatively small, and ζ is roughly estimated, we only need to find the circular frequency corresponding to the peak point on the experimental curve $A(\omega)$ which is recorded as ω_n^*, and ω_1 and ω_2 are obtained by drawing the graph, then the estimated damping ratio is about

$$\zeta\approx\frac{\omega_2-\omega_1}{2\omega_n^*} \tag{3.53}$$

Theoretically, the peak of $A(\omega)$ curve appears at $\omega=\omega_r=\omega_n\sqrt{1-2\zeta^2}$. Its peak is

$$A(\omega)_{max}=\frac{1}{2\zeta\sqrt{1-\zeta^2}}$$

And when ω is the lowest frequency ω_{min} in the measurement, $A(\omega_{min})$ should be approximately $A(0)=1$, so

$$\frac{A(\omega)_{max}}{A(\omega_{min})}=\frac{1}{2\zeta\sqrt{1-\zeta^2}}$$

When the actual ratio of $A(\omega)_{max}$ to $A(\omega_{min})$ is obtained from the experimental curve or from the experimental data, it is recorded as C, then

$$C=\frac{1}{2\zeta\sqrt{1-\zeta^2}}$$

Solve the above equation and remove the increasing root of $\zeta>1/\sqrt{2}$, and obtain the damping ratio ζ as

$$\zeta=\sqrt{\frac{1}{2}-\frac{1}{2}\sqrt{1-\frac{1}{C^2}}} \tag{3.54}$$

Then select ω corresponding to $A(\omega)_{max}$ according to the measurement data, and mark it as ω_r, then the natural frequency ω_n of the system is

$$\omega_n=\frac{\omega_r}{\sqrt{1-2\zeta^2}} \tag{3.55}$$

3.7.2 Step Response Method—Excitation

The main idea of the step response method is to input a unit step signal to the first-and second-order linear system under measurement. During the measurement, attention should be paid to observe the transition process before the transient part disappears and record the relevant data, and obtain the system based on them. The method of dynamic parameters is as follows.

其中,暂态部分为 $-e^{-\frac{t}{\tau}}$。

当 $t=\tau$ 时,它降为 $-1/e=-0.3679$,即过渡曲线升至 $1-1/e=0.6321$ 所需的时间便是一阶系统的时间常数 τ。此法要求精确测得一瞬间的数据,没有涉及全部响应过程。这看起来似乎简单,但很难操作。因此,可改用多点法。

②多点法。一阶系统对单位阶跃的响应的暂态部分为

$$e^{-t/\tau}=1-y(t)$$

两边取对数得

$$\ln(1-y(t))=-\frac{1}{\tau}\cdot t$$

由此可见,$\ln(1-y(t))$ 和 t 呈线性关系。

因此,根据试验数据 $y(t)$,可以作出 $\ln(1-y(t))$— t 曲线(图 3.26),图中曲线的斜率为

$$\frac{\ln(1-y(t))}{t}=\tan\alpha=-\frac{1}{\tau}$$

故

$$\tau=\left|\frac{1}{\tan\alpha}\right| \tag{3.56}$$

此法利用了全部数据,结果比较可靠。

(2)二阶系统阻尼比 ζ 和固有频率 ω_n。

①利用第 1 极值点。二阶系统在 $\zeta<1$ 时是欠阻尼系统,对单位阶跃的响应为

$$y(t)=1-\frac{e^{-\zeta\omega_n t}}{\sqrt{1-\zeta^2}}\sin(\omega_d t+\varphi_2) \tag{3.58}$$

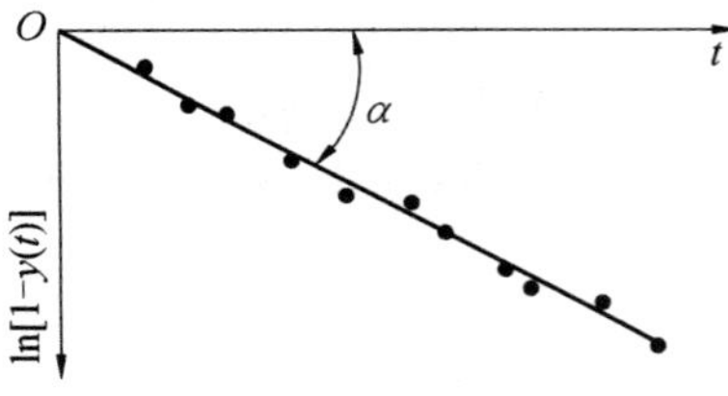

图 3.26 $\ln(1-y(t))$— t 曲线

其中,暂态部分为

$$y_1(t)=-\frac{e^{-\zeta\omega_n t}}{\sqrt{1-\zeta^2}}\sin(\omega_d t+\varphi_2)$$

是圆频率为 ω_d 的衰减振荡。按通常取极值的方法,在 $y_1'(t)=0$ 时可出现极值,根据 $y_1'(t)=0$的条件可得到

$$\tan(\omega_d t+\varphi_2)=\frac{\sqrt{1-\zeta^2}}{\zeta}$$

式中

$$\omega_d=\omega_n\sqrt{1-\zeta^2}$$

$$\tan\varphi_2=\frac{\sqrt{1-\zeta^2}}{\zeta}$$

因此,$y'(t)=0$ 这一条件等价于

(1) First-order system time constant τ.

① Single point method. The response of first-order system to a unit step is

$$y(t) = 1 - e^{-\frac{t}{\tau}}$$

In it, the transient part is $-e^{-\frac{t}{\tau}}$.

When $t = \tau$, it drops to $-1/e = -0.367\ 9$, that is, the time required for the transition curve to rise to $1 - 1/e = 0.632\ 1$ is the time constant τ of the first-order system. This method requires accurate measurement of data at a moment, not involving all the response processes. It seems simple, but it is difficult to operate. Therefore, the following method can be used instead.

② Multi-point method. The transient part of the response of a first-order system to a unit step is

$$e^{-t/\tau} = 1 - y(t)$$

Take the logarithm of both sides to get

$$\ln(1 - y(t)) = -\frac{1}{\tau} \cdot t$$

It can be seen that $\ln(1 - y(t))$ has a linear relationship with t.

Therefore, according to themeasurement data $y(t)$, a $\ln(1-y(t))$—t curve can be made (Figure 3.26), and the slope of the curve in the figure is

$$\frac{\ln(1 - y(t))}{t} = \tan\alpha = -\frac{1}{\tau}$$

Therefore

$$\tau = \left|\frac{1}{\tan\alpha}\right| \tag{3.56}$$

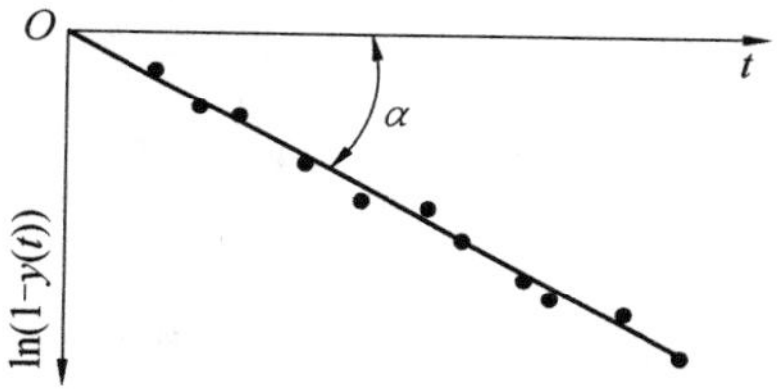

Figure 3.26　$\ln(1-y(t))$—t curve

This method uses all the data, and the result is relatively reliable.

(2) Second-order system damping ratio ζ and natural frequency ω_n.

① Using the first extreme point. The second-order system is an underdamped system when $\zeta < 1$, and the response to unit step is

$$y(t) = 1 - \frac{e^{-\zeta\omega_n t}}{\sqrt{1 - \zeta^2}}\sin(\omega_d t + \varphi_2) \tag{3.57}$$

In it, the transient part is

$$y_1(t) = -\frac{e^{-\zeta\omega_n t}}{\sqrt{1 - \zeta^2}}\sin(\omega_d t + \varphi_2)$$

It is a damped oscillation with a circular frequency of ω_d. According to the usual method of taking the extreme value, the extreme value can appear at $y_1'(t) = 0$, which can be obtained according to the condition of $y_1'(t) = 0$ that

$$\tan(\omega_d t + \varphi_2) = \frac{\sqrt{1 - \zeta^2}}{\zeta}$$

where

$$\omega_d = \omega_n\sqrt{1 - \zeta^2}$$

$$\tan\varphi_2 = \frac{\sqrt{1 - \zeta^2}}{\zeta}$$

Therefore, this condition $y'(t) = 0$ is equivalent to

$$\tan(\omega_n\sqrt{1 - \zeta^2}\,t + \varphi_2) = \tan\varphi_2$$

$$\tan(\omega_n\sqrt{1-\zeta^2}\,t+\varphi_2)=\tan\varphi_2$$

即相当于在 $\omega_n\sqrt{1-\zeta^2}\,t=n\pi(n=1,2,\cdots)$ 时，会使此正弦振荡出现极值。欠阻尼二阶装置的阶跃响应如图 3.27 所示。

出现第一极值的时刻为

$$t_p=\frac{\pi}{\omega_n\sqrt{1-\zeta^2}}=\frac{\pi}{\omega_d}$$

其极值为 y_{m1}，有

$$y_{m1}=1-\frac{e^{-\zeta\frac{\pi}{\sqrt{1-\zeta^2}}}}{\sqrt{1-\zeta^2}}\sin\left(\omega_d\cdot\frac{\pi}{\omega_d}+\varphi_2\right)$$

$$=1-\frac{e^{-\zeta\frac{\pi}{\sqrt{1-\zeta^2}}}}{\sqrt{1-\zeta^2}}\cdot\sin(\pi+\varphi_2)$$

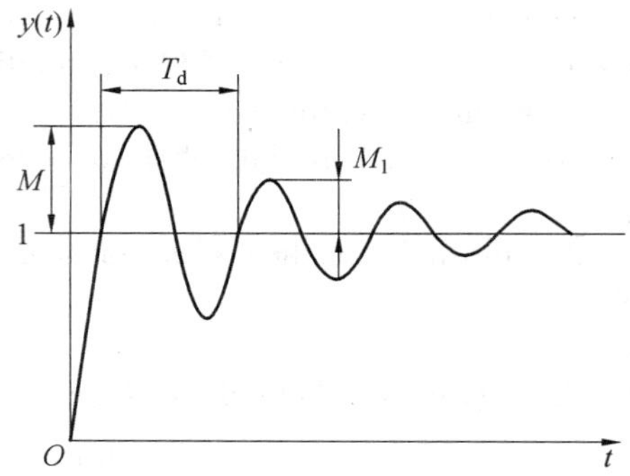

图 3.27　欠阻尼二阶装置的阶跃响应

因为

$$\sin(\pi+\varphi_2)=-\sqrt{1-\zeta^2}$$

故

$$y_{m1}=1+e^{-\frac{\zeta\pi}{\sqrt{1-\zeta^2}}}$$

与稳态输出 $(y(t)=1)$ 相比，其最大超调量为

$$M=e^{-\frac{\zeta\pi}{\sqrt{1-\zeta^2}}}$$

$$\zeta=\sqrt{\frac{1}{\left(\frac{\pi}{\ln M}\right)^2+1}}$$

可见，M 仅是 ζ 的函数，与系统固有频率 ω_n 无关，是二阶线性系统所固有的。欠阻尼二阶装置的 M—ζ 关系如图 3.28 所示。若试验中测得 M，则可从 M—ζ 曲线图中反查出系统的阻尼比 ζ。

②利用系列极值点。若响应中的暂态过程较长，可出现几个峰值，则任意两相邻超调量 M_i 和 M_{i+1} 出现的时间将分别为 t_{pi} 和 $t_{p(i+1)}$，其关系为

$$t_{p(i+1)}=t_{pi}+\frac{2\pi}{\omega_d}=t_{pi}+\frac{2\pi}{\omega_n\sqrt{1-\zeta^2}}$$

将 t_{pi} 和 $t_{p(i+1)}$ 代入式(3.58)中，可得

$$M_i=e^{-\frac{2\pi i\zeta}{\sqrt{1-\zeta^2}}}$$

$$M_{i+1}=e^{-\frac{2\pi(i+1)\zeta}{\sqrt{1-\zeta^2}}}$$

$$\ln\frac{M_i}{M_{i+1}}=\delta=\frac{2\pi\zeta}{\sqrt{1-\zeta^2}}$$

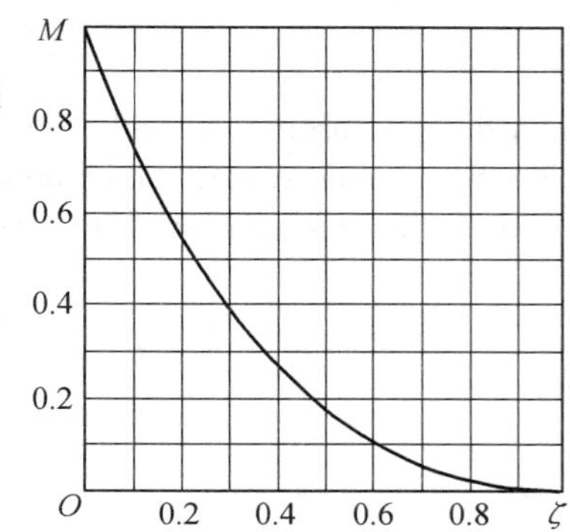

图 3.28　欠阻尼二阶装置的 M—ζ 关系

That is equivalent to when $\omega_n\sqrt{1-\zeta^2}t = n\pi$ ($n=1, 2, \cdots$), will make this sinusoidal oscillation appear extreme value. Step response of underdamped second-order device is shown in Figure 3.27.

The moment when the first extreme value appears is

$$t_p = \frac{\pi}{\omega_n\sqrt{1-\zeta^2}} = \frac{\pi}{\omega_d}$$

its extreme value is y_{m1}, we have

$$y_{m1} = 1 - \frac{e^{-\zeta\frac{\pi}{\sqrt{1-\zeta^2}}}}{\sqrt{1-\zeta^2}}\sin\left(\omega_d \cdot \frac{\pi}{\omega_d} + \varphi_2\right)$$

$$= 1 - \frac{e^{-\zeta\frac{\pi}{\sqrt{1-\zeta^2}}}}{\sqrt{1-\zeta^2}} \cdot \sin(\pi + \varphi_2)$$

Because of

$$\sin(\pi + \varphi_2) = -\sqrt{1-\zeta^2}$$

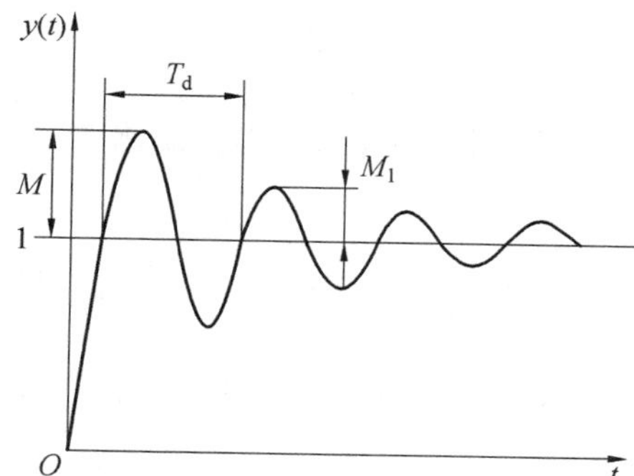

Figure 3.27　Step response of underdamped second-order device

Thus

$$y_{m1} = 1 + e^{-\frac{\zeta\pi}{\sqrt{1-\zeta^2}}}$$

Compared with the steady output ($y(t) = 1$), its maximum overshoot is

$$M = e^{-\frac{\zeta\pi}{\sqrt{1-\zeta^2}}}$$

$$\zeta = \sqrt{\frac{1}{\left(\frac{\pi}{\ln M}\right)^2 + 1}}$$

It can be seen that M is only a function of ζ, and has nothing to do with the natural frequency ω_n of the system. It is inherent to the second-order linear system. M—ζ relationship of underdamped second-order device is shown in Figure 3.28. If M is measured in the measurement, the damping ratio ζ of the system can be found in the M—ζ curve.

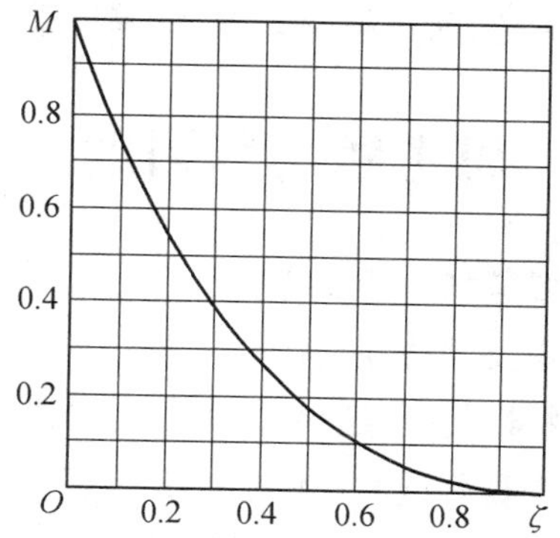

Figure 3.28　M—ζ relationship of underdamped second-order device

② Using series extreme points. If the transient process in the response is long and several peaks can appear, the time of occurrence of any two adjacent overshoots M_i and M_{i+1} will be t_{pi} and $t_{p(i+1)}$ respectively, and the relationship is

$$t_{p(i+1)} = t_{pi} + \frac{2\pi}{\omega_d} = t_{pi} + \frac{2\pi}{\omega_n\sqrt{1-\zeta^2}}$$

式中　δ——对数衰减率。

根据试验中所测得的 M_i 和 M_{i+1} 求出 δ 后，便可计算阻尼 ζ，即

$$\zeta=\frac{\delta}{\sqrt{\delta^2+4\pi^2}} \tag{3.57}$$

同理，当测得 M_i 和 M_{i+n}，求出 $\delta_n=\dfrac{2\pi n\zeta}{\sqrt{1-\zeta^2}}$后，有

$$\zeta=\frac{\delta_n}{\sqrt{\delta_n^2+4n^2\pi^2}} \tag{3.58}$$

若 $\zeta<0.3$，则$\sqrt{1-\zeta^2}\approx 1$，故上式可近似表示为

$$\zeta\approx\frac{\delta}{2\pi}$$

或

$$\zeta\approx\frac{\delta_n}{2\pi n}$$

若试验中能测得暂态过程（衰减振荡）的周期 T，根据

$$T=\frac{2\pi}{\omega_d}=\frac{2\pi}{\omega_n\sqrt{1-\zeta^2}}$$

可求出系统的固有频率，即

$$\omega_n=\frac{2\pi}{T\sqrt{1-\zeta^2}} \tag{3.59}$$

试验时，应根据系统可能产生的最大超调量来选择阶跃输入的幅值。超调量大时，幅值应小一些，并以它作为度量单位，所以超调量 M、M_i 等才是无量纲的。

思考题与练习题

3.1　一个理想的测试装置应具有什么特性？

3.2　线性系统中两个最重要的特性是什么？

3.3　描述测试装置动态特性的手段有哪几种？各自适应于什么样的场合？

3.4　测试装置可实现不失真测试的条件是什么？

3.5　在结构及工艺允许的条件下，为什么都希望将二阶测试装置的阻尼比定在 0.7 附近？

3.6　在选择不同的仪器时，如何在灵敏度、量程和稳定性方面进行考虑？

3.7　由传递函数为$H_1(s)=\dfrac{7}{1+5s}$和$H_2(s)=\dfrac{20\omega_n^2}{s^2+2\omega_n s+\omega_n^2}$的两个环节串联组成一个测试系统，该系统的总灵敏度是多少？

Substituting t_{pi} and $t_{p(i+1)}$ into formula(3.58), we get

$$M_i = e^{-\frac{2\pi i\zeta}{\sqrt{1-\zeta^2}}}$$

$$M_{i+1} = e^{-\frac{2\pi(i+1)\zeta}{\sqrt{1-\zeta^2}}}$$

$$\ln\frac{M_i}{M_{i+1}} = \delta = \frac{2\pi\zeta}{\sqrt{1-\zeta^2}}$$

where δ —the logarithmic decrement.

After calculating δ based on M_i and M_{i+1}, the damping ζ can be calculated, that is

$$\zeta = \frac{\delta}{\sqrt{\delta^2 + 4\pi^2}} \tag{3.58}$$

Similarly, when M_i and M_{i+n} are measured and $\delta_n = \frac{2\pi n\zeta}{\sqrt{1-\zeta^2}}$ is obtained, we have

$$\zeta = \frac{\delta_n}{\sqrt{\delta_n^2 + 4n^2\pi^2}} \tag{3.59}$$

If $\zeta < 0.3$, then $\sqrt{1-\zeta^2} \approx 1$, the above formula can be approximately expressed as

$$\zeta \approx \frac{\delta}{2\pi}$$

or

$$\zeta \approx \frac{\delta_n}{2\pi n}$$

If the period T of the transient process (damped oscillation) can be measured in the measurement, according to

$$T = \frac{2\pi}{\omega_d} = \frac{2\pi}{\omega_n\sqrt{1-\zeta^2}}$$

the natural frequency of the system can be calculated, that is

$$\omega_n = \frac{2\pi}{T\sqrt{1-\zeta^2}} \tag{3.60}$$

During the measurement, the amplitude of the step input should be selected according to the maximum overshoot that the system may produce. When the overshoot is larger, the amplitude should be smaller, and take it as the unit of measurement, so the overshoot M, M_i, etc. are dimensionless quantities.

Questions and Exercises

3.1　What should be the characteristics of an ideal measurement device?

3.2　What are the two most important characteristics of linear systems?

3.3　What are the methods for describing the dynamic characteristics of the measurement device? What kind of occasions do they adapt to?

3.4　What are the conditions for the measurement device to achieve non-distortion measurement?

3.5　Under the conditions allowed by the structure and process, why do we want to set the damping ratio of the second-order measurement device to be around 0.7?

3.6　How to consider sensitivity, range, and stability when choosing different instruments?

3.8 进行某次动态压力测量时，所采用的压电式力传感器的灵敏度为 80 pC/MPa，电荷放大器的灵敏度为 0.02 V/pC。若压力变化 20 MPa，为使记录笔在记录纸上的位移不大于 50 mm，则笔式记录仪的灵敏度应选多大？

3.9 图 3.29 所示为一测试系统的框图，试求该系统的总灵敏度。

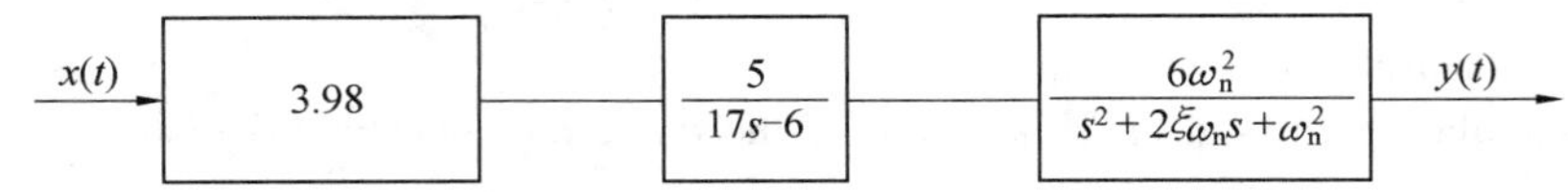

图 3.29 一测试系统的框图

3.10 用一个时间常数 $\tau=0.35$ s 的一阶测试系统测量周期分别为 5 s、1 s、2 s 的正弦信号，测量的幅值误差各是多少？

3.11 设用时间常数 $\tau=0.2$ s 的一阶装置测量正弦信号 $x(t)=\sin 4t+0.4\sin 40t$，试求其输出信号。

3.12 用一阶系统做 200 Hz 正弦信号的测量。如果要求振幅误差在 10% 以内，则时间常数应取多少？如用具有该时间常数的同一系统做 50 Hz 正弦信号的测试，则此时的振幅误差和相位差是多少？

3.13 将温度计从 20 ℃的空气中突然插入 100 ℃的水中。若温度计的时间常数 $\tau=2.5$ s，则 2 s 后的温度计指示值是多少？

3.14 用时间常数 $\tau=0.5$ s 的一阶装置进行测量，被测参数按正弦规律变化。若要求装置指示值的幅值误差小于 2%，则被测参数变化的最高频率是多少？如果被测参数的周期是 2 s 和 5 s，则幅值误差是多少？

3.15 求频率响应函数为 $\dfrac{3\ 155\ 072}{(1+0.01\mathrm{j}\omega)(1\ 577\ 536+176\mathrm{j}\omega-\omega^2)}$ 的系统对正弦输入 $x(t)=10\sin 62.8t$ 的稳态响应的均值显示。

3.16 一种力传感器可作为二阶系统处理。已知传感器的固有频率为 800 Hz，阻尼比为 0.14。当使用该传感器作频率为 500 Hz 和 1 000 Hz 正弦变化的外力测试时，其振幅和相位角各为多少？

3.17 求周期信号 $x(t)=0.6\cos(10t+30°)+0.3\cos(100t-60°)$ 通过传递函数为 $H(s)=1/(0.005s+1)$ 的测试装置后所得到的稳态响应。

3.18 在某发动机处于稳定工况下，对输出转矩进行了 10 次测量，得到如下测定值：14.3 N·m，14.3 N·m，14.5 N·m，14.3 N·m，13.8 N·m，14.0 N·m，14.4 N·m，14.5 N·m，14.3 N·m，14.0 N·m。试表达测量结果。

3.7　What is the total sensitivity of the system when two links with transfer functions $H_1(s)=\dfrac{7}{1+5s}$ and $H_2(s)=\dfrac{20\,\omega_n^2}{s^2+2\,\omega_n s+\omega_n^2}$ are connected in series to form a measurement system?

3.8　When we performe a certain dynamic pressure measurement, the sensitivity of the used piezoelectric force sensor is 80 pC/MPa, and the sensitivity of the charge amplifier is 0.02 V/pC. If the pressure changes by 20 MPa, how high should the sensitivity of the pen recorder be in order to keep the displacement of the recording pen on the recording paper no more than 50 mm?

3.9　Figure 3.29 shows a block diagram of a measurement system, try to find the total sensitivity of the system.

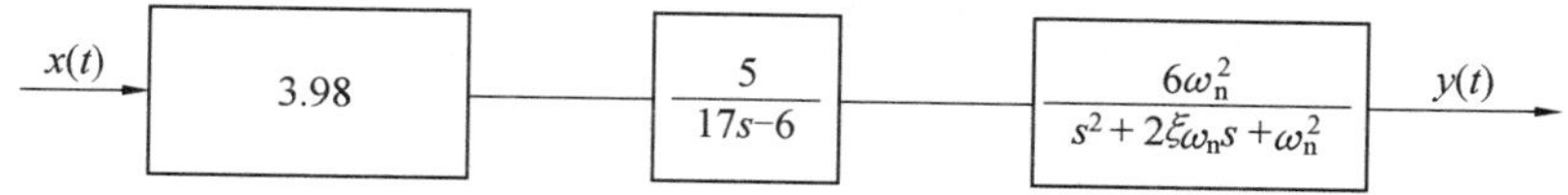

Figure 3.29　A block diagram of a measwement system

3.10　Use a first-order measurement system with a time constant $\tau = 0.35$ s to measure sine signals with periods of 5 s, 1 s, and 2 s, what is the amplitude error of the measurement?

3.11　Suppose a first-order device with time constant $\tau = 0.2$ s is used to measure a sine signal $x(t) = \sin 4t + 0.4\sin 40t$, and try to find its output signal.

3.12　Use a first-order system to measure a 200 Hz sine signal. If the amplitude error is required to be within 10% , what is the time constant? If we use the same system with this time constant to test a 50 Hz sine signal, what is the amplitude error and phase difference at this time?

3.13　The thermometer is suddenly inserted into the water at 100 ℃ from the air at 20 ℃. If the time constant of the thermometer is $\tau = 2.5$ s , what is the indicating value of the thermometer after 2 s?

3.14　We can use a first-order device with time constant $\tau=0.5$ s to measure, and the measured parameter changes according to the sine law. If the amplitude error of the device indication value is required to be less than 2% , what is the highest frequency of the measured parameter change? If the period of the measured parameter is 2 s and 5 s, what is the amplitude error?

3.15　Find the average value display of the steady-state response of the system with the frequency response function $\dfrac{3\,155\,072}{(1+0.01\mathrm{j}\omega)(1\,577\,536+176\mathrm{j}\omega-\omega^2)}$ to the sinusoidal input $x(t) = 10\sin 62.8t$.

3.16　A kind of force sensor can be treated as a second-order system. It is known that the natural frequency of the sensor is 800 Hz, and the damping ratio is 0.14. When using this sensor to test the external force with a sinusoidal frequency of 500 Hz and 1 000 Hz, what is the amplitude and phase angle?

3.17　Find the steady-state response obtained after the periodical signal $x(t) = 0.6\cos(10t + 30°) + 0.3\cos(100t - 60°)$ passes through the measurement device whose transfer function is $H(s) = 1/(0.005s + 1)$.

3.18　When an engine is in a stable working condition, the output torque is measured 10 times, and the following measured values are obtained: 14.3 N · m, 14.3 N · m, 14.5 N · m, 14.3 N · m, 13.8 N · m, 14.0 N · m, 14.4 N · m, 14.5 N · m, 14.3 N · m, 14.0 N · m. Try to express the measurement results.

第4章　传　感　器

【本章学习目标】

1. 掌握传感器的类型、作用；

2. 了解不同种类传感器的原理、特点、应用领域；

3. 掌握选用传感器的原则，能够根据被测量的特点来选用适合的传感器；

4. 了解常用的信号调理电路。

Chapter 4　Sensors

【Learning Objectives】

1. To be able to master the types and functions of sensors;

2. To be able to understand the principles, characteristics and application fields of different types of sensors;

3. To be able to master the principles to select sensors, and to select suitable sensors in accordance with the characteristics of the measured quantity;

4. To be able to understand frequently-used signal conditioning circuits.

4.1 概　　述

在信息时代的当今世界，现代信息技术的三大基础是传感器技术、通信技术和计算机技术。可以理解为传感器（感官）用来获取信息，通过通信技术（神经）进行传导，由计算机（大脑）处理并发出指令，这三者构成了信息技术系统。

信息处理首要的任务就是准确地获取信息。传感器作为测试系统的第一环节，是准确获取信息的首要环节，它直接作用于被测量，其作用一般是将非电被测量转为电量，从而得到非电量的量值。传感器是测量仪器与被测量之间的接口，处于测量装置的输入端，其性能直接影响着整个测量系统，对测量精确度起着决定性的作用。传感器可以认为是人类感官的延伸，借助传感器可以去测控那些人们无法用感官直接识别的事物。因此，可以说传感器是人们认识自然界事物的有力工具。

传感器在工程中应用十分普遍。例如，为实现对机械加工过程中零部件的生产加工、质量检测与控制，需要应用到多种传感器，如速度传感器、加速度传感器、力传感器、温度传感器、接近开关传感器、尺寸传感器、形状传感器等。机械手、机器人等设备为获得手臂末端的位置、姿态和受力，以及检测作业对象与作业环境的状态，应用了位移传感器、旋转/移动位置传感器、触觉传感器、力传感器、听觉传感器、视觉传感器、热觉传感器等。

在现代工业自动化生产过程中，需要传感器来监控生产过程中的各个参数，使得设备稳定工作在正常状态，从而保证产品统一良好的质量。没有传感器技术的支持，现代化生产也就无从谈起。在生产过程中，智能化程度和自动化程度越高，传感器的需求量就越大。

作为测量、控制与信息技术等领域的一个重要构成因素，传感器被视为当今科学技术发展的关键性因素之一，在测试技术与机械工程测量领域中有极其重要的作用。深入研究传感器原理和应用、研制开发新型传感器对于科学技术和生产过程中的自动控制和智能化发展，以及人类观测研究自然界的深度和广度都有重要的实际意义。

未来社会将是充满传感器的世界。传感器的水平是衡量一个国家综合经济实力和技术水平的标志之一，它的发展水平、生产能力和应用领域已成为一个国家科学技术进步的重要标志。可以说，谁支配了传感器，谁就支配了新时代。

4.1 Overview

In the information age of today ' s world, the three foundations of modern information technology are sensor technology, communication technology and computing technology. Therefore, information is acquired by sensors (sensory) , conducted through communication technology (meridian) , and processed and issued by computers (brain) , all which three parts constitute the information technology system.

The first and foremost step to make good use of information is how to accurately acquire information. As the first link of a measurement system, the sensor is the primary link for obtaining accurate information, acting directly on the measured quantity, whose function is generally to convert the non-electric measured quantity into electricity, thereby measuring the value of the non-electric. Being the interface between the measuring instrument device and the measured quantity, the sensor, whose performance directly affects the entire measuring system and plays a decisive role in the accuracy of the measurement, is at the input end of the measuring device. With the help of sensors considered as an extension of the human senses, we can measure and control things that people cannot directly recognize with the senses. Thus, sensors are recognized to be the powerful tools for people to understand things in nature.

Sensor applications are very common in engineering. For example, a variety of sensors, such as speed sensors, acceleration sensors, force sensors, temperature sensors, proximity switch sensors, size sensors, and shape sensors, etc. are applied to realize the production, and processing, quality inspection and control of parts in the machining process. To obtain the position, posture and force of the end of the arm, as well as to detect the status of the manipulating object and the working environment, equipment such as manipulators and robots are applied by displacement sensors, rotation/moving position sensors, tactile sensors, force sensors, auditory sensors, vision sensors and thermal sensor, etc.

In the modern industrial automated production process, sensors are needed to monitor various parameters in the production process, and to keep the equipment in a normal state in order to ensure the uniform and good quality of the products. None of modern production can happen without the support of sensor technology. In the production process, the higher the degree of intelligence and automation is, the greater the demand for sensors is.

Sensors, as an important constituent factor in the fields of measurement, control and information technology, are regarded as one of the key factors in the development of science and technology today and play an extremely important role in the field of measurement technology and mechanical engineering measurement. In-depth study of sensor principles and applications, and research and development of new sensors have important practical significance for the development of automatic control and intelligence in the process of science and technology and production, as well as the depth and breadth of human observation and research in the natural world.

The future society will be a world full of sensors. The level of the sensors is one of

4.1.1 传感器的定义与作用

传感器的概念来自“感觉(sense)”一词。生物能够获取外界信息,是因为具有相应的感觉器官。在研究和工程中,能补充或代替人类的感觉功能的传感器是必不可少的。

根据国家标准《传感器通用术语》(GB/T 7665—2005),将传感器(transducer/sensor)定义为“能感受被测量,并按照一定的规律转换成可用输出信号的器件或装置,通常由敏感元件和转换元件组成”。

传感器的输出信号为“可用输出信号”,“可用”是指便于传输、转换、处理、显示的信号,最常见的是电信号和光信号。一般来说,由于电信号是易于传输、检测和处理的物理量,因此有时也把将非电量转换成电量的器件或装置称为传感器。

传感器有两个作用:一是感应作用,即感受并拾取被测对象的信号;二是转换作用(又称变换作用),将被测信号转换成易于测量和传输的电信号(如电压、电阻、电流、电容、电感等),以便后续仪器接收和处理。当然,不是所有的传感器都有感应元件和变换元件之分,有些传感器中二者合为一体。

4.1.2 传感器的组成

传感器的组成按其定义由三部分组成:感应元件、变换元件和信号调理电路。一般还需外加辅助电源提供转换能量。传感器组成框图如图4.1所示。

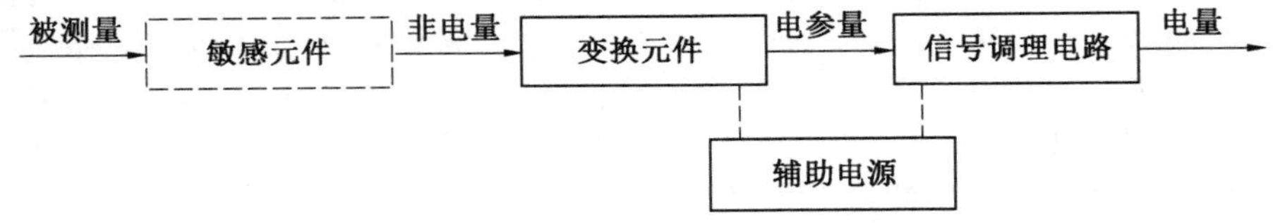

图4.1 传感器组成框图

敏感元件直接感受被测量(一般为非电量),并输出与被测量成确定关系的某一物理量的元件。在机械量(如力、压力、速度、位移等)测量中,常采用弹性元件作为感应元件,这种弹性元件又称弹性感应元件,它能够把被测量由一种物理状态转变为所需要的另一种物理量。

图4.2所示为一种气体压力传感器示意图。膜盒的下半部与壳体固接,上半部通过连杆与磁芯相连,磁芯置于两个电感线圈中,后者接入转换电路。膜盒是感应元件,其外部与大气压力相通,内部感受被测压力。当被测压力变化时,会引起膜盒上半部移动,即输出相应的位移量。

indicators to measure the comprehensive economic strength and technical level of a country. Its development level, production capacity and application fields have become an important indicator of a country's scientific and technological progress. It can be said that whoever dominates the sensor will dominate the new era.

4.1.1 The Definition and Function of Sensors

The concept of sensors comes from the term "sense". Living creatures can obtain information from the outside world because they have corresponding sensory organs. In research and engineering, sensors that can supplement or replace human sensory functions are essential.

According to the national standard "*General Terminology for Transducer*" (GB/T 7665-2005), a transducer/sensor is "a instrument or a device that senses a prescribed amount of measurement and converted it into usable output signals according to certain rule. It is usually composed of sensitive components and conversion components".

The output signal of the sensor is "available output signal". "Available" refers to a signal that is convenient for transmission, conversion, processing, and display. The most common ones are electrical signals and optical signals. Generally speaking, instruments or devices that can convert non-electricity into electricity are sometimes called sensors because electrical signals belong to physical quantities that are easy to transmit, detect, and process.

Sensors have two functions: one is the induction function, which is to sense and pick up the signal of the measured object; the other is the conversion function (or called the transformation function), which can convert the measured signal into an electrical signal that is easy to measure and transmit (such as voltage, resistance, current, capacitance, inductance) for the subsequent instrument to receive and process. Of course, not all sensors have sensing elements and transform elements, and some sensors combine the two into one.

4.1.2 The Composition of Sensors

According to its definition, the composition of sensors consists of three parts: sensing element, transform element, and signal conditioning circuit. Generally, an auxiliary power supply is required to provide conversion energy. Sensor composition block diagram is shown in Figure 4.1.

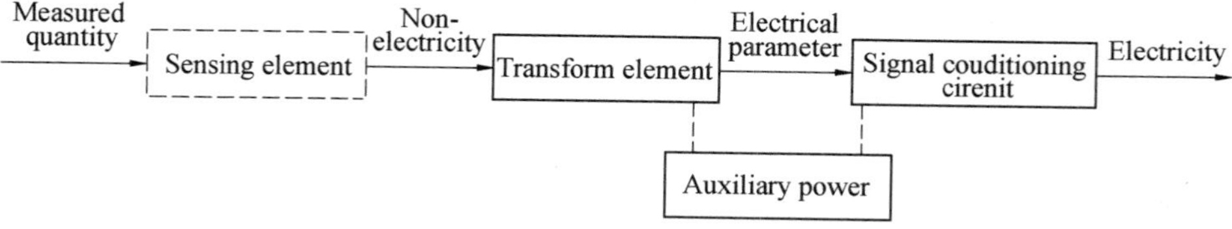

Figure 4.1 Sensor composition block diagram

The sensing element directly senses the measured quantity (generally non-electricity), and outputs a certain physical quantity that has a certain relationship with the measured quantity. In the measurement of mechanical quantities (such as force, pressure, speed, displacement), elastic elements are often used as sensing elements, which are also called elastic sensing elements, being able to transform the measured from one physical state into another physical quantity required.

敏感元件的输出就是变换元件的输入，变换元件把输入转换成电参量。在图4.2中，转换元件是可变电感线圈，它把输入的位移量转换成电感的变化。

电参量被输入到信号调理电路，进行放大、运算等处理，转换成易于进一步传输和处理的形式，从而获得被测值。

通常敏感元件与变换元件在结构上安装在一起，有些传感器是将二者合二为一，没有敏感元件、变换元件之分，甚至还有些新型的传感器将敏感元件、变换元件及信号调理电路集成为一个器件。

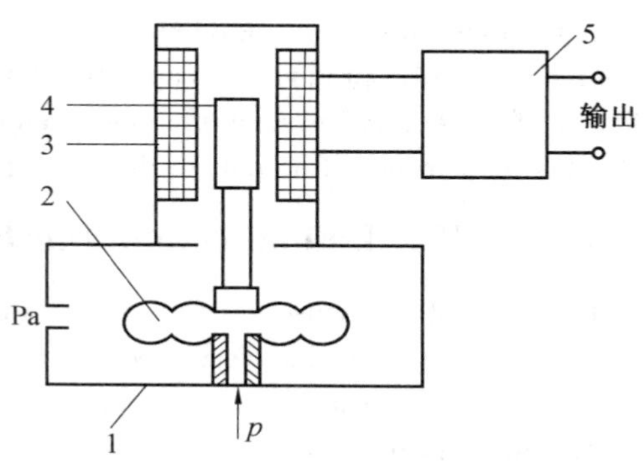

图4.2　气体压力传感器示意图

1—壳体；2—膜盒；3—电感线圈；4—磁芯；5—转换电路

4.1.3　传感器的分类

为便于研究、开发和选用，必须对传感器进行科学分类。工程中常用传感器的种类繁多，往往同一种被测量可以用不同类型的传感器来测量，而同一原理的传感器又可测量多种物理量。因此，传感器有许多种分类方法，概括起来，主要有下面几种分类方法。

1. 按变换原理分类

根据传感器对信息获取的不同变换原理，可将传感器分为电阻式传感器、电感式传感器、电容式传感器、压电式传感器、光电式传感器、磁电式传感器等。这种分类方法是从原理上认识输入与输出之间的变换关系。常见的按变换原理分类类型有电阻、电感、电容、磁电、热电、压电、光电（包括红外、光导纤维）、谐振、霍尔、超声、同位素、电化学、微波等。

2. 按被测物理量分类

工程应用中按被测量分类是最常见的分类方式，如力、速度、加速度、位移、温度、流量等物理量所用的传感器分别称为力传感器、速度传感器、加速度传感器、位移传感器、温度传感器、流量传感器等。

这种分类方法按用途进行分类，便于使用者根据测量对象来选型。需要注意的是，用途相同的传感器，其变换原理却不尽相同。例如，同是测量加速度用的传感器有电容式、应变式、压电式和电感式加速度传感器等。

3. 按信号变换特征分类

按信号变换特征的不同，传感器可分为两种：结构型和物性型。

Figure 4.2 shows a schematic diagram of a kind of gas pressure sensor. The lower half of the diaphragm capsule is fixedly connected to the shell, and the upper half is connected to the magnetic core through a connecting rod. The magnetic core is placed in two inductance coils, and the latter is connected to the conversion circuit. The diaphragm capsule is a sensing element, of which the outside is connected to atmospheric pressure, and the inside feels the measured pressure. When the measured pressure changes, it will cause the upper half of the diaphragm capsule to move, that is, to output the corresponding displacement.

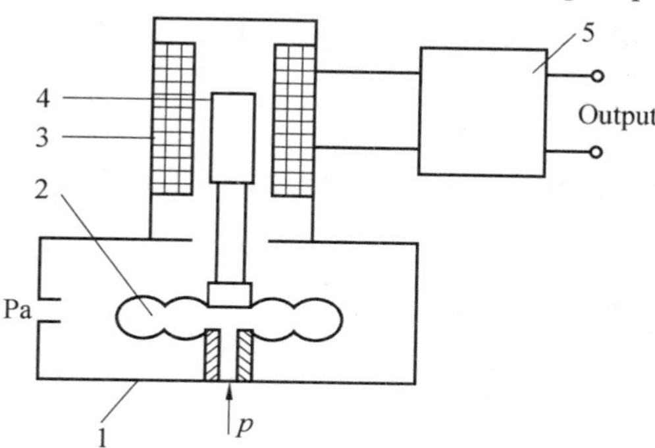

Figure 4.2 Schematic diagram of a gas pressure sensor
1—shell;2—diaphragm;3—inductance coil;4—magnetic core;5—conversion circuit

The output of sensing elements is the input of the transform element, which will convert the input into an electrical parameter. In Figure 4.2, the transducing element is a variable inductance coil, which will convert the input displacement into a change in inductance.

The electrical parameters are input to the signal conditioning circuit for amplifying, arithmetic and other processing, and converted into a form easy for further transmission and processing so as to obtain the measured value.

The sensing element and the transform element are usually installed together in structure. Some sensors combining the two into one have no distinction between the sensing element and the transform element. There are even some new types of sensors that integrate the sensing element, the transform element and the signal conditioning circuit to a device.

4.1.3 The Classification of Sensors

Sensors should be classified scientifically in order to facilitate research, development and selection. There are many types of sensors commonly used in engineering with the situation that often the same type of measurement can be measured with different types of sensors, and sensors of the same principle can measure a variety of physical quantities. Therefore, there are many classification methods for sensors, in summary, there are mainly the following classification.

1. Classification by Transformation Principles

In the light of the different transformation principles of information acquisition by sensors, sensors can be divided into resistive sensors, inductive sensors, capacitive sensors, piezoelectric sensors, photoelectric sensors, magnetoelectric sensors, etc. This classification method is to understand the transformation relationship between input and output in principle. The common classification types according to the transformation principle are resistance,

(1)结构型传感器。传感器利用自身结构(尺寸、形状、位置等)的变化,将外界被测参数变换为相应的电感、电阻、电容等物理量的变化。例如,电感式传感器由衔铁的位移引起自感或互感的变化;电容式传感器通过极板间距离发生变化而引起电容量的改变等。

(2)物性型传感器。利用感应器件材料本身物理特性的变化来实现信号的检测。例如,用水银温度计测温,是利用了水银的热胀冷缩现象;用光电传感器测速,是利用了光电器件本身的光电效应;压电式传感器是利用石英晶体的压电效应实现测量等。

4. 按能量传递方式分类

传感器是一种能量转换和传递的器件,按能量传递方式可将传感器分为两种:能量控制型和能量转换型。

(1)能量控制型传感器。需要从外部提供能量使传感器工作,由被测量的变化来控制外部能量的变化(图 4.3)。使用这种传感器时,必须加上外部电源,才能完成将上述电参量进一步转换成电量的过程,因此又称有源传感器。

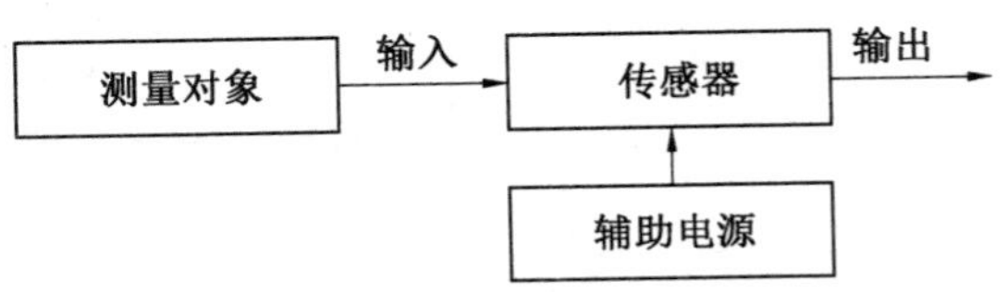

图 4.3　能量控制型传感器工作方式

电阻式、电容式、电感式传感器都属于能量控制型传感器。在感应到被测量后,它只改变自身的电参量(如电阻、电容、电感等),这类传感器本身不起变换能量的作用,但能对传感器提供的能量起控制作用。例如,电阻应变片接于电桥上,电桥工作能源由外部供给,而由被测量变化所引起的电阻变化去控制电桥输出电压。

(2)能量转换型传感器。这类传感器具有换能功能,它能将被测物理量直接转换成电量输出,无须外加电源,因此此类传感器又称无源传感器。例如,热电偶将被测温度直接转换为电量输出。光电池、压电式、磁电式、热电式等传感器均属能量转换型传感器。

5. 按输出信号方式分类

按被输出信号分类,可分为模拟式和数字式。模拟式传感器的输出为模拟量,数字式传感器的输出为数字量。

4.1.4　传感器采集信号的放大

传感器输出的信号往往很微弱,若不经放大,即使能够妥善地传输到终端,也很难驱动指示器、记录器或各种控制机构。由于传感器输出的信号大小和信号形式各不相同,传感器所处的环境条件和噪声对传感器的影响也不一样,因此所采用的放大电路的性能指

inductance, capacitance, magnetoelectricity, pyroelectricity, piezoelectricity, photoelectricity (including infrared, optical fiber), resonance, Hall, ultrasound, isotope, electrochemistry, microwave, etc.

2. Classification by the Measured Quantity

In engineering applications, classification by the measured quantity is the most common classification method. Sensors used for physical quantities such as force, speed, acceleration, displacement, temperature, flow, etc. are called force sensors, speed sensors, acceleration sensors, displacement sensors, temperature sensors, flow sensors, etc. respectively.

This method is classified according to purpose, convenient for users to select according to the measurement object. It should be noted that sensors with the same purpose have different conversion principles. For example, the same sensors used to measure acceleration include capacitive, strain-type, piezoelectric and inductive acceleration sensors.

3. Classification by Signal Transformation Characteristics

According to the difference in signal transformation characteristics, sensors are divided into two types: structural type and physical property type.

(1) Structure type sensor. The sensor uses changes in its own structure (size, shape, position, etc.) to transform the external measured parameters into corresponding changes in physical quantities such as inductance, resistance, and capacitance. For example, inductive sensors cause changes in self-inductance or mutual inductance by the displacement of the armature; capacitive sensors cause changes in capacitance through the changes in the distance between the plates, etc.

(2) Physical property type sensor. It uses changes in the physical properties of the sensing device material to achieve signal detection. For example, the principle of mercury thermometer used to measure temperature is to utilize the thermal expansion and contraction phenomenon of mercury; the principle of photoelectric sensor used to measure the speed is to use the photoelectric effect of the photoelectric device itself; and piezoelectric sensors use the piezoelectric effect of quartz crystals to achieve measurement, etc.

4. Classification by energy transfer method

The sensor is a device for energy conversion and transmission, according to the energy transmission mode, it can be divided into two types: energy control type and energy conversion type.

(1) Energy control type sensors. It is necessary to provide energy from the outside to make the sensor work, and the change in the external energy is controlled by the measured quantity change (Figure 4.3). When using this kind of sensor, an external power supply must be added to complete the process of further converting the above-mentioned electrical parameters into electricity, so it is also called an active sensor.

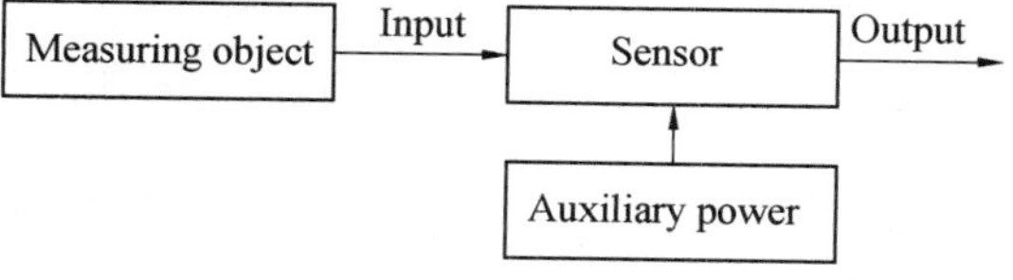

Figure 4.3 Working mode of energy control sensors

标和形式也不相同,这使得放大电路的种类多种多样。放大电路目前一般采用集成放大器构建,常用的测量装置有通用集成运算放大器、仪器放大器、斩波放大器、可编程增益放大器、互导放大器、电荷放大器、互阻抗放大器和载波放大器等。本节主要介绍仪器放大器的结构和特点,对放大器的讨论主要针对放大器的使用者。

1. 基本放大器

由集成运算放大器构成的放大电路包括三种基本形式:反相放大器、同相放大器和差动放大器。这些都是对电压放大的基本电路,如图4.4所示。

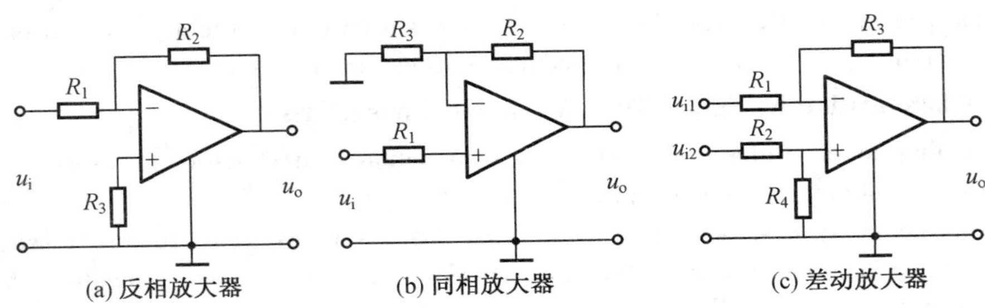

图4.4 电压放大的基本电路

反相放大电路性能稳定,但是输入阻抗(等于 R_1)较低,容易对传感器形成负载效应;同相放大电路输入阻抗高(理想情况下为无穷大),但易引入共模干扰;差动放大电路的输入阻抗(差模输入阻抗等于 R_1+R_2)也很难提高,而且电阻匹配困难。上述这些缺陷限制了这些基本放大电路在测量装置中的直接应用。

2. 仪器放大器

仪器放大器是一种能同时提供高输入阻抗和高共模抑制比的差动放大器。它具有以下特点:可以用单一电阻器调节增益、增益稳定、失调电压和失调电流低、漂移小、输出阻抗低。

仪器放大器可以由两个运算放大器构成,也可以由三个运算放大器构成。图4.5所示电路为基于三个运算放大器的仪器放大器。运算放大器 A_1 和 A_2 采用同相输入构成同相放大器,A_1 和 A_2 一起构成差动输入级。运算放大器 A_3 采用差动输入构成单端输出级。

输入级采用同相输入可以大幅提高电路的输入阻抗(差动输入电阻和共模输入电阻的理想值为无穷大,而实际值容易达到 $10^9\ \Omega$),减小电路对微弱输入信号的衰减(负载效应)。差动输入可以使电路只对差模信号放大,而对共模输入信号只起跟随作用,使得送到输出级输入端的差模信号与共模信号的幅值之比(即共模抑制比 CMRR)得到提高。

Resistive, capacitive, and inductive sensors are all energy control sensors. After sensing the measured quantity, it only changes its own electrical parameters (such as resistance, capacitance, inductance), which this type of sensor itself does not have the function of transforming energy, but it has controlling influence on the energy provided by the sensor. For example, the resistance strain gauges can be connected to the bridge, of which the working energy is supplied from the outside, and the resistance change caused by the measured quantity change controls the bridge's output voltage.

(2) Energy conversion sensor. With energy conversion function, energy conversion sensor can directly convert the measured physical quantity into electrical output without external power supply, so this type of sensor is also called a passive sensor. For example, a thermocouple directly converts the measured temperature into electrical output. Photocell, piezoelectric, magnetoelectric, pyroelectric and other sensors are all energy conversion sensors.

5. Classification by Output Signal Mode

It can be divided into analog type and digital type according to the output signal classification. The output of the analog sensor is an analog quantity, and the output of the digital sensor is a digital quantity.

4.1.4 Amplification of Sensor Acquisition Signals

With the signal output by the sensor often tending to be very weak, it is difficult to drive the indicator, recorder or various control mechanisms without amplification even if it can be properly transmitted to the terminal. Since the size and form of the signal output by the sensor are different, the environmental conditions of the sensor and the influence of noise on the sensor are also different, so the same as to the performance indicators and forms of the amplifying circuit used, leading to a wide variety of amplifying circuits. Amplification circuits are generally constructed with integrated amplifiers. Commonly used measuring devices include general integrated operational amplifiers, instrument amplifiers, chopper amplifiers, programmable gain amplifiers, capacitor feedback transimpedance amplifier, charge amplifiers, transimpedance amplifiers and carrier amplifiers, etc. This section mainly introduces the structure and characteristics of instrument amplifiers, and the discussion of amplifiers is mainly aimed at amplifier users.

1. Basic Amplifier

The amplifying circuit formed by the integrated operational amplifier includes three basic forms: inverting amplifier, non-inverting amplifier and differential amplifier. These are the basic circuits for voltage amplification, as shown in Figure 4.4.

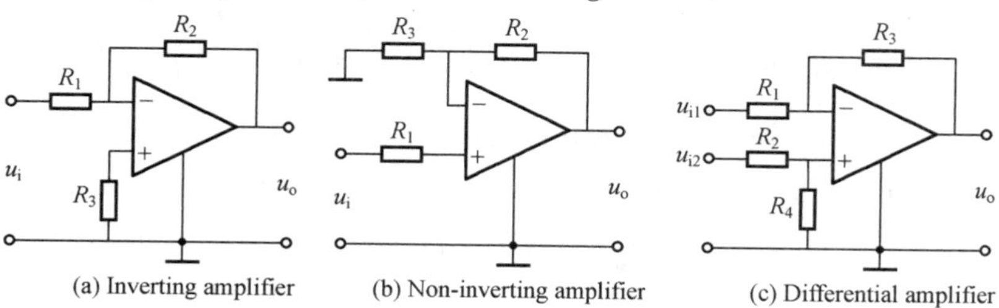

Figure 4.4 The basic circuit of voltage amplification

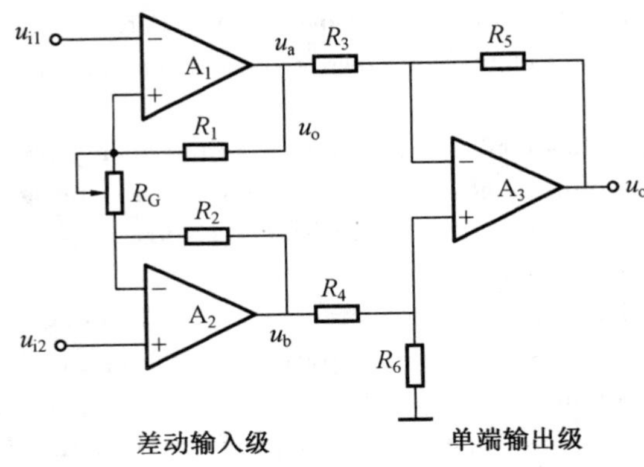

图 4.5　基于三个运算放大器的仪器放大器

这样，在以运算放大器 A_3 为核心部件组成的差动放大电路中，在 CMRR 要求不变的情况下，可明显降低对电阻 R_3 和 R_4、R_3 和 R_6 的精度匹配要求，从而使仪器放大器电差动输入级电路比简单的差动放大电路具有更好的共模抑制单端输出级能力。

设 A_1、A_2、A_3 都是理想运算放大器，则输入端无电流通过，所以 A_1 和 A_2 反相输入端的电压等于同相输入端的电压，并且流过电阻 R_1、R_G、R_2 的电流相等，则有

$$\frac{u_a - u_{i1}}{R_1} = \frac{u_{i1} - u_{i2}}{R_G} = \frac{u_{i2} - u_b}{R_2} \tag{4.1}$$

由式(4.1) 解得

$$\begin{cases} u_a = u_{i1} - \dfrac{R_1}{R_G}(u_{i2} - u_{i1}) \\ u_b = u_{i2} + \dfrac{R_2}{R_G}(u_{i2} - u_{i1}) \end{cases} \tag{4.2}$$

由式(4.2) 可知，如果 u_{i1} 和 u_{i2} 为共模输入电压 u_G，即 $u_{i1} = u_{i2} = u_c$，则 $u_a = u_b = u_c$。这说明，这种仪器放大器差动输入级对共模输入信号没有放大作用，只是跟随（即增益为 1）。

单端输出级的输出电压为

$$u_o = \frac{R_6}{R_4}\frac{R_5/R_3 + 1}{R_6/R_4 + 1}u_{i2} - \frac{R_5}{R_3}u_{i1} + \left(\frac{R_6}{R_4}\frac{R_5/R_3 + 1}{R_6/R_4 + 1}\frac{R_2}{R_G} + \frac{R_5}{R_3}\frac{R_1}{R_G}\right)(u_{i2} - u_{i1}) \tag{4.3}$$

由式(4.3) 可知，如果 u_{i1} 和 u_{i2} 为共模电压，即 $u_{i1} = u_{i2} = u_c$，则为消除这一共模输入电压，应该有 $u_o = 0$。令 $u_o = 0$，解得 $R_6/R_4 = R_5/R_3$。将此结果代入式(4.3)，可得

$$u_o = \frac{R_5}{R_3}\left(1 + \frac{R_1 + R_2}{R_G}\right)(u_{i2} - u_{i1}) \tag{4.4}$$

为减少电阻值的种类，一般取 $R_1 = R_2$，$R_3 = R_4 = R_5 = R_6$，则有

The performance of the inverting amplifier circuit is stable, but the input impedance (equal to R_1) is relatively low, which is easy to form a load effect on the sensor; the input impedance of the non-inverting amplifier circuit is high (ideally infinite), but it is easy to introduce common mode interference; the differential amplifier circuit input impedance (differential mode input impedance equal to R_1+R_2) is also hard to improve, and resistance matching is difficult. The above-mentioned shortcomings limit the direct application of these basic amplifying circuits in measuring devices.

2. Instrumentation Amplifier

Instrumentation amplifier is a differential amplifier that can provide high input impedance and high common-mode rejection ratio at the same time. It has the following characteristics: a single resistor being used to adjust the gain, the stable gain, the low offset voltage and offset current, the small drift, and the low output impedance.

The instrumentation amplifier can be composed of two operational amplifiers or three operational amplifiers. The circuit shown in Figure 4.5 is a instrumentation amplifier based on three operational amplifiers. Operational amplifiers A_1 and A_2 use non-inverting input to form a non-inverting amplifier, and A_1 and A_2 together form a differential input stage. Operational amplifier A_3 uses differential input to form a single-ended output stage.

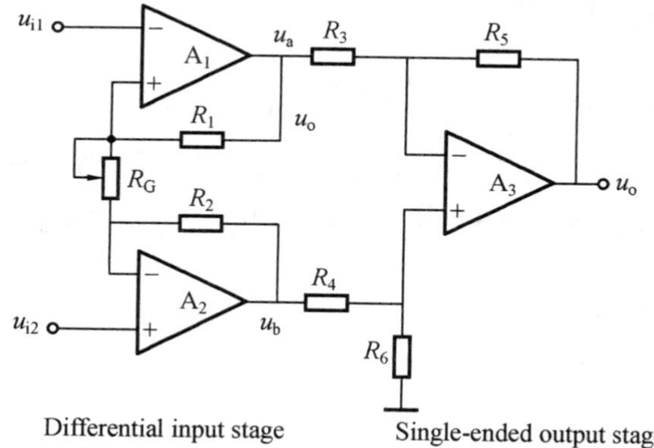

Figure 4.5 Instrumentation amplifier based on three operational amplifiers

The input stage adopts non-inverting input to greatly increase the input impedance of the circuit (the ideal value of differential input resistance and common mode input resistance is infinite, but the actual value is easy to reach $10^9\ \Omega$), and reduces the attenuation of the weak input signal (load effect). Differential input can make the circuit only amplify the differential mode signal, and only follow the common mode input signal, so that the ratio of the amplitude of the differential mode signal to the common mode signal sent to the input of the output stage (that is, the common mode rejection ratio CMRR) be improved. In this way, in the differential amplifier circuit composed of operational amplifier A_3 as the core component, under the condition that the CMRR requirements remain unchanged, the accuracy matching requirements for resistors R_3 and R_4, R_3 and R_6 can be significantly reduced, so that the electric defferential input stage circuit of instrument amplifier has better capability of common-mode rejection single-ended output stage than simple differential amplifier circuit.

Assuming that A_1, A_2, A_3 are all ideal operational amplifiers, there is no current through the input terminals, so the voltage at the inverting input terminals of A_1 and A_2 is equal to the

$$u_o = \left(1 + \frac{2R_1}{R_G}\right)(u_{i2} - u_{i1}) = G(u_{i2} - u_{i1}) \tag{4.5}$$

该仪器放大器的差模增益为

$$G = \frac{u_o}{u_{i2} - u_{i1}} = 1 + \frac{2R_1}{R_G} \tag{4.6}$$

由式(4.6)可见，通过改变 R_G 的值即可以调节放大器的差模增益，而共模抑制能力不受影响。

目前，市场上有很多公司生产的单片集成仪器放大器可用，如美国 ANALOG DEVICES 公司生产的 AD8221、AD8230 和美国 BURR – BROWN 公司推出的 INA114 等。

4.2 电阻式传感器

电阻式传感器能把被测量转换为电阻变化。导体中的电阻 R 与其电阻率 ρ 及长度 l 成正比，与截面积 A 成反比，即

$$R = \rho \frac{l}{A} \tag{4.7}$$

被测物理量引起 ρ、l、A 中任一个或几个量的变化，都可使电阻 R 改变。按其工作原理，电阻式传感器可分为变阻器式（电位器式）和电阻应变式传感器。

4.2.1 变阻器式传感器

1. 工作原理

变阻器式传感器又称电位器，其工作原理是将物体的位移转换为电阻的变化，即通过移动滑片与电阻线的接触点来改变接入电路中电阻线的长度，从而改变电阻。

2. 类型

根据 $R = \rho l/A$，变阻器式传感器可分为三种：线性位移型、非线性位移型和角位移型。三种变阻器式传感器如图 4.6 所示。

(1) 线性位移型（图 4.6(a)）。改变触点 C 的位置时，AC 间电阻值为 $R = k_L x$，k_L 为单位长度内的电阻值（Ω/m），灵敏度 $S = dR/dx = k_L$ = 常数。当导线分布均匀时，传感器的输出（电阻）与输入（线位移）呈线性关系。

(2) 非线性位移型（图 4.6(b)）。又称函数电位器，其骨架形状根据要求的输出 $f(x)$ 来决定。

(3) 角位移型（图 4.6(c)）。当导线分布均匀时，传感器的输入（角位移）与输出（电

voltage at the non-inverting input terminal, and the currents flowing through the resistors R_1, R_G, R_2 are equal, so

$$\frac{u_a - u_{i1}}{R_1} = \frac{u_{i1} - u_{i2}}{R_G} = \frac{u_{i2} - u_b}{R_2} \tag{4.1}$$

We can solve from the formula(4.1) that

$$\begin{cases} u_a = u_{i1} - \dfrac{R_1}{R_G}(u_{i2} - u_{i1}) \\ u_b = u_{i2} + \dfrac{R_2}{R_G}(u_{i2} - u_{i1}) \end{cases} \tag{4.2}$$

From formula (4.2), we can see that if u_{i1} and u_{i2} are the common mode input voltage u_G, that is, $u_{i1} = u_{i2} = u_c$, then $u_a = u_b = u_c$. This shows that the differential input stage of this instrument amplifier does not amplify the common-mode input signal, but only follows (that is, the gain is 1).

The output voltage of the single-ended output stage is

$$u_o = \frac{R_6}{R_4}\frac{R_5/R_3 + 1}{R_6/R_4 + 1}u_{i2} - \frac{R_5}{R_3}u_{i1} + \left(\frac{R_6}{R_4}\frac{R_5/R_3 + 1}{R_6/R_4 + 1}\frac{R_2}{R_G} + \frac{R_5}{R_3}\frac{R_1}{R_G}\right)(u_{i2} - u_{i1}) \tag{4.3}$$

From formula (4.3), we can see that if u_{i1} and u_{i2} are the common mode input voltage, that is, $u_{i1} = u_{i2} = u_c$, then in order to eliminate this common-mode input voltage, there should be $u_o = 0$. Let $u_o = 0$, we can get $R_6/R_4 = R_5/R_3$. Substitute this result into formula (4.3) to get

$$u_o = \frac{R_5}{R_3}\left(1 + \frac{R_1 + R_2}{R_G}\right)(u_{i2} - u_{i1}) \tag{4.4}$$

In order to reduce the type of resistance value, generally take $R_1 = R_2$, $R_3 = R_4 = R_5 = R_6$, then there are

$$u_o = \left(1 + \frac{2R_1}{R_G}\right)(u_{i2} - u_{i1}) = G(u_{i2} - u_{i1}) \tag{4.5}$$

The differential mode gain of the instrument amplifier is

$$G = \frac{u_o}{u_{i2} - u_{i1}} = 1 + \frac{2R_1}{R_G} \tag{4.6}$$

It can be seen from formula (4.6) that the differential mode gain of the amplifier can be adjusted by changing the value of R_G, and the common mode rejection capability is not affected.

At present, there are many monolithic integrated instrument amplifiers produced by companies on the market, such as the AD8221, AD8230 produced by ANALOG DEVICES in the United States, and the INA114 introduced by BURR-BROWN in the United States.

4.2 Resistive Sensors

Resistive sensors can convert the measured quantity into resistance changes. The resistance R of the conductor is proportional to its resistivity ρ and length l, and inversely proportional to the cross-sectional area A, that is

$$R = \rho\frac{l}{A} \tag{4.7}$$

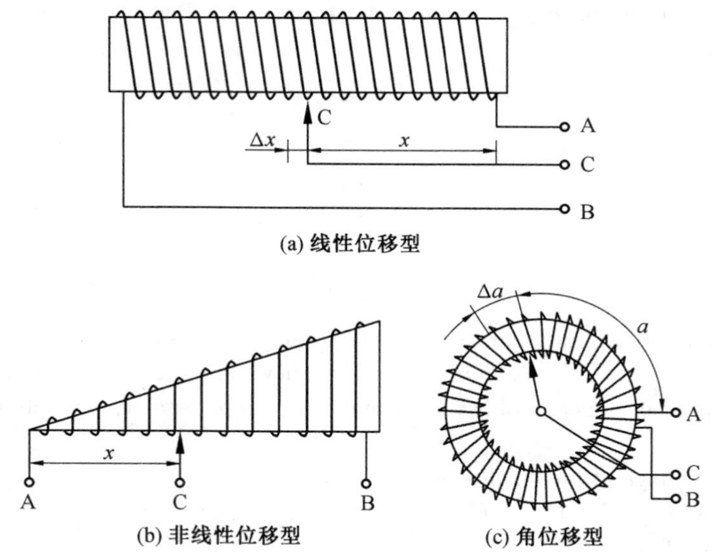

图 4.6　三种变阻器式传感器

阻）呈线性关系。其灵敏度 $S = \mathrm{d}R/\mathrm{d}\alpha = k_\alpha =$ 常数，α 为角位移，k_α 为单位弧度电阻值（Ω/rad）。

变阻式传感器还可按电阻体的材料分类，如线绕、合成碳膜、金属玻璃釉、有机实芯和导电塑料等类型。电性能主要决定于所用的材料。此外，还有用金属箔、金属膜和金属氧化膜制成电阻体的电位器，具有特殊用途。

电位器按使用特点区分，有通用、高精度、高分辨力、高阻、高温、高频、大功率等。

3. 特点

变阻器式传感器的优点如下。

（1）尺寸小、结构简单、价格低、质量轻、使用方便。

（2）可实现线性输出或任意函数特性输出，受环境因素（如湿度、温度、电磁场干扰等）影响小。

（3）性能稳定，输出信号大（一般无须再放大）。

其缺点为：若滑动触点有摩擦磨损的情况，则会影响其可靠性和寿命，而且会降低测量精度，所以分辨力较低，摩擦的同时会有噪声产生。

4.2.2　电阻应变式传感器

电阻应变式传感器是应用最广泛的传感器之一。电阻应变式传感器的感应元件是电阻应变片（简称应变片），能够将应变转换为电阻变化。将应变片粘贴在被测件表面，随

The measured physical quantity causes the change of any one or several of ρ, l, A, and the resistance R can be changed. According to its working principle, resistive sensors can be divided into rheostat type (potentiometer type) and resistance strain type sensors.

4.2.1 Rheostat Sensors

1. Working principle

The rheostat type sensor is also called a potentiometer whose working principle is to convert the displacement of an object into a change in resistance, that is, to change the length of the resistance line in the access circuit by moving the contact point between the slider and the resistance line, thereby changing the resistance.

2. Type

According to $R = \rho l/A$, rheostat sensors can be divided into three types: linear displacement type, nonlinear displacement type and angular displacement type. Three types of rheostat type sensors are shown in Figure 4.6.

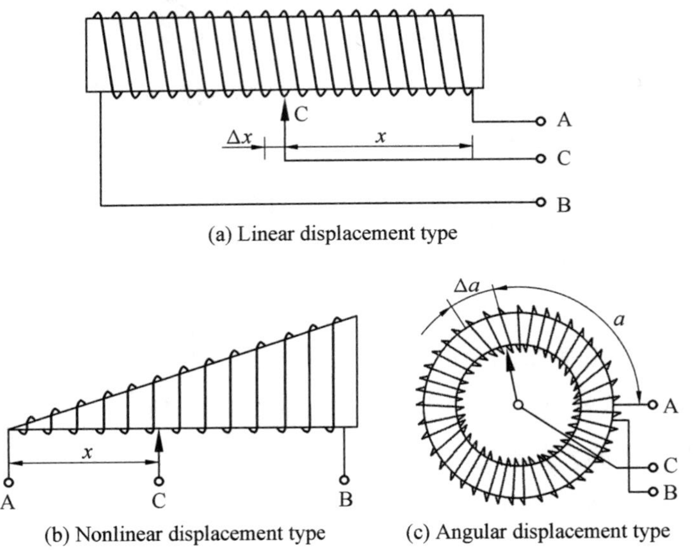

Figure 4.6 Three types of rheostat type sensors

(1) Linear displacement type (Figure 4.6(a)). When changing the position of contact C, the resistance between AC is $R = k_L x$, k_L is the resistance per unit length (Ω/m), and the sensitivity $S = dR/dx = k_L$ = constant. When the wires are evenly distributed, the output (resistance) of the sensor and the input (line displacement) have a linear relationship.

(2) Nonlinear displacement type (Figure 4.6(b)). Also known as function potentiometer, its skeleton shape is determined according to the required output $f(x)$.

(3) Angular displacement type (Figure 4.6(c)). When the wires are evenly distributed, the sensor's input (angular displacement) and output (resistance) have a linear relationship. The sensitivity $S = dR/d\alpha = k_\alpha$ = constant, α is the angular displacement, and k_α is the resistance value in radians (Ω/rad).

Rheostat sensors can also be classified according to the material of the resistor body, such

着被测件受力变形，应变片产生与被测件表面应变成比例的电阻变化，使用适当的测量电路和仪器就能测得被测件的应力或应变。应变片不仅能测应变，而且对能转化为应变变化的物理量，如力、压强、位移、加速度、转矩等，也可利用应变片来测量。

应变片具有以下优点。

(1) 电阻的变化与被测件的应变成线性关系。

(2) 适合动态测量，应变片尺寸小(国产应变片的最小栅长是0.178 mm)、质量轻(一般为0.1 ~ 0.2 g)、惯性小、频率响应好，可测0 ~ 500 kHz的动态应变。

(3) 测量范围广，一般测量范围为10 ~ 10^{-4} 量级的微应变，用高精度、高稳定性测量系统和半导体应变片可测出 10^{-2} 量级的微应变。

(4) 测量精度高，动态测试精度为1%，静态测试时达0.1%。

(5) 可在各种复杂或恶劣的环境中进行测量。例如，从 -270 ℃(液氦温度)的深冷温度到1 000 ℃的高温，从真空工作环境到几千个大气压的超高压状态，长时间地浸没于水下，如大的离心力、强烈振动、强磁场、放射性和化学腐蚀等恶劣环境中都可进行测量。

1. 工作原理——电阻应变效应

金属或半导体在外力作用下产生机械变形(伸长或缩短)时，引起金属或半导体的电阻值发生相应变化的物理现象称为电阻应变效应，电阻应变片是基于电阻应变效应工作的。

如果把应变片贴在弹性结构体上，当弹性体受外力作用而成比例地变形(在弹性范围内)时，应变片也随之变形，所以可通过应变片电阻的大小来检测外力的大小。

设应变片在不受外力作用时的初始电阻值为

$$R = \rho \frac{l}{A} \tag{4.8}$$

当应变片随弹性结构受力变形后，应变片受力变形示意图如图4.7所示，应变片金属丝的长度 l、截面积 A 都发生变化，电阻率 ρ 也会因晶格的变化而有所改变。l、A 和 ρ 三个因素的变化必然导致电阻值 R 的变化。设其变化为 dR，则有

$$\mathrm{d}R = \frac{\partial R}{\partial l}\mathrm{d}l + \frac{\partial R}{\partial \rho}\mathrm{d}\rho + \frac{\partial R}{\partial A}\mathrm{d}A$$

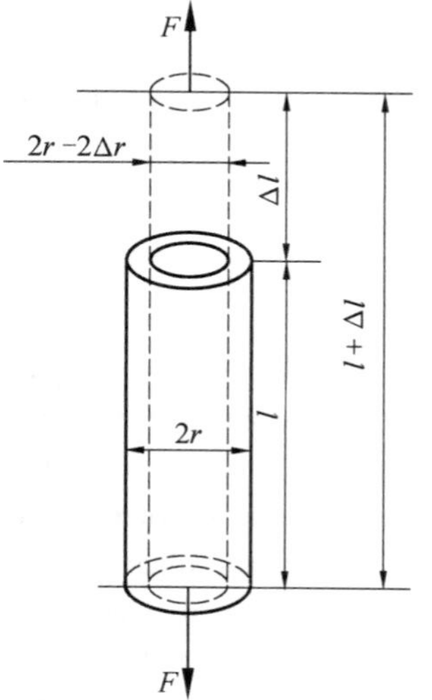

图4.7 应变片受力变形示意图

即

as wire wound, synthetic carbon film, metallic glass glaze, organic solid core and conductive plastic. The electrical performance is mainly determined by the material. In addition, there are potentiometers made of metal foil, metal film and metal oxide film into resistors, with special applications.

According to their characteristics, potentiometers are divided into types of general-purpose, high-precision, high-resolution, high-resistance, high-temperature, high-frequency, and high-power, etc.

3. Features

The advantages of rheostat sensors are as follows.

(1) Small size, simple structure, low price, light weight, and easy to use.

(2) It can realize linear output or arbitrary function characteristic output, and it is less affected by environmental factors (such as humidity, temperature, electromagnetic field interference).

(3) Stable performance, and large output signal (generally no need to amplify).

The disadvantages are: if the sliding contact has friction and wear, it will affect its reliability and life, and will reduce the measurement accuracy so that the resolution can tend to be low, and noise will be generated during friction.

4.2.2 Resistance Strain Type Sensors

Resistance strain type sensors are one of the most widely used sensors. The sensing element of the resistance strain sensor is a resistance strain gauge (strain gauge for short), which can convert strain into resistance change. We can paste the strain gauge on the surface of the measured part. As the measured part is deformed by force, the strain gauge will produce a resistance change proportional to the surface of the measured part. And stress or strain of the measured part can be measured by using appropriate measuring circuits and instruments. Strain gauges can not only measure strain, but also measure physical quantities that can be transformed into strain changes, such as force, pressure, displacement, acceleration, torque.

Strain gauges have the following advantages.

(1) The change of resistance and the strain of the measured part show a linear relationship.

(2) They are suitable for dynamic measurement. Due to small size (the minimum grid length of strain gauges is 0.178 mm), light weight (generally 0.1–0.2 g), small inertia, good frequency response, strain gauges can measure dynamic strains of 0–500 kHz.

(3) They prove wide measurement range where the general measurement range is micro strain of $10-10^{-4}$ magnitude. Micro strains of 10^{-2} magnitude can be measured with a high-precision, high-stability measurement system and semiconductor strain gauges.

(4) The measurement accuracy is high, the dynamicmeasurement accuracy is 1%, and the static measurement can reach 0.1%.

(5) They can be measured in various complex or harsh environments. For example, it can not only be measured from a cryogenic temperature of −270 °C (liquid helium temperature) to a high temperature of 1 000 °C, from a vacuum working environment to an ultra-high pressure state of thousands of atmospheres, but also be in a long-term immersion in water or in harsh environments such as large centrifugal force, strong vibration, strong magnetic field, radioactivity

$$dR = \frac{\rho}{A}dl + \frac{l}{A}d\rho - \frac{\rho l}{A^2}dA \tag{4.9}$$

式(4.9)两边都除以 R,并结合式(4.8),可得

$$\frac{dR}{R} = \frac{dl}{l} + \frac{d\rho}{\rho} - \frac{dA}{A}$$

若导体截面积为圆形,则式(4.9)变为

$$\frac{dR}{R} = \frac{dl}{l} + \frac{d\rho}{\rho} - 2\frac{dr}{r} \tag{4.10}$$

式中 $\frac{dl}{l}$——导体的轴向相对变形,称为纵向应变,即单位长度上的变化量,$\frac{dl}{l} = \varepsilon$;

$\frac{dr}{r}$——导体的径向相对变形,称为径向应变。

当导体纵向伸长时,其径向必然缩小,二者之间的关系为

$$\frac{dr}{r} = -\nu\frac{dl}{l} = -\nu\varepsilon$$

式中 ν——泊松比;

$\frac{d\rho}{\rho}$——导体电阻率的相对变化,与导体所受的轴向正应力有关。

$$\frac{d\rho}{\rho} = \lambda\sigma = \lambda E\varepsilon \tag{4.11}$$

式中 E——导线材料的弹性模量;

λ——压阻系数,与材质有关。

于是,式(4.10)可改写成

$$\frac{dR}{R} = \varepsilon + 2\nu\varepsilon + \lambda E\varepsilon = (1 + 2\nu + \lambda E)\varepsilon \tag{4.12}$$

当导体的材料确定后,E、λ 和 v 均为常数,则式(4.12)中的$(1 + 2\nu + \lambda E)$也为常数,这表明应变片电阻的相对变化率$\frac{dR}{R}$与应变 ε 之间呈线性关系,则应变片的灵敏度为

$$S = \frac{dR/R}{\varepsilon} = (1 + 2\nu + \lambda E) \tag{4.13}$$

因此,式(4.12)也可写成

$$\frac{dR}{R} = S \times \varepsilon \tag{4.14}$$

对于金属电阻应变片,其电阻的变化主要是由金属电阻丝的几何变形引起的。

由式(4.13)可知,其灵敏度 S 主要取决于$(1 + 2\nu)$项,λE 项则很小,可以忽略不计。

and chemical corrosion.

1. Working Principle—Resistance Strain Effect

When a metal or semiconductor is mechanically deformed (elongated or shortened) under the action of an external force, the physical phenomenon that causes the resistance value of the metal or semiconductor to change accordingly is called the resistance strain effect, on which the resistance strain gauge works based.

If a strain gauge is attached to the elastic structure, when the elastic body deforms proportionally (within the elastic range) by the external force, the strain gauge will also deform with it, so the magnitude of the external force can be detected by the resistance of the strain gauge.

Suppose the initial resistance value of the strain gauge when no external force is applied is

$$R = \rho \frac{l}{A} \tag{4.8}$$

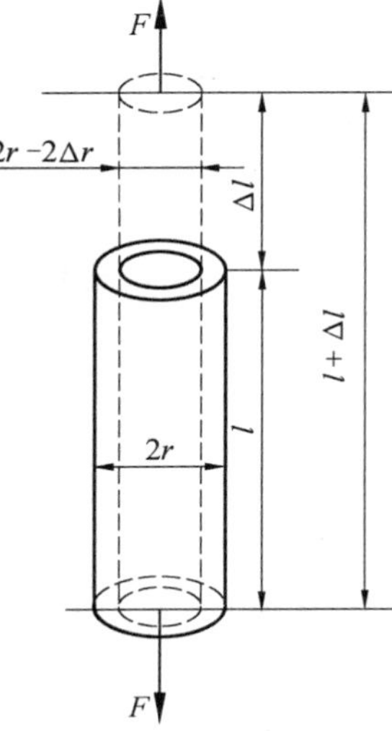

Figure 4.7 Schematic diagram of strain gauge stress and deformation

After the strain gauge is deformed by the elastic structure, schematic diagram of strain gauge stress and deformation is shown in Figure 4.7, the length l and the cross-sectional area A of the strain gauge wire will change, and the resistivity ρ will also change due to the change of the crystal lattice. The changes of the three factors of l, A and ρ will inevitably lead to the change of the resistance value R. Let the change be $\mathrm{d}R$, then there is

$$\mathrm{d}R = \frac{\partial R}{\partial l}\mathrm{d}l + \frac{\partial R}{\partial \rho}\mathrm{d}\rho + \frac{\partial R}{\partial A}\mathrm{d}A$$

that is

$$\mathrm{d}R = \frac{\rho}{A}\mathrm{d}l + \frac{l}{A}\mathrm{d}\rho - \frac{\rho l}{A^2}\mathrm{d}A \tag{4.9}$$

Divide both sides of formula(4.9) by R, and combine formula(4.8) , we get

$$\frac{\mathrm{d}R}{R} = \frac{\mathrm{d}l}{l} + \frac{\mathrm{d}\rho}{\rho} - \frac{\mathrm{d}A}{A}$$

If the conductor cross-sectional area is circular, then the formula (4.9) becomes

$$\frac{\mathrm{d}R}{R} = \frac{\mathrm{d}l}{l} + \frac{\mathrm{d}\rho}{\rho} - 2\frac{\mathrm{d}r}{r} \tag{4.10}$$

where $\frac{\mathrm{d}l}{l}$— the relative axial deformation of the conductor, called the longitudinal strain, which is the amount of change per unit length, $\frac{\mathrm{d}l}{l} = \varepsilon$;

$\frac{\mathrm{d}r}{r}$—the relative deformation of the conductor in the radial direction, called the radial strain.

When the conductor is elongated longitudinally, its radial direction will inevitably shrink. The relationship between the two is

金属电阻应变片的灵敏度 $S=1.7\sim3.6$。

常用电阻丝材料物理性能见表4.1。

表4.1 常用电阻丝材料物理性能

<table>
<tr><th rowspan="2">材料名称</th><th colspan="2">成　分</th><th rowspan="2">灵 敏 度
S_g</th><th rowspan="2">电阻率 ρ/
$(\Omega\cdot mm^2\cdot m^{-1})$</th><th rowspan="2">电阻温度系数/
$(\times10^{-6}$ ℃$)$</th><th rowspan="2">线胀系数/
$(\times10^{-6}$ ℃$)$</th></tr>
<tr><th>元素</th><th>百分比/%</th></tr>
<tr><td>康　铜</td><td>Cu
Ni</td><td>57
43</td><td>1.7 ~ 2.1</td><td>0.49</td><td>-20 ~ 20</td><td>14.9</td></tr>
<tr><td>镍铬合金</td><td>Ni
Cr</td><td>80
20</td><td>2.1 ~ 2.5</td><td>0.9 ~ 1.1</td><td>110 ~ 150</td><td>14.0</td></tr>
<tr><td>镍铬铝合金</td><td>Ni
Cr
Al
Fe</td><td>73
20
3 ~ 4
余量</td><td>2.4</td><td>1.33</td><td>-10 ~ 10</td><td>13.3</td></tr>
</table>

一般市售电阻应变片的标准阻值有60 Ω、120 Ω、350 Ω、600 Ω和1 000 Ω等，其中以120 Ω最为常用。应变片的尺寸可根据使用要求来选定。

对半导体应变片，由于其压阻系数 λ 及弹性模量 E 都较大，因此其灵敏度主要取决于 λE 项，而其几何变形引起的电阻变化则很小，可以忽略不计。半导体应变片的灵敏度 $S=60\sim170$，约为金属丝式应变片灵敏度的50 ~ 70倍。

2. 应变片的结构和种类

应变片主要分为两大类：金属电阻应变片和半导体应变片。常用的金属电阻应变片有三种：丝式、箔式和薄膜式。前两种为粘贴式应变片。粘贴式应变片由用苯酚、环氧树脂等绝缘材料浸泡过的玻璃基板制成，粘贴直径由金属丝或金属箔制成。应变片结构及其粘贴方式如图4.8所示。

（1）金属电阻应变片。

金属电阻应变片的结构如图4.8(a)所示，由绝缘的基片、电阻丝、覆盖层及引出线组成。

① 丝式应变片。

a. 回线式应变片。回线式应变片是将电阻丝绕制成敏感栅粘贴在绝缘基底上制成

$$\frac{dr}{r} = -\nu \frac{dl}{l} = -\nu\varepsilon$$

where ν —Poisson's ratio;

$\frac{d\rho}{\rho}$ —the relative change of conductor resistivity is related to the axial normal stress on the conductor.

$$\frac{d\rho}{\rho} = \lambda\sigma = \lambda E\varepsilon \tag{4.11}$$

where E—the elastic modulus of the wire material;

λ —the piezoresistive coefficient, which is related to the material.

Then formula (4.10) can be rewritten as

$$\frac{dR}{R} = \varepsilon + 2\nu\varepsilon + \lambda E\varepsilon = (1 + 2\nu + \lambda E)\varepsilon \tag{4.12}$$

When the conductor material is determined, E, λ and ν are all constants, then $(1 + 2\nu + \lambda E)$ in formula (4.12) is also a constant, which indicates that there is a linear relationship between the relative change rate of the strain gauge resistance $\frac{dR}{R}$ and the strain ε, and the sensitivity of the strain gauge is

$$S = \frac{dR/R}{\varepsilon} = (1 + 2\nu + \lambda E) \tag{4.13}$$

Thus, formula (4.12) can be expressed as

$$\frac{dR}{R} = S \times \varepsilon \tag{4.14}$$

For metal resistance strain gauges, the resistance change is mainly caused by the geometric deformation of the metal resistance wire.

It can be seen from formula (4.13) that the sensitivity S mainly depends on the item $(1 + 2\nu)$, while the item λE is very small and can be ignored. The sensitivity of the metal resistance strain gauge is $S=1.7-3.6$.

Physical properties of commonly used resistance wire materials are listed in Table 4.1.

Table 4.1 Physical properties of commonly used resistance wire materials

Material name	Ingredients		Sensitivity / S_g	Resistivity ρ / $(\Omega \cdot mm^2 \cdot m^{-1})$	Temperature coefficient of resistance/ $(\times 10^{-6}℃)$	Linear expansion coefficient/ $(\times 10^{-6}℃)$
	Element	Percentage /%				
Constantan	Cu Ni	57 43	1.7–2.1	0.49	−20–20	14.9
Nichrome	Ni Cr	80 20	2.1–2.5	0.9–1.1	110–150	14.0
Ni Cr Al alloy	Ni Cr Al Fe	73 20 3–4 Margin	2.4	1.33	−10–10	13.3

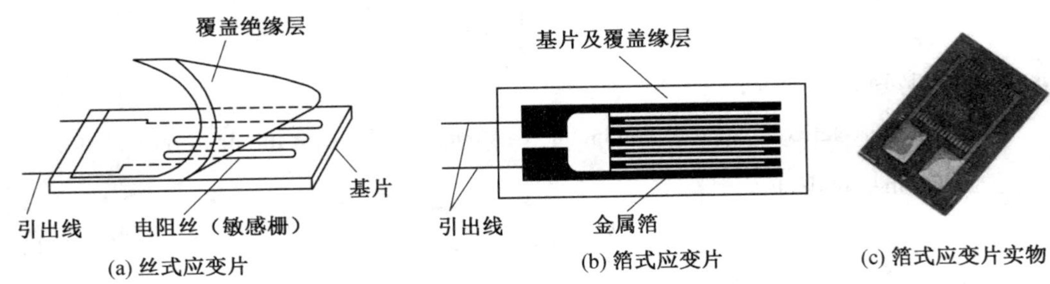

图 4.8　应变片结构及其粘贴方式

的，是一种常用的应变片，基底很薄（一般在 0.03 mm 左右），粘贴性能好，能保证有效地传递变形，引线多用直径 0.15 ~ 0.30 mm 的镀锡铜线与敏感栅相连。图 4.9(a) 所示为常见的回线式应变片构造图。

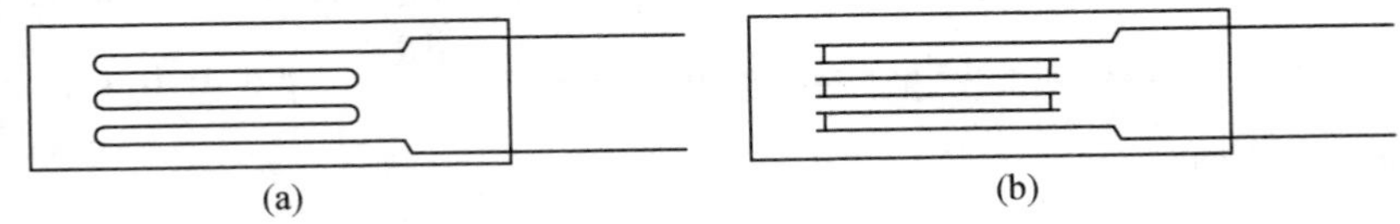

图 4.9　丝式应变片

b. 短接式应变片。短接式应变片的敏感栅平行安放，两端用直径为栅丝直径的 5 ~ 10 倍的镀银丝短接而构成，如图 4.9(b) 所示。短接式应变片的优点是克服了回线式应变片的横向效应。但由于焊点多，因此在冲击和振动试验条件下，易在焊接点处出现疲劳破坏，且制造工艺要求高。

② 箔式应变片。

这类应变片利用光刻腐蚀的方法，将箔材在绝缘基底下制成各种图形而成。箔片厚 1 ~ 10 μm，散热好，黏结情况好，传递被测件应变性能好。图 4.10 所示为常见的几种箔式应变片形式，其在常温下已逐步取代了回线式应变片。它的主要优点如下。

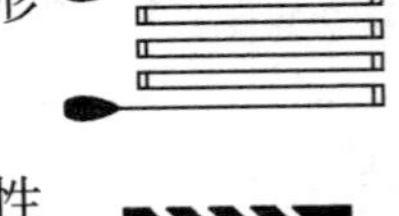

图 4.10　常见的几种箔式应变片形式

a. 能确保敏感栅尺寸正确、线条均匀，可制成任意形状以适应不同的测量要求。

b. 敏感栅薄而宽，黏结性能及传递被测件应变性能好。

c. 敏感栅界面为矩形，表面积与截面积之比远比圆断面的大，故黏合面积大。

d. 散热性能好，允许较大的工作电流通过，从而增大输出信号。

e. 敏感栅弯头横向效应可忽略，蠕变和机械滞后较小，疲劳寿命高。

Generally, the standard resistance values of commercially available resistance strain gauges are 60 Ω, 120 Ω, 350 Ω, 600 Ω, and 1 000 Ω, etc., of which 120 Ω is the most commonly used. The size of the strain gauge can be selected according to the requirements of use.

For semiconductor strain gauges, since the piezoresistance coefficient λ and elastic modulus E are large, its sensitivity mainly depends on λE, the resistance change caused by geometric deformation is very small and can be ignored. The sensitivity of the semiconductor strain gauge is $S=60-170$, about 50–70 times the sensitivity of wire strain gauges.

2. The structure and types of strain gauges

Strain gauges are mainly divided into two categories: metal resistance strain gauges and semiconductor strain gauges. Commonly used metal resistance strain gauges have three types: wire type, foil type, and thin film type. The first two are bonded strain gauges. The pasted strain gauge is made of a glass substrate soaked with insulating materials such as phenol and epoxy resin, and the pasting diameter is made of metal wire or metal foil. The structure of the strain gauge and its pasting method is shown in Figure 4.8.

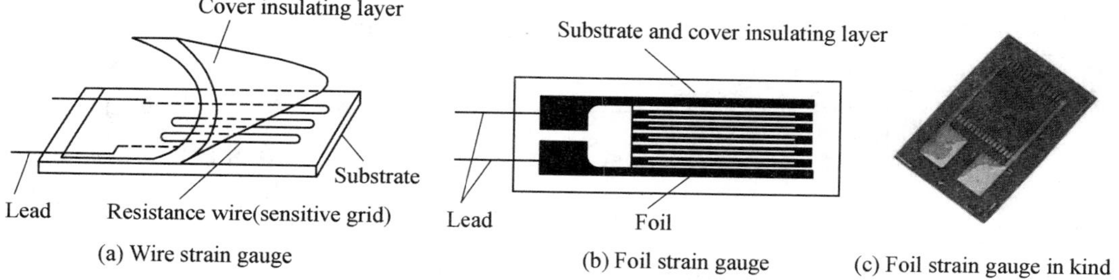

(a) Wire strain gauge (b) Foil strain gauge (c) Foil strain gauge in kind

Figure 4.8 The structure of the strain gauge and its pasting method

(1) Metal resistance strain gauge.

The structure of the metal resistance strain gauge is shown in Figure 4.8(a), which consists of an insulated substrate, resistance wire, covering layer and lead wires.

① Wire strain gauge.

a. Loop strain gauge. Loop strain gauge is made by winding resistance wire into a sensitive grid and pasting it on an insulating base. It is a commonly used strain gauge whose base is very thin (usually around 0.03 mm) and has good bonding performance, which can ensure effective to transfer deformation, the lead wire is mostly connected to the sensitive grid with tinned copper wire with a diameter of 0.15–0.30 mm. Figure 4.9(a) shows the structure diagram of a common loop strain gauge.

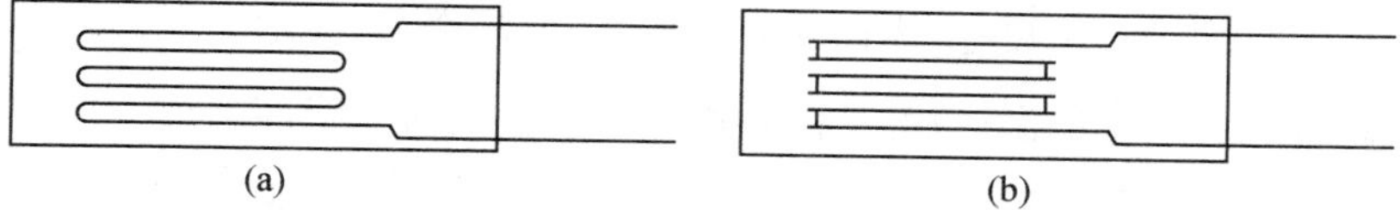

(a) (b)

Figure 4.9 Wire strain gauge

b. Short-connect strain gauge. The sensitive grids of the short-connect strain gauge are placed in parallel, and the two ends are short-circuited with silver-plated wires 5–10 times of the grid wire diameter, as shown in Figure 4.9(b). The advantage of short-connect strain

③ 薄膜式应变片。

金属薄膜式应变片是采用真空镀膜（如蒸发或沉积等）方式将金属在基底如表面有绝缘层的金属、有机绝缘材料或玻璃、石英、云母等无机材料上制成一层很薄的感应电阻膜（膜厚在0.1 μm以下）而构成的一种应变片。

（2）半导体应变片。

半导体应变片的工作原理基于半导体材料的压阻效应。所谓压阻效应，是指单晶半导体材料在沿某一轴向受到外力作用时，其电阻率ρ发生变化的现象。所有材料在某种程度上都具有压阻效应，但半导体的这种效应特别显著，能直接反映出很微小的应变。半导体应变片有三种：体型、薄膜型和扩散型（图4.11）。

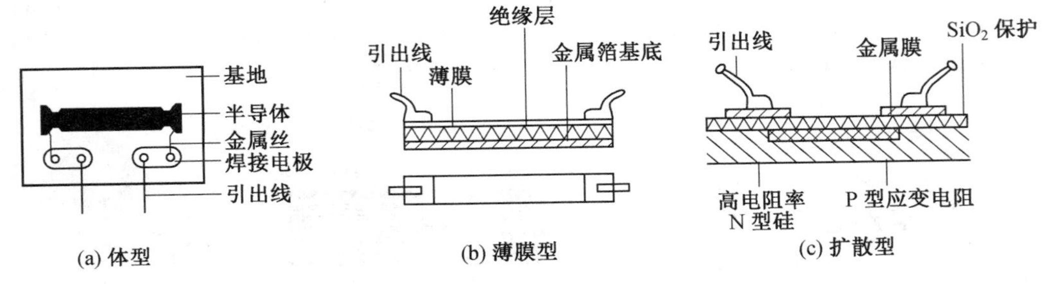

图4.11　半导体应变片

常见的半导体应变片是用锗或硅等半导体材料作为敏感栅，一般为单根状。半导体应变片的结构形式如图4.12所示。根据压阻效应，半导体把应变转换成电阻的变化。

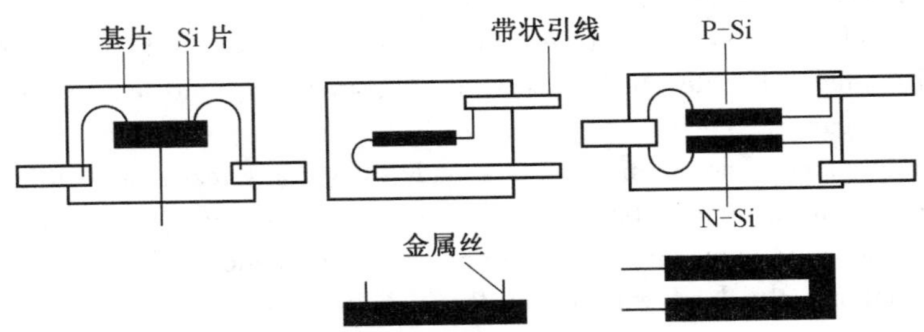

图4.12　半导体应变片的结构形式

半导体应变片的优点是尺寸、横向效应、机械滞后都很小，灵敏度高，因此输出也大。其缺点是电阻值和灵敏度的温度稳定性差，测量较大应变时非线性严重，灵敏度随受拉或受压而变，且分散度大，一般为3% ~ 5%。

3. 应变片的主要工作参数

（1）应变片的电阻值。应变片的电阻值指应变片没有安装且不受力的情况下，在室温时测定的电阻值。市售金属电阻应变片的电阻值已趋于标准化，主要规格有60 Ω、

gauge is able to overcome the lateral effect of the loop-type strain gauge. However, due to the large number of solder joints, fatigue damage is likely to occur at the solder joints under impact and vibration measurement conditions, and the manufacturing process requirements are high.

② Foil strain gauge.

This type of strain gauge will form various patterns on the foil under the insulating base by using methods of photolithography and etching. The thickness of the foil is about 1 - 10 μm, with good heat dissipation, good adhesion, and excellent strain transmission performance of the measured part. Figure 4. 10 shows several common foil-type strain gauges, which have gradually replaced loop-type strain gauges at room temperature. Its main advantages are as follows.

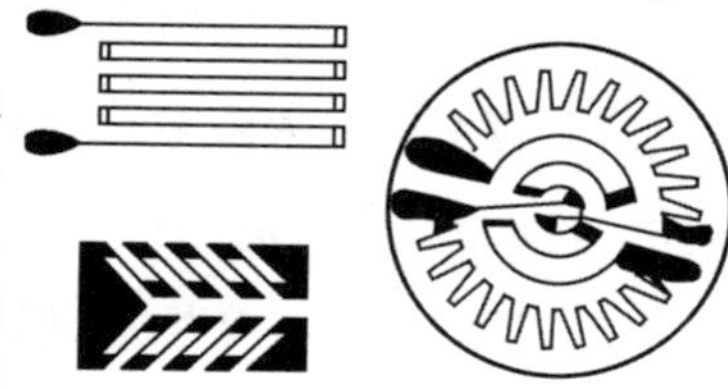
Figure 4. 10 Several common strain gauges

a. They can ensure the correct size of the sensitive grid, uniform lines, and they can be made into any shape to meet different measurement requirements.

b. The sensitive grid is thin and wide, with good adhesion and strain transmission performance.

c. The sensitive grid interface is rectangular, and the ratio of surface area to cross-sectional area is much larger than that of round cross-section, hence the large bonding area.

d. The heat dissipation performance is good, with allowing a larger working current to pass, thereby increasing the output signal.

e. The lateral effect of the sensitive grid elbow can be ignored, the creep and mechanical hysteresis are small, and the fatigue life is high.

③ Thin film strain gauges.

The metal thin film strain gauge is a kind of strain gauge made up of a thin induction resistance film (the film thickness is less than 0. 1 μm) on the substrate by vacuum coating (such as evaporation or deposition). The substrate mainly refers to the metal with insulating layer on the surface, organic insulating material or inorganic material such as glass, quartz, mica, etc. , on the surface.

(2) Semiconductor strain gauges.

The working principle of semiconductor strain gauges is based on the piezoresistive effect of semiconductor materials. The so-called piezoresistive effect refers to the phenomenon that the resistivity ρ of a single crystal semiconductor material will change when an external force is applied along a certain axis. All materials have a piezoresistive effect to some extent, but this effect of semiconductors is particularly significant with directly reflecting very small strains. There are three types of semiconductor strain gauges: body type, thin film type, and diffusion type (Figure 4. 11).

Common semiconductor strain gauges use semiconductor materials such as germanium or silicon as sensitive gates, which are generally single-rooted. The structure of the semiconductor strain gauges is shown in Figure 4. 12. On basis of the piezoresistive effect, the semiconductor converts strain into a change in resistance.

The advantages of semiconductor strain gauges are small size, lateral effects, and small mechanical hysteresis, high sensitivity, and large output. The disadvantage is that they have poor temperature stability of resistance and sensitivity, nonlinearity is serious when measuring

120 Ω、350 Ω、600 Ω 和 1 000 Ω 等，其中 120 Ω 用处最广。应变片在相同的工作电流下，电阻值越大，允许的工作电压越大，可提高测量灵敏度。

（2）机械滞后。在恒定的温度环境下，对已粘贴应变片的试件加载、卸载过程中，同一载荷下应变的最大差数称为机械滞后。造成此现象的原因很多，如应变片本身特性差、试件的材质差、黏结剂选择不当、固化不良、黏结技术不佳、部分脱落或黏结层太厚等。

（3）蠕变。当温度恒定，已粘贴应变片的试件承受长时间恒定的外力时，应变片的应变值随时间的变化称为蠕变。蠕变主要是由胶层引起的，如黏结剂种类选择不当、粘贴层较厚或固化不充分及在黏结剂接近软化温度下进行测量等。

（4）应变极限。应变的线性（灵敏系数为常数）特性只有在一定的应变限度范围内才能保持。当试件输入的真实应变超过某一限值时，应变的输出特性将出现非线性。在恒温条件下，使非线性误差达到 10% 时的真实应变值称为应变极限 $\varepsilon_{\mathrm{lim}}$。

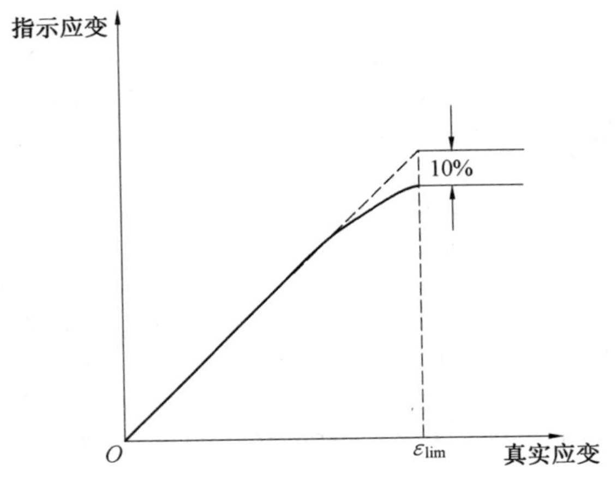

图 4.13　应变极限

（5）疲劳寿命。已粘贴应变片的试件在一定的交变应力作用下，应变片可连续工作而不致产生疲劳损坏的循环次数称为疲劳寿命。疲劳寿命的循环次数与动载荷的特性及大小密切相关。一般情况下，循环次数可达 10^6 ~ 10^7。

（6）绝缘电阻。应变片引线和粘贴应变片的试件之间的电阻值称为绝缘电阻。此值常作为应变片黏结层固化程度和是否受潮的标志。绝缘电阻下降会带来零点漂移和测量误差，尤其是绝缘电阻的不稳定会导致测试失败。

（7）零点漂移。在温度恒定、试件不受力的条件下，应变片的应变值随时间的变化称为零点漂移（简称零漂）。这是应变片的绝缘电阻过低及通过电流而发热等原因造成的。

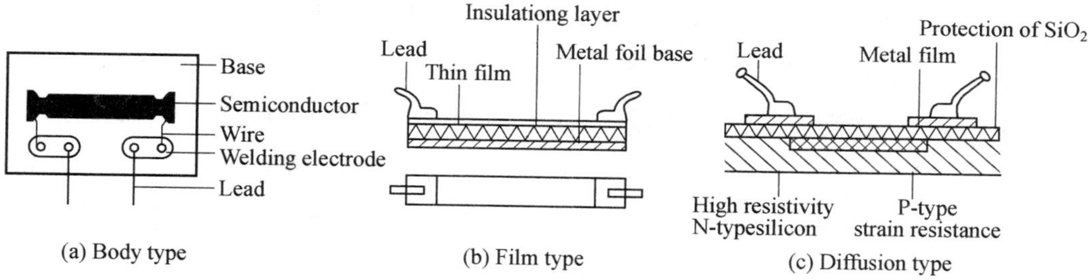

Figure 4. 11 Semiconductor strain gauges

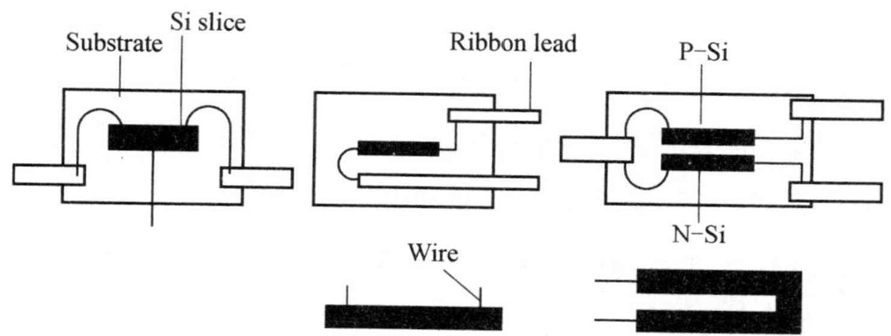

Figure 4. 12 The structure of the semiconductor strain gauges

large strains, sensitivity changes with tension or compression, and large dispersion, generally around 3% and 5%.

3. The Main Working Parameters of Strain Gauges

(1) The resistance value of strain gauges. The resistance value of strain gauges refers to the resistance value measured at room temperature when the strain gauge is not installed and is not stressed. The resistance value of commercially available metal resistance strain gauges has tended to be standardized which its main specifications are 60 Ω, 120 Ω, 350 Ω, 600 Ω and 1 000 Ω, among which 120 Ω is the most widely used. Under the same working current, the larger the resistance value of the strain gauges, the larger the allowable working voltage, which can improve the measurement sensitivity.

(2) Mechanical hysteresis. In a constant temperature environment, the maximum difference in strain under the same load during loading and unloading of a specimen with a strain gauge attached is called mechanical hysteresis. There are many reasons for this phenomenon, such as poor characteristics of the strain gauge itself, poor material of the measurement piece, improper adhesive selection, poor curing, poor bonding technology, partial fall off or too thick bonding layer.

(3) Creep. When the temperature is constant, and the specimen with the strain gauge is subjected to a constant external force for a long time, the change of the strain value of the strain gauge with time is called creep. Creep is mainly caused by the adhesive layer, such as improper selection of the type of adhesive, thick adhesive layer or insufficient curing, and measurement when the adhesive is close to the softening temperature.

(4) Strain limit. The linearity of the strain (its sensitivity coefficient is constant) can only be maintained within a certain strain limit. When the true strain input by the specimen exceeds a certain limit, the output characteristics of the strain will appear nonlinear. Under constant

(8) 最大工作电流。允许通过应变片而不影响其工作特性的最大电流值称为最大工作电流。该电流与外界条件有关,一般为几十毫安,箔式应变片有的可达500 mA。当进行静态测量时,为提高测量精度,流过应变片的电流应小一些;当进行短期动态测量时,为增大输出功率,电流可大一些。工作电流大,输出信号也大,灵敏度高。但工作电流超过 I_{max} 值,会使应变片过热,灵敏系数会变化,零点漂移和蠕变会增加,甚至烧毁应变片。工作电流的大小由试件的导热性能及敏感栅形状和尺寸等决定。

4. 调理电路 —— 直流电桥

因为电阻应变式传感器的电阻的变化不能直接产生可检测并显示的信号,所以需要一种方法将这些传感器的参量变化转换成电压或电流输出。电桥电路就能够起到这样的作用,将电阻转化为电信号,便于放大、显示或记录。

按照供电电源的不同,电桥电路分为交流电桥和直流电桥。按照输出方式,电桥电路可分为不平衡桥式电路与平衡桥式电路。

直流电桥采用直流恒压源或恒流源供电,在电阻应变式传感器测试中应用广泛,是本节讲述的重点。交流电桥则采用幅值和频率恒定的交流电源供电,适用于电容式和电感式传感器,将在后续的电感式传感器章节中详细讲述。

直流电桥的基本形式是惠斯通电桥(纯电阻电桥),如图4.14所示。图中,U_r 是直流供电电压,U_o 是电桥输出电压。电阻式传感器,如电阻应变计、热电阻、热敏电阻等,可以接在电桥的一个桥臂或多个桥臂上。在输出开路时,这种电桥电路相当于两个分压电路的并联,R_1 和 R_2 构成一个分压电路,R_3 和 R_4 构成另一个分压电路。在输出开路时,根据分压定律,可得

$$U_b = \frac{R_1}{R_1 + R_2} U_r \tag{4.15}$$

$$U_d = \frac{R_4}{R_3 + R_4} U_r \tag{4.16}$$

因此,电桥的开路输出电压为

$$U_o = U_b - U_d = \frac{R_1}{R_1 + R_2} U_r - \frac{R_4}{R_3 + R_4} U_r = \frac{R_1 R_3 - R_2 R_4}{(R_1 + R_2)(R_3 + R_4)} U_r \tag{4.17}$$

注意,式(4.17)是在电桥输出不带负载的情况下得到的,所以为空载(或开路)输出电压表达式。

当电桥的输出电压为0时,电桥处于平衡状态。由式(4.17)可知,直流电桥的平衡条件为

temperature conditions, the true strain value when the nonlinear error reaches 10% is called the strain limit $\varepsilon_{\lim}$.

(5) Fatigue life. Under a certain alternating stress, the number of cycles that the specimen with the strain gauge pasted can work continuously without fatigue damage, which is called the fatigue life. The number of cycles of fatigue life is closely related to the characteristics and magnitude of the dynamic load. In general, the number of cycles can reach $10^6 - 10^7$.

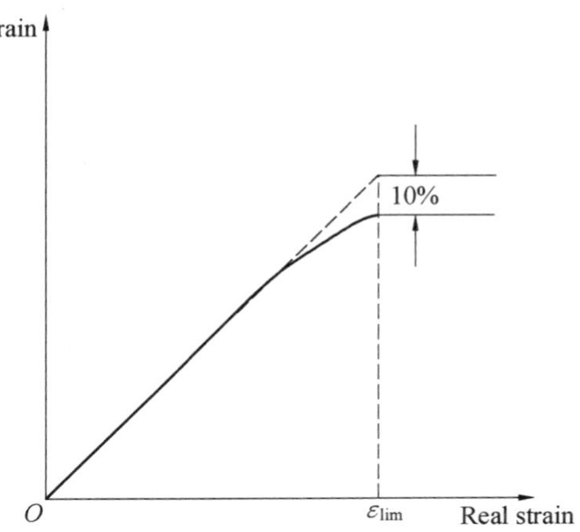

Figure 4.13 Strain limit

(6) Insulation resistance. The resistance between the lead of the strain gauge and the specimen to which the strain gauge is attached is called the insulation resistance. This value is often used as a sign of the degree of curing of the strain gauge bonding layer and whether it is damp. The decrease of insulation resistance will bring zero drift and measurement error, especially the instability of insulation resistance will lead to measurement failure.

(7) Zero drift. Under the conditions of constant temperature and no force on the specimen, the change of the strain value of the strain gauge with time is called zero point drift (referred to as zero drift). This is caused by the low insulation resistance of the strain gauge and heat generated by the current passing through it.

(8) Maximum operating current. The maximum operating current is referred to the maximum current value allowed to pass through the strain gauge without affecting its operating characteristics. The current is related to external conditions, generally tens of milliamperes, and some of the foil strain gauges can reach 500 mA. When performing static measurement, the current flowing through the strain gauge should be smaller in order to improve the measurement accuracy; when performing short-term dynamic measurement, the current can be larger in order to increase the output power. The working current is large, the output signal is also large, and the sensitivity is high. However, if the operating current exceeds the $I_{\max}$ value, the strain gauge will overheat, the sensitivity coefficient will change, zero drift and creep will increase, and the strain gauge will even be burned. The size of the working current is determined by the thermal conductivity of the measurement piece and the shape and size of the sensitive grid.

4. Conditioning Circuit—DC Bridge

Because the resistance change of the resistance strain sensor cannot directly produce a detectable and displayable signal, a method is needed to convert the parameter change of these sensors into a voltage or current output. The bridge circuit can play such a role, converting the resistance into an electrical signal for amplifying, displaying or recording.

According to the power supply, the bridge circuit is divided into AC bridge and DC bridge. According to the output mode, the bridge circuit can be divided into unbalanced bridge circuit and balanced bridge circuit.

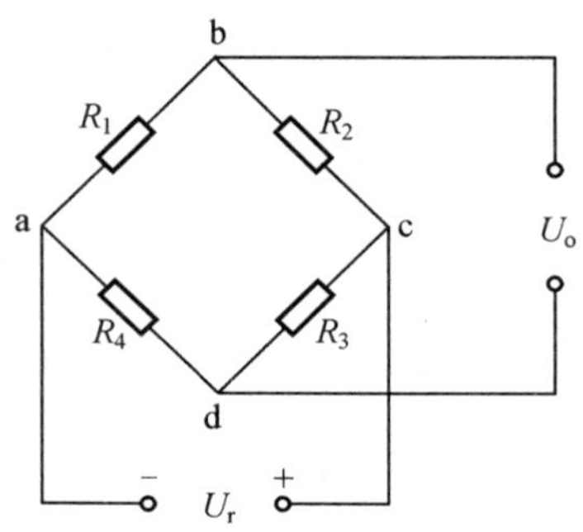

图 4.14　直流电桥的基本形式

$$R_1R_3 = R_2R_4$$

或

$$R_1/R_2 = R_4/R_3 \tag{4.18}$$

即当相对两臂电阻乘积相等或相邻两臂电阻比值相等时，电桥达到平衡。

由式(4.17)可知，如果电桥初始状态平衡，则电桥任意一个桥臂上电阻发生变化，电桥都会失去平衡，从而使输出电压 U_o 不再为 0。因此，U_o 的大小反映了电阻的变化量。如果在电桥的输出端 b 和 d 之间接入显示装置(如电压表)，则显示装置的读数反映了桥臂上电阻的变化量，因此反映了被测量的大小。这种测量电阻变化的方法称为偏位测量法。如果在电桥的一个桥臂上或在两个桥臂之间设置一个可变电阻器，当电桥某个桥臂或几个桥臂上的电阻发生变化导致电桥失去平衡时，通过调节可变电阻器使电桥重新回到平衡状态(显示装置归零)，则可变电阻的变化值反映了被测电阻或被测量的变化值。这种在电桥平衡状态下通过读取可变电阻器的变化值(在可变电阻器的刻度盘上读取)来反映被测量大小的方法称为零位测量法。

零位测量法和偏位测量法都得到了实际应用。例如，静态电阻应变仪采用的就是零位测量法，而动态电阻应变仪则采用了偏位测量法。

偏位测量法需要一个经校准的显示或记录装置。由式(4.17)可知，由于电桥的输出电压与电源电压 U_r 成正比，因此电源电压 U_r 的变化会带来测量误差。在偏位测量法中，随着桥臂电阻的变化，b 和 d 之间产生瞬变输出电压，该电压可用示波器显示或记录仪器记录，所以能测量快速动态现象。因此，偏位测量法测量效率高，适用于静、动态测量。

零位测量法需要一个经校准的可变电阻器。由于读数是在电桥处于平衡的状态下进行的，此时电桥输出电压为 0，电源电压的变化不会带来测量误差，因此零位测量法的测量精度较高。但零位测量法在获取读数之前，需要调节平衡电阻使显示装置指向零位。如果手动调节，则调节过程需要很长时间。即使使用一个仪器伺服机构进行自动调节，所

Due to adopting DC constant voltage source or constant current source to supply power, the DC bridge is widely used in resistance strain sensor measurement, which is the focus of this section. The AC bridge powered by an AC power supplying with constant amplitude and frequency is suitable for capacitive and inductive sensors, and will be described in detail in the subsequent chapter about inductive sensor.

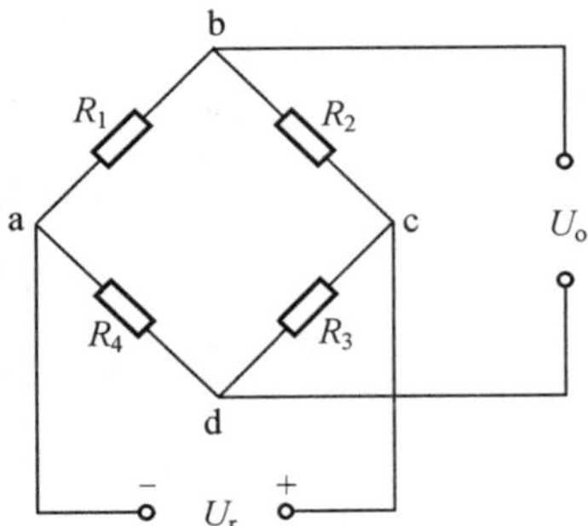

Figure 4.14 Basic form of DC bridge

The basic form of DC bridge is Wheatstone bridge (pure resistance bridge), as shown in Figure 4.14. In the figure, U_r is the DC supply voltage, U_o is the bridge output voltage. Resistive sensors, such as resistance strain gauges, thermal resistors, thermistors, can be connected to one or more bridge arms of the bridge. When the output is open, this bridge circuit is equivalent to the parallel connection of two voltage dividers. R_1 and R_2 form a voltage divider circuit, and R_3 and R_4 form another voltage divider circuit. When the output is open, according to voltage division principle, we get

$$U_b = \frac{R_1}{R_1 + R_2} U_r \tag{4.15}$$

$$U_d = \frac{R_4}{R_3 + R_4} U_r \tag{4.16}$$

Therefore, the open-circuit output voltage of the bridge is

$$U_o = U_b - U_d = \frac{R_1}{R_1 + R_2} U_r - \frac{R_4}{R_3 + R_4} U_r = \frac{R_1 R_3 - R_2 R_4}{(R_1 + R_2)(R_3 + R_4)} U_r \tag{4.17}$$

Note that formula (4.17) is obtained when the bridge output is not loaded, so it is the no-load (or open circuit) output voltage expression.

When the output voltage of the bridge is 0, the bridge is in equilibrium. From formula (4.17), it can be seen that the balance condition of the DC bridge is

$$R_1 R_3 = R_2 R_4$$

or

$$R_1 / R_2 = R_4 / R_3 \tag{4.18}$$

That is, when the product of the resistance of the two arms is equal or the resistance ratio of the two adjacent arms is equal, the bridge reaches equilibrium.

It can be seen from formula (4.17) that if the initial state of the bridge is balanced, the resistance on any of the bridge arms of the bridge will change, and the bridge will lose balance, so that the output voltage U_o is no longer 0. Therefore, the size of U_o reflects the change in resistance. If a display device (such as a voltmeter) is connected between the output terminals b and d of the electric bridge, the reading of the display device reflects the change in resistance on the bridge arm, thus reflecting the measured size. This method of measuring resistance changes is called offset measurement. If a variable resistor is set on a bridge arm or between two bridge arms, when the resistance of a bridge arm or several bridge arms changes leading to imbalance, we can adjust the variable resistor to make the bridge return to a balanced state (the display device returns to zero), and the change value of the variable resistance reflects the measured resistance or the measured change value. This method of reflecting the measured size by reading the change value of the variable resistor (read on the dial of the variable resistor) in

需时间也比测量很多快变的变量所容许的时间长很多。因此,零位测量法测量效率低,只适合静态或变化较慢的物理量的测量。在给定情况下,选择零位测量法还是偏位测量法取决于该测量对响应速度和漂移等指标的要求。

下面讨论图4.14所示直流电桥的电压灵敏度。电桥的电压灵敏度定义为单位电阻的相对变化量所引起的输出电压大小。由式(4.17)可知,电桥的输出电压与桥臂上的电阻值不是正比关系,因此电压灵敏度不是常数。对式(4.17)求微分得

$$\mathrm{d}U_o = U_r\left[\frac{R_2}{(R_1+R_2)^2}\mathrm{d}R_1 - \frac{R_1}{(R_1+R_2)^2}\mathrm{d}R_2 + \frac{R_4}{(R_3+R_4)^2}\mathrm{d}R_3 - \frac{R_3}{(R_3+R_4)^2}\mathrm{d}R_4\right] \tag{4.19}$$

由式(4.19)可知,电桥的输出电压与各桥臂电阻的变化量并非线性关系,这给测量带来了非线性误差。但如果电阻的变化值相对于原始阻值很小,则由于$\frac{R_2}{(R_1+R_2)^2}$、$\frac{R_1}{(R_1+R_2)^2}$、$\frac{R_4}{(R_3+R_4)^2}$和$\frac{R_3}{(R_3+R_4)^2}$近似为常数,因此电桥的输出电压与各桥臂电阻的变化近似呈线性关系。例如,在电阻应变计中,$\Delta R/R$很少超过1%。由于$\Delta R/R$很小这种情况具有实用价值,因此下面针对这种情况对电桥的电压灵敏度进行讨论。

在$\Delta R/R$很小,且电桥初始处于平衡状态的条件下,式(4.19)可近似表示为

$$U_o = U_r\left[\frac{R_2}{(R_1+R_2)^2}\Delta R_1 - \frac{R_1}{(R_1+R_2)^2}\Delta R_2 + \frac{R_4}{(R_3+R_4)^2}\Delta R_3 - \frac{R_3}{(R_3+R_4)^2}\Delta R_4\right] \tag{4.20}$$

设$\Delta R_2=\Delta R_3=\Delta R_4=0$,代入式(4.20)中,可得

$$U_o = U_r\frac{R_2}{(R_1+R_2)^2}\Delta R_1 = U_r\frac{R_1R_2}{(R_1+R_2)^2R_1}\Delta R_1 \tag{4.21}$$

在这种情况下,电桥的电压灵敏度近似为

$$S = \frac{U_o}{\Delta R_1/R_1} = U_r\frac{R_1R_2}{(R_1+R_2)^2} \tag{4.22}$$

式(4.20)~(4.22)虽然是近似关系式,说明了电桥电路存在非线性,但在某些情况下,如电阻应变计,只要合理地将电阻应变计布置在不同的桥臂上,这种近似关系会很好地得到满足,甚至可以完全消除非线性。下面就电阻应变计在直流电桥电路中的配置情况进行讨论,其他电阻式传感器在直流电桥电路中的配置可以以此为参考。

电阻应变计经常采用的是等臂电桥。所谓等臂电桥,就是当电桥处于平衡状态时,各桥臂的电阻值相等。设等臂电桥各桥臂的初始电阻值$R_1=R_2=R_3=R_4=R_0$,$R_i(i=1,2,$

the balanced state of the bridge is called the null method of measurement.

Both the null method of measurement and the offset measurement method have been applied in practice. For example, the static resistance strain gauge uses the null method of measurement, while the dynamic resistance strain gauge uses the offset measurement method.

The offset measurement method requires a calibrated display or recording device. It can be seen from formula (4.17) that since the output voltage of the bridge is proportional to the power supply voltage U_r, the change of the power supply voltage U_r will bring measurement errors. In the offset measurement method, as the bridge arm resistance changes, a transient output voltage that can be displayed by an oscilloscope or recorded by a recording instrument is generated between b and d, so fast dynamic phenomena can be measured. Therefore, the offset measurement method has high measurement efficiency and is suitable for static and dynamic measurement.

A calibrated variable resistor is required in the null method of measurement. Since the reading is performed when the bridge is in a balanced state, the output voltage of the bridge is 0 at this time, and the change of the power supply voltage will not bring measurement errors, so the zero measurement method has a higher measurement accuracy. However, if we use the null method of measurement, we need to adjust the balance resistance to point the display device to the zero position before taking the reading. If it is adjusted manually, the adjustment process will take a long time. Even if an instrument servo mechanism is used for automatic adjustment, the required time is much longer than the allowed time to measure many fast-varying variables. Therefore, the null method of measurement of low measurement efficiency is only suitable for the measurement of static or slow-changing physical quantities. Whether choosing the null method of measurement or offset measurement method in a given situation depends on the measurement's requirements for indicators such as response speed and drift.

The voltage sensitivity of the DC bridge shown in Figure 4.14 is discussed below. The voltage sensitivity of the bridge is defined as the output voltage caused by the relative change of unit resistance. From formula(4.17), it can be seen that the output voltage of the bridge is not proportional to the resistance value on the bridge arm, so the voltage sensitivity is not constant. Differentiate formula(4.17) to get

$$\mathrm{d}U_o = U_r\left[\frac{R_2}{(R_1+R_2)^2}\mathrm{d}R_1 - \frac{R_1}{(R_1+R_2)^2}\mathrm{d}R_2 + \frac{R_4}{(R_3+R_4)^2}\mathrm{d}R_3 - \frac{R_3}{(R_3+R_4)^2}\mathrm{d}R_4\right] \tag{4.19}$$

It can be seen from formula (4.19) that the output voltage of the bridge has a non-linear relationship with variation of the resistance of each bridge arm, which will bring a non-linear error to the measurement. However, if the change value of the resistance is small relative to the orginal resistance value, since $\frac{R_2}{(R_1+R_2)^2}$, $\frac{R_1}{(R_1+R_2)^2}$, $\frac{R_4}{(R_3+R_4)^2}$ and $\frac{R_3}{(R_3+R_4)^2}$ are approximately constants, the output voltage of the bridge is approximately linear with the changes in the resistance of each bridge arm. For example, in a resistance strain gauge, $\Delta R/R$ is very small, this situation has practical value, the voltage sensitivity of the bridge will be discussed below for this situation.

Under the condition that $\Delta R/R$ is very small and the bridge is initially in equilibrium, formula (4.19) can be approximately expressed as

3,4）由 R_0 变到 $R_0+\Delta R_i$，则电桥空载输出电压为

$$U_o=\frac{R_1R_3-R_2R_4}{(R_1+R_2)(R_3+R_4)}U_r=\frac{(R_0+\Delta R_1)(R_0+\Delta R_3)-(R_0+\Delta R_2)(R_0+\Delta R_4)}{(R_0+\Delta R_1+R_0+\Delta R_2)(R_0+\Delta R_3+R_0+\Delta R_4)}U_r$$

$$=\frac{R_0(\Delta R_1-\Delta R_2+\Delta R_3-\Delta R_4)+\Delta R_1\Delta R_3-\Delta R_2\Delta R_4}{(2R_0+\Delta R_1+\Delta R_2)(2R_0+\Delta R_3+\Delta R_4)}U_r \tag{4.23}$$

当 $\Delta R_i \ll R_0(i=1,2,3,4)$ 时，忽略分子中的二次微项和分母中的微项，得到

$$U_o=\frac{1}{4R_0}(\Delta R_1-\Delta R_2+\Delta R_3-\Delta R_4)U_r \tag{4.24}$$

式(4.24)为等臂电桥的自动加减特性表达式。该式表明，相对两臂电阻的变化为相加关系，而相邻两臂电阻的变化为相减关系。此式为在使用等臂电桥电路时合理布置电阻应变计、提高灵敏度、减小非线性、减小温度误差提供了理论依据。

例如，在图4.15所示悬臂梁式电阻应变式传感器中，在等截面悬臂梁同一个横截面的上下两个表面各粘贴了两片电阻应变计，上表面为 R_1 和 R_3，下表面为 R_2 和 R_4。只要按图中所标序号将四个应变计分别接入图4.15所示电桥电路的四个桥臂上，就不仅提高了电桥的电压灵敏度，也实现了温度补偿，并减小了非线性。设四个电阻应变计的特性参数相同，原始阻值 $R_1=R_2=R_3=R_4=R_0$，R_m 是被测力引起的电阻应变计的电阻变化，ΔR_t 是温度变化引起的电阻应变计的电阻变化，则有

$$\Delta R_1=\Delta R_3=\Delta R_m+\Delta R_t \tag{4.25}$$

$$\Delta R_2=\Delta R_4=-\Delta R_m+\Delta R_t \tag{4.26}$$

代入式(4.23)中，可得

$$U_o=\frac{R_0(\Delta R_1-\Delta R_2+\Delta R_3-\Delta R_4)+\Delta R_1\Delta R_3-\Delta R_2\Delta R_4}{(2R_0+\Delta R_1+\Delta R_2)(2R_0+\Delta R_3+\Delta R_4)}U_r$$

$$=\frac{2R_0(\Delta R_1-\Delta R_2)+\Delta R_1^2-\Delta R_2^2}{(2R_0+\Delta R_1+\Delta R_2)^2}U_r$$

$$=\frac{2R_0[(\Delta R_m+\Delta R_t)-(-\Delta R_m+\Delta R_t)]+(\Delta R_m+\Delta R_t)^2-(-\Delta R_m+\Delta R_t)^2}{[2R_0+(\Delta R_m+\Delta R_t)+(-\Delta R_m+\Delta R_t)]^2}U_r$$

$$=\frac{\Delta R_m}{R_0+\Delta R_t}U_r$$

$$\approx\frac{\Delta R_m}{R_0}U_r=S_0\varepsilon U_r \tag{4.27}$$

$$U_o = U_r\left[\frac{R_2}{(R_1 + R_2)^2}\Delta R_1 - \frac{R_1}{(R_1 + R_2)^2}\Delta R_2 + \frac{R_4}{(R_3 + R_4)^2}\Delta R_3 - \frac{R_3}{(R_3 + R_4)^2}\Delta R_4\right] \tag{4.20}$$

Suppose $\Delta R_2 = \Delta R_3 = \Delta R_4 = 0$, substituting formula (4.20) to get

$$U_o = U_r \frac{R_2}{(R_1 + R_2)^2}\Delta R_1 = U_r \frac{R_1 R_2}{(R_1 + R_2)^2 R_1}\Delta R_1 \tag{4.21}$$

In this case, the voltage sensitivity of the bridge is approximately

$$S = \frac{U_o}{\Delta R_1 / R_1} = U_r \frac{R_1 R_2}{(R_1 + R_2)^2} \tag{4.22}$$

Although formulas (4.20)–(4.22) are approximate relational expressions, illustrating the non-linearity of the bridge circuit, in some cases, for example, resistance strain gauges whose approximate relationship will be well satisfied, and even the nonlinearity can be completely eliminated as long as the resistance strain gauges are reasonably arranged in different on the bridge arm. The configuration of the resistance strain gauges in the DC bridge circuit is discussed below. The configuration of other resistive sensors in the DC bridge circuit can be used as a reference.

Equal arm bridge is often used in resistance strain gauges. It means that the resistance value of each bridge arm is equal when the bridge is in a balanced state. Suppose the initial resistance value $R_1 = R_2 = R_3 = R_4 = R_0$, and $R_i (i = 1,2,3,4)$ changes from R_0 to $R_0 + \Delta R_i$, then the no-load output voltage of the bridge is

$$\begin{aligned} U_o &= \frac{R_1 R_3 - R_2 R_4}{(R_1 + R_2)(R_3 + R_4)}U_r = \frac{(R_0 + \Delta R_1)(R_0 + \Delta R_3) - (R_0 + \Delta R_2)(R_0 + \Delta R_4)}{(R_0 + \Delta R_1 + R_0 + \Delta R_2)(R_0 + \Delta R_3 + R_0 + \Delta R_4)}U_r \\ &= \frac{R_0(\Delta R_1 - \Delta R_2 + \Delta R_3 - \Delta R_4) + \Delta R_1 \Delta R_3 - \Delta R_2 \Delta R_4}{(2R_0 + \Delta R_1 + \Delta R_2)(2R_0 + \Delta R_3 + \Delta R_4)}U_r \end{aligned} \tag{4.23}$$

When $\Delta R_i \ll R_0 (i = 1,2,3,4)$, ignoring the quadratic differential term in the numerator and the differential term in the denominator, we get

$$U_o = \frac{1}{4R_0}(\Delta R_1 - \Delta R_2 + \Delta R_3 - \Delta R_4) U_r \tag{4.24}$$

Formula (4.24) is the automatic addition and subtraction characteristic expression of the equal-arm bridge. This formula shows that the relative resistance changes of the two arms are in an additive relationship, while the changes in the adjacent two arms' resistance are in a subtractive relationship. This formula provides a theoretical basis for reasonably arranging resistance strain gauges to improve sensitivity, reduce non-linearity and temperature errors when using equal-arm bridge circuits.

For example, in the cantilever beam resistance strain sensor shown in Figure 4.15, two resistance strain gauges are attached to the upper and lower surfaces of the same cross-section of the cantilever beam of constant cross-section. The upper surface is R_1 and R_3, and the lower surface for R_2 and R_4. As long as the four strain gauges are connected to the four bridge arms of the bridge circuit shown in Figure 4.15 according to the serial numbers shown in the figure, it not only improves the voltage sensitivity of the bridge, but also realizes temperature compensation and reduces non-linearity. Suppose that the characteristic parameters of the four resistance strain gauges are the same. The original resistance value $R_1 = R_2 = R_3 = R_4 = R_0$, R_m is the resistance change of the resistance strain gauges caused by the measured force, and ΔR_t

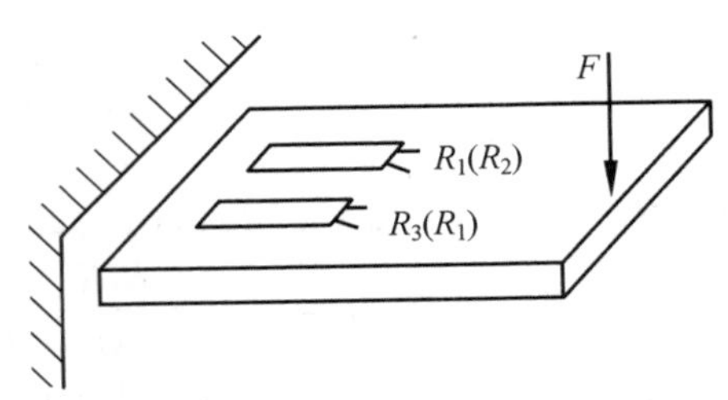

图 4.15　悬臂梁式电阻应变式传感器

可见，温度误差影响要小得多，实现了温度互补偿，线性也得到很大的改善。如果不考虑电阻应变计的温度误差，则电桥输出电压与电阻的变化满足准确的线性关系。

电阻式传感器在直流电桥电路中可以采用单臂、半桥双臂和全桥四臂等配置方式。下面以电阻应变计在等臂电桥中的配置为例，说明传感器在直流电桥电路中各种配置方式的特点。电阻应变计在电桥电路中的配置的基本原则包括：能提高灵敏度；能实现温度补偿；减小非线性误差。

以悬臂梁式电阻应变式传感器为例，电阻应变计在电桥电路中三种配置方式的输出电压和电压灵敏度见表4.2。四个桥臂上的原始阻值为$R_1=R_2=R_3=R_4=R_0$。表中的输出电压表达式和电压灵敏度表达式是在不考虑温度引起的电阻变化的条件下推出的。各种配置方式特点总结如下。

（1）单臂配置方式存在较大的非线性误差。在高精度测量中，需要采用非线性补偿措施，温度误差大，需要单独的温度补偿措施。

（2）半桥双臂配置方式消除了非线性误差。如果将温度引起的电阻变化考虑进来，其非线性误差也比单臂配置方式小得多。这种配置方式基本消除了温度误差，实现了温度互补偿，电压灵敏度是单臂配置方式的2倍。

（3）全桥四臂配置方式消除了非线性误差。如果将温度引起的电阻变化考虑进来，其非线性误差也比单臂配置方式小得多。这种配置方式基本消除了温度误差，实现了温度互补偿，电压灵敏度是单臂配置方式的4倍，是半桥双臂配置方式的2倍。

表 4.2　电阻应变计在电桥电路中的三种配置方式的输出电压和电压灵敏度

配置方式	单臂	半桥双臂	全桥四臂
电阻应变计在悬臂梁上的布置	R_1　F	R_1　F　R_2	$R_1(R_3)$　F　$R_2(R_4)$

is the resistance variation of the resistance strain gauges caused by temperature changes, then

$$\Delta R_1 = \Delta R_3 = \Delta R_m + \Delta R_t \tag{4.25}$$

$$\Delta R_2 = \Delta R_4 = - \Delta R_m + \Delta R_t \tag{4.26}$$

Substituting into the formula (4.23), we get

$$\begin{aligned}
U_o &= \frac{R_0(\Delta R_1 - \Delta R_2 + \Delta R_3 - \Delta R_4) + \Delta R_1 \Delta R_3 - \Delta R_2 \Delta R_4}{(2R_0 + \Delta R_1 + \Delta R_2)(2R_0 + \Delta R_3 + \Delta R_4)} U_r \\
&= \frac{2R_0(\Delta R_1 - \Delta R_2) + \Delta R_1^2 - \Delta R_2^2}{(2R_0 + \Delta R_1 + \Delta R_2)^2} U_r \\
&= \frac{2R_0[(\Delta R_m + \Delta R_t) - (-\Delta R_m + \Delta R_t)] + (\Delta R_m + \Delta R_t)^2 - (-\Delta R_m + \Delta R_t)^2}{[2R_0 + (\Delta R_m + \Delta R_t) + (-\Delta R_m + \Delta R_t)]^2} U_r \\
&= \frac{\Delta R_m}{R_0 + \Delta R_t} U_r \\
&\approx \frac{\Delta R_m}{R_0} U_r = S_0 \varepsilon U_r
\end{aligned} \tag{4.27}$$

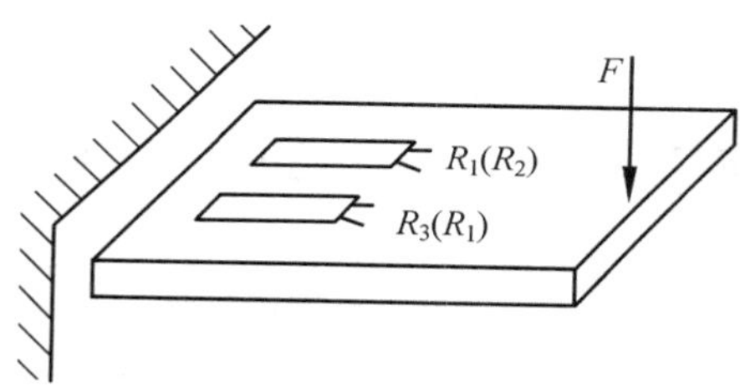

Figure 4.15 The cantilever beam resistance strain sensor

It can be seen that the influence of temperature error is much smaller, the temperature mutual compensation can be realized, and the linearity is also greatly improved. Not considering the temperature error of the resistance strain gauges, the output voltage of the bridge and the change of the resistance meet an accurate linear relationship.

Resistive sensors can be configured in single-arm, half-bridge double arms and full-bridge four arms configurations in the DC bridge circuit. The following takes the configuration of the resistance strain gauge in the equal-arm bridge as an example to illustrate the characteristics of various configurations of the sensor in the DC bridge circuit. The basic principles of the configuration of the resistance strain gauge in the bridge circuit include: the improvement of the sensitivity; the realization of temperature compensation; the reduction of the non-linear error.

The output voltage and voltage sensitivity of the three configurations of the resistance strain gauges in the bridge circuit are listed in Table 4.2, taking the cantilever beam resistance strain sensor as an example. The original resistance value on the four bridge arms is $R_1 = R_2 = R_3 = R_4 = R_0$. The output voltage expression and voltage sensitivity expression in the table are deduced without considering the resistance change caused by temperature. The characteristics of various configuration methods are summarized as follows.

(1) The single-arm configuration has a large nonlinear error. In high-precision measurement, nonlinear compensation measures are required, and the temperature error is large, thus separate temperature compensation measures are required.

(2) The half-bridge double arms configuration can eliminate non-linear errors. If the resistance change caused by temperature is taken into consideration, the nonlinear error is

续表 4.2

配置方式	单臂	半桥双臂	全桥四臂
电阻应变计在电桥电路中的配置	$R_1\pm\Delta R$, R_2, R_3, R_4, U_o, $-\ U_r\ +$	$R_1\pm\Delta R$, $R_2\mp\Delta R$, R_3, R_4, U_o, $-\ U_r\ +$	$R_1\pm\Delta R$, $R_2\mp\Delta R$, $R_3\pm\Delta R$, $R_4\mp\Delta R$, U_o, $-\ U_r\ +$
电桥输出电压	$U_o=\dfrac{\Delta R}{2(2R_0+\Delta R)}U_r\approx\dfrac{1}{4}\dfrac{\Delta R}{R_0}U_r$	$U_o=\dfrac{1}{2}\dfrac{\Delta R}{R_0}U_r$	$U_o=\dfrac{\Delta R}{R_0}U_r$
电压灵敏度	$S=\dfrac{U_o}{\Delta R/R_0}=\dfrac{1}{4}U_r$	$S=\dfrac{U_o}{\Delta R/R_0}=\dfrac{1}{2}U_r$	$S=\dfrac{U_o}{\Delta R/R_0}=U_r$
特点	线性差，需要单独设置温补措施	线性好，温度误差互补偿	线性好，温度误差互补偿

前面对直流电桥电路的分析都是建立在开路输出的情况下的。实际上，电桥的输出总是要接到后续电路或测量装置中。这时，后续的测量装置等便成为电桥的负载。在这种情况下，开路输出电压表达式即式(4.17)便不再准确成立。

设电桥负载电阻为 R_L，则电桥与负载相连的等效电路如图4.16所示。根据戴维南定理，可以求得电桥电路带负载时的输出电压为

$$U_o=\frac{R_1R_3-R_2R_4}{(R_1+R_2)(R_3+R_4)}\frac{1}{1+R_e/R_L}U_r \tag{4.28}$$

式中　R_e——电桥的等效电阻，$R_e=\dfrac{R_1R_2}{R_1+R_2}+\dfrac{R_3R_4}{R_3+R_4}$。

由式(4.28)可知，负载电阻 R_L 将使输出信号衰减，引起更严重的非线性，产生负载效应。衰减量取决于负载电阻 R_L 相对于等效电阻 R_e 的大小。尽管如此，由于多数电压测量装置(如数字电压表、示波器、电压记录仪器等)都含有输入电阻为1 MΩ或更高的输入放大器，可近似认为空载，因此通常在实际使用中，前述的空载输出表达式及由此得到的一些分析结果都是能很好地近似满足的。但在高精度测量中，可能需要采取适当的线性化措施。

much smaller than that of the single-arm configuration. This configuration method can basically eliminate temperature errors and realize temperature mutual compensation, and the voltage sensitivity is twice that of the single-arm configuration.

(3) The full-bridge four arms configuration can eliminate nonlinear errors. The nonlinear error is much smaller than that of the single-arm configuration if the resistance change caused by temperature is taken into consideration. This configuration can basically eliminate temperature errors and realize temperature mutual compensation. The voltage sensitivity is four times that of the single-arm configuration and twice that of the half-bridge double arms configuration.

Table 4.2 The output voltage and voltage sensitivity of the three configurations of the resistance strain gauges in the bridge circuit

Configuration method	Single-arm	Half-bridge double arms	Full-bridge four arms
The arrangement of the resistance strain gauge on the cantilever beam	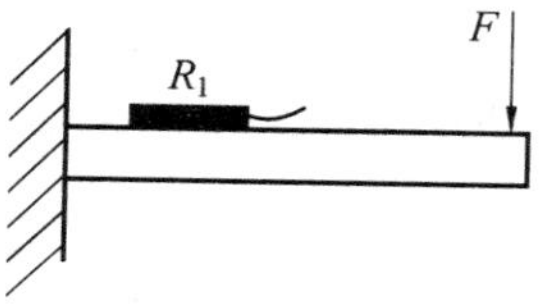 F, R_1	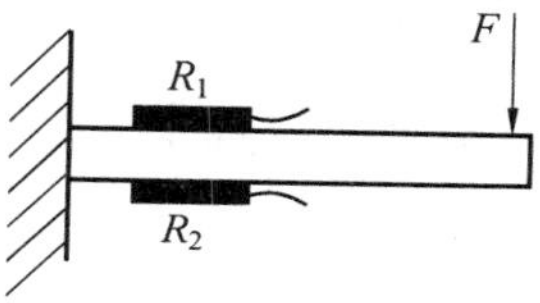 F, R_1, R_2	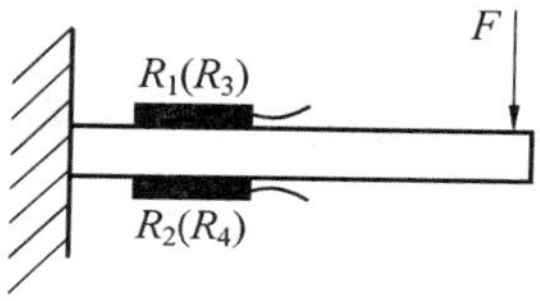 F, $R_1(R_3)$, $R_2(R_4)$
The configuration of resistance strain gauge in the bridge circuit	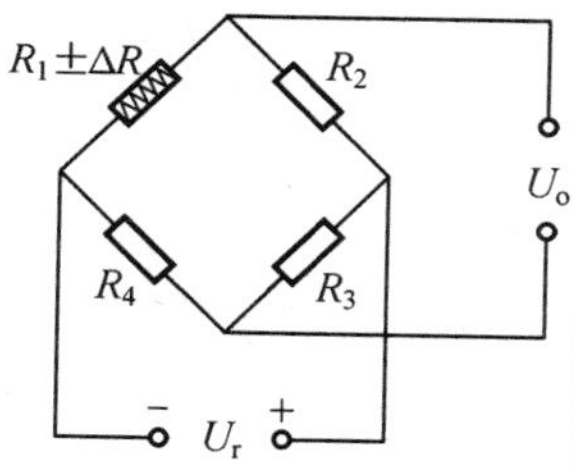 $R_1 \pm \Delta R$, R_2, R_3, R_4, U_o, $-\ U_r\ +$	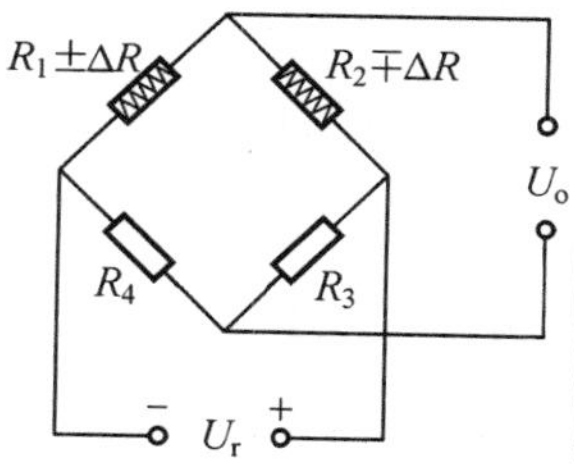 $R_1 \pm \Delta R$, $R_2 \mp \Delta R$, R_3, R_4, U_o, $-\ U_r\ +$	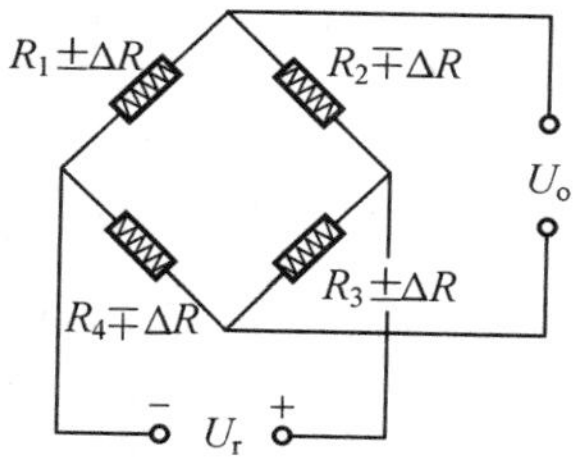 $R_1 \pm \Delta R$, $R_2 \mp \Delta R$, $R_3 \pm \Delta R$, $R_4 \mp \Delta R$, U_o, $-\ U_r\ +$
Bridge output voltage	$U_o = \dfrac{\Delta R}{2(2R_0 + \Delta R)} U_r \approx \dfrac{1}{4} \dfrac{\Delta R}{R_0} U_r$	$U_o = \dfrac{1}{2} \dfrac{\Delta R}{R_0} U_r$	$U_o = \dfrac{\Delta R}{R_0} U_r$
Voltage sensitivity	$S = \dfrac{U_o}{\Delta R / R_0} = \dfrac{1}{4} U_r$	$S = \dfrac{U_o}{\Delta R / R_0} = \dfrac{1}{2} U_r$	$S = \dfrac{U_o}{\Delta R / R_0} = U_r$
Features	Poor linearity, and temperature compensation measures need to be set separately	Good linearity, temperature error mutual compensation	Good linearity, temperature error mutual compensation

The previous analysis of the DC bridge circuit is based on the case of open-circuit output. In fact, the output of the bridge is always connected to the subsequent circuit or measuring device. At this time, the subsequent measuring devices will become the load of the bridge. In

图4.14和图4.16给出的是最简单的电桥电路。通常为方便用户使用,需要附加另外一些装置,如增加调零(调平衡)电阻器、灵敏度调节电阻器、标定电阻器等。商业用传感器也可能包括附加的温度敏感电阻器,实现温度补偿。

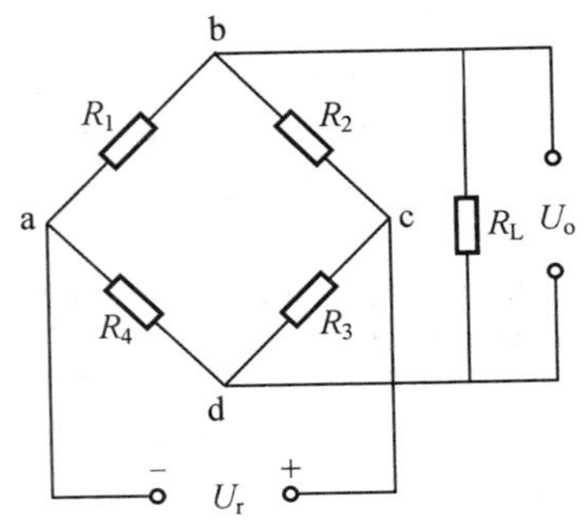

图4.16　电桥与负载相连的等效电路

直流电桥具有下列特点:高稳定度直流电源较易获得;对连接导线要求较低;平衡电路简单。但在进行静态或变化缓慢信号的测量时,后需接直流放大器,而直流放大器往往存在着零漂、接地电位影响大等缺陷。因此,要选择性能良好的直流放大器。

上述电桥是在不平衡条件下工作的,它的缺点是当电源电压不稳定或环境温度有变化时,都会引起电桥输出的变化,从而产生测量误差。为此,在某些情况下采用平衡电桥(图4.17)。设被测量等于0时,电桥处于平衡状态,此时指示仪表G及可调电位器H指零。当某一桥臂随被测量变化时,电桥会失去平衡。调节电位器H,改变电阻R_5触点位置,可使电桥重新平衡,电表G指针回零。电位器H上的标度与桥臂电阻值的变化成比例,故H的指示值可以直接表达被测量的数值。这种测量方法的特点是在电表G指零时再读值,因此称为零位测量法。

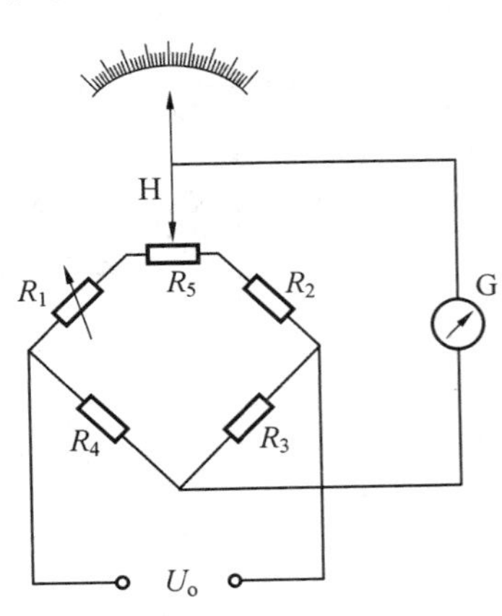

图4.17　平衡电桥

由于平衡电桥最终的输出为0,因此测量误差取决于可调电位器的精确度,而与电桥电源电压无关。

一般静态应变仪往往采用这种平衡电桥,并通过手动实现平衡。在电子电位差计或$X-Y$记录仪中,通常是用伺服电动机来调整电位器的位置,实现自动调平衡的。

5. 电阻应变片的温度误差及补偿

温度变化引起应变片的电阻变化与被测件应变所造成的电阻变化几乎具有相同的数

this case, the open-circuit output voltage expression, that is, formula(4.17), is no longer accurate.

Assuming that the load resistance of the bridge is R_L, the equivalent circuit of the bridge and the load is shown in Figure 4.16. According to Thevenin's theorem, the output voltage of the bridge circuit with load can be obtained as

$$U_o = \frac{R_1R_3 - R_2R_4}{(R_1 + R_2)(R_3 + R_4)} \frac{1}{1 + R_e/R_L} U_r \tag{4.28}$$

where R_e—the equivalent resistance of the bridge, $R_e = \frac{R_1R_2}{R_1 + R_2} + \frac{R_3R_4}{R_3 + R_4}$.

From formula (4.28), it can be seen that the load resistance R_L will attenuate the output signal, causing more serious non-linearity and generating load effects. The amount of attenuation depends on the size of the load resistance R_L relative to the equivalent resistance R_e. Nevertheless, since most voltage measurement devices (such as digital voltmeters, oscilloscopes, voltage recording instruments) contain input amplifiers with an input resistance of 1 MΩ or higher, they can be regarded as no-load. Therefore the aforementioned no-load output expressions and some of the analysis results obtained from them can be approximately satisfied in usually actual use. However, it may be necessary to take appropriate linearization measures in high-precision measurement.

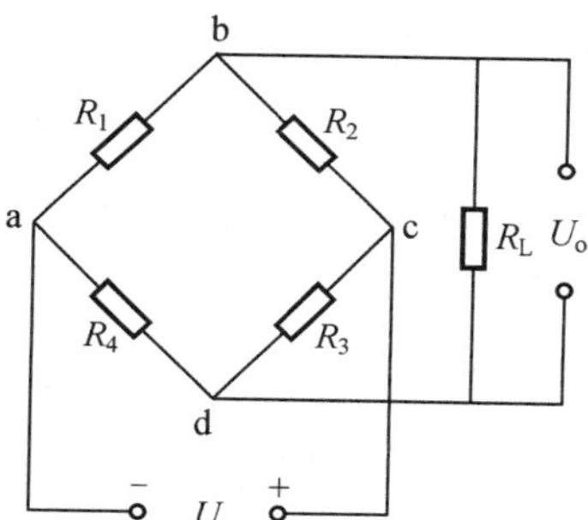

Figure 4.16 Equivalent circuit of the bridge and the load

Figure 4.14 and Figure 4.16 show the simplest bridge circuit. Generally, additional devices are needed for the convenience of users, such as adding zero-regulator(balance adjustment) resistors, sensitivity adjustment resistors, calibration resistors. Commercial sensors may also include additional temperature-sensitive resistors to achieve temperature compensation.

The DC bridge has the following characteristics: high-stability DC power is easier to obtain; lower requirements for connecting wires; simple balance circuit. However, in the measurement of static or slow-changing signals, a DC amplifier is required afterwards, and DC amplifiers often have defects such as zero drift and large ground potential. Therefore, a DC amplifier with good performance should be selected.

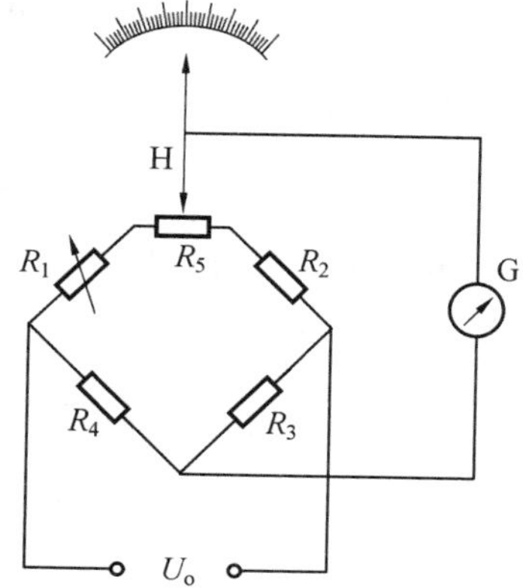

Figure 4.17 Balanced bridge

The above-mentioned electric bridge works under unbalanced conditions, whose disadvantage is that when the power supply voltage is unstable or the ambient temperature changes, it will cause changes in the output of the electric bridge, with resulting in measurement errors. For this reason, a balanced bridge is used in some cases (Figure 4.17). When the measured value is equal to 0, the bridge is in a balanced state, at this time, the indicator G and the adjustable potentiometer H point to zero. When a certain bridge arm changes with the measured quantity, the bridge will lose balance. Adjust the potentiometer H

量级。若不采取措施消除温度的影响，测量精度将无法保证。温度补偿方法有两大类：桥路补偿和应变片自补偿。其中，桥路补偿法应用较多。桥路补偿法可在较大温度范围内对粘贴普通应变片的各种材料试件进行温度误差补偿，如图4.18所示。

工作片 R_1 通常是作为平衡电桥的一个臂测量应变的，R_2 为补偿片。因此，桥路补偿法有时又称补偿片法。工作片 R_1 粘贴在需要测量应变的试件上，补偿片 R_2 粘贴在一块不受力的、与试件相同的材料上。这块材料自由地放在试件上或附近，R_1 和 R_2 的环境温度相同（图4.18）。当温度发生变化时，工作片 R_1 和补偿片 R_2 的电阻都发生变化。R_1 与 R_2 为同类应变片，又贴在相同的材料上，因此 R_1 和 R_2 的电阻变化相同，即 $\Delta R_1=\Delta R_2$。由于 R_1 和 R_2 分别接入电桥的相邻两桥臂，因此温度变化引起的电阻变化 ΔR_1 和 ΔR_2 的作用相互抵消。根据电桥理论，其输出电压 U_{sc} 与温度变化无关。

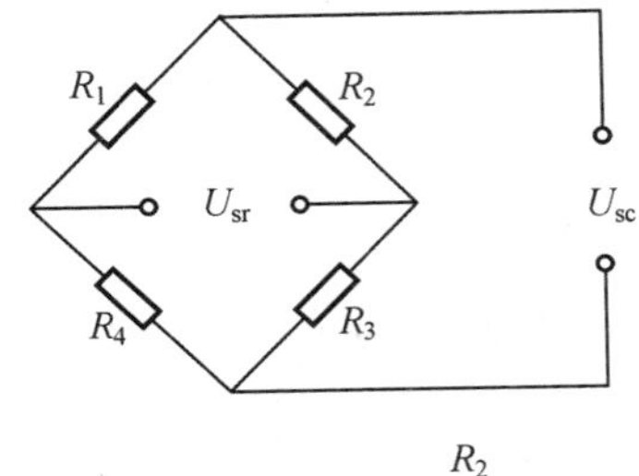

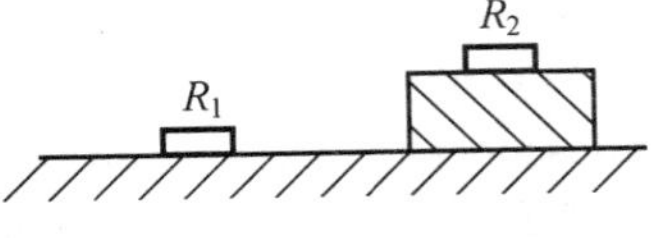

图4.18　桥路补偿法

桥路补偿法的优点是方便简单，在常温下补偿效果较好；缺点是在温度变化梯度较大时，工作片与补偿片很难保证处于完全一致的温度情况，因此影响补偿效果。

若采用差动电桥，可起自动补偿作用。差动电桥补偿法如图4.19所示，在测弹性梁弯曲应变时，将两个应变片分别贴在上下对称位置上，接入图4.18所示电路中，构成差动电桥。当梁处于上下温度一致的同一环境中时，R_1 与 R_2 由温度变化引起的电阻变化又与图4.18作用一样，相互抵消。

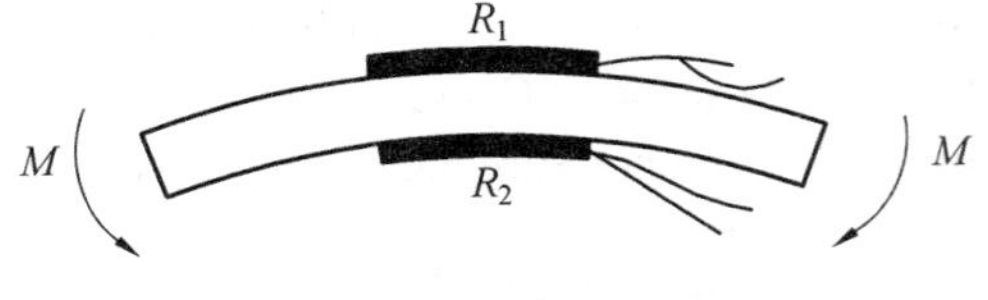

图4.19　差动电桥补偿法

6. 应变式传感器

若应变片贴在弹性体上，则力的作用及可转换成作用力的物理量，如扭矩、位移、速度、加速度等，均可使弹性体产生应变，引起贴片电阻变化，经电桥输出电压。在弹性范围内，应变 ε 与应力 σ 及作用力 F 的关系为

$$\varepsilon=\sigma/E=F/E\cdot S$$

式中　E——弹性模量，kgf/mm^2（1 kgf = 9.806 65 N）；

A——弹性体截面积，mm^2。

由应变片灵敏系数 $S=\dfrac{\Delta R}{R}/\varepsilon$ 有 $\dfrac{\Delta R}{R}=S\cdot\varepsilon$，则电桥输出电压为

and change the position of the resistor R_5 contact point to rebalance the bridge and return the G pointer of the meter to zero. The scale on the potentiometer H is proportional to the change in the resistance of the bridge arm, so the indicated value of H can directly express the measured value. The characteristic of this measurement method is to read the value when the meter G is zero, so it is called null method of measurement.

Since the final output of the balanced bridge is 0, the measurement error depends on the accuracy of the adjustable potentiometer, and has nothing to do with the bridge power supply voltage.

General static strain gauges often use this kind of balance bridge to balance manually. In electronic potentiometers or $X-Y$ recorders, servo motors are usually used to adjust the position of the potentiometer to achieve automatic balance adjustment.

5. Temperature Error and Compensation of Resistance Strain Gauges

The resistance change of the strain gauge caused by the temperature change is almost the same order of magnitude as the resistance change caused by the strain of the device undermea-surement. If measures are not taken to eliminate the influence of temperature, the measurement accuracy will not be guaranteed. There are two types of temperature compensation methods: bridge circuit compensation and strain gauge self-compensation. The bridge circuit compensation method is widely used. The bridge circuit compensation method can compensate the temperature error of various material specimens pasted with ordinary strain gauges in a larger temperature range, as shown in Figure 4.18.

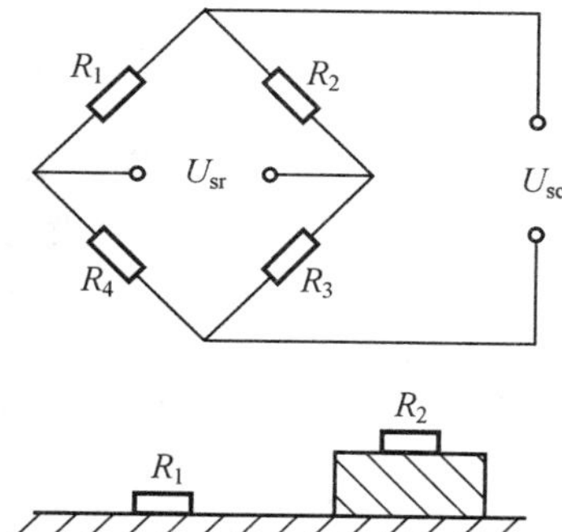

Figure 4. 18 Bridge circuit compensation method

Work sensor piece R_1 is usually used as an arm of a balanced bridge to measure strain, and R_2 is a compensation piece. Therefore, the bridge circuit compensation method is sometimes called the compensating gauges method. The work sensor piece R_1 is pasted on the measurement piece that needs to measure the strain, and the compensation sensor piece R_2 is pasted on a non-stressed material that is the same as the measurement piece. This material is freely placed on or near the measurement piece, and the ambient temperature of R_1 and R_2 is the same (Figure 4.18). When the temperature changes, the resistances of the work sensor piece R_1 and the compensation sensor piece R_2 both change. R_1 and R_2 are similar strain gauges, and they are attached to the same material, so the resistance changes of R_1 and R_2 are the same, that is, $\Delta R_1 = \Delta R_2$. Since R_1 and R_2 are respectively connected to two adjacent bridge arms of the electric bridge, the effects of resistance changes ΔR_1 and ΔR_2 caused by temperature changes cancel each other out. According to the bridge theory, its output voltage U_{sc} has nothing to do with temperature changes.

The advantage of the bridge compensation method is that it is convenient and simple, and the compensation effect is better at room temperature. The disadvantage is that when the temperature gradient is large, it is difficult for the work sensor piece and the compensation sensor piece to ensure that the temperature is exactly the same, which affects the compensation effect.

$$U_o = \frac{U}{4} \cdot \frac{\Delta R}{R} = \frac{U}{4} S \cdot \varepsilon = \frac{U}{4} S \cdot \frac{F}{EA}$$

从而建立起 U_o—F 关系。梁类弹性体最为常用。在外力作用下，梁弯曲变形，可在上下表面贴应变片，应力状态相反，接入电桥相邻臂组成差动电桥。

常用应变式传感器的弹性体结构如图 4.20 所示。其中，图 4.20(a) 所示为膜片式压力应变传感器；图 4.20(b) 所示为圆柱式力应变传感器；图 4.20(c) 所示为圆环力应变传感器；图 4.20(d) 所示为转矩应变传感器；图 4.20(e) 所示为八角环车削测力仪，可用来同时测量三个互相垂直的力(走刀抗力 F_x、吃刀抗力 F_y 及主削力 F_z)；图 4.20(f) 所示为弹性梁应变加速度计。

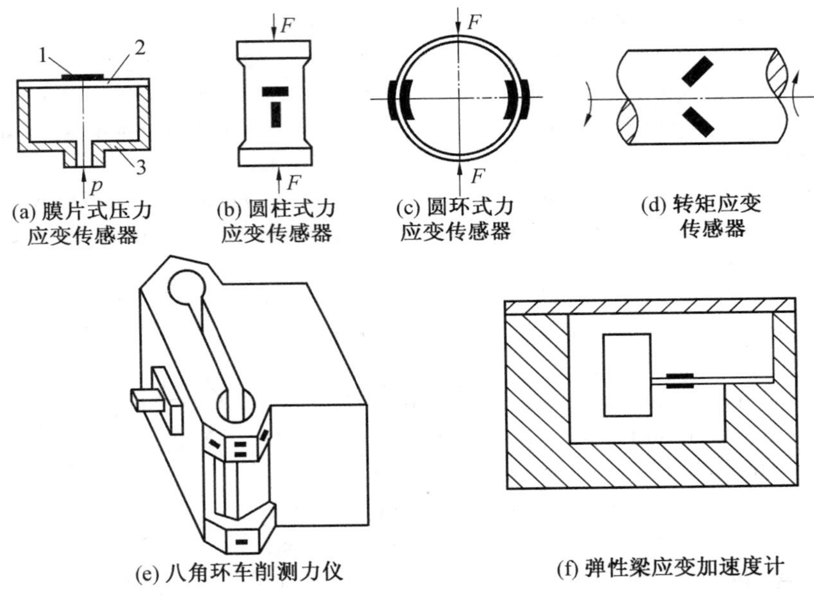

图 4.20　常用应变式传感器的弹性体结构

1— 应变片；2— 膜片；3— 壳体

将应变片贴于弹性体上，可制成多种用途的应变式传感器，用以测量各种能使弹性体产生应变的物理量，如位移、加速度、压力、流量等。这时，被测物理量使弹性体产生与之成正比的应变，这个应变再由应变片转换成其自身电阻的变化。根据应变效应可知，应变片电阻的相对变化与应变片所感受的应变成比例，从而通过电阻与应变、应变与被测量的关系测得被测物理量的大小。

7. 压阻式传感器

半导体应变片典型结构如图 4.21 所示。它的工作原理是基于半导体材料的压阻效应。所谓压阻效应，是指单晶半导体材料在沿某一轴向受到外力作用时，其电阻率 ρ 发生

If a differential bridge is used, it will be used for automatic compensation. Differential bridge compensation method is shown in Figure 4. 19, when measuring the elastic beam bending strain, stick two strain gauges on the upper and lower symmetrical positions respectively, and connect them to the circuit as shown in Figure 4. 18 to form a differential bridge. When the beam is in the same environment where the upper and lower temperatures are the same, the resistance changes caused by the temperature changes of R_1 and R_2 are the same as those in Figure 4. 18, which cancel each other out.

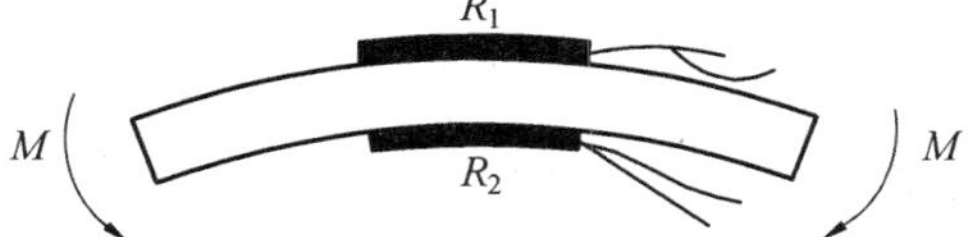

Figure 4. 19 Differential bridge compensation method

6. Strain Sensors

If the strain gauge is attached to the elastic body, the action of force and the physical quantity that can be converted into force, such as torque, displacement, speed, acceleration, can cause the elastic body to strain, causing the patch resistance to change, and outputting voltage through the bridge. In the elastic range, the relationship between strain ε and stress σ and force F is

$$\varepsilon = \sigma/E = F/E \cdot S$$

where E—elastic modulus, kgf/ mm^2(1 kgf=9. 806 65 N);

A—the cross-sectional area of the elastomer, mm^2.

Since the strain gauge sensitivity coefficient $S = \frac{\Delta R}{R}/\varepsilon$, we get $\frac{\Delta R}{R} = S \cdot \varepsilon$, then the output voltage of the bridge is

$$U_o = \frac{U}{4} \cdot \frac{\Delta R}{R} = \frac{U}{4} S \cdot \varepsilon = \frac{U}{4} S \cdot \frac{F}{EA}$$

Thus, the relationship $U_o - F$ has been established. Beam-type elastomers are the most commonly used. Under the action of external force, the beam is bent and deformed, and strain gauges can be attached to the upper and lower surfaces. The stress state is opposite, and the adjacent arms of the bridge are connected to form a differential bridge.

The elastic body structure of commonly used strain sensors is shown in Figure 4. 20. In it, Figure 4. 20(a) shows a diaphragm pressure strain sensor; Figure 4. 20(b) shows a cylindrical force strain sensor; Figure 4. 20(c) shows a ring force strain sensor; Figure 4. 20(d) shows a torque strain sensor; Figure 4. 20(e) shows an octagonal ring turning dynamometer, which can be used to simultaneously measure three mutually perpendicular forces (the feed resistance F_x , the cutting resistance F_y , and the main cutting force F_z); and Figure 4. 20(f) shows a elastic beam strain accelerometer.

Attaching the strain gauges to the elastic body can be made into a multi-purpose strain sensor to measure various physical quantities that can cause the elastic body to strain, such as displacement, acceleration, pressure, flow. At this time, the measured physical quantity causes the elastic body to produce a strain proportional to it, and this strain is converted into a change in its own resistance by the strain gauges. According to the strain effect, the relative change of the strain gauge resistance is proportional to the strain felt by the strain gauge, so that

变化的现象。

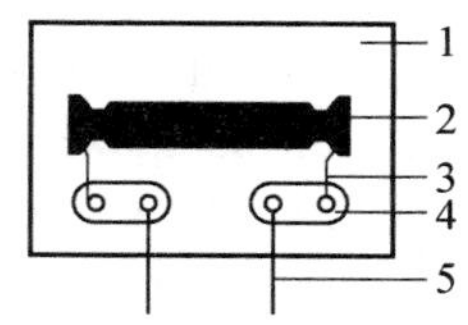

图 4.21　半导体应变片典型结构

1— 胶膜衬底;2—P - Si;3— 内引线;4— 焊接板;5— 外引线

若单晶半导体在外力作用下,则原子点阵排列规律会发生变化,导致载流子迁移率及载流子浓度的变化,从而引起电阻率的变化。

根据式(4.12),$(1+2\nu)\varepsilon$ 项是由几何尺寸变化引起的,$\lambda E\varepsilon$ 是由电阻率变化引起的。对半导体而言,后者远大于前者,它是半导体应变片电阻变化的主要部分,故式(4.12)可简化为

$$\frac{\mathrm{d}R}{R} \approx \lambda E\varepsilon \tag{4.29}$$

这样,半导体应变片灵敏度为

$$S_g = \frac{\mathrm{d}R/R}{\varepsilon} \approx \lambda E \tag{4.30}$$

这一数值为金属丝电阻应变片的 50 ~ 70 倍。

几种常用半导体材料特性见表 4.3。从表中可以看出,材料和载荷施加方向不同,则压阻效应和灵敏度也不同。

表 4.3　几种常用半导体材料特性

材料	电阻率 ρ /($\Omega\cdot$cm)	弹性模量 E /($\times 10^7$ N $\cdot$ cm^{-2})	灵敏度	晶向
P 型硅	7.8	1.87	175	[111]
N 型硅	11.7	1.23	- 132	[100]
P 型锗	15.0	1.55	102	[111]
N 型锗	16.6	1.55	- 157	[111]
N 型锗	1.5	1.55	- 147	[111]
P 型锑化铟	0.54		- 45	[100]
P 型锑化铟	0.01	0.745	30	[111]
N 型锑化铟	0.013		- 74.5	[100]

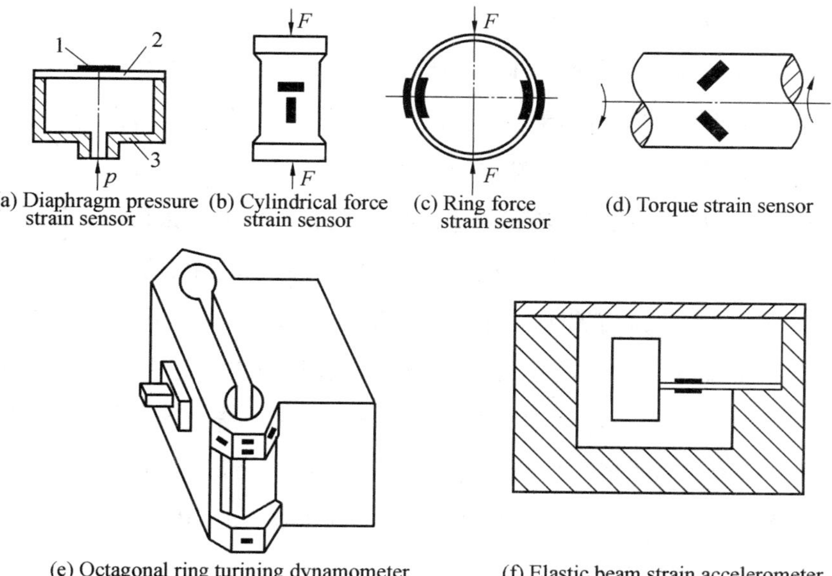

(a) Diaphragm pressure strain sensor (b) Cylindrical force strain sensor (c) Ring force strain sensor (d) Torque strain sensor

(e) Octagonal ring turining dynamometer (f) Elastic beam strain accelerometer

Figure 4.20 The elastic body structure of commonly used strain sensors
1—Strain gauge; 2—Diaphragm; 3—Shell

the measured physical quantity can be measured by the relationship between the resistance and the strain, and the strain and the measured.

7. Piezoresistive Sensors

The typical structure of a semiconductor strain gauge is shown in Figure 4.21. Its working principle is based on the piezoresistive effect of semiconductor materials. The so-called piezoresistive effect refers to the phenomenon that the resistivity ρ of a single crystal semiconductor material will change when an external force is applied along a certain axis.

If the single crystal semiconductors are under the external force, the arrangement of the atomic lattice will change, leading to changes in carrier mobility and carrier concentration, which will cause changes in resistivity.

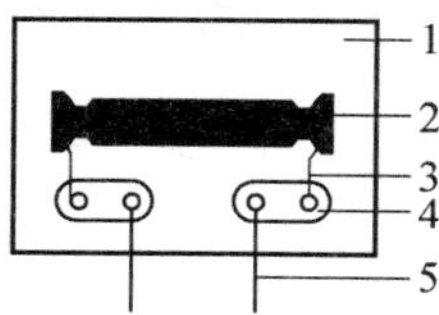

Figure 4.21 The typical structure of a semiconductor strain gauge
1—Adhesive film substrate;2—P-Si;3—Inner lead;4—Welding board;5—Outer lead

According to formula (4.12), $(1+2\nu)\varepsilon$ is caused by the change of geometric dimensions, and $\lambda E\varepsilon$ is caused by the change of resistivity. For semiconductors, the latter is far greater than the former. It is the main part of the resistance change of the semiconductor strain gauge, so the formula (4.12) can be simplified as

$$\frac{\mathrm{d}R}{R} \approx \lambda E\varepsilon \tag{4.29}$$

In this way, the sensitivity of the semiconductor strain gauge is

半导体应变片最突出的优点是灵敏度高，这为它的应用提供了有利条件。另外，由于机械滞后小、横向效应小及它本身体积小等特点，因此扩大了半导体应变片的使用范围。其最大缺点是温度稳定性能差、灵敏度分散度大（受晶向、杂质等因素的影响）及在较大应变作用下非线性误差大等，这些缺点也给使用带来一定困难。

图4.22(a)是压阻式压力传感器的结构示意图。压阻芯片采用周边固定的硅杯结构封装在外壳内。在一块圆形的单晶硅膜片上布置四个扩散电阻，两片位于受压应力区，另外两片位于受拉应力区，它们组成一个全桥测量电路，如图4.22(c)所示。硅膜片用一个圆形硅杯固定，如图4.22(b)所示，这种结构使硅膜片与固定支撑构成一体，既可提高灵敏度，减少非线性误差和滞后，又便于集成化和批量生产。硅膜片的两边有两个压力腔，一个是与被测压力相连的高压腔，另一个是低压腔，接参考压力，通常与大气压相通。当存在压差时，膜片产生变形，使两对电阻的阻值发生变化，进而电桥失去平衡，其输出电压反映膜片两边承受的压差大小。

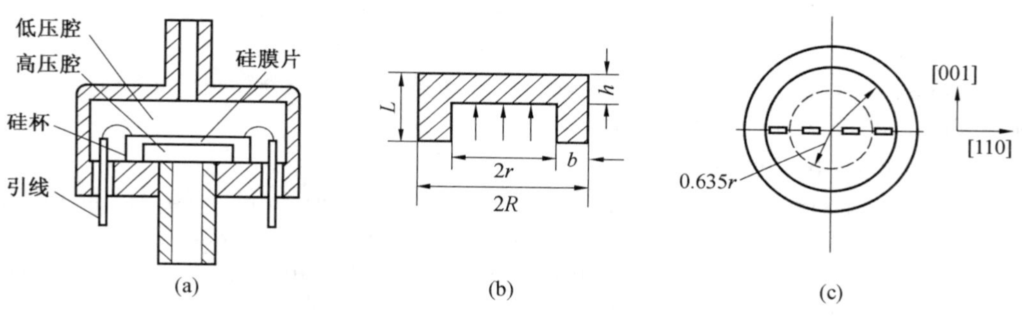

图4.22　压阻式压力传感器

4.3　电感式传感器

4.3.1　概述

电感式传感器的敏感元件是电感线圈，它利用电磁感应原理把被测量（如位移、压力、流量、振动等）转换成线圈的自感或互感的变化，再通过测量电路转换为电流或电压的变化量输出，实现被测非电量到电量的转换。

电感式传感器具有以下特点：结构简单，无活动电触点，因此工作可靠、寿命长；灵敏度和分辨力高，能测出0.01 μm的位移变化，传感器的输出信号强，电压灵敏度可达数百毫伏每毫米位移；线性度高，重复性好，在一定位移范围（几十微米至数毫米）内，非线性误差为0.05% ~0.1%；能实现信号的远距离传输、记录、显示和控制，在工业自动控制系

$$S_g = \frac{\mathrm{d}R/R}{\varepsilon} \approx \lambda E \tag{4.30}$$

This value is 50–70 times of the wire resistance strain gauges.

The characteristics of several commonly used semiconductor materials are listed in Table 4.3. It can be seen from the table that different materials and load application directions have different piezoresistive effects and sensitivity.

Table 4.3 Characteristics of several commonly used semiconductor materials

Material	Resistivity ρ /($\Omega \cdot$ cm)	Elastic modulus E/ ($\times 10^7$ N $\cdot$ cm^{-2})	Sensitivity	Crystal orientation
P-type silicon	7.8	1.87	175	[111]
N-type silicon	11.7	1.23	−132	[100]
P-type germanium	15.0	1.55	102	[111]
N-type germanium	16.6	1.55	−157	[111]
N-type germanium	1.5	1.55	−147	[111]
P-type indium antimonide	0.54		−45	[100]
P-type indium antimonide	0.01	0.745	30	[111]
N-type indium antimonide	0.013		−74.5	[100]

The most prominent advantage of the semiconductor strain gauge is its high sensitivity which provides favorable conditions for its application. In addition, due to the small mechanical hysteresis, small lateral effects, and its own small size, the use of semiconductor strain gauges has been expanded. Its biggest shortcomings are poor temperature stability, large sensitivity dispersion (the influence of crystal orientation, impurities and other factors), and large nonlinear errors under the action of large strains, whose shortcomings also bring certain difficulties to use.

Figure 4.22 (a) is the structure diagram of apiezoresistive pressure sensor. The piezoresistive chip with a fixed silicon cup structure around the periphery is encapsulated in a shell. On a circular single crystal silicon diaphragm, four diffusion resistors are arranged, two of which are located in the compressive stress area, and the other two of which are located in the tensile stress area. They form a full-bridge measurement circuit, as shown in Figure 4.22(c). The silicon diaphragm shown in Figure 4.22(b) is fixed with a circular silicon cup, whose structure makes the silicon diaphragm and the fixed support integral. It not only improves sensitivity, reduces nonlinear errors and hysteresis, but also facilitates integration and mass production. There are two pressure chambers on both sides of the silicon diaphragm, one is a high pressure chamber connected to the measured pressure, and the other is a low pressure chamber connected to the reference pressure, which is usually connected with the atmospheric pressure. When there is a pressure difference, the diaphragm is deformed to change the resistance of the two pairs of resistors, and then the bridge is unbalanced, whose output voltage reflects the pressure difference on both sides of the diaphragm.

统中应用广泛;无输入时存在零位输出电压,可引起测量误差;对激励电源的频率和幅值稳定性要求较高;频率响应较低,不适用于快速、高频动态测量。

电感式传感器种类很多,按照转换所依据的物理效应的不同,可将电感式传感器分为两种:自感型(可变磁阻式)和互感型(差动变压器式)。

4.3.2 自感型——变磁阻式电感传感器

1. 可变磁阻式

可变磁阻式自感型电感传感器是利用线圈自感的变化来实现测量的,它由三部分组成:线圈、铁芯和衔铁。铁芯和衔铁由导磁材料(硅钢片或坡莫合金)制成,在铁芯和衔铁之间有空气隙 δ,传感器的运动部分与衔铁相连。当被测量变化使衔铁产生位移时,会引起磁路中磁阻变化,从而导致线圈的电感发生变化。只要测出该电感的变化,就能确定衔铁位移的大小和方向。

由电工学原理可知,线圈的电感(自感量)L 为

$$L=\frac{W^2}{R_m} \tag{4.31}$$

式中 W——线圈匝数;

R_m——磁路的总磁阻。

由式(4.31)可知,若电感线圈的匝数 W 一定,则当磁阻 R_m 变化时,自感量 L 将随之改变,根据 L 可以求出被测位移 x。因此,自感型电感传感器又称变磁阻式传感器。

图4.23所示的磁路的总磁阻由两部分组成:空气隙的磁阻、衔铁和铁芯的磁阻。则有

$$R_m=\frac{L}{\mu A_1}+\frac{2\delta}{\mu_0 A_0} \tag{4.32}$$

式中 L——磁路中铁芯和衔铁的长度,m;

μ——软铁的磁导率,H/m;

μ_0——空气的磁导率,$\mu_0=4\pi\times10^{-7}$ H/m;

A——铁芯导磁截面积,m^2;

A_0——空气隙导磁截面积,m^2。

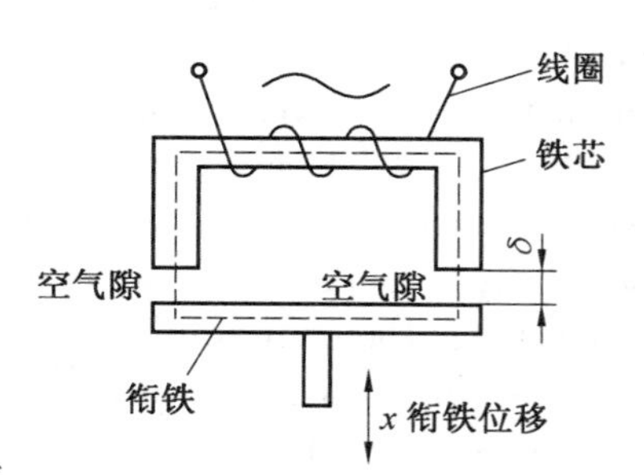

图4.23 可变磁阻式传感器原理图

由于铁芯和衔铁的磁导率 μ 远大于空气的磁导率 μ_0,即铁芯和衔铁的磁阻远小于空气隙的磁阻,因此磁

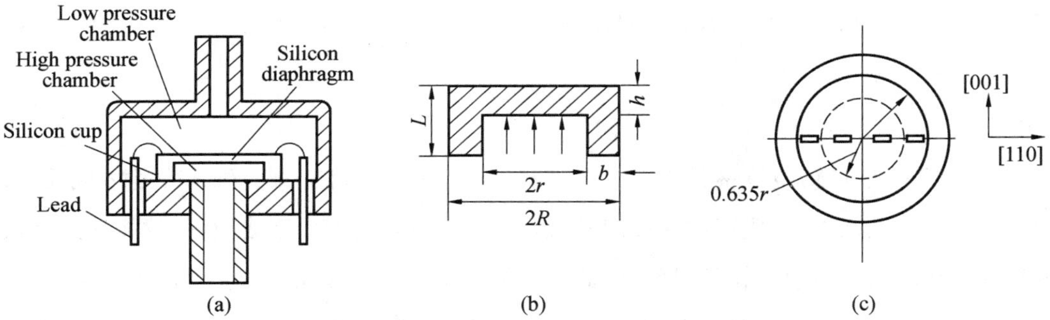

Figure 4.22 Piezoresistive pressure sensor

4.3 Inductive Sensors

4.3.1 Overview

The sensing element of the inductive sensor is an inductance coil, which uses the principle of electromagnetic induction to convert the measured quantity (such as displacement, pressure, flow, vibration) into the change of the self-inductance or mutual inductance of the coil, and then converts the output of the change of current or voltage through the measuring circuit to realize the conversion of the measured non-electricity to the electric quantity.

Inductive sensors have the following characteristics: simple structure, no movable electrical contacts, so reliable work and long life; high sensitivity and resolution, and can measure displacement changes of 0.01 μm, the output signal of the sensor is strong, and the voltage sensitivity can reach hundreds of millivolts per millimeter displacement; high linearity, good repeatability, the nonlinear error is 0.05% –0.1% within a certain displacement range (tens of microns to several millimeters); it can realize long-distance transmission, recording, display and control of signals, and is widely used in industrial automatic control systems; there is a zero output voltage without input, which can cause measurement errors; it has high requirements for the frequency and amplitude stability of the excitation power supply; low frequency response, not suitable for fast, high-frequency dynamic measurement.

There are many types of inductive sensors. Inductive sensors can be divided into two types according to the different physical effects on which the conversion is based: self-inductance (variable reluctance) and mutual inductance (differential transformer).

4.3.2 Self-Inductance—Variable Magnetoresistive Inductance Sensor

1. Variable Reluctance Type

Variable reluctance self-inductance type inductive sensor which consists of three parts: coil, iron core and armature uses the change of coil self-inductance to achieve measurement. The core and armature are made of magnetic material (silicon steel sheet or permalloy). There is an air gap δ between the core and the armature. The moving part of the sensor is connected to the armature. When the measured quantity change causes the armature to move and causes the magnetic resistance in the magnetic circuit to change, it will lead to a change in the inductance

路中的总磁阻可只考虑空气隙的磁阻这一项,故 $R_m \approx \dfrac{2\delta}{\mu_0 A_0}$。将此式代入式(4.31)可得

$$L = \frac{W^2\mu_0 A_0}{2\delta} \tag{4.33}$$

式(4.33)是自感型传感器的工作原理表达式。由该式可知,自感 L 与气隙 δ 成反比,而与气隙导磁截面积 A_0 成正比。被测量只要能够改变空气隙厚度或面积,就能达到将被测量的变化转换成自感变化的目的,由此可构成间隙变化型和面积变化型两种类型。

图 4.24(a)所示为间隙变化型电感传感器,W、μ_0 及 A_0 都不可变,δ 可变。当工件直径变化引起衔铁移动时,磁路中气隙的磁阻将发生变化,从而引起线圈电感的变化,由此可判断衔铁的位移(即被测工件直径的变化)值。

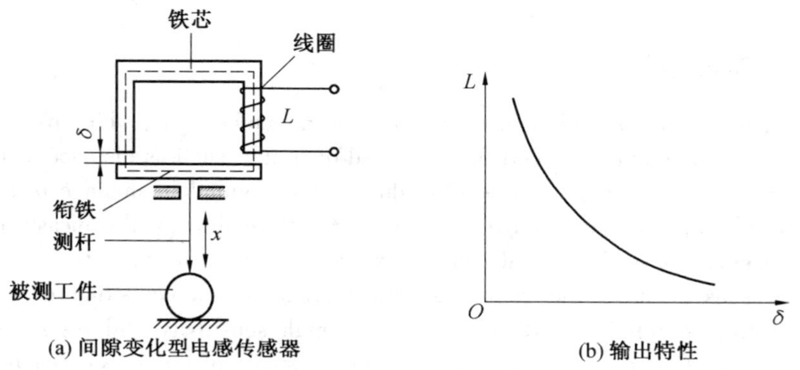

图 4.24　间隙变化型变磁阻式传感器结构及其输出特性

由式(4.33)可知,L—δ 的关系是双曲线关系,即非线性关系(图 4.24(b)),其灵敏度为

$$S = \frac{\mathrm{d}L}{\mathrm{d}\delta} = -\frac{W^2\mu_0 A_0}{2\delta^2} = -\frac{L}{\delta} \tag{4.34}$$

为保证传感器的线性度,限制非线性误差,间隙变化型电感传感器多用于微小位移的测量。实际应用中,一般取 $\Delta\delta/\delta_0 \leqslant 0.1$,位移测量范围为 0.001 ~ 1 mm。

图 4.25(a)所示为面积变化型变磁阻式传感器,W、μ_0 及 δ 均不变,铁芯和衔铁之间的相对覆盖面积(即磁通截面)随被测量的变化而改变,从而改变磁阻。由于磁路截面积变化了 ΔA,因此传感器的电感改变 ΔL,实现了被测参数到电参量 ΔL 的转换。由式(4.33)可知,L—A_0(输出—输入)成线性(图 4.25(b)),其灵敏度为

$$S = \frac{\mathrm{d}L}{\mathrm{d}A} = \frac{W^2\mu_0}{2\delta_0} = 常数 \tag{4.35}$$

这种传感器自由行程限制小,示值范围较大,线性度良好,灵敏度为常数(但灵敏度

of the coil. As long as the change in inductance is measured, the magnitude and direction of the armature displacement can be determined.

According to the principle of electrical engineering, the inductance (self-inductance) L of the coil is

$$L = \frac{W^2}{R_m} \tag{4.31}$$

where W—the number of turns of the coil;

R_m—the total magnetic resistance of the magnetic circuit.

It can be seen from formula (4.31) that if the number of turns W of the inductance coil is constant, then when the magnetic resistance R_m changes, the self-inductance L will change accordingly, and the measured displacement x can be obtained from L. Therefore, the self-inductance type inductive sensor is also known as variable reluctance sensor.

The total reluctance of the magnetic circuit shown in Figure 4.23 consists of two parts: the reluctance of the air gap, the reluctance of the armature and the iron core. So we have

$$R_m = \frac{L}{\mu A_1} + \frac{2\delta}{\mu_0 A_0} \tag{4.32}$$

where L—the length of the core and armature in the magnetic circuit, m;

μ—the magnetic permeability of soft iron, H/m;

μ_0—permeability of air, $\mu_0 = 4\pi \times 10^{-7}$ H/m;

A—core magnetic cross-sectional area, m^2;

A_0—air gap magnetic cross-sectional area, m^2.

Since the magnetic permeability of the core and armature μ is much greater than the permeability of air μ_0, that is, the magnetic resistance of the core and armature is much smaller than that of the air gap, so the total reluctance in the magnetic circuit can only consider the reluctance of the air gap, so $R_m \approx \frac{2\delta}{\mu_0 A_0}$. Substituting it into formula (4.31), we get

$$L = \frac{W^2 \mu_0 A_0}{2\delta} \tag{4.33}$$

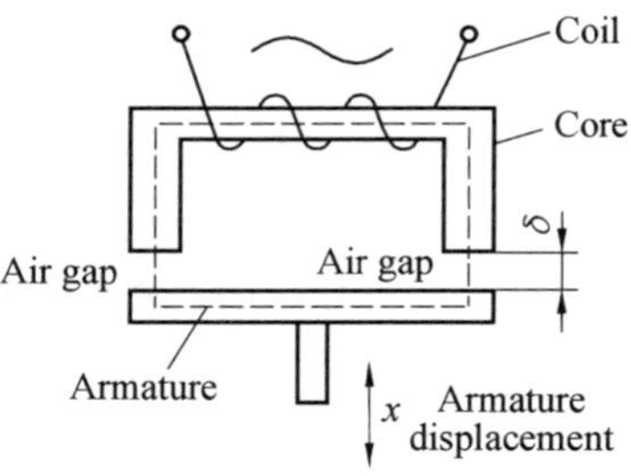

Figure 4.23 Schematic diagram of variable reluctance sensors

Formula (4.33) is an expression of the working principle of a self-inductive sensor. From this formula, the self-inductance L is inversely proportional to the air gap δ, and proportional to the air gap magnetic cross-sectional area A_0. As long as the thickness or area of the air gap can be changed, the measured quantity change can be converted into a self-inductance change, constituting gap change type and variable area type.

Figure 4.24(a) shows a gap change type inductive sensor, W, μ_0 and A_0 are not variable, and δ is variable. When the change in the diameter of the workpiece causes the armature to move, the magnetic resistance of the air gap in the magnetic circuit will change, with causing the change of the coil inductance, and the displacement of the armature (that is, the change in the diameter of the measured workpiece) can be judged.

From formula (4.33), it is known that the relationship of $L - \delta$ is a hyperbolic relationship, that is, a non-linear relationship (Figure 4.24(b)), the sensitivity is

较低)。若将衔铁做成转动式,则可测量角位移。

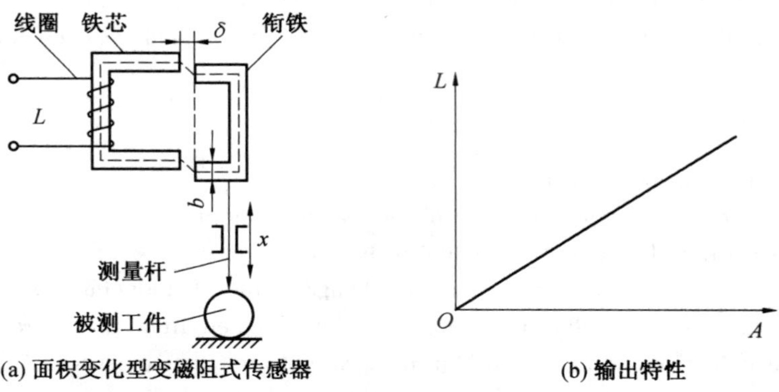

图 4.25　面积变化型变磁阻式传感器结构及其输出特性

图 4.26 所示为单螺管线圈型变磁阻式传感器结构。当铁芯在线圈中运动时,将改变磁阻,使线圈自感发生变化。这种传感器结构简单、制造容易,但灵敏度低,适用于较大位移(数毫米)测量。

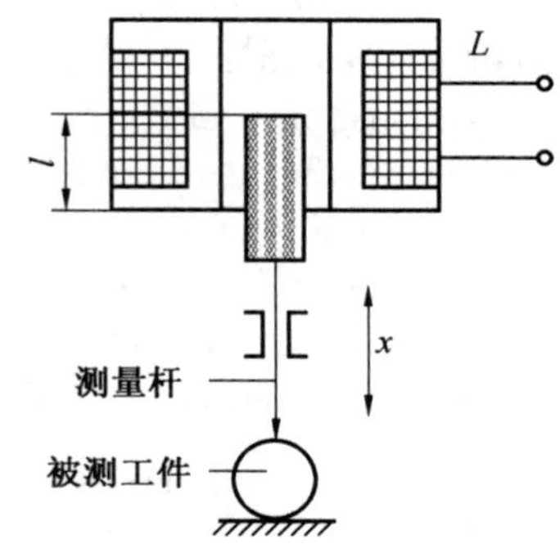

图 4.26　单螺管线圈型变磁阻式传感器结构

实际应用中,常将两个完全相同的线圈与一个共用的活动衔铁结合在一起,构成差动结构。图 4.27 所示为间隙变化型差动式电感传感器的结构和输出特性。

当衔铁位于气隙的中间位置时,$\delta_1 = \delta_2$,两线圈的电感相等($L_1 = L_2 = L_0$),总的电感 $L_1 - L_2 = 0$;当衔铁偏离中间位置时,一个线圈的线圈电感增加为 $L_1 = L_0 + \Delta L$,另一个线圈的电感减小为 $L_2 = L_0 - \Delta L$,总的电感变化量为

$$L_1 - L_2 = (+\Delta L) - (-\Delta L) = 2\Delta L$$

于是,差动式电感传感器的灵敏度 S 为

$$S = \frac{\mathrm{d}L}{\mathrm{d}\delta} = -2\frac{L}{\delta} \tag{4.36}$$

与式(4.34)相比可知,差动结构比单边式传感器的灵敏度提高 1 倍。从图 4.27(b)

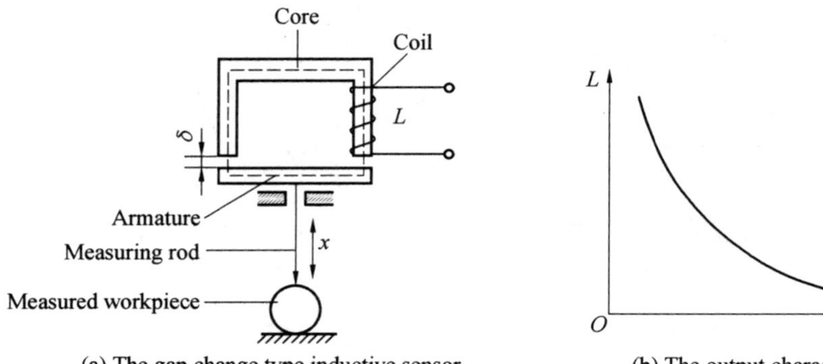

(a) The gap change type inductive sensor (b) The output characteristics

Figure 4.24 The structure and output characteristics of the gap change type variable magnetoresistive sensor

$$S = \frac{\mathrm{d}L}{\mathrm{d}\delta} = -\frac{W^2\mu_0 A_0}{2\delta^2} = -\frac{L}{\delta} \tag{4.34}$$

In order to ensure the linearity of the sensor and limit the non-linear error, the gap change type inductive sensor is mostly used for the measurement of small displacement. In practical applications, $\Delta\delta/\delta_0 \leqslant 0.1$ is generally taken, and the displacement measurement range is 0.001–1 mm.

Figure 4.25(a) shows the variable magnetoresistive sensor with variable, W, μ_0 and δ are unchanged, and the relative coverage area (that is, the magnetic flux cross-section) between the core and the armature changes with the measured change, thereby changing the magnetic resistance. Because the cross-sectional area of the magnetic circuit changes by ΔA, the inductance of the sensor changes by ΔL to realize the conversion of the measured parameter to the electrical parameter ΔL. From formula(4.33), we can see that $L - A_0$ (output-input) is linear (Figure 4.25(b)). Its sensitivity is

$$S = \frac{\mathrm{d}L}{\mathrm{d}A} = \frac{W^2\mu_0}{2\delta_0} = \text{constant} \tag{4.35}$$

This kind of sensor has a small free travel limit, a large display range, good linearity, and a constant sensitivity (but the sensitivity is low). If the armature is made into a rotating type, the angular displacement can be measured.

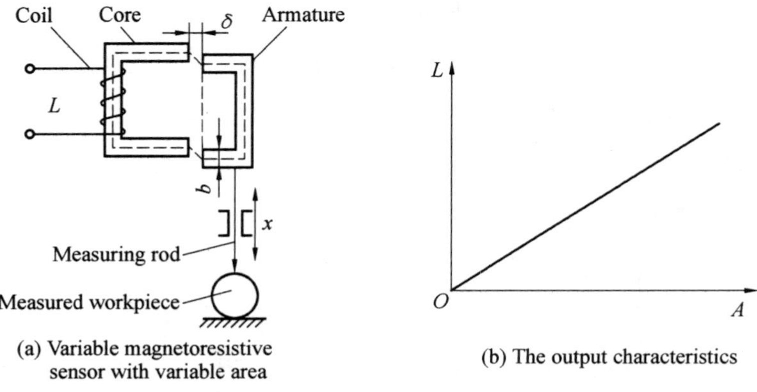

(a) Variable magnetoresistive sensor with variable area (b) The output characteristics

Figure 4.25 Structure and output characteristics of variable magnetoresistive sensor with variable area

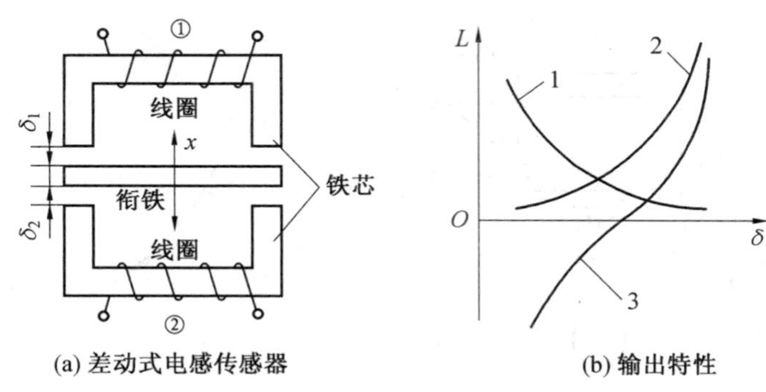

(a) 差动式电感传感器　　(b) 输出特性

图 4.27　间隙变化型差动式电感传感器的结构和输出特性

1— 线圈 ① 的输出曲线;2— 线圈 ② 的输出曲线;3— 差动式传感器的输出特性

中可见,其输出线性度改善很多。面积变化型和螺线管型也可以构成差动结构(图 4.28)。图 4.28(b) 所示的差动螺丝管型结构中,其总电感的变化是单一螺管型电感变化量的 2 倍,可以部分消除磁场不均匀造成的非线性,测量范围为 0 ~ 300 μm,最高分辨率可达 0.5 μm。

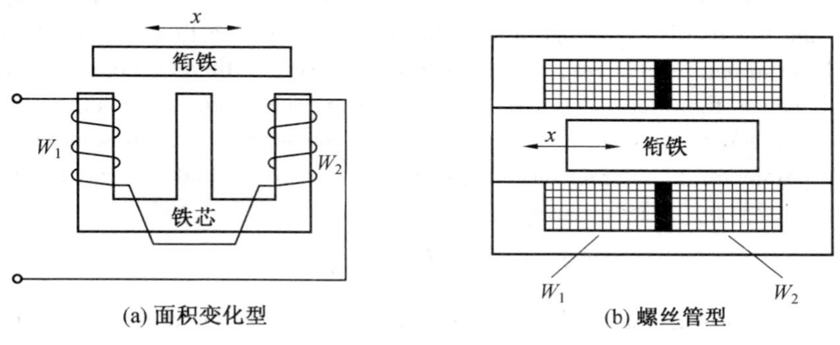

(a) 面积变化型　　(b) 螺丝管型

图 4.28　面积变化型、螺线管型差动式传感器

2. 调理电路 —— 交流电桥、调幅电路

前面有关电阻应变式传感器的章节中讲到了调理电路中直流电桥的原理和应用。当电桥电路工作在交流状态下时,电路中各种寄生电容和电感等都要起作用,此时就需要复数阻抗对电路进行分析。与直流电桥不同的是,交流电桥采用幅值和频率恒定的交流电源供电,适用于电容式和电感式传感器。

一些静态或缓变的被测量经过传感器得到的信号多为缓变信号,采用直流放大器直接放大处理这样的信号会遇到两个难题 —— 放大器的零漂和外界低频干扰。为解决这些问题,经常先是将直流或缓变信号变成较高频率的交变信号,然后采用交流放大器进行放大,最后再从高频交变信号中将直流或缓变信号提取出来。这些过程一般称为调制和

Figure 4.26 shows the structure of single solenoid coil type variable magnetoresistive sensor. When the iron core moves in the coil, it will change the magnetic resistance and change the self-inductance of the coil. This kind of sensor is simple in structure and easy to manufacture, but has low sensitivity and is suitable for large displacement (several millimeters) measurement.

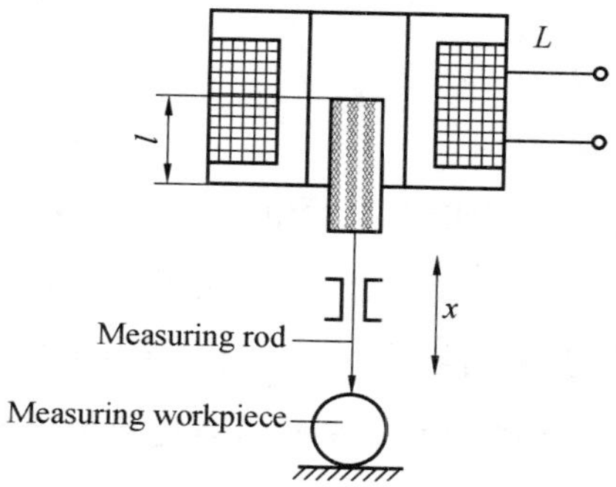

Figure 4.26 Structure of single solenoid coil type variable magnetoresistive sensor

In practical applications, two identical coils are often combined with a common movable armature to form a differential structure. Figure 4.27 shows the structure and output characteristics of a differential inductive sensor with a changing gap.

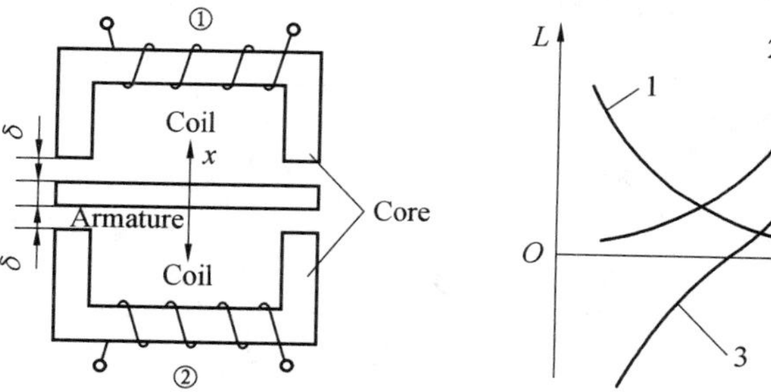

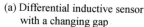

(a) Differential inductive sensor with a changing gap

(b) The output characteristics

Figure 4.27 The structure and output charactreristics of a differential inductive sensor with a changing gap
1—Outpt curve of coil ①; 2—Output curve of coil ②; 3—The output characteristics of the differential sensor

When the armature is in the middle of the air gap, $\delta_1 = \delta_2$, the inductance of the two coils are equal ($L_1 = L_2 = L_0$), the total inductance is $L_1 - L_2 = 0$; when the armature deviates from the middle position, the coil inductance of one coil increases to $L_1 = L_0 + \Delta L$, and the inductance of the other coil decreases to $L_2 = L_0 - \Delta L$, the total inductance change is

$$L_1 - L_2 = (+\Delta L) - (-\Delta L) = 2\Delta L$$

So the sensitivity S of the differential inductive sensor is

$$S = \frac{\mathrm{d}L}{\mathrm{d}\delta} = -2\frac{L}{\delta} \tag{4.36}$$

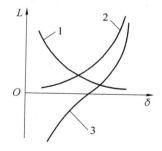

Compared with formula (4.34), it can be seen that the sensitivity of the differential structure is 1 times higher than that of the unilateral sensor. It can be seen from Figure 4.27 (b) that the output linearity has improved a lot. Variable area type and solenoid type can also

解调。根据载波被调制的参数不同,调制可分为三种模式:调幅(AM)、调频(FM)和调相(PM)。这三种模式分别是载波的幅值 A、频率 f 和相位 P 各自随调制信号的幅值做线性变化的过程。三种模式的已调制波分别称为调幅波、调频波和调相波。

本节详细讨交流电桥的原理和应用,同时介绍电感式传感器测量装置中调幅及其解调的基本方法和原理。

(1) 交流电桥。

交流电桥电路结构与直流电桥相似,交流电桥的基本形式如图 4.29 所示。其分析过程与直流电桥类似,但要注意以下区别:交流电桥由于工作在交流状态,电路中各种寄生电容和电感等都要起作用,因此采用复数阻抗进行分析比较方便;交流电桥的平衡既要考虑阻抗模平衡,又要考虑阻抗相角平衡,故平衡装置比直流电桥复杂,需要可变电阻器和可变电抗器两套平衡装置,且平衡过程复杂;一般采用高频正弦波供电,其输出为调幅波,故需要解调电路;使用时要注意导线、元器件等各种分布电容、电感及寄生电容、电感等的影响。

交流电桥不仅适用于电阻式传感器,也适用于电容式传感器和电感式传感器。

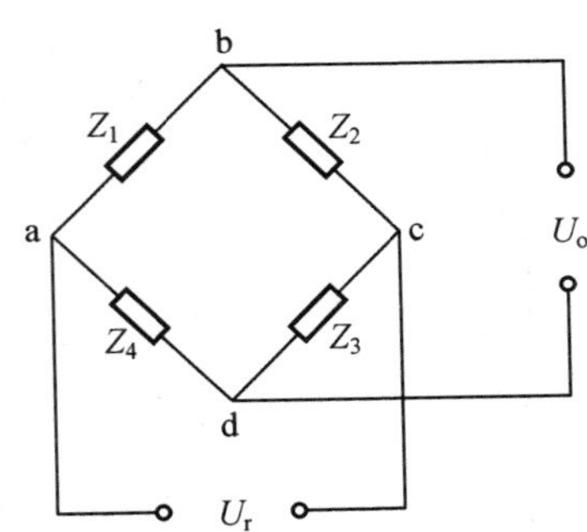

图 4.29　交流电桥的基本形式

若将交流电桥的阻抗、电流及电压用复数表示,则图 4.29 所示交流电桥的输出电压为

$$\dot{U}_o = \frac{Z_1Z_3 - Z_2Z_4}{(Z_1 + Z_2)(Z_3 + Z_4)}\dot{U}_r \tag{4.37}$$

各桥臂的复数阻抗为

$$Z_i = Z_{0i}e^{j\varphi_i}, \quad i = 1,2,3,4 \tag{4.38}$$

式中　Z_{0i}—— 复数阻抗的模;

φ_i—— 复数阻抗的相角。

交流电桥的平衡条件为

form a differential structure (Figure 4.28). In the differential solenoid structure shown in Figure 4.28(b), the total inductance change is twice that of a single solenoid inductance, which can partially eliminate the non-linearity caused by the uneven magnetic field. The measurement range is 0–300 μm, the highest and the resolution can reach 0.5 μm.

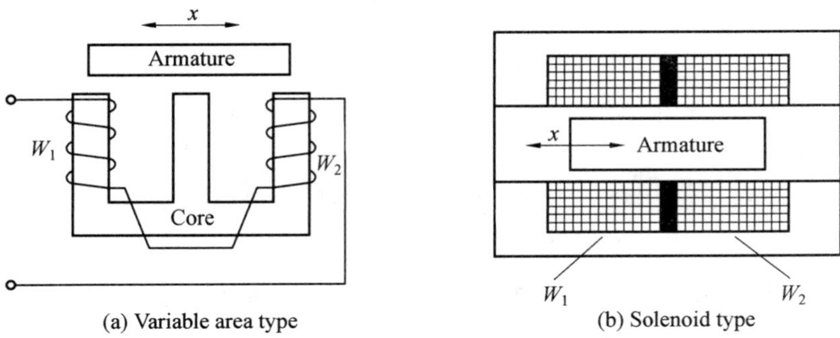

(a) Variable area type (b) Solenoid type

Figure 4.28 Variable area type, solenoid type differential sensor

2. Conditioning Circuit—AC Bridge, Amplitude Modulation Circuit

The principle and application of the DC bridge in the conditioning circuit were discussed in the previous chapter of the resistance strain sensor. When the bridge circuit is working in AC state, all kinds of parasitic capacitances and inductances in the circuit must take effect with complex impedance being needed to analyze the circuit. Different from the DC bridge, the AC bridge uses an AC power supply with constant amplitude and frequency, suitable for capacitive and inductive sensors.

Some static signals that are measured through the sensor are mostly slowly varying signals. Using a DC amplifier to directly amplify and process such signals will encounter two problems—zero drift of the amplifier and external low-frequency interference. In order to solve these problems, the DC or slow-changing signal is often first converted into a higher frequency alternating signal, and then an AC amplifier is used for amplification, and finally the DC or the slow-changing signal is extracted from the high-frequency alternating signal. These processes are generally called modulation and demodulation. According to the difference of modulated parameters of the carrier, modulation can be divided into three modes: amplitude modulation (AM), frequency modulation (FM) and phase modulation (PM). They are the processes in which the amplitude A, frequency f and phase P of the carrier vary linearly with the amplitude of the modulated signal. The modulated waves of the three modes are called amplitude-modulated wave, frequency-modulated wave and phase-modulated wave.

This section discusses the principles and applications of AC bridges in detail, and have an induction into the basic methods and principles of amplitude modulation and demodulation in inductive sensor measurement devices.

(1) AC bridge.

The circuit structure of an AC bridge is parallel to that of a DC bridge, the basic form of AC bridge is shown in Figure 4.29. The analysis process is similar to that of a DC bridge, but the following differences should be noted: since the AC bridge works in AC state, various parasitic capacitances and inductances in the circuit must play a role, so it is more convenient to use complex impedance for analysis; the balance of AC bridge must consider both impedance

$$Z_1 Z_3 = Z_2 Z_4 \tag{4.39}$$

即

$$Z_{01} Z_{03} e^{j(\varphi_1+\varphi_3)} = Z_{02} Z_{04} e^{j(\varphi_2+\varphi_4)} \tag{4.40}$$

根据复数相等的条件可得

$$\begin{cases} Z_{01} Z_{03} = Z_{02} Z_{04} \\ \varphi_1 + \varphi_3 = \varphi_2 + \varphi_4 \end{cases} \tag{4.41}$$

式(4.41)表明,交流电桥平衡要满足两个条件:两相对桥臂阻抗模的乘积相等;其阻抗相角的和相等。

由于交流电桥平衡必须满足模和相角的两个条件,因此桥臂结构需采取不同的组合方式,以满足相对桥臂阻抗相角之和相等这一条件。

图 4.30 所示为一种常见的电容电桥,电桥中两相邻桥臂为纯电阻 R_2 和 R_3,而另两相邻桥臂为电容 C_1 和 C_4。其中,R_1 和 R_4 可视为电容介质损耗的等效电阻。根据式(4.39)的平衡条件,有

$$\left(R_1 + \frac{1}{j\omega C_1}\right) R_3 = \left(R_4 + \frac{1}{j\omega C_4}\right) R_2 \tag{4.42}$$

展开得

$$R_1 R_3 + \frac{R_3}{j\omega C_1} = R_2 R_4 + \frac{R_2}{j\omega C_4}$$

根据复数相等的条件(实部、虚部分别相等),可得

$$\begin{cases} R_1 R_3 = R_2 R_4 \\ \dfrac{R_3}{C_1} = \dfrac{R_2}{C_4} \end{cases} \tag{4.43}$$

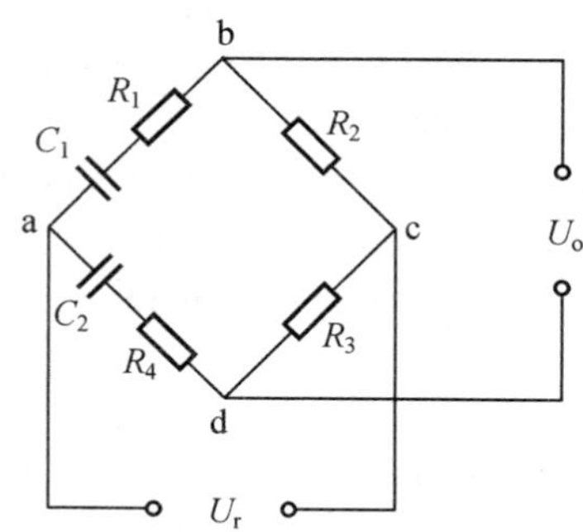

图 4.30 电容电桥

由式(4.43)可知,为达到电桥平衡,必须同时调节电容与电阻两个参数,使之分别取

mode balance and impedance phase angle balance, therefore the balancing device is more complicated than the DC bridge, requiring two sets of balancing devices, a variable resistor and a variable reactor, and the balancing process is complicated; generally, high-frequency sine wave is used for power supply, and the output is an amplitude modulation wave, so a demodulation circuit is required; when using it, we should pay attention to the influence of various distributed capacitance and inductance, and parasitic capacitance and inductance such as wires and components.

The AC bridge is not only suitable for resistive sensors, but also for capacitive sensors and inductive sensors.

If the impedance, current and voltage of the AC bridge are represented by complex numbers, the output voltage of the AC bridge shown in Figure 4.29 is

$$\dot{U}_o = \frac{Z_1 Z_3 - Z_2 Z_4}{(Z_1 + Z_2)(Z_3 + Z_4)} \dot{U}_r \tag{4.37}$$

The complex impedance of each bridge arm is

$$Z_i = Z_{0i} e^{j\varphi_i}, \quad i = 1,2,3,4 \tag{4.38}$$

where Z_{0i} —the modulus of complex impedance;

φ_i —the phase angle of the complex impedance.

Figure 4.29 Basic form of AC bridge

The balance condition of the AC bridge is

$$Z_1 Z_3 = Z_2 Z_4 \tag{4.39}$$

that is

$$Z_{01} Z_{03} e^{j(\varphi_1 + \varphi_3)} = Z_{02} Z_{04} e^{j(\varphi_2 + \varphi_4)} \tag{4.40}$$

According to the condition that the plural numbers are equal, we get

$$\begin{cases} Z_{01} Z_{03} = Z_{02} Z_{04} \\ \varphi_1 + \varphi_3 = \varphi_2 + \varphi_4 \end{cases} \tag{4.41}$$

Formula (4.41) shows that the AC bridge balance must meet two conditions: the product of the impedance modes of the two opposing bridge arms is equal; the sum of the impedance phase angles is equal.

Since the AC bridge balance must meet the two conditions of mode and phase angle, the bridge arm structure needs to adopt different combinations to meet the condition that the sum of the relative bridge arm impedance phase angles is equal.

Figure 4.30 shows a common capacitive bridge. Two adjacent bridge arms in the bridge are pure resistors R_2 and R_3, while the other two adjacent bridge arms are capacitors C_1 and C_4, where R_1 and R_4 can be regarded as the equivalent resistance of the dielectric loss of the capacitor. According to the balance condition of formula (4.39), there is

$$\left(R_1 + \frac{1}{j\omega C_1}\right) R_3 = \left(R_4 + \frac{1}{j\omega C_4}\right) R_2 \tag{4.42}$$

Expand it, we get

$$R_1 R_3 + \frac{R_3}{j\omega C_1} = R_2 R_4 + \frac{R_2}{j\omega C_4}$$

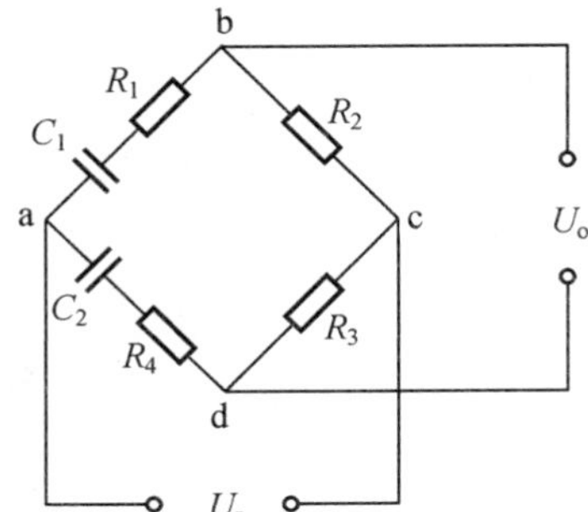

Figure 4.30 Capacitance bridge

According to the condition that complex numbers are equal (the real part and imaginary part

得电阻和容抗的平衡。

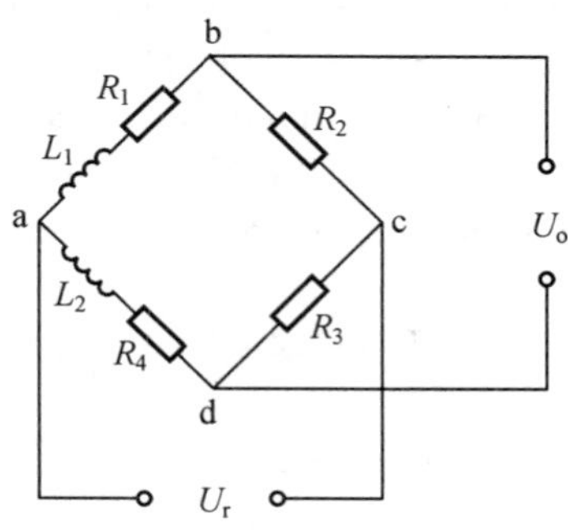

图 4.31　电感电桥

图 4.31 所示为一种常见的电感电桥,两相邻桥臂为电感 L_1 和 L_4 与电阻 R_2 和 R_3。根据式(4.41),电桥平衡条件为

$$R_1R_3 = R_2R_4 \tag{4.44}$$

$$L_1R_3 = L_4R_2 \tag{4.45}$$

由于交流电桥的平衡必须同时满足阻抗模平衡与阻抗相角平衡两个条件,因此相比于直流电桥,其平衡调节要复杂得多。即使是纯电阻交流电桥,电桥导线之间形成的分布电容及电阻本身的寄生电容也会产生影响,相当于在各桥臂的电阻上并联了一个电容。纯电阻交流电桥的分布电容图如图 4.32 所示。为此,在调电阻平衡时还需进行电容的调平衡。图 4.33 所示为一种动态电阻应变仪采用的纯电阻电桥,平衡调节由变阻器 R_3 和差动可变电容器 C 配合进行,而且需要反复调节,才能最终达到平衡。早期的电阻应变仪采用的是人工手动调平衡,所以要花费很长时间;目前的电阻应变仪广泛使用微控制器,调平衡一般都是由微控制器自动完成的。

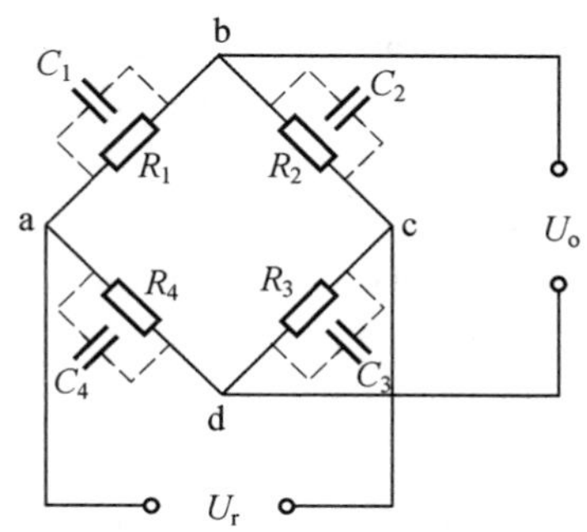

图 4.32　纯电阻交流电桥的分布电容图

在交流电桥的使用中,影响交流电桥测量精度及误差的因素比直流电桥要多得多,如电容的泄漏电阻、无感电阻的残余电抗、电桥各元件之间的互感耦合、元件间及元件对地

are equal respectively), we can get

$$\begin{cases} R_1R_3 = R_2R_4 \\ \dfrac{R_3}{C_1} = \dfrac{R_2}{C_4} \end{cases} \tag{4.43}$$

It can be seen from formula (4.43) that in order to achieve the bridge balance, the two parameters of capacitance and resistance must be adjusted at the same time, so that the resistance and capacitive reactance are balanced respectively.

Figure 4.31 shows a common inductance bridge. Two adjacent bridge arms are inductance L_1 and L_4, and resistance R_2 and R_3. According to formula (4.41), the bridge balance condition is

$$R_1R_3 = R_2R_4 \tag{4.44}$$

$$L_1R_3 = L_4R_2 \tag{4.45}$$

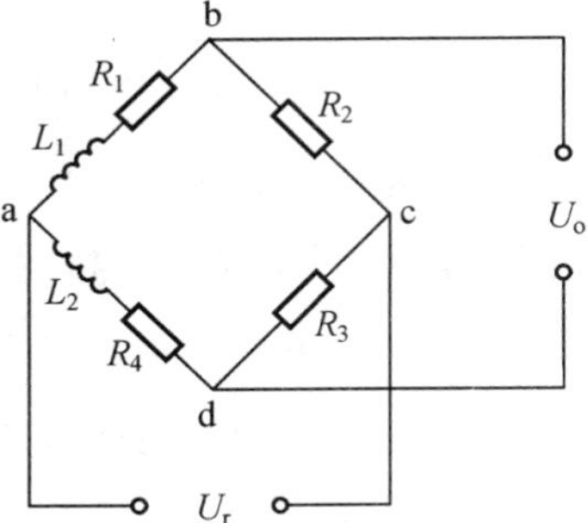

Figure 4.31 Inductance bridge

Since the balance of the AC bridge must meet the two conditions of impedance mode balance and impedance phase angle balance at the same time, its balance adjustment is much more complicated than that of a DC bridge. Even if it is a pure resistance AC bridge, the distributed capacitance formed between the bridge wires and the parasitic capacitance of the resistance itself will have an impact, which is equivalent to connecting a capacitor in parallel to the resistance of each bridge arm. Distributed capacitance diagram of pure resistance AC bridge is shown in Figure 4.32. For this reason, it is necessary to adjust the capacitance when adjusting the resistance balance. Figure 4.33 shows a pure resistance bridge used in a dynamic resistance strain gauge. The balance adjustment is performed by the cooperation of the rheostat R_3 and the differential variable capacitor C, and it needs to be adjusted repeatedly to finally achieve balance. Early resistance strain gauges used manual balance adjustment, so it took a long time; the current resistance strain gauges widely use microcontrollers, and the balance adjustment is usually done automatically by the microcontroller.

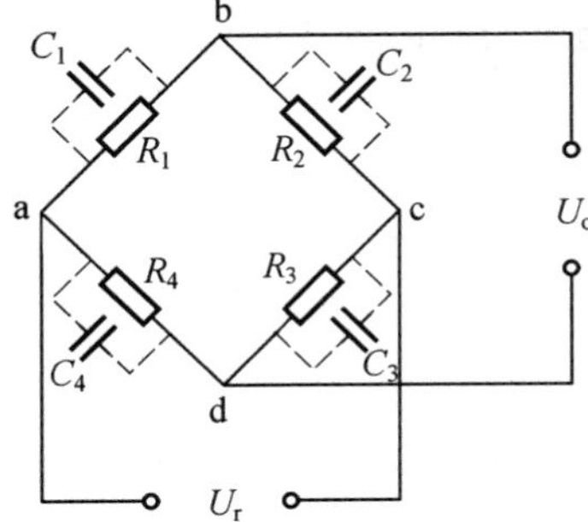

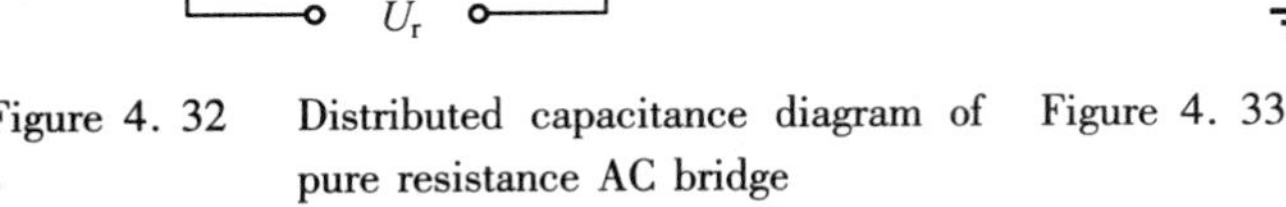

Figure 4.32 Distributed capacitance diagram of pure resistance AC bridge

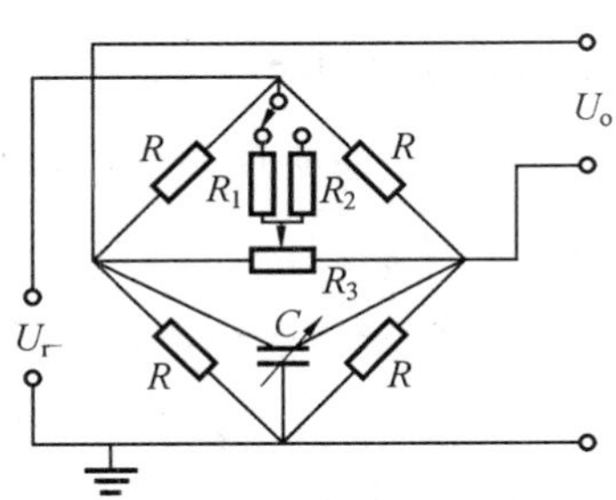

Figure 4.33 Pure resistance bridge used in a dynamic resistance strain gauge

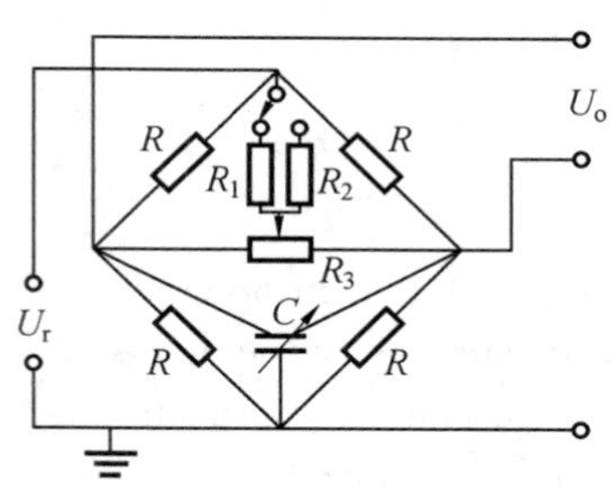

图 4.33　动态电阻应变仪采用的纯电阻电桥

之间的分布电容、邻近交流电路对电桥的感应影响等。对此，应尽可能地采取适当措施加以消除。另外，对交流电桥的激励电源，要求其电压波形和频率必须具有很好的稳定性，否则将影响到电桥的平衡。当电源电压波形畸变时，其中也包含了高次谐波。对于基波来说，电桥达到了平衡；而对于高次谐波来说，电桥不一定能平衡，因此会有高次谐波的不平衡电压输出。

交流电桥一般采用音频交流电源(5 ~ 10 kHz) 作为电桥电源。这样，电桥输出将为调幅波，后接交流放大电路简单而无零漂，并且解调电路和滤波电路容易去除干扰而获得有用信号。

(2) 调幅原理及其解调。

调制就是利用输入信号控制高频交流载波的某个参数，使该参数随输入信号而变化，从而实现将直流或缓变信号变成高频交变信号的过程。解调是从高频交变已调信号中提取原输入信号的过程。

线性调制模型如图 4.34 所示。图中，M 是调制器；$x(t)$ 是输入信号，即要检测或传输的信号，称为调制信号或调制波；$y(t)$ 是高频载波，通常由载波发生器(又称高频振荡器)提供；$S(t)$ 是调制器的输出，称为已调波。

调制的目的包括：实现频率变换，使信息能在给定的信道内传输(即满足工作频带要求)；实现信道多路复用，使信息能充分利用给定的信道频带；提高抗干扰能力。

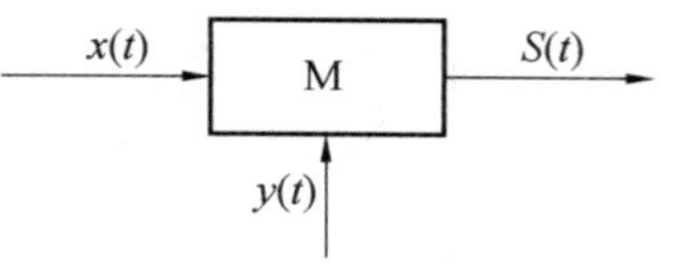

图 4.34　线性调制模型

调幅就是用调制信号对载波的幅值进行调制，使载波的幅值随调制信号的大小而成比例变化。已调波(调幅波) 的幅值反映了调制信号的大小。调幅可以采用不同的方法实现，如平方律调幅、斩波调幅、模拟乘法器调幅等。本节主要介绍测量装置常见的模拟乘法器调幅原理。例如，交流电桥、差动变压器式传感器本身就相当于一个乘法器，输出的都是调幅波。在交流电桥中，供电电压为载波，桥臂上元件值的变化量为调制信号。在差动变压器中，一次绕组

In the use of AC bridges, there are many more factors that affect the accuracy and error of AC bridges than DC bridges, such as the leakage resistance of capacitors, residual reactance of non-inductive resistance, mutual inductance coupling between the components of the bridge, the distributed capacitance between the components and the ground, and the inductive influence of the adjacent AC circuit on the bridge, which should be eliminated as much as possible by appropriate measures. In addition, for the excitation power supply of the AC bridge, the voltage waveform and frequency must have good stability, otherwise it will affect the balance of the bridge. When the power supply voltage waveform is distorted, it also contains high-order harmonics. For the fundamental wave, the bridge is balanced; but for the higher harmonics, the bridge may not be able to balance, so there will be an unbalanced voltage output of the higher harmonics.

The AC bridge will generally use audio AC power (5 - 10 kHz) as the bridge power supply. In this way, the bridge output will be an amplitude modulated wave, followed by an AC amplifier circuit that is simple without zero drift, and the demodulation circuit and filter circuit are easy to remove interference to obtain useful signals.

(2) Principle of amplitude modulation and its demodulation.

Modulation is to control a certain parameter of the high-frequency AC carrier by using the input signals, making the parameter change with the input signals so as to realize the process of changing the DC or slowly changing signals into high-frequency alternating signals. Demodulation is the process of extracting the original input signals from the high-frequency alternating modulated signals.

Linear modulation model is shown in Figure 4.34. In the figure, M is the modulator; $x(t)$ is the input signal, that is, the signal to be detected or transmitted, called the modulation signal or modulated wave; $y(t)$ is the high-frequency carrier, usually provided by the carrier generator (or high-frequency oscillator); $S(t)$ is the output of the modulator, called the modulated wave.

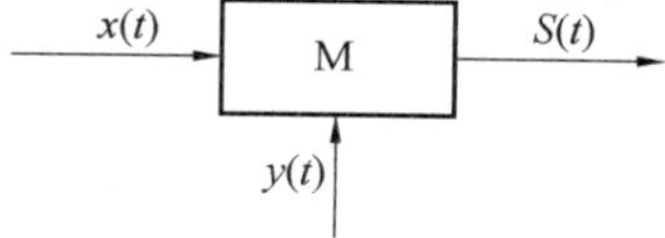

Figure 4.34 Linear modulation model

The purpose of modulation includes: achieving frequency conversion to transmit information in a given channel (that is, to meet the requirements of the working frequency band); achieving channel multiplexing to make information take full advantage of the given channel frequency band; improving anti-interference ability.

Amplitude modulation is to modulate the amplitude of the carrier with a modulation signal, making the amplitude of the carrier change proportionally with the size of the modulation signal. It is recognized that the amplitude of the modulated wave (amplitude modulation wave) reflects the magnitude of the modulation signal. The amplitude modulation can be achieved in different ways, such as square law amplitude modulation, chopping amplitude modulation, and analog multiplier amplitude modulation. This section mainly introduces the principle of analog multiplier amplitude modulation common in measuring devices. For example, the AC bridge and differential transformer type sensor itself is equivalent to a multiplier, and its output is all

的供电电压为载波,铁芯的位移信号为调制信号。

模拟乘法器调幅就是用调制信号 $x(t)$ 与高频载波相乘,使载波的幅值随调制信号 $x(t)$ 的大小而变。载波可以是正弦(或余弦)波,也可以是方波。图 4.35 所示为模拟乘法器调幅原理模型图。图中,$y(t)$ 是载波信号,$x(t)$ 是调制信号,$x_a(t)$ 是调幅信号(已调波)。设

$$y(t) = \cos 2\pi f_0 t \tag{4.46}$$

式中　f_0—— 载波频率。

因此,有

$$x_a(t) = x(t)y(t) = x(t)\cos 2\pi f_0 t \tag{4.47}$$

图 4.35　模拟乘法器调幅原理模型图

调幅信号的波形如图 4.36 所示。由图可见,调幅波的轮廓线与调制信号的波形相同,即调幅波的幅值反映了调制信号的大小。注意,在调制波有正有负,即调制波分布在零线两侧的情况下,在调制波过零点处,调幅波反相(图 4.36(a))。这一反相在解调时需要同步解调器或相敏解调器,否则不能正确解调。如果调制波在零线一侧,调制波过零点时,调幅波没有反相(图 4.36(b))。这时,可以不使用同步解调或相敏解调器,而使用简单的包络检波解调器即可实现解调。

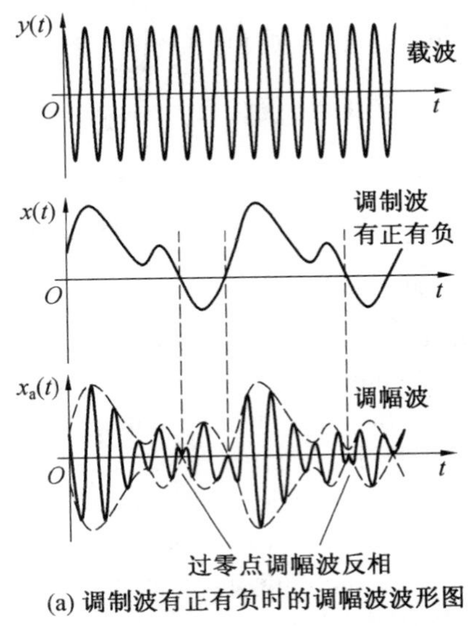

(a) 调制波有正有负时的调幅波波形图

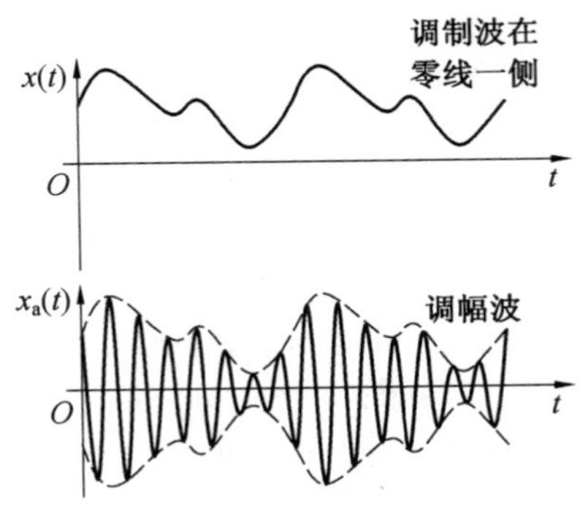

(b) 调制波在零线一侧时的调幅波波形图

图 4.36　调幅信号的波形

载波的频谱为

amplitude modulated waves. In an AC bridge, the power supply voltage is the carrier, and the change in the component value on the bridge arm proves the modulation signal. In a differential transformer, the supply voltage of the primary winding is the carrier, and the displacement signal of the core is the modulation signal.

The analog multiplier amplitude modulation is to multiply the modulation signal $x(t)$ and the high-frequency carrier so that the amplitude of the carrier varies with the magnitude of the modulation signal $x(t)$. The carrier wave can be a sine (or cosine) wave or a square wave. Figure 4.35 shows the principle model diagram of analog multiplier amplitude modulation. In the figure, $y(t)$ is the carrier signal, $x(t)$ is the modulation signal, and $x_a(t)$ is the amplitude modulation signal (modulated wave). Suppose

$$y(t) = \cos 2\pi f_0 t \tag{4.46}$$

where f_0— carrier frequency.

x(t) ⊗ $x_a(t)$
y(t)

Figure 4.35 Principle model diagram of analog multiplier amplitude modulation

Therefore, we have

$$x_a(t) = x(t)y(t) = x(t)\cos 2\pi f_0 t \tag{4.47}$$

The waveform of the amplitude modulation signal is shown in Figure 4.36. It can be seen from the figure that the contour line of the amplitude modulation wave is the same as the waveform of the modulation signal, that is, the amplitude of the amplitude modulation wave reflects the magnitude of the modulation signal. Note that when the modulating wave is positive or negative, that is, when the modulating wave is distributed on both sides of the zero line, the AM wave is inverted at the zero-crossing point of the modulating wave(Figure 4.36(a)). This inversion will require a synchronous demodulator or phase-sensitive demodulator during demodulation, otherwise, it cannot be demodulated correctly. If the modulating wave is on the side of the zero line, the AM wave is not inverted when the modulating wave crosses the zero point(Figure 4.36(b)). In this case, you can use a simple envelope detection demodulator instead of a synchronous demodulator or a phase-sensitive demodulator to achieve demodulation.

The frequency spectrum of the carrier is

$$Y(f) = F(\cos 2\pi f_0 t) = \frac{1}{2}\delta(f - f_0) + \frac{1}{2}\delta(f + f_0) \tag{4.48}$$

Suppose the frequency spectrum of modulating wave $x(t)$ is

$$F[x(t)] = X(f) \tag{4.49}$$

According to the convolution characteristics of the Fourier transform, the spectrum of the amplitude modulated wave is

$$\begin{aligned} X_a(f) &= X(f) * Y(f) = X(f) * \left[\frac{1}{2}\delta(f - f_0) + \frac{1}{2}\delta(f + f_0)\right] \\ &= \frac{1}{2}X(f) * \delta(f - f_0) + \frac{1}{2}X(f) * \delta(f + f_0) \end{aligned}$$

According to the convolution characteristics of the impulse function, we get

$$X_a(f) = \frac{1}{2}X(f - f_0) + \frac{1}{2}X(f + f_0) \tag{4.50}$$

$$Y(f) = F(\cos 2\pi f_0 t) = \frac{1}{2}\delta(f - f_0) + \frac{1}{2}\delta(f + f_0) \tag{4.48}$$

设调制波 $x(t)$ 的频谱为

$$F[x(t)] = X(f) \tag{4.49}$$

根据傅里叶变换的卷积特性,可得调幅波的频谱为

$$\begin{aligned} X_a(f) &= X(f) * Y(f) = X(f) * \left[\frac{1}{2}\delta(f - f_0) + \frac{1}{2}\delta(f + f_0)\right] \\ &= \frac{1}{2}X(f) * \delta(f - f_0) + \frac{1}{2}X(f) * \delta(f + f_0) \end{aligned}$$

根据脉冲函数的卷积特性,可得

$$X_a(f) = \frac{1}{2}X(f - f_0) + \frac{1}{2}X(f + f_0) \tag{4.50}$$

由式(4.50)可知,在频域上,调幅的结果是将调制信号的频谱沿频率轴平移至载波频率 f_0 处,且频谱密度幅值减半,如图 4.36 所示。因此,调幅过程就相当于频谱“搬移”过程。因为载波频率较高,所以这样通过调幅便可将信号的频率提高,然后在接收端通过解调器再恢复原信号。由调幅波频谱图可知,为保证调幅波频谱不至于混叠,载波信号的频率必须高于调制信号中的最高谐波频率 f_m。为减小放大电路可能引起的失真,也为在解调滤波时有效地滤掉高频成分,调制信号的频宽相对载波频率来说应越小越好,这就要求载波频率越高越好。一般要求载波频率 $f_0 > (5 \sim 10)f_m$。载波频率也会受到放大电路截止频率的限制。

调幅波的解调有多种方法,常用的有同步解调和包络检波。

同步解调又称相干解调。因为这种解调方法既可以恢复调制信号的幅度信息,也可以恢复调制信号的相位信息,能够识别信号的正、负极性,所以这种解调方法又称相敏解调。同步解调的具体实现方案多种多样,下面以采用模拟乘法器进行解调为例说明同步解调的过程。

采用模拟乘法器进行解调包含以下两个过程:首先将调幅波信号 $x_a(t)$ 和与载波信号 $y(t)$ 同频同相的参考信号 $v_r(t)$ 相乘;然后再用低通滤波器对得到的信号滤波(图 4.37)。在测量装置中,参考信号可以直接取自调幅时所用的载波振荡器。

设 $v_r(t) = y(t) = \cos 2\pi f_0 t$,则

$$\begin{aligned} x_p(t) &= x_a(t)v_r(t) = x(t)y^2(t) = x(t)\cos^2 2\pi f_0 t \\ &= \frac{x(t)}{2} + \frac{1}{2}x(t)\cos 4\pi f_0 t \end{aligned} \tag{4.51}$$

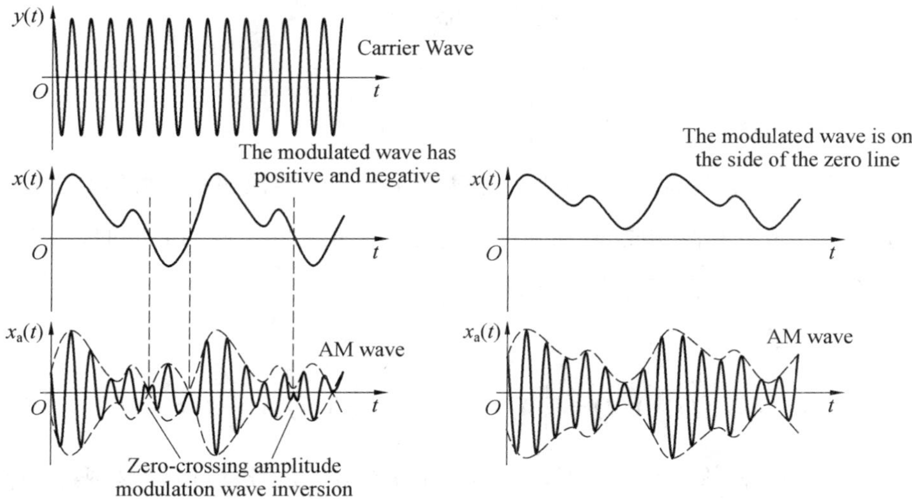

(a) Oscillogram of amplitude modulated wave with the modulating wave being positive or negative

(b) Oscillogram of AM wave with the modulating wave being on the side of the zero line

Figure 4.36 The waveform of the amplitude modulation signal

It can be seen from formula (4.50) that in the frequency domain, the result of amplitude modulation is to shift the frequency spectrum of the modulated signal to the carrier frequency f_0 along the frequency axis, and the amplitude of the spectral density is halved, as shown in Figure 4.36. Therefore, the amplitude modulation process is equivalent to the frequency spectrum "shifting" process. Because the carrier frequency is higher, the frequency of the signal can be increased by amplitude modulation, and then the original signal can be restored by the demodulator at the receiving end. It can be seen from the spectrum diagram of the AM wave that the frequency of the carrier signal must be higher than the highest harmonic frequency f_m in the modulation signal in order to ensure that the spectrum of the AM wave is not aliased. To reduce the distortion that may be caused by the amplifier circuit, and to effectively filter out high-frequency components during demodulation and filtering, the bandwidth of the modulation signal should be as small as possible relative to the carrier frequency, which requires that the carrier frequency should be as high as possible. It generally requires carrier frequency $f_0 > (5-10)f_m$. The carrier frequency will also be limited by the cut-off frequency of the amplifying circuit.

There are many methods for demodulation of AM waves whose commonly used are synchronous demodulation and envelope detection.

Synchronous demodulation is also called coherent demodulation, because this demodulation method can recover both the amplitude information of the modulated signal and the phase information of the modulated signal, and also can identify the positive and negative polarity of the signal, hence phase sensitive demodulation. There are various specific implementation schemes for synchronous demodulation. The following takes the demodulation with an analog multiplier as an example to illustrate the process of synchronous demodulation.

Demodulation using an analog multiplier includes the following two processes: first multiply the AM wave signal $x_a(t)$ with the reference signal $v_r(t)$ of the same frequency and phase of the same carrier signal $y(t)$; and then use a low-pass filter to filter the obtained signal

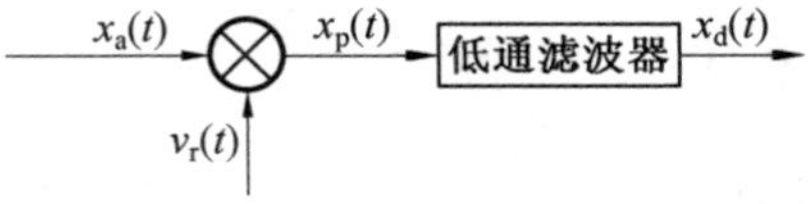

$x_a(t)$ → ⊗ → $x_p(t)$ → 低通滤波器 → $x_d(t)$；$v_r(t)$

图 4.37　调幅波同步解调原理

由式(4.51)可见,调幅波信号 $x_a(t)$ 与载波信号 $y(t)$ 再次相乘后得到的信号是由原信号(幅度减半)成分和高频成分(最后一个等号右边第二项)叠加而成的,只要用低通滤波器将该高频成分滤掉,即可得到原信号。

$x_p(t)$ 的频谱为

$$X_p(f) = X_a(f) * Y(f) = \frac{1}{4}X(f - 2f_0) + \frac{1}{2}X(f) + \frac{1}{4}X(f + 2f_0) \tag{4.52}$$

由式(4.52)可见,在频域上,调幅波信号 $x_a(t)$ 与载波信号 $y(t)$ 再次相乘后得到信号 $x_p(t)$ 的频谱,即调幅波的频谱沿频率轴平移,相当于调制信号的频谱再次"搬移",结果中包含调制信号的频谱信息 $X(f)/2$。由于 $x_p(t)$ 信号中包含高频边频带$(2f_0 - f_m, 2f_0 + f_m)$,因此用低通滤波器将其滤掉即可得到调制信号的频谱。调幅及同步解调过程中信号频谱的变化过程如图 4.38 所示。

解调后的调制信号幅度虽然发生了变化,但这种变化只是比例缩放,不会引起时域波形的变化,即不会引起调制信号失真,且幅度的变化很容易用放大器来补偿。

采用模拟乘法器进行同步解调对调制信号是否在零线一侧没有要求,但解调时需要与载波同频同相的参考信号,实现上比较复杂,还需要性能良好的线性乘法器。实现同步解调可以采用集成的载波放大器。例如,NE5521、AD598 和 AD698 都是单片集成载波放大器。这些载波放大器是用来完成交流放大、调幅波解调和低通滤波功能的电路,其中还包括必需的振荡器。

包络检波又称非相干解调或非同步整流低通滤液解调,采用先整流检波再低通滤波的方法。调幅波包络检波解调原理方框图如图 4.39 所示。但是,这种方法不能识别调制信号的正、负极性。为正确地解调,要求调制信号在零线一侧(或全大于零,或全小于零)。如果调制信号不在零线一侧,则需要加一直流偏压调整到零线一侧。当调制信号中加上了直流偏压解调后,需要准确地减去该直流偏压,最简单的方法是采用隔直电容。

图 4.40 所示为半波整流解调电路、全波整流解调电路和带隔直电容的半波整流解调电路。

包络检波电路具有简单、稳定、成本低的优点。但由于二极管具有一定的电压降,因此要求调幅波的幅值大于二极管的导通阈值。

(Figure 4.37). In the measuring device, the reference signal can be directly taken from the carrier oscillator used in amplitude modulation.

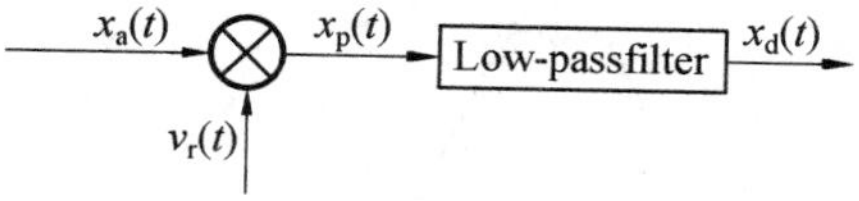

Figure 4.37 Principle of synchronous demodulation of AM wave

Suppose $v_r(t)=y(t)=\cos 2\pi f_0 t$, then

$$\begin{aligned} x_p(t) &= x_a(t)v_r(t)=x(t)y^2(t)=x(t)\cos^2 2\pi f_0 t \\ &= \frac{x(t)}{2}+\frac{1}{2}x(t)\cos 4\pi f_0 t \end{aligned} \tag{4.51}$$

It can be seen from formula(4.51) that the signal obtained by multiplying the AM wave signal $x_a(t)$ and the carrier signal $y(t)$ again is composed of the original signal (amplitude halved) component and high frequency component (the second item on the right of the last equal sign). As long as the high-frequency components are filtered out with a low-pass filter, the original signal can be obtained.

The spectrum of $x_p(t)$ is

$$X_p(f)=X_a(f)*Y(f)=\frac{1}{4}X(f-2f_0)+\frac{1}{2}X(f)+\frac{1}{4}X(f+2f_0) \tag{4.52}$$

It can be seen from formula(4.52) that in the frequency domain, the spectrum of the signal $x_p(t)$ obtained by multiplying the AM wave signal $x_a(t)$ and the carrier signal $y(t)$ again, that is, the spectrum of the AM wave shifts along the frequency axis, which is equivalent to the frequency spectrum of the modulated signal "move" again, with the result containing the spectrum information $X(f)/2$ of the modulated signal. Because the signal $x_p(t)$ contains the high frequency sideband $(2f_0-f_m,\ 2f_0+f_m)$, the frequency spectrum of the modulated signal can be obtained by filtering it out with a low-pass filter. The changing process of signal spectrum in amplitude modulation and synchronous demodulation is shown in Figure 4.38.

Although the amplitude of the modulated signal after demodulation changes, this change is only a scaling and will not cause a change in the time-domain waveform, that is, it will not cause distortion of the modulated signal, and the amplitude change can be easily compensated by an amplifier.

Synchronous demodulation using an analog multiplier does not require whether the modulated signal is on the zero side or not, but a reference signal with the same frequency and phase as the carrier is required for demodulation, which is more complicated to implement and requires a linear multiplier with good performance. An integrated carrier amplifier can be used to achieve synchronous demodulation. For example, NE5521, AD598 and AD698 are all monolithic integrated carrier amplifiers. These carrier amplifiers are circuits used to complete ACamplification, AM demodulation, and low-pass filtering function, in which also includes the necessary oscillators.

Envelope detection is also called non-coherent demodulation or demodulation of non-synchronous rectification low-pass filtrate, adopting the method of first rectifying detector and then low-pass filtering. Block diagram of the principle of amplitude modulation wave envelope detection and demodulation is shown in Figure 4.39. However, this method cannot identify the

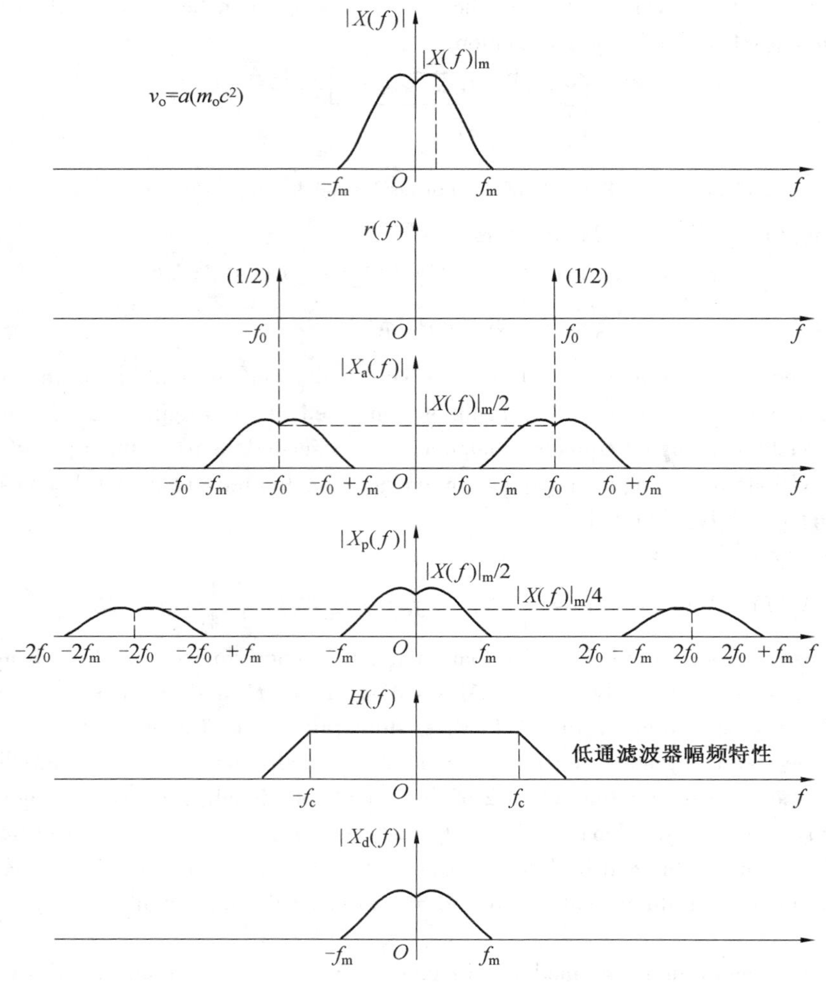

图 4.38　调幅及同步解调过程中信号频谱的变化过程

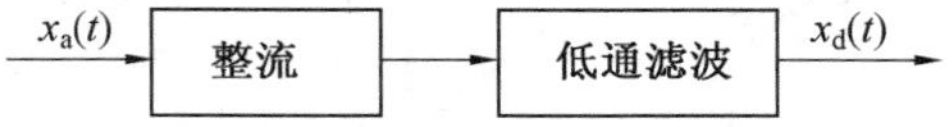

图 4.39　调幅波包络检波解调原理方框图

4.3.3　互感型——差动变压器式电感传感器

互感型传感器实质上是一个具有可动铁芯的变压器，其原理为变压器原理，且次级线圈以差动方式连接，故互感型传感器又称差动变压器式传感器。

这种传感器利用电磁感应中的互感现象，如图 4.41 所示。当线圈 W_1 输入交流电流 i_1 时，线圈 W_2 产生感应电动势 e_{12}，其大小与电流 i_1 的变化率成正比，即

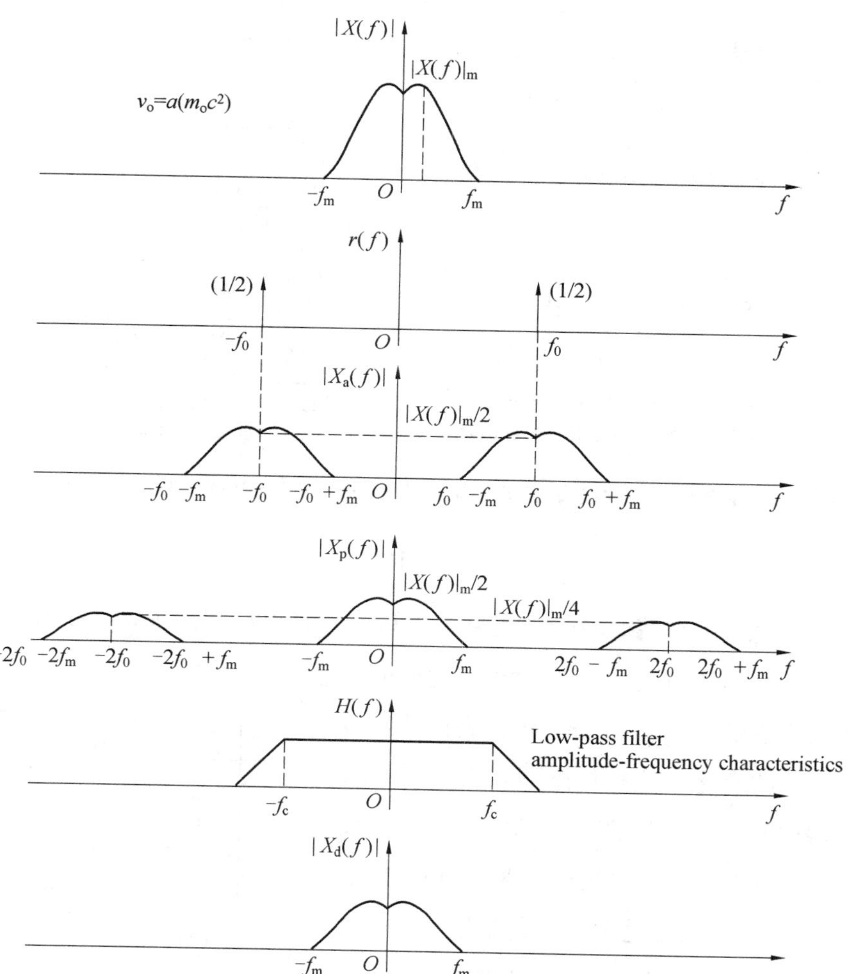

Figure 4.38 The changing process of signal spectrum in amplitude modulation and synchronous demodulation

positive and negative polarity of the modulated signal. In order to demodulate correctly, the modulated signal is required to be on the side of the zero line (either all greater than zero, or all less than zero). If this is not the case, it is necessary to add a DC bias to adjust it to the side of the neutral line. After a DC bias voltage is added to the modulation signal for demodulation, the DC bias voltage needs to be accurately subtracted, a DC blocking capacitor proving the simplest method.

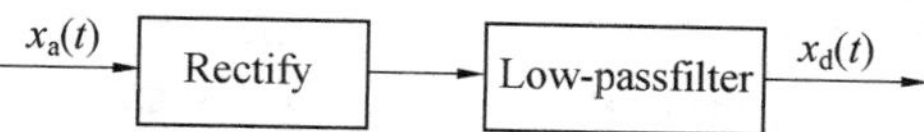

Figure 4.39 Block diagram of the principle of amplitude modulation wave envelope detection and demodulation

Figure 4. 40 shows a half-wave rectification and demodulation circuit, a full-wave rectification and demodulation circuit, and a half-wave rectification and demodulation circuit with blocking capacitors.

The envelope detection circuit has the advantages of simple, stable and low cost.

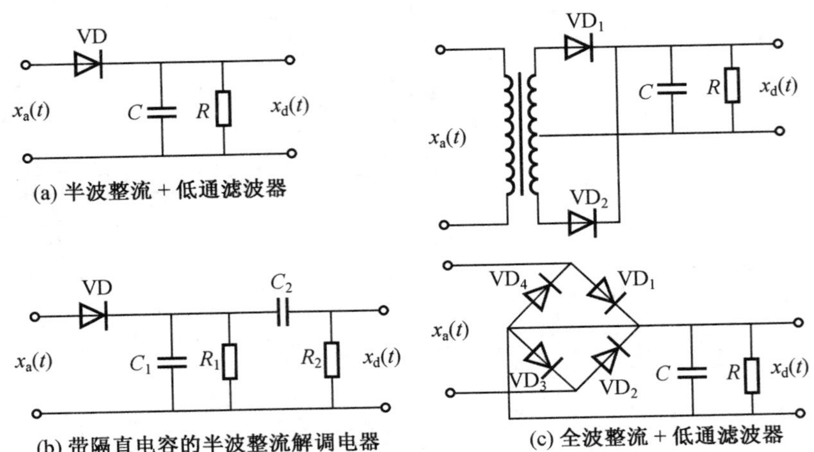

图 4.40　半波整流解调电路、全波整流解调电路和带隔直电容的半波整流解调电路

$$e_{12} = - M \frac{\mathrm{d}i_1}{\mathrm{d}t} \tag{4.53}$$

式中　M—— 比例系数,称为互感,H,其大小与两线圈相对位置及周围介质的导磁能力等因素有关,它表明两线圈之间的耦合程度。

互感型传感器利用这一原理,将被测位移量转换成线圈互感的变化。这种传感器实质上就是一个变压器,其初级线圈接入稳定交流电源,次级线圈感应产生一输出电压。当被测参数使互感 M 变化时,次级线圈输出电压也产生相应变化。由于常常采用两个次级线圈组成差动式,因此又称差动变压器式传感器。实际应用较多的是螺管形差动变压器,其工作原理如图 4.42(a)、(b) 所示。变压器由初级线圈 W 和两个参数完全相同的次级线圈 W_1、W_2 组成。线圈中心插入圆柱铁芯 P,次级线圈 W_1 及 W_2 反极性串联。当初级线圈 W 加上交流电压时,次级线圈 W_1 和 W_2 分别产生感应电势 e_1 及 e_2,其大小与铁芯位置有关。当铁芯在中心位置时,$e_1 = e_2$,输出电压 $e_0 = 0$;当铁芯向上运动时,$e_1 > e_2$;当铁芯向下运动时,$e_1 < e_2$。随着铁芯逐渐偏离中心位置,e_0 逐渐增大,其输出特性如图 4.42(c) 所示。

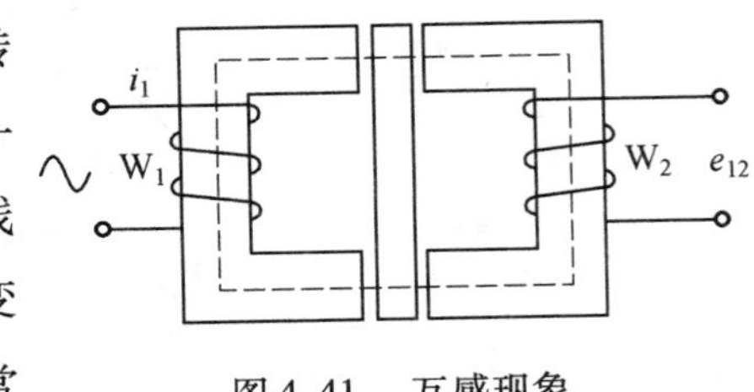

图 4.41　互感现象

差动变压器的输出电压是交流量,其幅值与铁芯位移成正比。其输出电压若用交流电压表指示,则输出值只能反映铁芯位移大小,不能反映移动的方向性。交流电压输出存在一定的零点残余电压。零点残余电压是两个次级线圈结构不对称、初级线圈铜损电阻、铁磁材质不均匀、线圈间分布电容等原因造成的。因此,即使在铁芯处于中间位置时,输

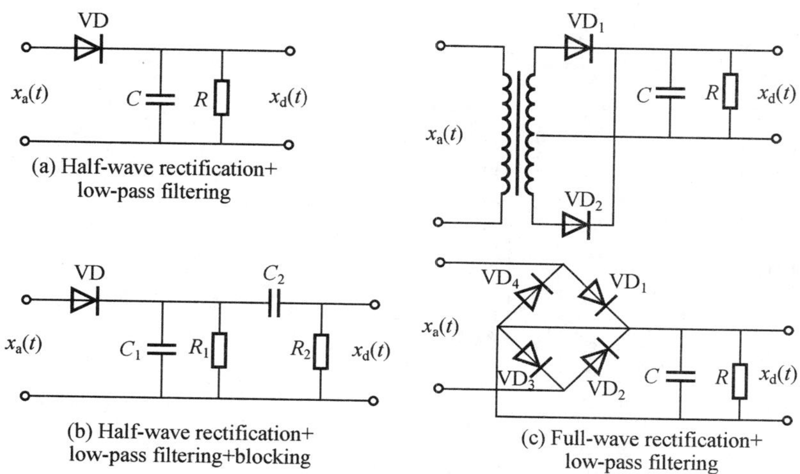

Figure 4.40 A half-wave rectification and demodulation circuit, a full-wave rectification and demodulation circuit, and a half-wave rectification and demodulation circuit with blocking capacitors

However, since the diode has a certain voltage drop, the amplitude of the amplitude modulation wave is required to be greater than the turn-on threshold of the diode.

4.3.3 Mutual Inductance Type—Differential Transformer Type Inductive Sensor

The mutual inductance sensor is essentially a transformer with a movable iron core, whose principle is the transformer principle, and the secondary coils are connected in a differential manner, so it is also called differential transformer sensor.

This kind of sensor uses the mutual inductance phenomenon in electromagnetic induction, as shown in Figure 4.41. When coil W_1 inputs alternating current i_1, coil W_2 generates induced electromotive force e_{12}, whose magnitude is proportional to the rate of change of current i_1, namely

$$e_{12} = - M \frac{di_1}{dt} \tag{4.53}$$

where M—proportionality coefficient, called mutual inductance, H, whose size is related to factors such as the relative position of the two coils and the magnetic permeability of the surrounding medium, indicating the degree of coupling between the two coils.

The mutual inductance sensor uses this principle to convert the measured displacement into the change of coil mutual inductance. This kind of sensor is essentially a transformer, whose primary coil is connected to a stable AC power supply, and the secondary coil induces an output voltage. When the measured parameter changes the mutual inductance M, the output voltage of the secondary coil also

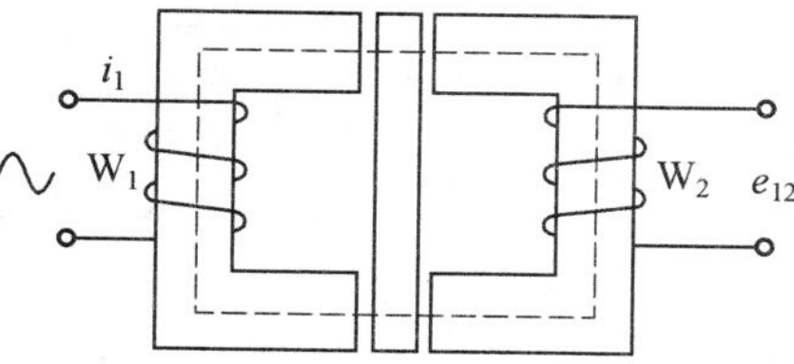

Figure 4.41 Mutual inductance phenomenon

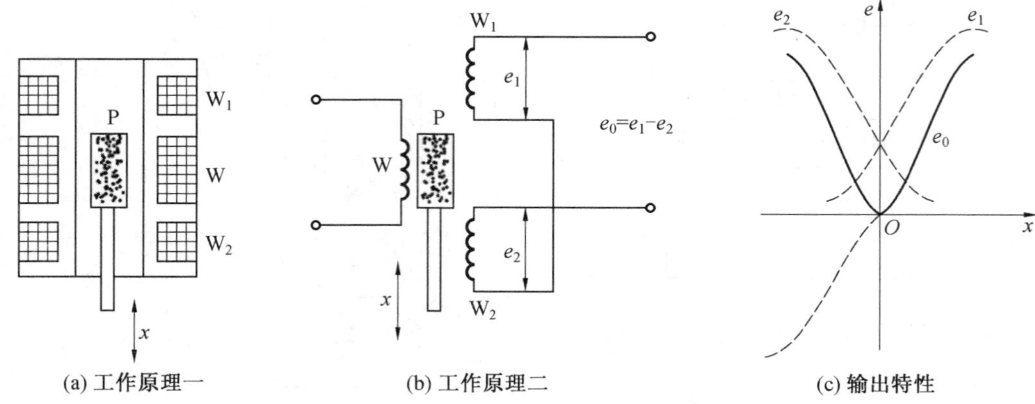

图 4.42　差动变压器式传感器工作原理

出也不为零。为此,差动变压器式传感器的后接电路形式需要采用既能反映铁芯位移方向性,又能补偿零点残余电压的差动直流输出电路。

图 4.43 所示为一种用于小位移测量的差动相敏检波电路工作原理。在没有输入信号时,铁芯处于中间位置,调节电阻 R,使零点残余电压减小;当有输入信号时,铁芯上移或下移,其输出电压经交流放大、相敏检波、滤波后得到直流输出,由表头指示输入位移量大小和方向。

差动变压器式电感传感器具有精确度高(高到 0.1 μm 数量级)、线性范围大(可扩大到 ±100 mm)、稳定度好和使用方便等特点,被广泛应用于直线位移的测量。但其实际测量频率上限受制于传感器中所包含的机械结构。借助弹性元件,可以将压力、质量等物理量转换成位移的变化,故也将这类传感器用于压力、质量等物理量的测量。

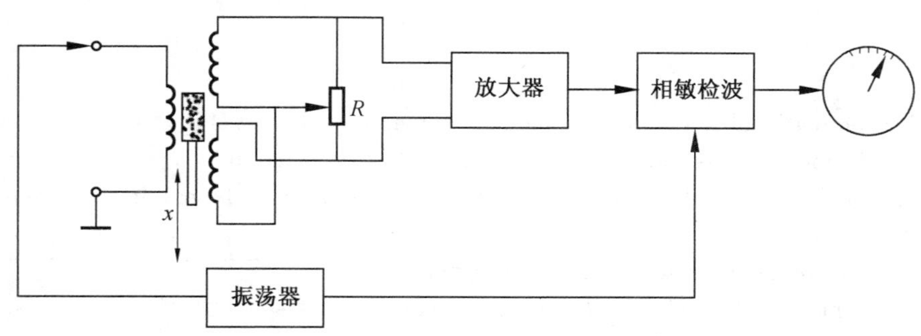

图 4.43　用于小位移测量的差动相敏检波电路工作原理

changes accordingly. Since two secondary coils are often used to form a differential type, it is also called a differential transformer sensor. The most practical application is the screw-type differential transformer, and its working principle is shown in Figure 4.42(a) and (b). The transformer is composed of a primary coil W and two secondary coils W_1, W_2 with exactly the same parameters. The center of the coil is inserted into the cylindrical core P, and the secondary coils W_1 and W_2 are connected in series with reverse polarity. When the primary coil W is applied with an alternating voltage, the secondary coils W_1 and W_2 generate induced potentials e_1 and e_2 respectively, whose size is related to the position of the iron core. When the iron core is in the center position, $e_1 = e_2$, the output voltage $e_0 = 0$; when the iron core moves upward, $e_1 > e_2$; when it moves downward, $e_1 < e_2$. As the core deviates from the center position gradually, e_0 gradually increases, and its output characteristics are shown in Figure 4.42(c).

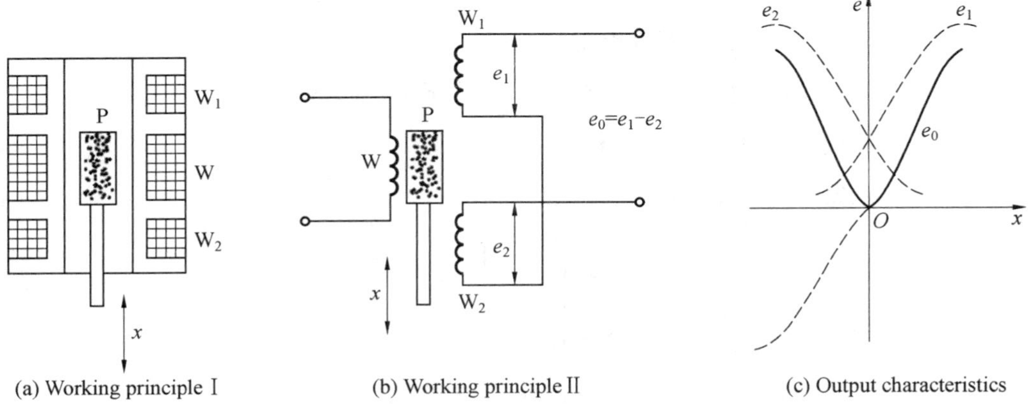

Figure 4.42 Working principle of differential transformer sensor

The output voltage of the differential transformer is AC, and its amplitude is proportional to the core displacement. If the output voltage is indicated by an AC voltmeter, the output value can only reflect the core displacement, not the direction of movement. The AC voltage output has a certain zero residual voltage. The zero residual voltage is caused by the asymmetricalstructure of the two secondary coils, the copper loss resistance of the primary coil, the uneven ferromagnetic material, and the distributed capacitance between the coils. Therefore, even when the core is in the middle position, the output is not zero. For this reason, the post-connection circuit form of the differential transformer type sensor needs to adopt a differential DC output circuit that not only reflects the directionality of the core displacement, but also compensates for the zero-point residual voltage.

Figure 4.43 shows the working principle of a differential phase-sensitive detection circuit for small displacement measurement. Without input signal, the iron core is in the middle position, adjust the resistance R to reduce the zero residual voltage; when there is an input signal, the core moves up or down, and the output voltage is amplified by AC, phase-sensitive detection, and filtered to obtain a DC output. The meter indicates the magnitude and direction of the input displacement.

Differential transformer-type inductive sensors have the characteristics of high accuracy (up to the order of 0.1 μm), large linear range (expandable to ±100 mm), good stability,

4.4 电容式传感器

4.4.1 概述

1. 工作原理

电容式传感器是将被测物理量转换为电容量变化的装置。它实质上是一个具有可变参数的电容器。两块平行金属板构成的电容器(图 4.44),其电容量 C 为

$$C = \frac{\varepsilon \varepsilon_0 A}{\delta} \tag{4.54}$$

式中 ε—— 极板间介质的相对介电常数,空气介质 $\varepsilon = 1$;

ε_0—— 真空的介电常数,$\varepsilon_0 = 8.85 \times 10^{-12}$ F/m;

A—— 极板相互覆盖的面积,m^2;

δ—— 极板间距离,简称极距,m。

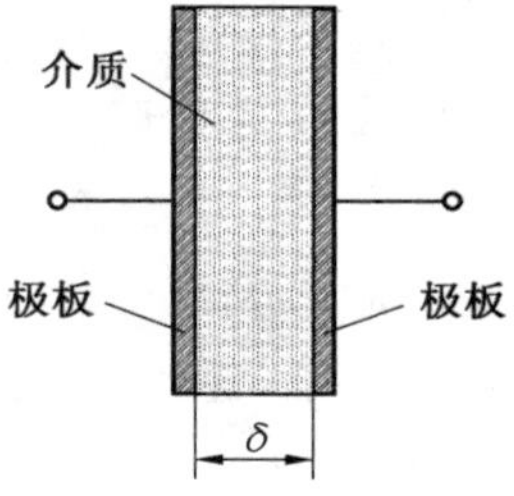

图 4.44　平板电容器

由式(4.54) 知,当被测量使式中的 δ、A 和 ε 变化时,都将引起电容器电容量 C 的变化。

2. 类型

在实际应用中,通常限定式(4.54) 中 δ、A 和 ε 三个参数中的两个保持不变,只改变其中的一个参数,使电容产生变化。根据被变化的三个参数的不同,电容式传感器可分为三类:极距变化型、面积变化型和介质变化型。

4.4.2 极距变化型电容传感器

极距变化型电容传感器的结构和特性如图 4.45 所示。当电容器两平行板的重合面积及介质不变,动板因受被测量控制而移动时,极板间距 δ 发生了改变,引起电容器电容量的变化,从而达到将被测参数转换成电容量变化的目的。若电容器的极板面积为 A,初

and convenient use, which are widely used in linear displacement measurement. However, the upper limit of the actual measurement frequency is restricted by the mechanical structure contained in the sensor. With the help of elastic elements, physical quantities such as pressure and mass can be converted into changes in displacement, so this type of sensor is also used for the measurement of physical quantities such as pressure and mass.

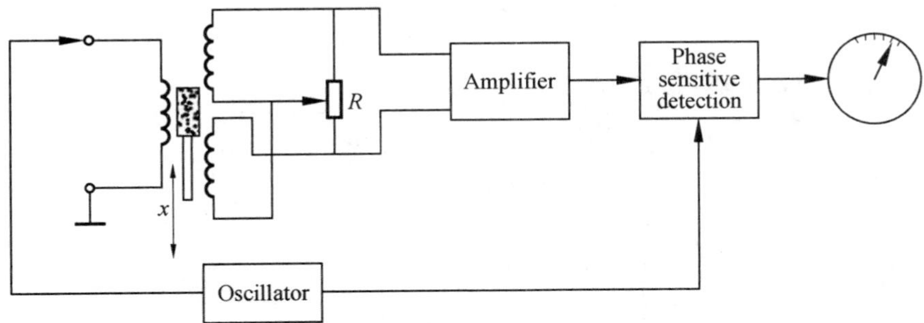

Figure 4.43 The working principle of a differential phase-sensitive detection circuit for small displacement measurement

4.4 Capacitive Sensors

4.4.1 Overview

1. Working Principle

A capacitive sensor is a device that converts the measured physical quantity into a change in capacitance. It is essentially a capacitor with variable parameters. A capacitor composed of two parallel metal plates (Figure 4.44), its capacitance C is

$$C = \frac{\varepsilon\varepsilon_0 A}{\delta} \tag{4.54}$$

where ε—the relative permittivity of the medium between the plates, the air medium $\varepsilon = 1$;

ε_0—the dielectric constant of vacuum, $\varepsilon_0 = 8.85 \times 10^{-12}$ F/m;

A—the area covered by the plates, m^2;

δ—the distance between the plates, referred to as the electrode distance, m.

From formula (4.54), when the measured δ, A and ε in the formula change, it will cause the change of the capacitance C of the capacitor.

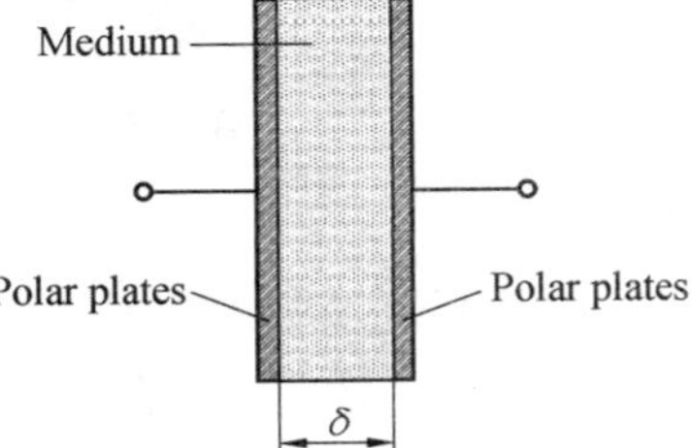

Figure 4.44 Flat capacitor

2. Type

In practical applications, two of the three parameters δ, A and ε in formula (4.54) are usually limited to remain unchanged, and only one parameter is changed to change the capacitance. According to the difference of the three parameters being changed, capacitive sensors can be divided into three types: electrode distance variable type, electrode plate area type and variable dielectric type.

始极距为δ_0，极板间介质的介电常数为ε，根据式(4.54)，如果两极板重合及极间介质不变，则电容量C与极距δ呈非线性关系。当极距有一微小变化量$\mathrm{d}\delta$时，引起电容的变化量$\mathrm{d}C$为

$$\mathrm{d}C=-\varepsilon\varepsilon_0 A\frac{1}{\delta^2}\mathrm{d}\delta$$

由此可以得到传感器灵敏度为

$$S=\frac{\mathrm{d}C}{\mathrm{d}\delta}=-\varepsilon\varepsilon_0 A\frac{1}{\delta^2} \tag{4.55}$$

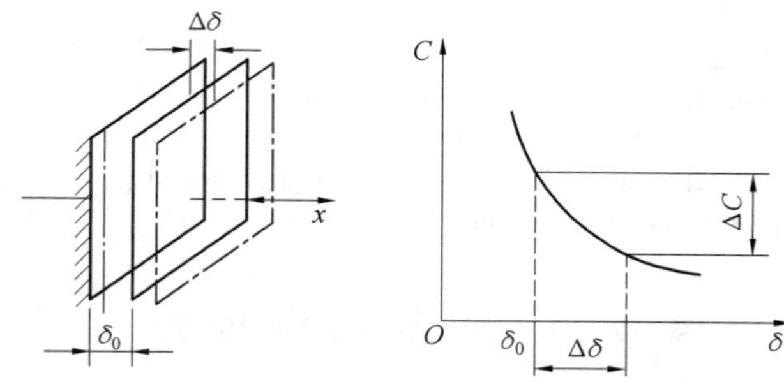

图4.45　极距变化型电容传感器的结构和特性

可以看出，灵敏度S与极距平方成反比，极距越小，灵敏度越高。灵敏度S不是常数，说明一定存在非线性误差。

设初始电容为C_0，则电容器的初始电容量为

$$C_0=\frac{\varepsilon A}{\delta_0} \tag{4.56}$$

当间隙δ_0减小$\Delta\delta$时，电容量增加ΔC，即

$$C=C_0+\Delta C=\frac{\varepsilon A}{\delta_0-\Delta\delta}=C_0\frac{1}{1-\dfrac{\Delta\delta}{\delta_0}}$$

$$\Delta C=C_0\frac{1}{1-\dfrac{\Delta\delta}{\delta_0}}-C_0=C_0\left(\frac{1}{1-\dfrac{\Delta\delta}{\delta_0}}-1\right)=\frac{C_0\dfrac{\Delta\delta}{\delta_0}}{1-\dfrac{\Delta\delta}{\delta_0}}$$

$$\frac{\Delta C}{C_0}=\frac{\Delta\delta}{\delta_0}\left(1-\frac{\Delta\delta}{\delta_0}\right)^{-1}=\frac{\Delta\delta}{\delta_0}\left[1+\frac{\Delta\delta}{\delta_0}+\left(\frac{\Delta\delta}{\delta_0}\right)^2+\left(\frac{\Delta\delta}{\delta_0}\right)^3+\cdots+\left(\frac{\Delta\delta}{\delta_0}\right)^n\right]$$

由式(4.56)可知，极距变化型电容传感器的输入(被测参数引起的极距变化$\Delta\delta$)与输出(电容的变化ΔC)之间的关系是非线性的，由非线性引起的误差为

$$\Delta=\left(\frac{\Delta\delta}{\delta_0}\right)^2+\left(\frac{\Delta\delta}{\delta_0}\right)^3+\cdots+\left(\frac{\Delta\delta}{\delta_0}\right)^n \tag{4.57}$$

4.4.2 Variable Electrode Distance Capacitance Sensor

The structure and characteristics of electrode distance variable type capacitance sensor are shown in Figure 4.45. When the overlapping area and medium of the two parallel plates of the capacitor remain unchanged, and the movable plate moves due to the control of the measured quantity, the plates distance δ changes, causing the capacitance of the capacitor to change, thus achieving the purpose of converting the measured parameter into a change in capacitance. If the plate area of the capacitor is A, the initial electrodes distance is δ_0, and the dielectric constant of the medium between the plates is ε, according to formula (4.54), if the overlapping area of the two plates and the inter-electrode medium remain unchanged, then the capacitance C and the distance δ have a nonlinear relationship. When there is a slight change in the distance $\mathrm{d}\delta$, the change in capacitance $\mathrm{d}C$ is

$$\mathrm{d}C = -\varepsilon\varepsilon_0 A \frac{1}{\delta^2}\mathrm{d}\delta$$

From this, the sensor sensitivity can be obtained that

$$S = \frac{\mathrm{d}C}{\mathrm{d}\delta} = -\varepsilon\varepsilon_0 A \frac{1}{\delta^2} \tag{4.55}$$

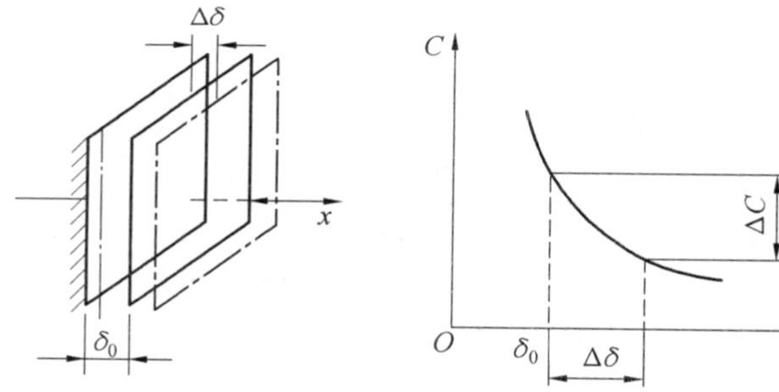

Figure 4.45 The structure and characteristics of electrode distance variable type capacitance sensor

It can be seen that the sensitivity S is inversely proportional to the square of the plates distance, and the smaller the plates distance, the higher the sensitivity. The sensitivity S is not a constant, indicating that there must be a nonlinear error.

Suppose the initial capacitance is C_0, then the initial capacitance of the capacitor is

$$C_0 = \frac{\varepsilon A}{\delta_0} \tag{4.56}$$

When the gap δ_0 decreases $\Delta\delta$, the capacitance increases ΔC, that is

$$C = C_0 + \Delta C = \frac{\varepsilon A}{\delta_0 - \Delta\delta} = C_0 \frac{1}{1 - \dfrac{\Delta\delta}{\delta_0}}$$

$$\Delta C = C_0 \frac{1}{1 - \dfrac{\Delta\delta}{\delta_0}} - C_0 = C_0\left(\frac{1}{1 - \dfrac{\Delta\delta}{\delta_0}} - 1\right) = \frac{C_0 \dfrac{\Delta\delta}{\delta_0}}{1 - \dfrac{\Delta\delta}{\delta_0}}$$

$$\frac{\Delta C}{C_0} = \frac{\Delta\delta}{\delta_0}\left(1 - \frac{\Delta\delta}{\delta_0}\right)^{-1} = \frac{\Delta\delta}{\delta_0}\left[1 + \frac{\Delta\delta}{\delta_0} + \left(\frac{\Delta\delta}{\delta_0}\right)^2 + \left(\frac{\Delta\delta}{\delta_0}\right)^3 + \cdots + \left(\frac{\Delta\delta}{\delta_0}\right)^n\right]$$

From formula (4.56), it can be seen that the relationship between the input (the change

但当($\Delta\delta/\delta_0$) $\ll 1$ 时,可略去高次项而认为是线性的。此时,有

$$\frac{\Delta C}{C_0}=\frac{\Delta\delta}{\delta_0}$$

它的灵敏度可近似为

$$S=\frac{\mathrm{d}(\Delta C)}{\mathrm{d}(\Delta\delta)}\approx\frac{C_0}{\delta_0}=\varepsilon A_0 \tag{4.58}$$

显然,要减小非线性误差,必须缩小测量范围 $\Delta\delta$。一般取测量范围为0.1 μm至数百微米。对于精密的电容传感器,取 $\Delta\delta/\delta_0 < 0.01$。

极距变化型电容传感器的特点是:动态特性好;灵敏度和精度较高(可达纳米级);适用于较小位移(1 nm ~ 1 μm) 的精密测量。由于存在原理上的非线性误差,因此相应的测量电路比较复杂。

图 4.46 所示为差动式极距变化型电容传感器,两电容器的变化量大小相等,符号相反。利用后接的转换电路(如电桥等) 可以检出两电容器电容量的差值,该差值与活动极板的移动量 $\Delta\delta$ 有一一对应关系。

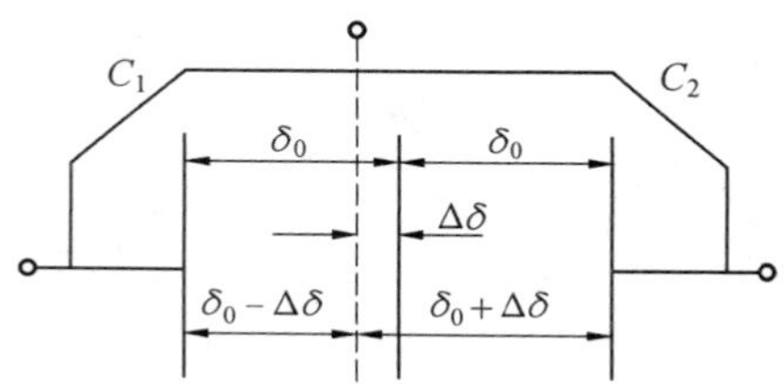

图 4.46　差动式极距变化型电容传感器

采用差动式原理后,传感器的灵敏度提高了 1 倍,非线性得到了很大的改善,某些因素(如环境温度变化、电源电压波动等) 对测量精度的影响也得到了一定的补偿。

4.4.3　面积变化型电容传感器

1. 直线位移型

面积变化型电容传感器(平面线位移型) 如图 4.47 所示,当动板沿 x 方向移动时,相互覆盖面积发生了变化,电容量随之改变,其输出特性为

$$C=\frac{\varepsilon bx}{\delta} \tag{4.59}$$

式中　b—— 极板宽度;

x—— 位移;

δ—— 极板间距。

of the plates distance caused by the measured parameter $\Delta\delta$) and the output (the change of the capacitance ΔC) of the capacitance sensor with a change in plates distance is non-linear, and the error caused by the non-linearity is

$$\Delta = \left(\frac{\Delta\delta}{\delta_0}\right)^2 + \left(\frac{\Delta\delta}{\delta_0}\right)^3 + \cdots + \left(\frac{\Delta\delta}{\delta_0}\right)^n \tag{4.57}$$

But when ($\Delta\delta/\delta_0$) $\ll 1$, the higher-order term can be omitted and it is considered linear. At this time, we have

$$\frac{\Delta C}{C_0} = \frac{\Delta\delta}{\delta_0}$$

Its sensitivity can be approximated as

$$S = \frac{d(\Delta C)}{d(\Delta\delta)} \approx \frac{C_0}{\delta_0} = \varepsilon A_0 \tag{4.58}$$

In order to reduce the nonlinear error, it is obviously necessary to reduce the measurement range $\Delta\delta$. Generally, the measurement range is 0.1 μm to hundreds of microns. For precision capacitive sensors, take $\Delta\delta/\delta_0 < 0.01$.

The characteristics of electrode distance variable type capacitance sensor are: good dynamic characteristics; high sensitivity and accuracy (up to nanometer level) ; and suitable for precise measurement of small displacements (1 nm–1 μm). Due to the non-linear error in principle, the corresponding measurement circuit is relatively complicated.

Figure 4. 46 shows a differential electrode distance variable type capacitance sensor, the changes of the two capacitors are equal in magnitude and opposite in sign. The difference between the capacitances of the two capacitors can be detected by the subsequent conversion circuit (such as a bridge) , and there is a one-to-one correspondence between the difference and the amount of movement $\Delta\delta$ of the movable plate.

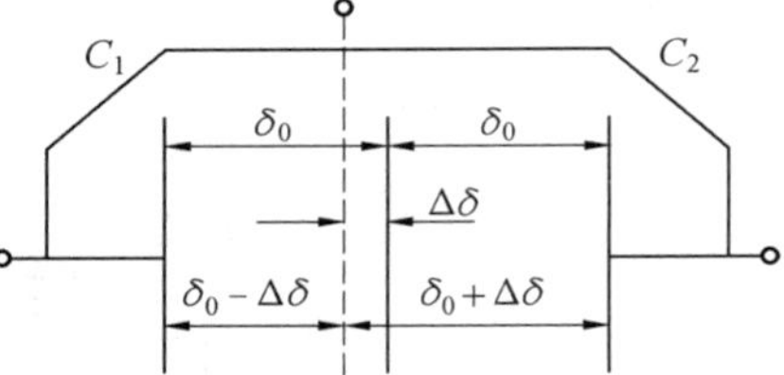

Figure 4.46 Differential electrode distance variable type capacitance sensor

Afterwe adopting the differential principle, the sensor sensitivity has been doubled, the nonlinearity has been greatly improved, and the influence of certain factors (such as environmental temperature changes, power supply voltage fluctuations) on the measurement accuracy has also been compensated to a certain extent.

4.4.3 Variable Electrode Plate Area Type Capacitance Sensor

1. Linear Displacement Type

Variable electrode plate area type capacitive sensor (plane linear displacement type) is shown in Figure 4.47, when the moving plate moves along thex direction, the mutual coverage area changes, and the capacitance changes accordingly, whose output characteristic is

$$C = \frac{\varepsilon b x}{\delta} \tag{4.59}$$

where b —the width of the plate;
x —displacement;
δ —the distance between the plates.

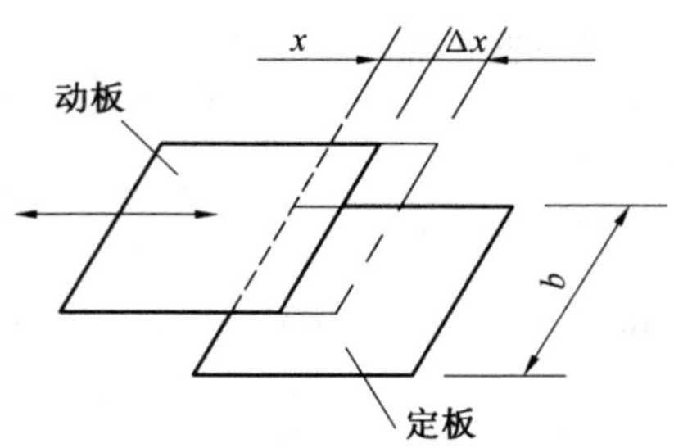

图 4.47　面积变化型电容传感器(平面线位移型)

其灵敏度为

$$S = \frac{\mathrm{d}C}{\mathrm{d}x} = \frac{\varepsilon b}{\delta} = 常数 \tag{4.60}$$

图 4.48 所示为面积变化型电容传感器(单边圆柱体线位移型),动板(圆柱)与定板(圆筒)相互覆盖,其电容量为

$$C = \frac{2\pi \varepsilon x}{\ln(D/d)} \tag{4.61}$$

式中　d—— 圆柱外径;

D—— 圆筒孔径。

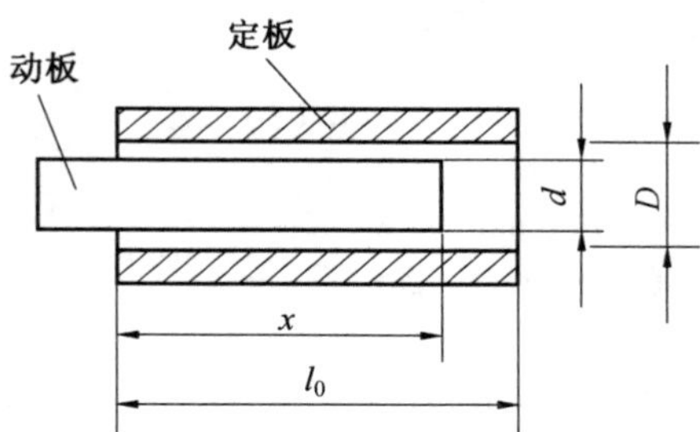

图 4.48　面积变化型电容传感器(单边圆柱体线位移型)

当覆盖长度 x 变化时,电容量 C 随之改变,其灵敏度为

$$S = \frac{\mathrm{d}C}{\mathrm{d}x} = \frac{2\pi \varepsilon}{\ln(D/d)} = 常数 \tag{4.62}$$

可见,面积变化型线位移传感器的输出(电容的变化 ΔC)与其输入(由被测量引起的极板覆盖面积的改变)呈线性关系。

2. 角位移型

图 4.49 所示为面积变化型电容传感器(角位移型)。当动板转动一角度时,与定板之间的覆盖面积就发生变化,导致电容量随之改变。覆盖面积为

$$A = \frac{\alpha r^2}{2}$$

Its sensitivity is

$$S = \frac{\mathrm{d}C}{\mathrm{d}x} = \frac{\varepsilon b}{\delta} = \text{constant} \tag{4.60}$$

Figure 4.47 Variable electrode plate area type capacitive sensor (plane linear displacement type)

Figure 4. 48 shows a variable electrode plate area type capacitive sensor (unilateral cylinder linear displacement type), the movable plate (cylinder) and the fixed plate (cylinder) cover each other, and its capacitance is

$$C = \frac{2\pi\varepsilon x}{\ln(D/d)} \tag{4.61}$$

where d —outer diameter of cylinder;
D —cylinder aperture.

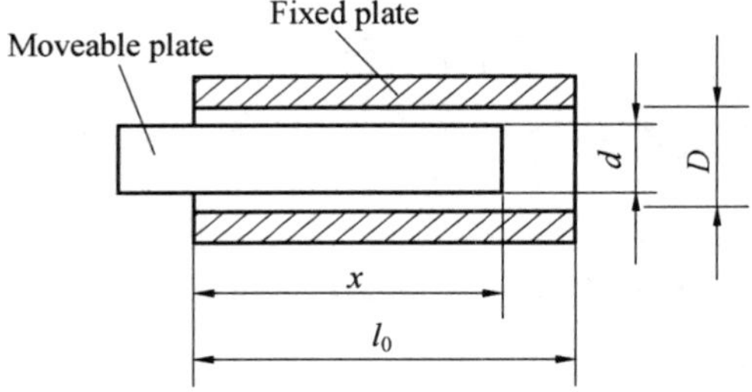

Figure 4.48 Variable electrode plate area type capacitive sensor (unilateral cylinder linear displacement type)

When the cover length x changes, the capacitance C changes accordingly, and its sensitivity is

$$S = \frac{\mathrm{d}C}{\mathrm{d}x} = \frac{2\pi\varepsilon}{\ln(D/d)} = \text{constant} \tag{4.62}$$

It can be seen that the output of the variable electrode plate area type linear displacement sensor (the change in capacitance ΔC) and its input (the change in the coverage area of the electrode plate caused by the measurement) are in a linear relationship.

2. Angular Displacement Type

Figure 4. 49 shows the variable electrode plate area type capacitive sensor (angular displacement type). When the movable plate rotates by an angle, the coverage area between it and the fixed plate changes, causing the capacitance to change accordingly. Coverage area is

$$A = \frac{\alpha r^2}{2}$$

where α —the center angle corresponding to the coverage area;
r —the radius of the plate.

式中　α—— 覆盖面积对应的中心角；

r—— 极板半径。

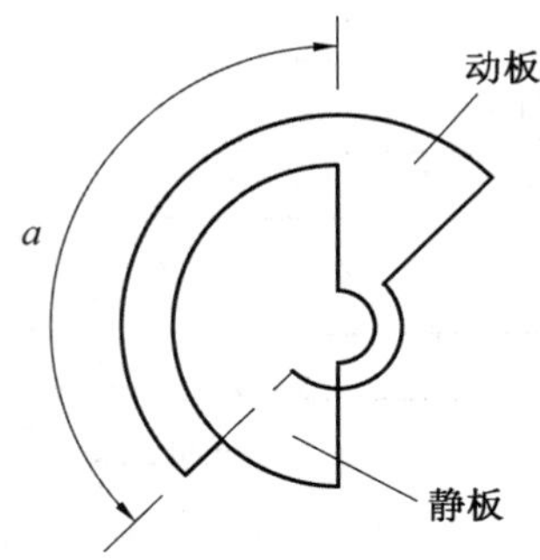

图 4. 49　面积变化型电容传感器(角位移型)

因此，电容量为

$$C = \frac{\varepsilon \alpha r^2}{2\delta} \tag{4.63}$$

其灵敏度为

$$S = \frac{\mathrm{d}C}{\mathrm{d}\alpha} = \frac{\varepsilon r^2}{2\delta} = 常数 \tag{4.64}$$

可见，角位移型电容传感器的输入(被测量引起的极板角位移 $\Delta\alpha$)与输出(电容的变化 ΔC)呈线性关系。

改变极板的形状和数量，面积变化型电容传感器可变为差动型，提高灵敏度。面积变化型电容传感器的其他形式如图 4. 50 所示。

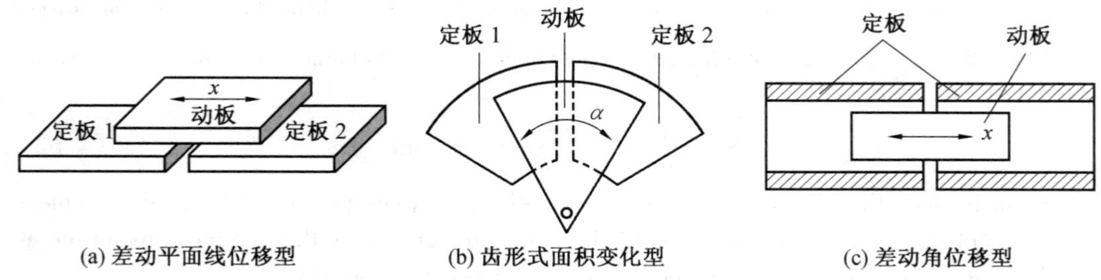

图 4. 50　面积变化型电容传感器的其他形式

面积变化型电容传感器在理想情况下灵敏度为常数，不存在非线性误差，但实际上受电场边缘效应的影响仍存在一定的非线性误差，且灵敏度较低。

4.4.4　介质变化型电容传感器

被测参数使介电常数发生变化，从而引起电容量的变化，称为介质变化型电容式传感

Therefore, the capacitance is

$$C = \frac{\varepsilon \alpha r^2}{2\delta} \tag{4.63}$$

Its sensitivity is

$$S = \frac{dC}{d\alpha} = \frac{\varepsilon r^2}{2\delta} = \text{constant} \tag{4.64}$$

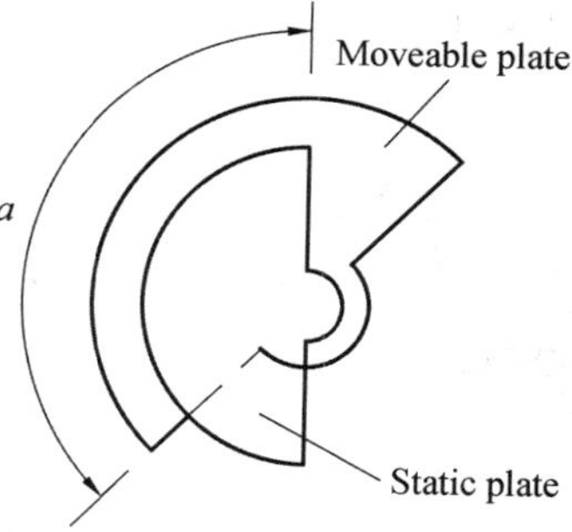

Figure 4.49 Variable electrode plate area type capacitive sensor (angular displacement type)

It can be seen that the input (the plate angular displacement $\Delta\alpha$ caused by the measured quantity) of the angular displacement capacitive sensor and the output (the change in capacitance ΔC) are in a linear relationship.

By changing the shape and number of the plates, the variable electrode plate area type capacitive sensor can be changed to a differential type, which can improve the sensitivity. Other forms of variableelectrode plate area capacitive sensor are shown in Figure 4.50.

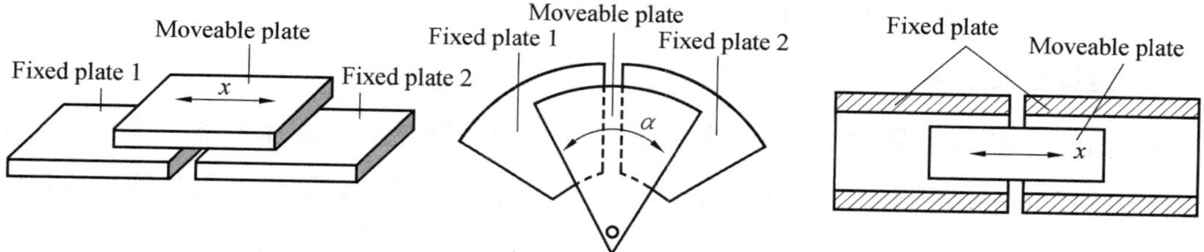

(a) Differential plane linear displacemen type (b) Tooth form area variable type (c)Differential angular displacement type

Figure 4.50 Other forms of variableelectrode plate area capacitive sensor

The variable electrode plate area type capacitance sensor has a constant sensitivity under ideal conditions and does not have a nonlinear error, but in fact, there is still a certain nonlinear error due to the influence of the edge effect of the electric field with low sensitivity.

4.4.4 Variable Dielectric Type Capacitive Sensor

The measured parameter causes the dielectric constant to change, causing the change of the capacitance, which is called the variable dielectric type capacitive sensor. This kind of sensor is mostly used to measure the thickness of materials, liquid level, temperature, and humidity, etc. Figure 4.51 shows the application of dielectric change type capacitive sensor. Figure 4.51(a) shows that a medium layer (such as paper, movie film) passes between two fixed plates. When the thickness, temperature, or humidity of the dielectric layer changes, its dielectric constant changes with causing a change in capacitance. Figure 4.51(b) shows a

器。这种传感器多用来测量材料的厚度、液体的液面、温度和湿度等。图 4.51 所示为介质变化型电容传感器应用。图4.51(a) 所示为在两固定极板间有一介质层(如纸张、电影胶片等) 通过,当介质层的厚度、温度或湿度发生变化时,其介电常数发生变化,引起电容量的变化。图4.51(b) 所示为一种电容式液面计,当液面位置发生变化时,两电极的浸入高度也发生变化,引起电容量的变化。

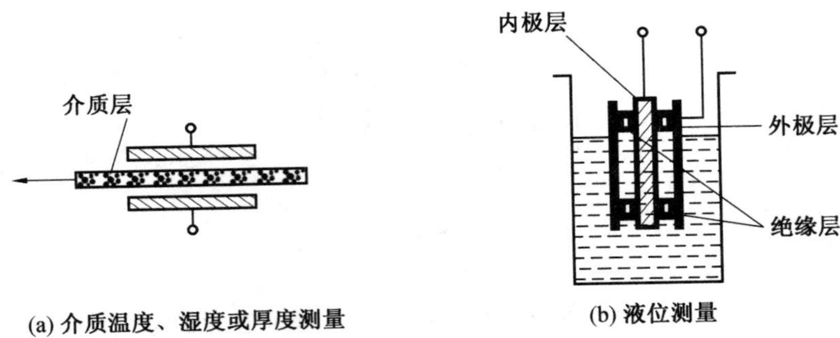

(a) 介质温度、湿度或厚度测量　　(b) 液位测量

图 4.51　介质变化型电容传感器应用

4.4.5　电容传感器的测量电路

电容传感器把被测位移量转换成电容量,还需要后续测量电路将电容量转换成电压、电流或频率信号。常用的测量电路有以下几种。

1. 运算放大器电路

由前述已知,变极距型电容传感器的极距变化与电容量呈非线性关系。这一缺点使电器传感器的应用受到一定限制,而采用运算放大器电路可得到输出电压与输入位移的线性关系。

运算放大器电路如图 4.52 所示,C_0 为固定电容,C_x 为反馈电容且为电容式传感器。根据运算放大器的运算关系,有

$$\dot{U}_{sc}=\frac{Z_x}{Z_0}\dot{U}_{sr}=-\frac{C_0}{C_x}\dot{U}_{sr}$$

图 4.52　运算放大器电路

将 $C_x=\varepsilon A/x$ 代入上式,得输出特性为

$$\dot{U}_{sc}=-\frac{C_0x}{\varepsilon A}\dot{U}_{sr} \tag{4.65}$$

2. 桥式电路

图 4.53 所示为电容式传感器桥式测量电路。图 4.53(a) 所示为单臂接法的桥式测

capacitive liquid level gauge. When the position of the liquid level changes, the immersion height of the two electrodes also changes, causing a change in capacitance.

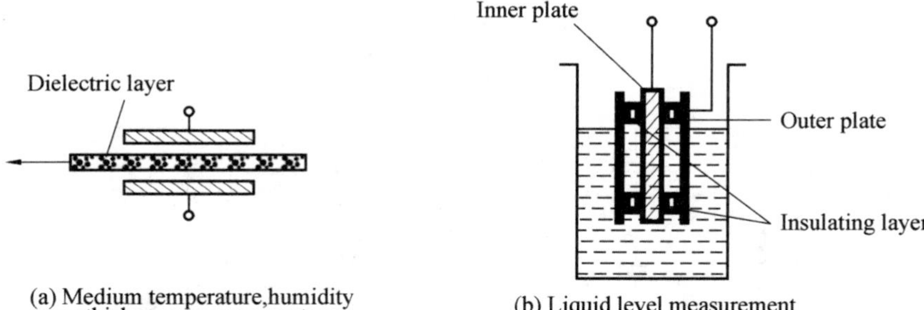

Figure 4.51 Application of dielectric change type capacitive sensor

4.4.5 The Measurement Circuit of the Capacitance Sensor

The capacitance sensor will convert the measured displacement into capacitance, and a subsequent measuring circuit will be needed to convert the capacitance into voltage, current, or frequency signals. The commonly used measurement circuits are as follows.

1. Operation Amplifier Circuit

It is known from the foregoing that the change of the electrodes distance of the variable electrode distance capacitance sensor has a non-linear relationship with the capacitance. This shortcoming limits the application of electrical sensors to a certain extent, however, the use of operational amplifier circuits can obtain the linearity between the output voltage and the input displacement relationship.

Operation amplifier circuit is shown in Figure 4.52, C_0 is a fixed capacity, and C_x is a feedback capacity and a capacitive sensor. According to the operational relationship of the operational amplifier, we get

$$\dot{U}_{sc} = \frac{Z_x}{Z_0}\dot{U}_{sr} = -\frac{C_0}{C_x}\dot{U}_{sr} \tag{4.31}$$

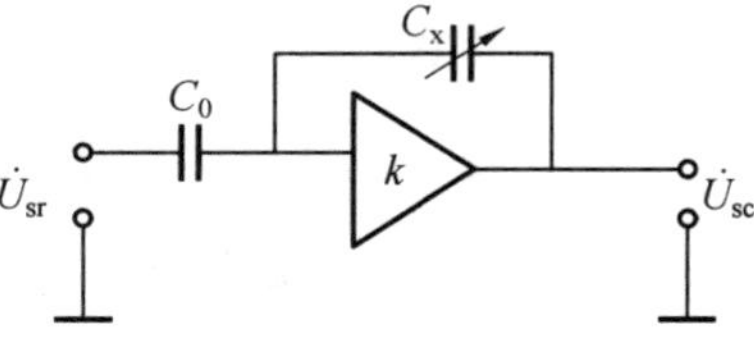

Figure 4.52 Operation amplifier circuit

Substituting $C_x = \varepsilon A/x$ into the above formula, the output characteristic is

$$\dot{U}_{sc} = -\frac{C_0 x}{\varepsilon A}\dot{U}_{sr} \tag{4.65}$$

2. Bridge circuit

Figure 4.53 shows the capacitive sensor bridge measurement circuit. Figure 4.53(a) shows a single-arm connection bridge measurement circuit. The high-frequency power supply is connected to a diagonal line of the capacitor bridge via a transformer. Capacitors C_1, C_2, C_3 and C_x constitute the four arms of the capacitor bridge, and C_x is a capacitive sensor. When the AC bridge is balanced, we have

$$\frac{C_1}{C_2} = \frac{C_x}{C_3}, U_{sc} = 0$$

When C_x changes, $U_{sc} \neq 0$ has an output voltage.

量电路，高频电源经变压器接到电容桥的一条对角线上，电容 C_1、C_2、C_3、C_x 构成电容桥的四臂，C_x 为电容传感器。交流电桥平衡时，有

$$\frac{C_1}{C_2}=\frac{C_x}{C_3}, U_{sc}=0$$

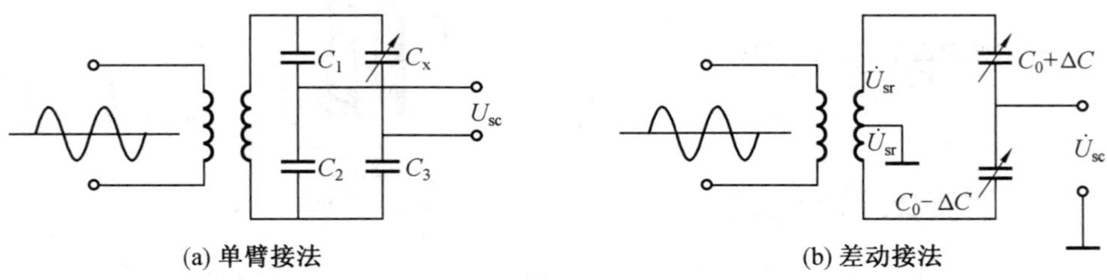

图 4.53　电容式传感器桥式测量电路

当 C_x 改变时，$U_{sc}\neq 0$ 存在输出电压。

在图 4.53(b) 所示的电路中，接有差动电容传感器，其空载输出电压为

$$\dot{U}_{sc}=\frac{(C_0-\Delta C)-(C_0+\Delta C)}{(C_0+\Delta C)+(C_0-\Delta C)}\dot{U}_{sr}=\frac{\Delta C}{C_0}\dot{U}_{sr} \tag{4.66}$$

式中　$\dot{U}_{sr}$—— 工作电压；

C_0—— 电容传感器平衡状态的电容值；

ΔC—— 电容传感器的电容变化值。

3. 谐振电路

谐振电路及其工作特性如图 4.54 所示。电容传感器的电容 C_x 作为谐振电路(L_2、$C_2//C_x$ 或 C_2+C_x) 调谐电容的一部分。此谐振回路通过电压耦合，从稳定的高频振荡器中获得振荡电压。当传感电容 C_x 发生变化时，谐振回路的阻抗发生相应变化，并被转换成电压或电流输出，经过放大和检波即可得到输出。为获得较好的线性，一般工作点应选择在谐振曲线一边的准线性区域内。这种电路比较灵敏，但工作点不易选好，其变化范围也较窄，传感器连接电缆的杂散电容影响也较大。

4. 驱动电缆

在使用电容传感器时需要注意以下两个方面：一是电容传感器的电容量很小，一般只有几十或几百皮法，测量时电容量的变化更小，常在 1 pF 以下；二是传感器的板极与周围元件之间及连接电缆都存在着寄生电容，其电容值很大且不稳定，这就使测量精确度受到严重影响，甚至无法工作。为此，必须采取适当的技术措施来减小或消除寄生电容的影响。常用的措施有：缩短传感器和测量电路之间的电缆，甚至将测量电路的一部分和传感

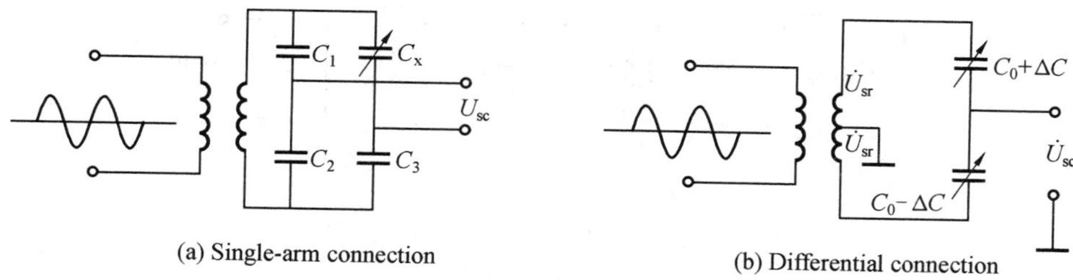

Figure 4.53 Capacitive sensor bridge measurement circuit

In the circuit shown in Figure 4.53(b), a differential capacitance sensor is connected, and its no-load output voltage is

$$\dot{U}_{sc}=\frac{(C_0-\Delta C)-(C_0+\Delta C)}{(C_0+\Delta C)+(C_0-\Delta C)}\dot{U}_{sr}=\frac{\Delta C}{C_0}\dot{U}_{sr} \tag{4.66}$$

where $\dot{U}_{sr}$ —working voltage;

C_0—capacitance value of the capacitive sensor in the balanced state;

ΔC —the capacitance change value of the capacitance sensor.

3. Resonant Circuit

Resonant circuit and its operating characteristics are shown in Figure 4.54. The capacitance C_x of the capacitance sensor is used as part of the tuning capacitance of the resonant circuit (L_2, $C_2//C_x$ or C_2+C_x). This resonant tank obtains an oscillating voltage from a stable high-frequency oscillator through voltage coupling. When the sensing capacitance C_x changes, the impedance of the resonant circuit changes correspondingly and will be converted into a voltage or current output, and the output can be obtained after amplification and detection. In order to obtain better linearity, the general operating point should be selected in the quasi-linear region on one side of the resonance curve. This kind of circuit is more sensitive, but the operating point is not easy to choose, with narrow the range of change, and the influence of the stray capacitance of the sensor connecting cable being also greater.

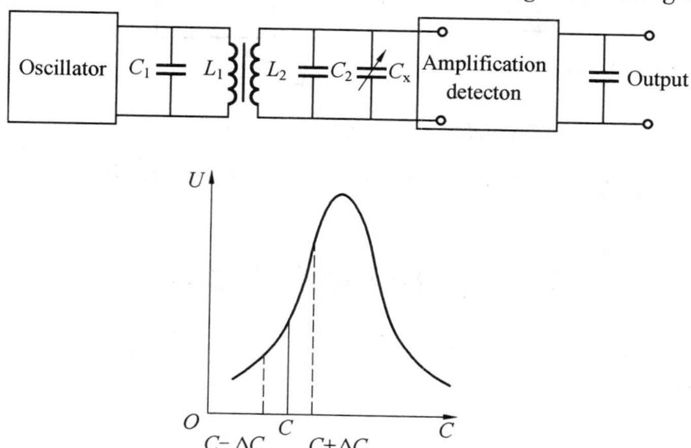

Figure 4.54 Resonant circuit and its operating characteristics

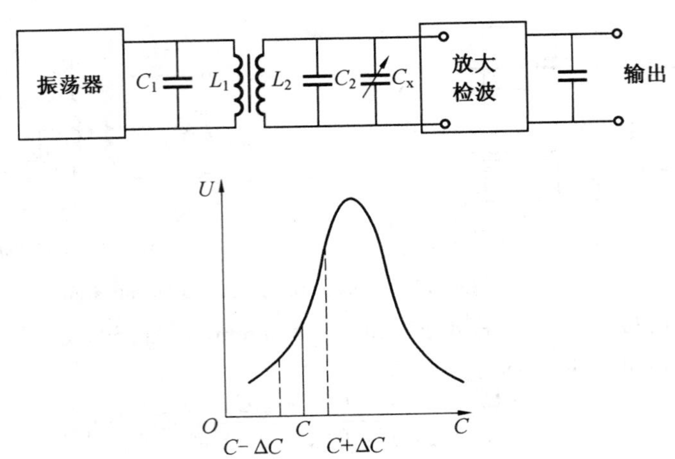

图 4.54　谐振电路及其工作特性

器做成一体或采用专用驱动电缆。

图4.55所示为驱动电缆的工作原理,它采用双层屏蔽电缆。其中,用一个增益为1的放大器,放大器输入端接于芯线,输出端接于内屏蔽线,用芯线的电位来驱动内屏蔽线的电位。当放大器严格保持增益为1和相移为0时,内屏蔽线和芯线等电位,可以免除芯线和内屏蔽线之间的容性漏电流,从而消除二者之间寄生电容的影响。若放大器增益非1或相移非0,芯线和内屏蔽线的电位仍会有差别。

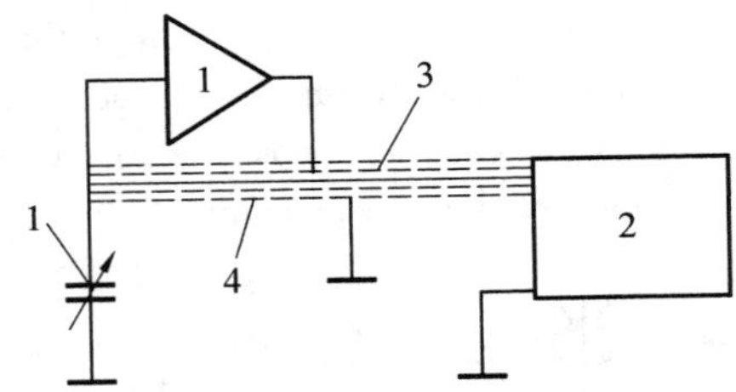

图 4.55　驱动电缆的工作原理

1— 传感器;2— 测量电路;3— 内屏蔽;4— 外屏蔽

4.5　电涡流式传感器

4.5.1　概述

金属导体在交变磁场中,会在导体内产生感应电流,此电流在导体内是闭合的(呈涡旋状)电涡流,这种效应称为电涡流效应。电涡流式传感器就是基于金属导体在交变磁场中的电涡流效应原理制成的传感器。

4. Drive Cable

We shall need to pay attention to two aspects when using capacitive sensors: firstly, the capacitance of a capacitive sensor is very small, generally only tens or hundreds of picofarad, and the capacitance change during measurement is even smaller, often below 1 pF; secondly, there are parasitic capacitances between the sensor's plate and surrounding components and connecting cables, and the capacitance value is very large and unstable. This severely affects the accuracy of the measurement and even fails to work. For this reason, appropriate technical measures should be taken to reduce or eliminate the influence of parasitic capacitance. Commonly used measures are: to shorten the cable between the sensor and the measurement circuit, and even to make a part of the measurement circuit and the sensor into one or use a dedicated drive cable.

Figure 4.55 shows the working principle of the drive cable. It uses double-layer shielded cables. With an amplifier with a gain of 1 being used, the input end is connected to the core wire, and the output end is connected to the inner shield wire. The potential of the core wire is used to drive the potential of the inner shield wire. When the amplifier strictly maintains the gain at 1 and the phase shift at 0, the inner shield wire and the core wire are at the same potential, which can avoid the capacitive leakage current between the core wire and the inner shield wire, thereby eliminating the influence of the parasitic capacitance between the two. If the amplifier gain is not 1 or the phase shift is not 0, the potential of the core wire and the inner shield wire is still different.

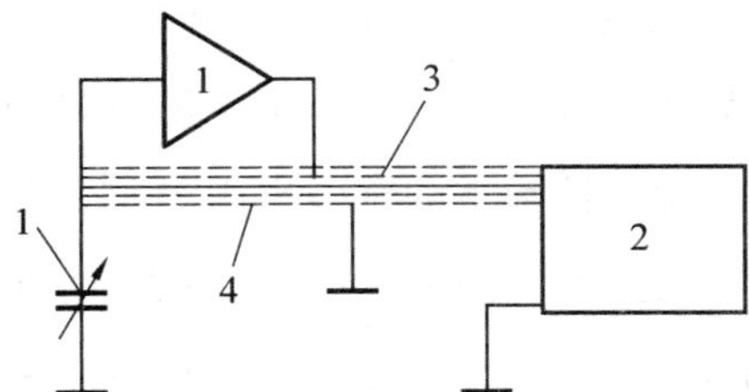

Figure 4.55 Working principle of the drive cable
1—sensor; 2—measurement circuit; 3—inner shield; 4—outer shield

4.5 Eddy Current Sensors

4.5.1 Overview

In an alternating magnetic field, a metal conductor will induce current in the conductor that is a closed (vortex-like) eddy current in the conductor, whose effect is called the eddy current effect. The eddy current sensor is a sensor made based on the principle of the eddy current effect of a metal conductor in an alternating magnetic field.

The generation of eddy currents inevitably consumes a part of the magnetic field energy so that the impedance of the coil that generates the magnetic field changes. The measuring principle of the eddy current sensor is that the measured non-electricity causes the change of the coil impedance through the eddy current effect.

The eddy current sensor is essentially a coil-metal conductor system. In the system, the

电涡流的产生必然要消耗一部分磁场能量，从而使产生磁场的线圈阻抗发生变化。电涡流式传感器的测量原理为被测非电量通过电涡流效应引起线圈阻抗的变化。

电涡流传感器实质是一个线圈 - 金属导体系统。在系统中，线圈的阻抗是一个多元函数，与金属导体的性质（电阻率ρ、磁导率μ、厚度h等）、线圈的几何参数、线圈与金属之间的距离x、线圈电流的激励频率ω等有关，即阻抗$Z=f(\rho,\mu,h,x,\omega,\cdots)$。因此，可把线圈作为传感器的敏感元件，通过其阻抗的变化实现被测参数的测量。例如，仅改变参数x（其余参数不变）时，可测位移、转速、厚度、振动；只改变ρ或μ时，可测量导体表面缺陷、裂纹、硬度和强度，可用于探伤、材质鉴别。

电涡流式传感器有两种类型：高频反射式和低频透射式。其中，高频反射式电涡流传感器应用较为广泛。

4.5.2 高频反射式电涡流传感器

1. 工作原理

高频反射式电涡流传感器的工作原理如图 4.56 所示。当传感器线圈通以高频交变电流I_1时，线圈周围空间形成交变磁场该磁场，作用于金属板。由于集肤效应（交变电流通过导体时，感应作用引起导体截面上电流分布不均匀，越接近导体表面，电流密度越大，这种现象称为集肤效应，集肤效应使导体的有效电阻增加，交流电的频率越高，集肤效应越显著），因此高频磁场不能透过有一定厚度h的金属板，而仅作用于表面的薄层内，并在这薄层中产生电涡流I_2。I_2又产生新的交变磁场Φ_2（反抗原磁场Φ_1），从而引起线圈等效阻抗Z发生变化。

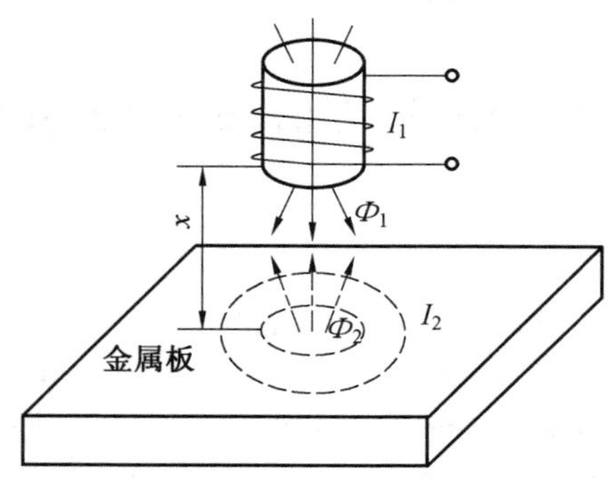

图 4.56 高频反射式电涡流传感器的工作原理

若激励线圈和金属导体材料确定，则线圈的阻抗Z就是线圈与金属板之间的距离x的单值函数，即$Z=f(x)$。当距离x发生变化时，线圈的阻抗就发生变化，从而达到以传感

impedance of the coil is a multivariate function, which is related to the properties of the metal conductor (resistivity ρ, permeability μ, thickness h, etc.), the geometric parameters of the coil, the distance x between the coil and the metal, and the excitation frequency of the coil current ω, and so on. That is, the impedance $Z = f(\rho, \mu, h, x, \omega, \cdots)$. Therefore, the coil can be used as the sensitive element of the sensor, and the measured parameter can be measured through the change of its impedance. For example, when only changing parameter x (other parameters remain unchanged), it can measure displacement, speed, thickness, and vibration; when only changing ρ or μ, it can measure conductor surface defects, cracks, hardness and strength, which can be used for flaw detection and material identification.

There are two types of eddy current sensors: high frequency reflection type and low frequency transmission type, of which high frequency reflection type eddy current sensors are widely used.

4.5.2 High-Frequency Reflective Eddy Current Sensor

1. Working Principle

The working principle of the high-frequency reflective eddy current sensor is shown in Figure 4.56. When the sensor coil is energized with a high-frequency alternating current I_1, an alternating magnetic field is formed in the space around the coil. This magnetic field acts on the metal plate. Due to the skin effect (when the alternating current passes through the conductor, the current distribution on the cross section of the conductor is uneven due to the induction effect that the closer to the surface of the conductor, the greater the current density, whose phenomenon is called the skin effect which increases the effective resistance of the conductor, thus the higher the frequency of alternating current, the more pronounced the skin effect), the high-frequency magnetic field cannot pass through the metal plate with a certain thickness h, but only acts on the thin layer on the surface, and generates eddy current I_2 in this thin layer. I_2 generates a new alternating magnetic field Φ_2 (anti-antigen magnetic field Φ_1), which causes the equivalent impedance Z of the coil to change.

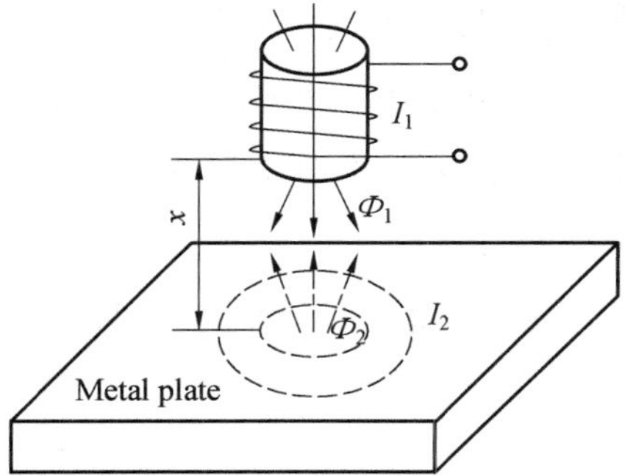

Figure 4.56 Working principle of the high-frequency reflective eddy current sensor

If the excitation coil and the metal conductor material are determined, the impedance Z of the coil will be a single value function of the distance x between the coil and the metal plate, that is, $Z = f(x)$. When the distance x changes, the impedance of the coil will change so as to achieve the purpose of detecting the displacement by the impedance change of the sensor coil. This is the principle of the eddy current sensor to measure the displacement.

器线圈的阻抗变化来检测位移量的目的，这就是电涡流传感器测位移的原理。

根据线圈－导体系统的电磁作用，若不考虑电涡流分布的不均匀性，可以演算得到导体中的电涡流 I_2 与距离 x 的关系为

$$I_2 = I_1\left(1 - \frac{x}{\sqrt{x^2 - r_{os}^2}}\right) \tag{4.67}$$

式中　I_1——线圈的激励电流；

r_{os}——线圈的外半径。

由式(4.67)可以画出电涡流强度与 x/r_{os} 的关系曲线，如图4.57所示。曲线表明，电涡流随着 x/r_{os} 的增加而迅速减小。在导体与线圈间的距离 $x > r_{os}$ 处，所产生的电涡流已经很微弱了。为产生相当强的电涡流效应，应使 $x/r_{os} < 1$，一般取 $x/r_{os} = 0.05 \sim 0.15$。

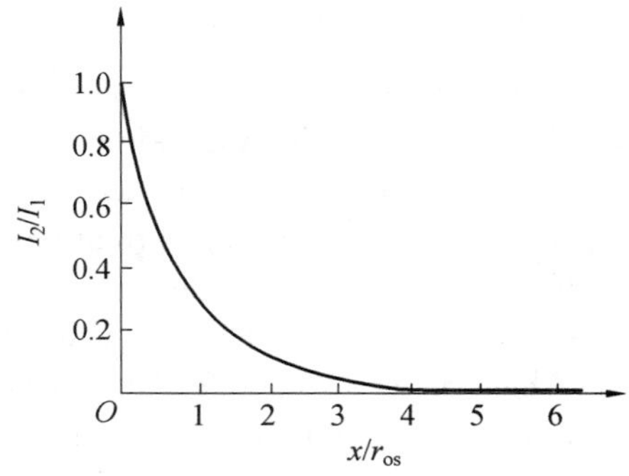

图4.57　电涡流强度与 x/r_{os} 的关系曲线

2. 测量电路、调频原理及解调

涡电流式传感器的测量电路一般有阻抗分压式调幅电路和调频电路。

测量时，为提高灵敏度，将已知电容 C 与传感器线圈并联(一般在传感器内)组成LC并联谐振回路。传感器线圈等效电感的变化使并联谐振回路的谐振频率发生变化，将被测量变换为电压或电流信号输出。

分压式调幅度电路的工作原理如图4.58所示。传感器线圈 L 和电容 C 组成并联谐振回路，其谐振频率为

$$f = \frac{1}{2\pi\sqrt{LC}} \tag{4.68}$$

电路中由振荡器提供稳定的高频信号电源。当谐振频率与该电源频率相同时，输出电压 u 最大。在测量时，传感器线圈阻抗随 δ 而改变，LC回路失谐。输出信号 $u(t)$ 频率虽然仍为振荡器的工作频率，但幅值随 δ 而变化，它相当于一个调幅波。此调幅波经放大、检波、滤波后即可以得到气隙 δ 的动态变化信息。分压式调幅电路的谐振曲线及输出

According to the electromagnetic effect of the coil-conductor system, if the inhomogeneity of the eddy current distribution is not considered, the relationship between the eddy current I_2 in the conductor and the distance x can be calculated as

$$I_2 = I_1\left(1 - \frac{x}{\sqrt{x^2 - r_{os}^2}}\right) \tag{4.67}$$

where I_1— the excitation current of the coil;

r_{os} —the outer radius of the coil.

From formula(4.67), the relationship curve between eddy current intensity and x/r_{os} can be drawn, as shown in Figure 4.57. The curve shows that the eddy current decreases rapidly as x/r_{os} increases. At the distance $x > r_{os}$ between the conductor and the coil, the generated eddy current is already very weak. In order to produce a relatively strong eddy current effect, $x/r_{os} < 1$ should be used, generally $x/r_{os} = 0.05 - 0.15$.

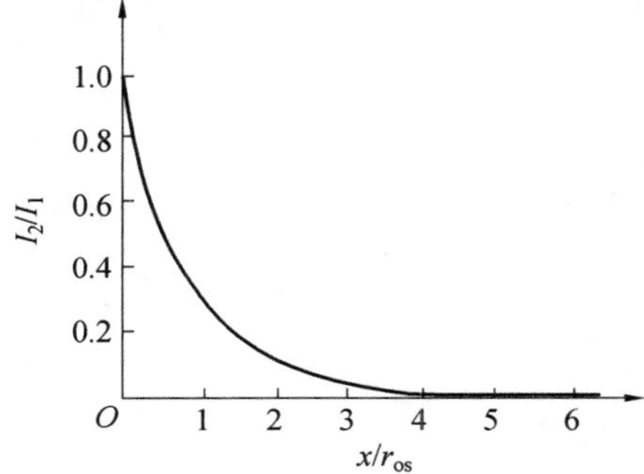

Figure 4.57 The relationship curve between eddy current intensity and x/r_{os}

2. Measuring Circuit, FM Principle and Demodulation

The measuring circuit of the eddy current sensor generally has an impedance divider type amplitude modulation circuit and a frequency modulation circuit.

When measuring, in order to improve the sensitivity, the known capacitance C is connected in parallel with the sensor coil (usually in the sensor) to form an LC parallel resonant circuit. The change of the equivalent inductance of the sensor coil causes the resonant frequency of the parallel resonant circuit to change, which will be measured and converted into a voltage or current signal output.

The working principle of the voltage divider amplitude modulation circuit is shown in Figure 4.58. The sensor coil L and the capacitor C form a parallel resonant circuit, and its resonant frequency is

$$f = \frac{1}{2\pi\sqrt{LC}} \tag{4.68}$$

The oscillator provides a stable high-frequency signal power supply in the circuit. When the resonance frequency is the same as the power supply frequency, the output voltage u is the largest. When measuring, the impedance of the sensor coil changes with δ, and the LC loop is detuned. Although the frequency of the output signal $u(t)$ is still the operating frequency of the oscillator, the amplitude changes with δ. It is equivalent to an amplitude modulation wave. After the amplitude modulation wave is amplified, detected, and filtered, the information of the

特性如图 4. 59 所示。

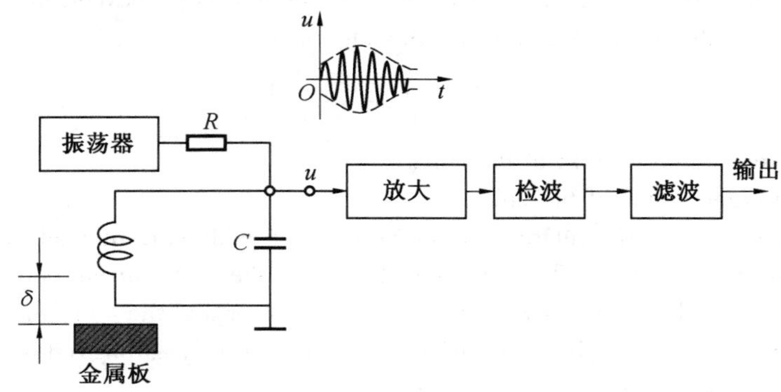

图 4. 58　分压式调幅电路的工作原理

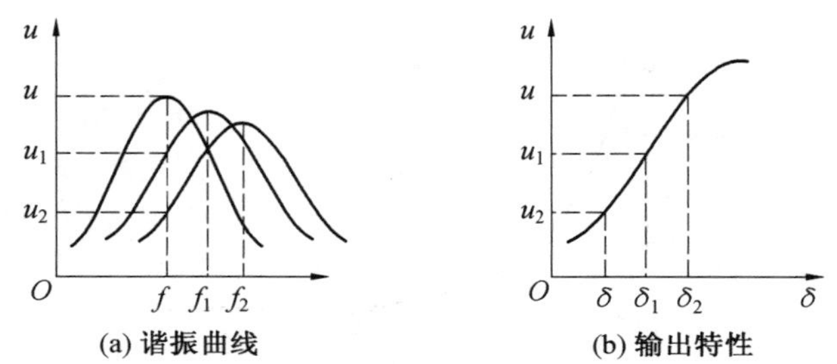

图 4. 59　分压式调幅电路的谐振曲线及输出特性

调频电路的工作原理如图 4. 60 所示。这种方法也是把传感器线圈接入 LC 振荡回路。其与调幅法不同之处是以回路的谐振频率作为输出量。当金属板至传感器之间的距离 δ 发生变化时,将引起线圈电感变化,从而使振荡器的振荡频率 f 发生变化,再通过鉴频器进行频率 - 电压转换,即可得到与 δ 成比例的输出电压。

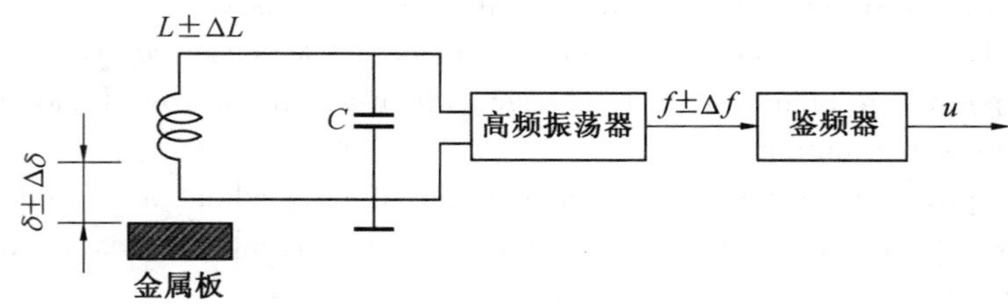

图 4. 60　调频电路的工作原理

4. 3. 2 节中讲到调幅电路的原理及其解调方法,这里继续补充调频电路的原理及其解调。

调频是利用调制信号对载波的频率进行调制,使载波的频率随调制信号的大小而成比例变化,但幅值保持不变,已调波(调频波) 频率的变化量反映了调制信号的大小。载

dynamic change of the air gap δ can be obtained. The resonance curve and output characteristics of the voltage divider type amplitude modulation circuit are shown in Figure 4.59.

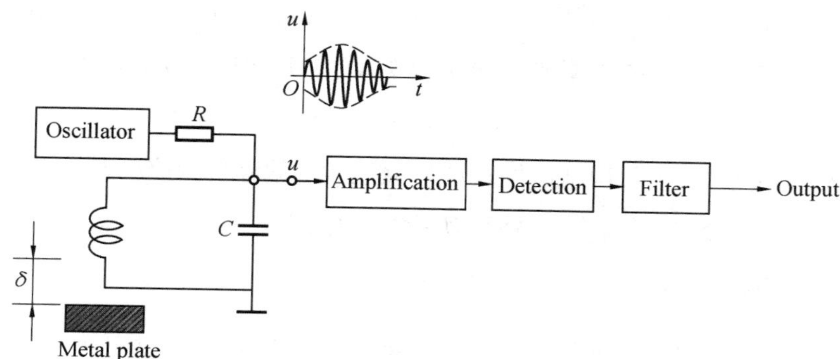

Figure 4.58 Working principle of the voltage divider amplitude modulation circuit

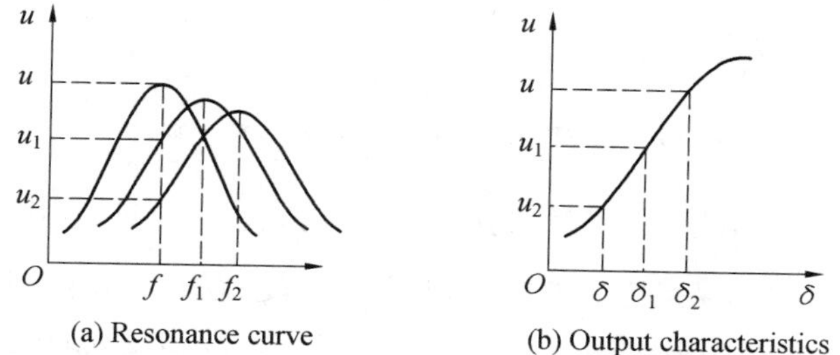

Figure 4.59 The resonance curve and output characteristics of voltage divider type amplitude modulation circuit

The working principle of the FM circuit is shown in Figure 4.60. This method also connects the sensor coil to the LC oscillating circuit. The difference from the amplitude modulation method is that the resonant frequency of the circuit is used as the output. When the distance δ between the metal plate and the sensor changes, it will cause the coil inductance to change, so that the oscillation frequency f of the oscillator changes, and then the frequency-voltage conversion is carried out through the frequency discriminator, the output voltage proportional to δ can be obtained.

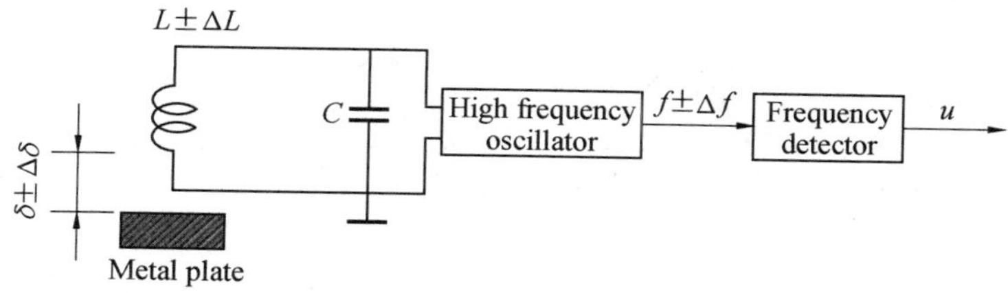

Figure 4.60 The working principle of FM circuit

The principle of the amplitude modulation circuit and its demodulation method are mentioned in section 4.3.2, here we continue to supplement the principle of the frequency modulation circuit and its demodulation.

Frequency modulation is the use of modulating signals to modulate the frequency of the

波可以采用正弦(或余弦)波或方波等。调频信号具有抗干扰能力强、便于远距离传输、不易失真等特点,也很容易采用数字技术和计算机处理。

设载波 $y(t)=A\cos\omega_0 t$。其中,角频率 ω_0 为载波频率,是没有受到调制时的载波频率。如果保持振幅 A 为常数,令载波瞬时角频率 $\omega(t)$ 随调制信号 $x(t)$ 做线性变化,则有

$$\omega(t)=\omega_0+kx(t) \tag{4.69}$$

式中 k——比例常数,表示调制信号 $x(t)$ 对载波频率偏移量的控制能力,是单位调制信号产生的频率偏移量,故称为调频灵敏度。

调频信号的瞬时相角为

$$\varphi(t)=\int_0^t\omega(\tau)\mathrm{d}\tau+\varphi_0=\omega_0 t+k\int_0^t x(\tau)\mathrm{d}\tau+\varphi_0 \tag{4.70}$$

式中 φ_0——调频信号的起始相角。

因此,调频信号可以表示为

$$x_{\mathrm{f}}(t)=A\cos\left[\omega_0 t+k\int_0^t x(\tau)\mathrm{d}\tau+\varphi_0\right] \tag{4.71}$$

图4.61所示为调频信号频率与调制信号大小的关系。可见,在$0\sim t_1$区间,调制信号 $x(t)=0$,调频信号的频率保持原始的中心频率 ω_0 不变;在 $t_1\sim t_5$ 区间,调频波 $x_{\mathrm{f}}(t)$ 的瞬时频率随着调制信号的大小而变化;在 $t_1\sim t_2$ 和 $t_3\sim t_4$ 区间,调制信号 $x(t)<0$,所以调频信号的频率小于 ω_0;在 $t_2\sim t_3$ 和 $t_4\sim t_5$ 区间,调制信号 $x(t)>0$,所以调频信号的频率大于 ω_0;在 $t\geqslant t_5$ 后,调制信号 $x(t)=0$,调频信号的频率又恢复到原始的中心频率 ω_0。

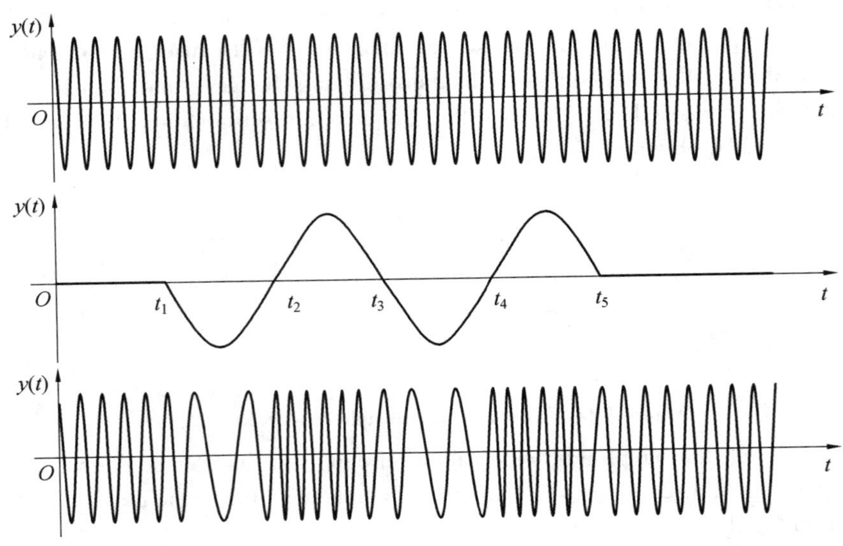

图4.61 调频信号频率与调制信号大小的关系

carrier wave, so that the frequency of the carrier wave changes proportionally with the size of the modulating signal, but the amplitude remains unchanged, which the amount of change in the frequency of the modulated wave (FM wave) reflects the magnitude of the modulated signal. The carrier wave can be a sine (or cosine) wave or a square wave. FM signals not only have the characteristics of strong anti-interference ability, convenient long-distance transmission, and not easy to be distorted, but also are easy to adopt digital technology and computer processing.

Suppose carrier $y(t) = A\cos \omega_0 t$, where the angular frequency ω_0 is the frequency of the carrier, which is the carrier frequency when it is not modulated. If the amplitude A is kept constant and the instantaneous angular frequency $\omega(t)$ of the carrier changes linearly with the modulation signal $x(t)$, then we get

$$\omega(t) = \omega_0 + kx(t) \tag{4.69}$$

where k —proportional constant, which represents the control ability of the modulation signal $x(t)$ to the carrier frequency offset, and is the frequency offset generated by the unit modulation signal, so it is called the frequency modulation sensitivity.

The instantaneous phase angle of the FM signal is

$$\varphi(t) = \int_0^t \omega(\tau)\mathrm{d}\tau + \varphi_0 = \omega_0 t + k\int_0^t x(\tau)\mathrm{d}\tau + \varphi_0 \tag{4.70}$$

where φ_0 —the initial phase angle of the FM signal.

Thus, the FM signal can be expressed as

$$x_f(t) = A\cos\left[\omega_0 t + k\int_0^t x(\tau)\mathrm{d}\tau + \varphi_0\right] \tag{4.71}$$

Figure 4.61 shows the relationship between the frequency of the FM signal changes and the size of the modulation signal. It can be seen that in the interval $0-t_1$, the frequency of the modulation signal $x(t) = 0$ and the FM signal keeps the original center frequency ω_0; in the interval $t_1 - t_5$, the instantaneous frequency of the FM wave $x_f(t)$ changes with the magnitude of the modulation signal; in the interals $t_1 - t_2$ and $t_3 - t_4$, modulate signal $x(t) < 0$, so the frequency of FM signal is less than ω_0; in the intervals $t_2 - t_3$ and $t_4 - t_5$, modulate signal $x(t) > 0$, so the frequency of FM signal is greater than ω_0; after $t \geqslant t_5$, modulate signal $x(t) = 0$, the frequency of the FM signal is restored to the original center frequency ω_0.

(1) The realization method of frequency modulation.

For different sensors, there are many specific schemes for realizing frequency modulation. For parametric sensors such as resistance, capacitance, and inductance, the optional frequency modulation schemes include variable oscillators (including harmonic oscillators, relaxation oscillators, CMOS oscillators) and voltage frequency converters. For power-generating sensors (such as eddy current sensors), voltage-frequency converters must be used.

① Harmonic oscillator. The output of the harmonic oscillator is a sine wave. Place capacitive, inductive or resistive sensors in the RC harmonic oscillator or LC harmonic oscillator as a frequency component of the oscillator. When the capacitance, inductance or resistance of the sensor changes with the measurement, the sensor's impedance changes, resulting in a change in the oscillation frequency of the oscillator and then the output of the FM signal.

RC harmonic oscillator relies on RC phase shift network or Wien bridge to work, it is wise to use Wien bridge, because it is more stable. Figure 4.62 shows the basic structure and the e-

(1) 调频的实现方法。

针对不同的传感器,实现调频的具体方案很多。对于电阻、电容、电感等参量型传感器,可选的调频方案有可变振荡器(包括谐波振荡器、弛缓振荡器、CMOS 振荡器等)和电压频率变换器等。对于发电型传感器(如电涡流式传感器),则要采用电压 - 频率变换器。

① 谐波振荡器。谐波振荡器输出的是正弦波。将电容式、电感式或电阻式传感器放置在 RC 谐波振荡器或 LC 谐波振荡器中作为振荡器的一个选频元件。当传感器的电容、电感或电阻随被测量而变时,传感器的阻抗发生变化,从而导致振荡器的振荡频率也随之发生变化,输出调频信号。

RC 谐波振荡器依靠 RC 移相网络或维恩电桥进行工作,最好是采用维恩电桥,因为它更稳定。图 4.62 所示为维恩电桥的基本结构和等效框图。图中,A_d 和 A_c 分别是运算放大器的开环差模放大倍数和开环共模放大倍数,Z_1 是 R_1 和 C_1 的串联等效阻抗,Z_2 是 R_2 和 C_2 的并联等效阻抗。如果运算放大器满足 $A_d \gg A_c$,则输出电压为

$$\dot{U}_o = A_d \dot{U}_o \left(\frac{Z_2}{Z_1 + Z_2} - \frac{R_3}{R_3 + R_4} \right) \tag{4.72}$$

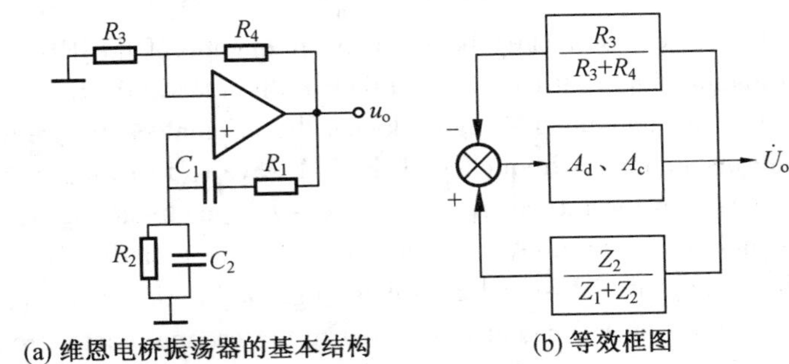

图 4.62　维恩电桥的基本结构和等效框图

当满足下列条件时,电路将产生振荡,即

$$\frac{R_3}{R_4} = \frac{Z_2}{Z_1} = \frac{R_2}{1 + j\omega R_2 C_2} \frac{j\omega C_1}{1 + j\omega R_1 C_1} \tag{4.73}$$

为满足上述条件,有

$$\frac{R_4}{R_3} = \frac{R_1}{R_2} + \frac{C_2}{C_1} \tag{4.74}$$

振荡频率为

$$f_0 = \frac{1}{2\pi \sqrt{R_1 R_2 C_1 C_2}} \tag{4.75}$$

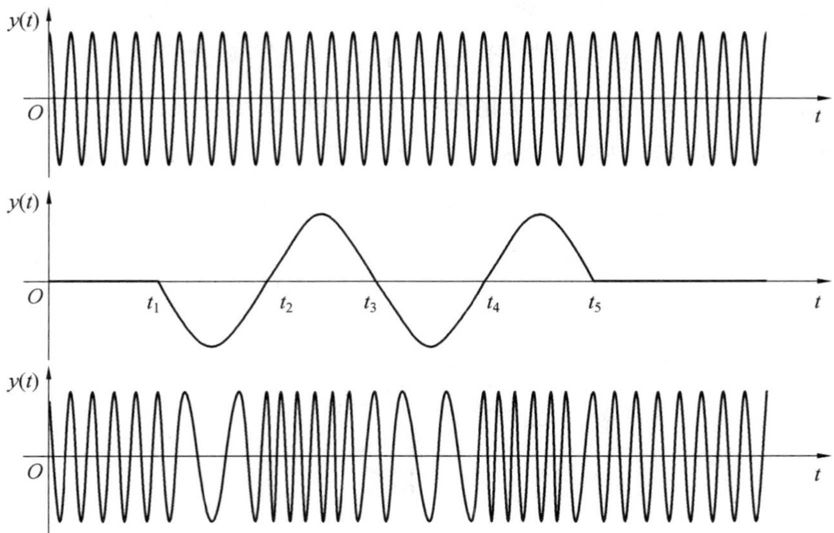

Figure 4.61 The relationship between the frequency of the FM signal and the size of the modulation signal

quivalent block diagram of Wien bridge. In the figure, A_d and A_c are the open-loop differential mode amplification and open-loop common mode amplification of the operational amplifier, respectively, Z_1 is the series equivalent impedance of R_1 and C_1 , and Z_2 is the parallel equivalent impedance of R_2 and C_2 . If the operational amplifier satisfies $A_d \gg A_c$, then the output voltage is

$$\dot{U}_o = A_d \dot{U}_o \left(\frac{Z_2}{Z_1 + Z_2} - \frac{R_3}{R_3 + R_4} \right) \tag{4.72}$$

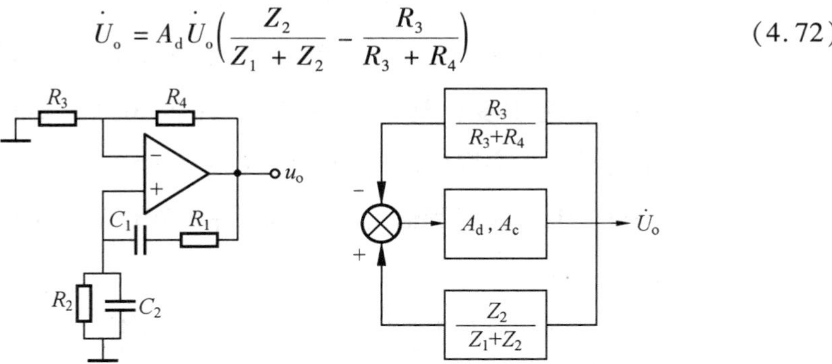

(a) Basic structure of Wien bridge oscillator (b) Equivalent block diagram

Figure 4.62 The basic structure and the equivalent block diagram of Wien bridge oscillator

When the following conditions are met, the circuit will oscillate, that is

$$\frac{R_3}{R_4} = \frac{Z_2}{Z_1} = \frac{R_2}{1 + j\omega R_2 C_2} \frac{j\omega C_1}{1 + j\omega R_1 C_1} \tag{4.73}$$

To meet the above conditions, we get

$$\frac{R_4}{R_3} = \frac{R_1}{R_2} + \frac{C_2}{C_1} \tag{4.74}$$

The oscillation frequency is

传感器可以是 Z_1 或 Z_2 的任一部分(电阻或电容)。为保证在起动时形成振荡,R_3 或 R_4 的选择取决于 u_o。当 u_o 很小时,需要大的增益,以对运算放大器输入端频率为 f_0 的任何扰动进行放大。一旦 u_o 达到足够大的幅度,便要求降低增益,以防出现输出饱和。图 4.63 所示为实用维恩电桥振荡器。当 u_o 很小时,负反馈电阻 $R_4 = R_4'$。但当 u_o 大到一定程度时,两个二极管轮流在各自相应的半周期内导通,$R_4 = R_4' \mathbin{/\!/} R_4''$,增益下降。例如,可以选择 $R_4' = 2.1R_3$ 和 $R_4'' = 10R_4'$。

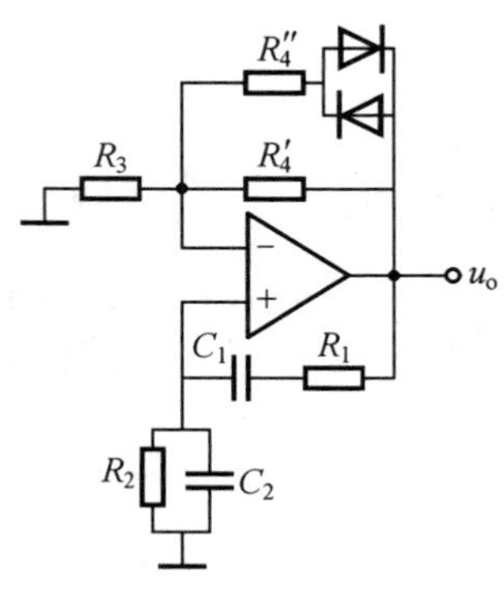

图 4.63　实用维恩电桥振荡器

② 电压 - 频率变换器。本节讲述的电涡流式传感器就是使用该电路作为测量电路。频率变换器从输入电压或电流给出与一般的逻辑电平(TTL 或 CMOS) 兼容的脉冲串、方波信号或二者兼备,信号的频率与模拟输入量成线性比例关系。压控振荡器也是一种电压 - 频率变换器,但其变化范围非常有限(最多为 100 : 1),线性较差。然而,它们可以工作在远高于电压 - 频率变换器的 10 MHz 常规极限频率之上。单片电压 - 频率变换器能给出 10 ~ 100 MHz 的满量程输出频率,其频率的变化范围为 1 ~ 10 000,相当于 A/D 转换器 13 位的分辨力。

许多电压 - 频率变换器都根据电荷平衡技术进行工作,基于电荷平衡技术的电压 - 频率变换器的简化电路如图 4.64 所示。该电路由以下几部分组成:积分器、比较器、精密单稳态脉冲发生器、输出级,以及具有高的时间稳定性和温度稳定性的开关电流源。当输入电压 u_i 为正电压时,电容 C 以正比于输入量的速率充电,并在积分器的输出端给出负斜率的斜坡电压 u_o。比较器测定这个电压达到预置电平的时间,并启动单稳态脉冲发生器输出一个具有固定幅度和持续时间 T_d 的脉冲。数字缓冲器(图中用一个简单的集电极开路的 NPN 型晶体管表示) 将该脉冲反馈至电流输出端。该脉冲还控制积分器通过电流恒定的电流源 I_d(通常为 1 mA) 放电,电容器泄放的电荷量为 I_dT_d。如果输入电压仍然存在,则经过一段由输入电压大小决定的时间之后,输入电压将对泄放的电荷量进行补偿。

$$f_0 = \frac{1}{2\pi\sqrt{R_1 R_2 C_1 C_2}} \tag{4.75}$$

The sensor can be either part of Z_1 or Z_2 (resistance or capacitance). In order to ensure the formation of oscillation at the start, the choice of R_3 or R_4 will depend on u_o. When u_o is very small, a large gain is required to amplify any disturbance at the input end of the operational amplifier with the frequency f_0. Once u_o reaches a sufficiently large amplitude, it will be required to reduce the gain to prevent output saturation. Figure 4.63 shows a practical Wien bridge oscillator. When u_o is very small, negative feedback resistance $R_4 = R_4'$. But when u_o reaches a certain level, the two diodes turn on in their respective half cycles, and $R_4 = R_4' \mathbin{/\!/} R_4''$, the gain drops. For example, we can have a choice of $R_4' = 2.1R_3$ and $R_4'' = 10R_4'$.

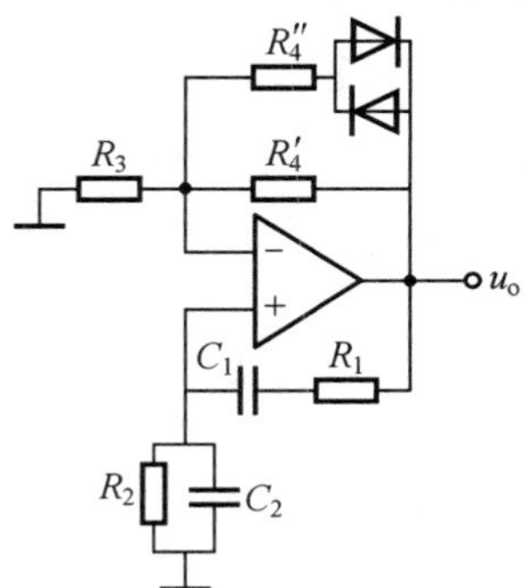

Figure 4.63 A practical Wien bridge oscillator

② Voltage-frequency converter. The eddy current sensor described in this section uses this circuit as a measuring circuit. The frequency converter gives a pulse train, a square wave signal, or both from the input voltage or current compatible with general logic levels (TTL or CMOS), and the frequency of the signal is linearly proportional to the analog input. The voltage-controlled oscillator is also a voltage-frequency converter, but the range of variation is very limited (up to 100 : 1), and the linearity is poor. However, they can work well above the 10 MHz conventional limit frequency of the voltage-frequency converter. The single-chip voltage-frequency converter can give a full-scale output frequency from 10–100 MHz, and the frequency range is 1–10 000, which is equivalent to the 13-bit resolution of an A/D converter.

Many voltage-frequency converters work based on charge balancing technology, the simplified circuit of voltage-frequency converter based on charge balance technology is shown in Figure 4.64. The circuit is composed of the following parts: integrator, comparator, precision monostable pulse generator, output stage, and switching current source with high time stability and temperature stability. When the input voltage u_i is a positive voltage, the capacitor C is charged at a rate proportional to the input quantity, and a ramp voltage u_o with a negative slope is given at the output of the integrator. The comparator measures the time when the voltage reaches the preset level, and starts the monostable pulse generator to output a pulse with a fixed amplitude and duration T_d. The digital buffer (in the figure represented by a simple open-collector NPN transistor) feeds the pulse to the output of the circuit. The pulse also controls the integrator to discharge through the current source I_d (usually 1 mA), with a constant current, and the amount of charge discharged by the capacitor is $I_d T_d$. If the input voltage still exists, the input voltage will compensate for the amount of charge discharged after a period of time de-

这个过程经一段时间 T 后将重复，使得

$$IT = I_d T_d \tag{4.76}$$

$$f = \frac{1}{T} = \frac{I}{I_d T_d} \tag{4.77}$$

注意，C 和比较器阈值都不会影响 f。关键参数是单稳态脉冲的持续时间和放电电流值，二者都必须十分稳定。

在类似于图 4.64 所示的电压频率变换器中，输入放大器的相加点可以接入电流，允许通过在该点上添加一个电流 I_0 来移动输出范围（0 Hz 对应于 0 V）。通过用一个电阻衰减器对输入电压进行分压，可以减小输出范围，以便使最高频率低于电路所能提供的频率。另外，也可以用数字计数器对输出频率进行分频。由于衰减器易受电阻器温度系数的影响，因此通常首选的是数字计数器。

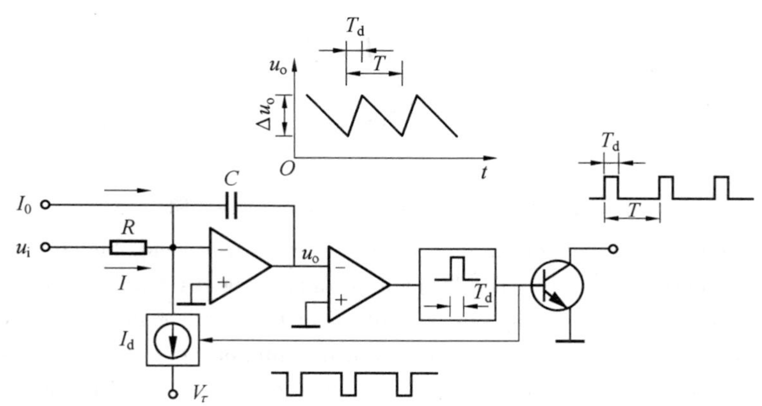

图 4.64　基于电荷平衡技术的电压 - 频率变换器的简化电路

电压 - 频率变换器成高度线性，具有高的分辨率和抗噪能力，但其变换速度较慢。

（2）鉴频的实现方法。

调频信号的解调过程是从调频信号中恢复调制信号的过程，称为鉴频，用于实现鉴频的电路称为鉴频器。实现鉴频的方案有很多，可分为直接鉴频和间接鉴频。直接鉴频就是直接检测调频信号中的频率，然后还原出调制信号；间接鉴频是将调频信号转换成另一种形式的信号，然后还原出调制信号。

① 直接鉴频法。调频信号是等幅振荡信号，通过将调频信号整形成与数字信号电平兼容的方波信号，很容易将调频信号转换成数字信号，这样就可以利用数字计数器和计时器实现计数器直接鉴频。频率通常通过在已知时间间隔内对周期计数来测量。图 4.65 所示为频率计的基本框图，设调频信号的频率为 f_x，T_0 是计数时间，这个时间间隔（选通时

termined by the input voltage. This process will repeat after a period of time T , making

$$IT = I_d T_d \tag{4.76}$$

$$f = \frac{1}{T} = \frac{I}{I_d T_d} \tag{4.77}$$

Note that neither C nor the comparator threshold will affect f_o . The key parameters are the duration of the monostable pulse and the discharge current value, both of which must be very stable.

In a voltage frequency converter similar to that shown in Figure 4.64, the addition point of the input amplifier can be connected to the current, which allows moving the output range by adding a current I_0 to this point (0 Hz corresponds to 0 V) . By dividing the input voltage with a resistive attenuator, we can reduce the output range so that the highest frequency can be lower than the frequency that the circuit provides. In addition, a digital counter can also be used to divide the output frequency. Since the attenuator is susceptible to the temperature coefficient of the resistor, a digital counter is usually the first choice.

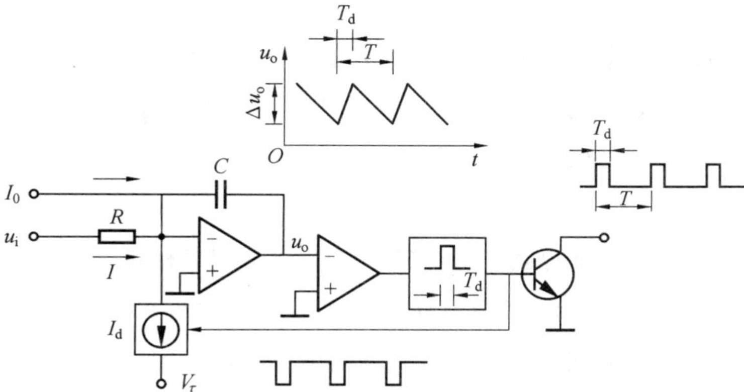

Figure 4.64 Simplified circuit of voltage-frequency converter based on charge balance technology

The voltage-frequency converter proves highly linear, with high resolution and noise immunity, but the conversion speed is relatively slow.

(2) Implementation methods of frequency discrimination.

The demodulation process of the FM signal is the process of recovering the modulated signal from the FM signal, which is called frequency discrimination, and the circuit used to implement frequency discrimination is called a frequency discriminator. There are many schemes for realizing frequency discrimination, which are divided into direct frequency discrimination andindirect frequency discrimination. Direct frequency discrimination is to directly detect the frequency in the frequency modulation signal, and then restore the modulation signal. Indirect frequency discrimination is to convert the FM signal into another form of signal, and then restore the modulated signal.

① Direct frequency discrimination method. FM signal is a kind of equal-amplitude oscillation signal by shaping the FM signal into a square wave signal compatible with the digital signal level. It is easy to convert the FM signal into a digital signal so that digital counters and timers can be used to realize direct frequency discrimination of counters. Frequency is usually measured by counting cycles in a known time interval. Figure 4. 65 shows the basic block

间）通常从精确的时钟经由分频器获得的，则在 T_0 时间内，计数值 N 为

$$N = f_x T_0 \tag{4.78}$$

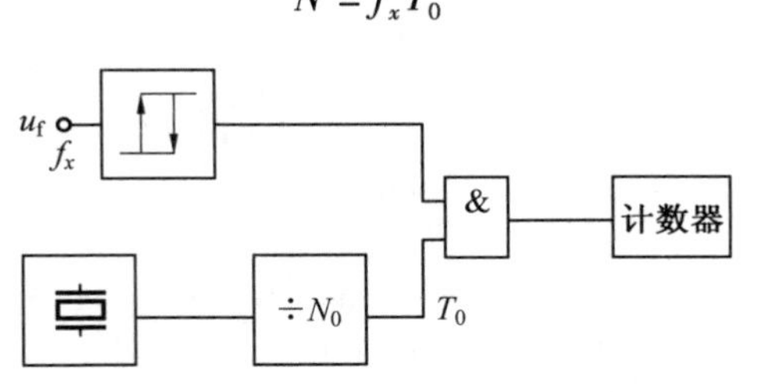

图 4.65　频率计的基本框图

由于输入信号和内部时钟不同步，因此能够恰好在下一个输入转换到达之前或恰好在到达之后停止计数。这意味着不确定度为 1 个数，通常描述为实际结果是 $(N \pm 1)$ 个数。

由于这种测量方法的分辨力是 1 个数，因此分辨率是 $1/N$，N 越大，分辨率越高。然而，加大 N 意味着延长测量时间，这对于低频尤其明显。例如，以低于 0.1% 的不确定度测量 10 kHz 需要 $N = 1\ 000$，由于每个输入周期要持续 100 μs，因此测量时间将是 100 ms。

微控制器一般不包括能提供按图 4.65 的方式测量频率时基的分频器。如果用微控制器（单片计算机）对调频信号计数，则可利用微控制器的两个可编程序计数器，其中一个计数器用来对经历的时间计数，另一个计数器则用来对输入脉冲计数。

② 间接鉴频法。间接鉴频可以采用振幅鉴频法和相位鉴频法等。振幅鉴频法是将等幅的调频信号变换成振幅也随瞬时频率变化的调频 - 调幅波，然后通过包络检波器获得原调制信号。振幅鉴频的实现方法有直接时域微分法、斜率鉴频法等。

斜率鉴频法利用在所需频率范围内具有线性幅频特性的网络完成鉴频，可以采用单失谐回路，也可以采用双失谐回路。这两种鉴频网络利用调谐回路幅频特性曲线的倾斜部分鉴频，因此称为斜率鉴频法。另外，由于是利用调谐回路的失（离）谐状态，因此又称失（离）谐回路法。

图 4.66 所示为一种采用变压器耦合的单失谐回路实现斜率鉴频的电路。图 4.66(a) 中，L_1 和 L_2 是变压器耦合的一次、二次绕组，它们与 C_1 和 C_2 组成并联谐振回路。谐振回路对简谐输入信号的稳态响应幅值 U_{am} 随输入信号频率变化的关系曲线如图 4.66(b) 中的左上所示。

调频波经过 C_1、L_1 耦合，加于 C_2、L_2 组成的谐振回路上。将等幅调频波 u_f 输入，在回路的谐振频率 f_r 处，L_1、L_2 中的耦合电流最大，二次侧输出电压 u_a 也最大。当 u_f 频率偏离

diagram of the frequency counter. Let the frequency of the FM signal be f_x and T_0 is the counting time, this time interval (gating time) is usually obtained from an accurate clock through a frequency divider, and the count value N during T_0 time is

$$N = f_x T_0 \tag{4.78}$$

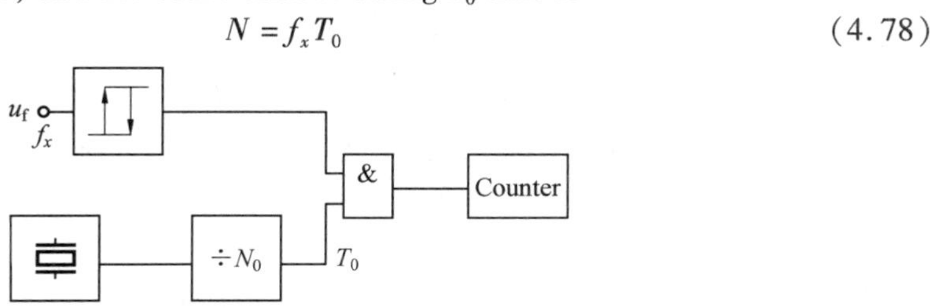

Figure 4.65 Basic block diagram of frequency counter

Since the input signal is not synchronized with the internal clock, we can stop counting just before the next input transition arrives or just after it arrives. This means the uncertainty is a number, which is usually described as the actual result being ($N \pm 1$) number.

Since the resolution of this measurement method is one number, the resolution is $1/N$. The larger the N, the higher the resolution. However, increasing N means extending the measurement time, which is especially prominent for low frequencies. For example, measuring 10 kHz with an uncertainty of less than 0.1% requires $N = 1\ 000$. Since each input period lasts for 100 μs, the measurement time will be 100 ms.

Microcontrollers generally do not include a frequency divider that can provide a time base for measuring frequency in the manner shown in Figure 4.65. If using a microcontroller (single-chip computer) to count the frequency modulation signal, we can use the two programmable counters of the microcontroller, one of which is used to count the elapsed time, and the other one is used to count the input pulses.

② Indirect frequency discrimination method. Indirect frequency discrimination can be used as the form such as amplitude frequency discrimination method or phase frequency discrimination method. The amplitude frequency discrimination method transforms the constant-amplitude FM signal into a frequency-modulated wave whose amplitude changes with the instantaneous frequency, and then obtains the original modulated signal through an envelope detector. The realizition of amplitude frequency discrimination includes the methods such as direct time domain differentiation method and slope frequency discrimination method.

The slope frequency discrimination method uses a network with linear amplitude-frequency characteristics within the required frequency range to complete frequency discrimination, which a single detuning loop or a double detuning loop can be used. These two frequency discrimination networks use the slope part of the tuning loop amplitude-frequency characteristic curve to discriminate, so they are called slope frequency discrimination methods. In addition, it is also called the detuning (out-of-harmonic) loop method, because the detuning (out-of-harmonic) state of the tuning loop is used.

Figure 4.66 shows a circuit that uses a transformer-coupled single detuning loop to achieve slope frequency discrimination. In Figure 4.66(a), L_1 and L_2 are transformer-coupled primary and secondary windings, they form a parallel resonant circuit with C_1 and C_2. The steady-state response amplitude U_{am} of the resonant circuit to the harmonic input signal varies with the frequency of the input signal. The curve is shown in the upper left of Figure 4.66(b).

f_r 时，u_a 也随之下降。u_a 的频率（即调频波的频率）和 u_f 保持一致，但其幅值与调制信号成正比，如图 4.66（b）的右上所示。通常利用谐振回路幅频特性曲线的亚谐振区近似直线的一段实现频率－电压变换。当被测量（如位移）为零值时，调频信号的中心频率 f_0 对应特性曲线上升部分近似直线段的中点（这要求 $f_r > f_0$）。将 u_a 经由 VD、C、R 组成的包络检波电路，即可得到与调制信号成比例的输出电压 u_o，如图 4.66（b）的右下所示，从而实现斜率鉴频。

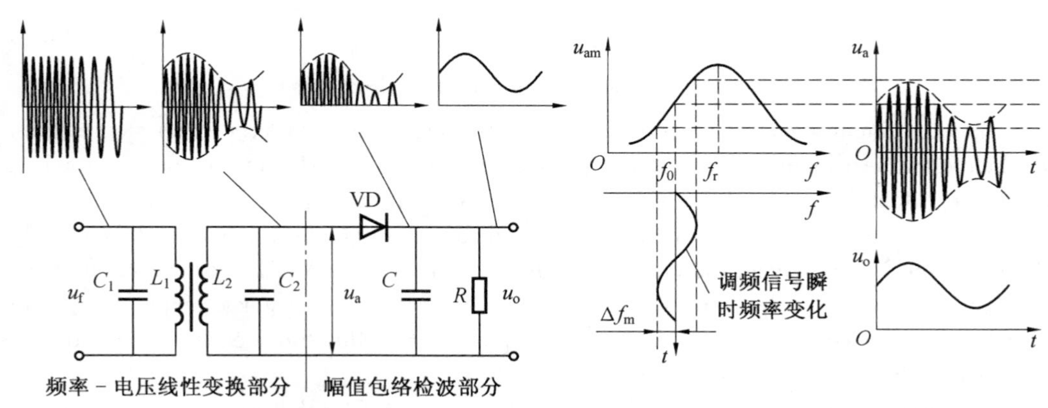

(a) 鉴频器　　(b) 失谐回路电压幅值－频率特性曲线

图 4.66　采用变压器耦合的单失谐回路实现鉴频的电路

图 4.66（a）所示电路的缺点是线性范围较窄。为改善其线性，可以采用双失谐回路鉴频器，也可以采用集成频率－电压转换器。一些集成频率－电压转换器通过使用数字方法，避免了滤波器响应较慢的问题。

4.5.3　低频透射式电涡流传感器

金属导体内电涡流的贯穿深度与传感器线圈激励电流的频率有关：频率越低，贯穿深度越厚。因此，采用低频电流激励时，可以测量金属导体的厚度。

图 4.67（a）所示为低频透射式电涡流测厚仪原理。发射线圈 W_1 和接收线圈 W_2 分别置于被测金属板的两边。当低频（1 000 Hz 左右）电压加到线圈 W_1 的两端时，线圈 W_1 中即流过一个同频率的交流电流，并在其周围产生交变磁场。如果两线圈间不存在被测金属板，W_1 的磁场就能直接贯穿线圈 W_2。于是，W_2 的两端会产生交变感应电动势 e_2。当 W_1 与 W_2 之间放置一个金属板时，W_1 产生的交变磁场在金属板中会产生涡流 i，这个涡流损耗了 W_1 的部分磁场能量，使其贯穿金属板后耦合到 W_2 的磁通量减少，从而引起感应电势 e_2 的下降。

金属板的厚度 h 越大，涡流损耗的磁场能量也越大，e_2 就越小。因此，e_2 的大小就反

The FM wave is coupled through C_1 and L_1 and added to the resonant circuit composed of C_2 and L_2. Input the constant-amplitude frequency modulation wave u_f, at the resonant frequency f_r of the loop, the coupling current in L_1 and L_2 is the largest, and the secondary side output voltage u_a is also the largest. When u_f frequency deviates from f_r, u_a will also decrease. The frequency of u_a (that is, the frequency of the FM wave) and u_f remain the same , but its amplitude is proportional to the modulation signal, as shown in the upper right of Figure 4.66(b). The frequency-voltage conversion is usually achieved by using the approximate straight section of the sub-resonant region of the amplitude-frequency characteristic curve of the resonant circuit. When the measured (such as displacement) is zero, the center frequency f_0 of the FM signal corresponds to the midpoint of the rising part of the characteristic curve approximate to the straight line (this requires $f_r > f_0$). Pass u_a through the envelope detection circuit composed of VD, C and R to obtain the output voltage u_o proportional to the modulation signal, as shown in the lower right of Figure 4.66(b), so as to achieve slope frequency discrimination.

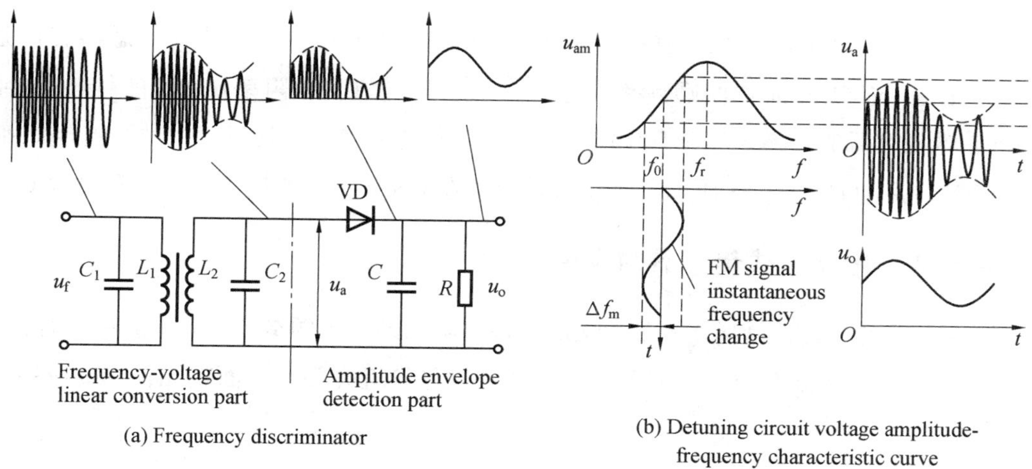

Figure 4.66 A circuit that uses a transformer-coupled single detuning loop to achieve slope frenquency discrimination

The disadvantage of the circuit shown in Figure 4.66(a) is that the linear range is narrow. In order to improve its linearity, a double detuning loop discriminator can be used, or an integrated frequency-voltage converter can be used. Some integrated frequency-voltage converters avoid the problem of slow filter response by using digital methods.

4.5.3 Low-Frequency Transmission Eddy Current Sensor

The penetration depth of the eddy current in the metal conductor is related to the frequency of the excitation current of the sensor coil: the lower the frequency, the thicker the penetration depth. Therefore, when using low-frequency current excitation, the thickness of metal conductors can be measured.

Figure 4.67(a) shows the principle of a low-frequency transmission eddy current thickness gauge. The transmitting coil W_1 and the receiving coil W_2 are respectively placed on both sides of the metal plate to be measured. When a low frequency (about 1 000 Hz) voltage

映了金属板厚度 h 的大小,如图 4.67(b) 所示。这就是低频透射式涡流传感器测厚的原理。

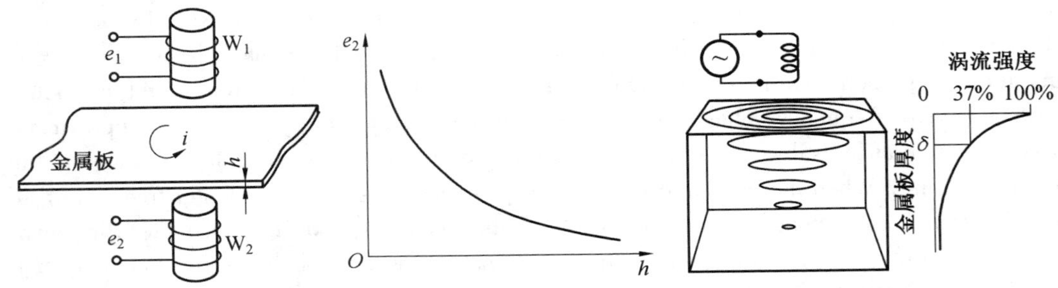

(a) 低频透射式涡流传感器原理 (b) 接收线圈感应电势与厚度的关系 (c) 涡流强度与金属板厚度的关系

图 4.67 低频透射式电涡流测厚仪

接收线圈的电压 e_2 随被测材料的厚度按负指数幂的规律减小。为使传感器有较宽的测量范围,应选用较低的激励频率 f,通常 $f \approx 1$ kHz。在测电阻率 ρ 较小的材料(如纯铜)时,选择较低的频率 f(等于 500 Hz);而当测电阻率 ρ 较大的材料(如黄铜、铝)时,则选用频率较高的 f(等于 2 kHz),从而保证传感器在测量不同材料时的线性度和灵敏度。

4.5.4 电涡流式传感器的应用

电涡流传感器已广泛应用于工业生产和科学研究的各个领域,可以测量转速、厚度、振动、位移、温度等参数,还可以进行无损探伤和制作接近开关。下面就几种主要应用做简略介绍。

1. 振幅测量

电涡流式传感器可以非接触地测量各种振动的幅值,特别适合做低频振动测量。

图 4.68 所示为电涡流式传感器的工程应用实例。图 4.68(a) 所示为径向振摆测量;图 4.68(b) 所示为零件尺寸测量。

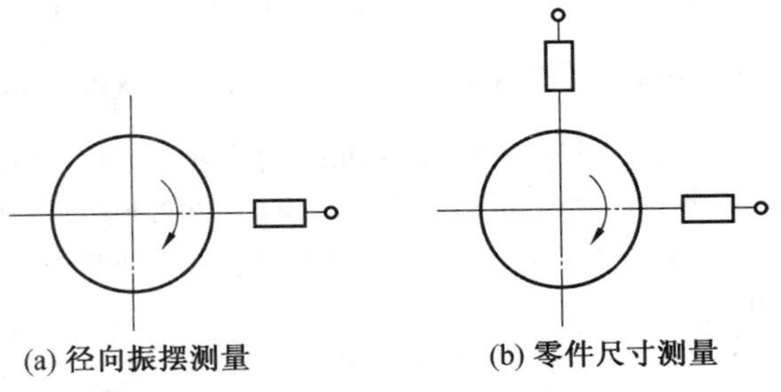
(a) 径向振摆测量 (b) 零件尺寸测量

图 4.68 电涡流式传感器的工程应用实例

is applied to both ends of the coil W_1, an alternating current of the same frequency flows through the coil W_1 and an alternating magnetic field is generated around it. If there is no measured metal plate between the two coils, the magnetic field of W_1 can directly penetrate coil W_2, so the alternating induced electromotive force e_2 will be generated at both ends of W_2. When a metal plate is placed between W_1 and W_2, the alternating magnetic field generated by W_1 will generate an eddy current i in the metal plate. This eddy current loses part of the magnetic field energy of W_1, making the magnetic flux of W_2 coupled to the metal plate after passing through the metal plate is reduced, causing the induced potential e_2 to drop.

The greater the thickness h of the metal plate, the greater the magnetic field energy lost by the eddy current, and the smaller the e_2. Therefore, the size of e_2 reflects the size of the thickness h of the metal plate, as shown in Figure 4.67(b). This is the principle of thickness measurement with low-frequency transmission eddy current sensor.

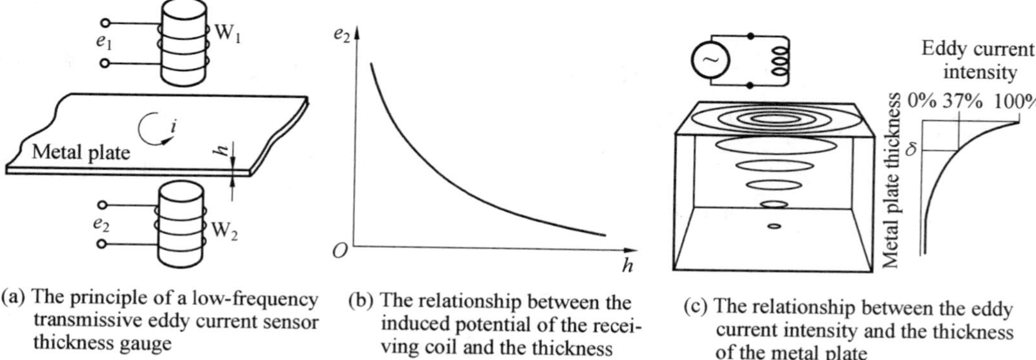

(a) The principle of a low-frequency transmissive eddy current sensor thickness gauge
(b) The relationship between the induced potential of the receiving coil and the thickness
(c) The relationship between the eddy current intensity and the thickness of the metal plate

Figure 4.67 Low-frequency transmission type eddy current thickness gauge

The voltage e_2 of the receiving coil decreases with the thickness of the material to be measured according to the rule of negative exponential power. In order to make the sensor have a wider measurement range, a lower excitation frequency f should be selected, usually $f \approx$ 1 kHz. When measuring materials with small resistivity ρ (such as pure copper), choose a lower frequency f (equal to 500 Hz); and when measuring materials with large resistivity ρ (such as brass, aluminum), choose a higher frequency f (equal to 2 kHz), so as to ensure the linearity and sensitivity of the sensor when measuring different materials.

4.5.4 The Application of Eddy Current Sensors

Eddy current sensors have been widely used in various fields of industrial production and scientific research, which can measure speed, thickness, vibration, displacement, temperature and other parameters, and can also perform nondestructive inspection and make proximity switches. The following is a brief introduction to several main applications.

1. Amplitude Measurement

The eddy current sensor can measure the amplitude of various vibrations in a non-contact manner, and is particularly suitable for low-frequency vibration measurement.

Figure 4.68 shows the engineering application examples of eddy current sensors. Figure 4.68(a) shows the radial runout measurement; Figure 4.68(b) shows parts dimension measurement.

2. 尺寸测量

电涡流传感器可以测量试件的几何尺寸(图 4. 69)。被测工件通过传送线时,若几何尺寸不合格(过大或偏小) 的工件通过电涡流传感器,传感器会输出不同的信号。

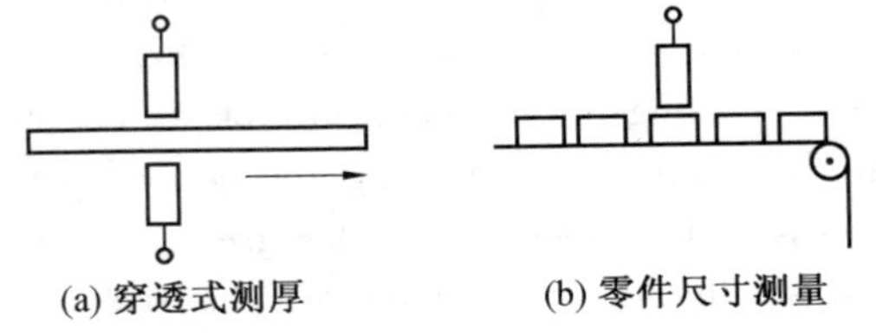

图 4. 69　电涡流式传感器用于尺寸测量

3. 转速测量

电涡流式转速计的工作原理如图 4. 70 所示。在旋转体(转轴或飞轮) 上开一个或数个槽或齿,旁边安装电涡流传感器,轴转动时便能检出传感器与轴表面的间隙(周期性) 变化,于是传感器的输出也发生周期性变化,经放大、整形后,成为周期性的脉冲信号,然后可由频率计计数并指示频率值,即转速。

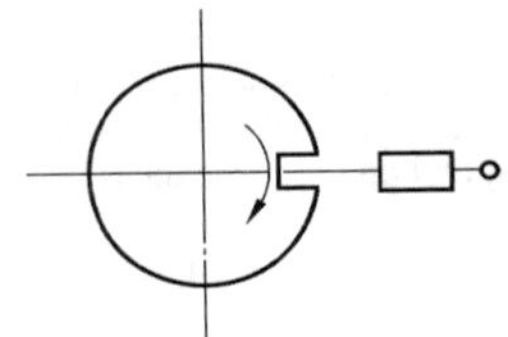

图 4. 70　电涡流式转速计

脉冲信号频率与轴的转速成正比,即

$$n = \frac{f}{z} \times 60 \tag{4.79}$$

式中　n—— 轴的转速;

f—— 脉冲信号频率;

z—— 转轴上的槽数或齿数。

4. 金属零件表面裂纹检查(电涡流探伤)

用电涡流传感器可以探测金属导体表面或近表面裂纹、热处理裂纹及焊缝裂纹等缺陷。电涡流探伤如图 4. 71 所示。测试时,传感器贴近零件表面,传感器与被测金属零件之间的距离保持不变。当遇到裂纹时,金属的电阻率、磁导率会产生变化,裂缝处也有位移量的改变,会使电涡流传感器等效电路中的涡流反射电阻和涡流反射电感发生变化,导致线圈的阻抗改变,输出信号电压也随之发生改变。

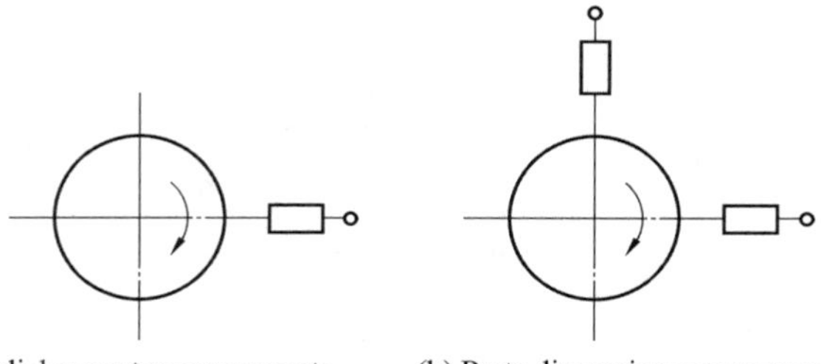

(a) Radial runout measurement (b) Parts dimension measurement

Figure 4.68 The engineering appliocation examples of eddy current sensors

2. Size Measurement

The eddy current sensor can measure the geometric size of themeasurement piece (Figure 4.69). When the measured workpiece passes through the transmission line, if the workpiece with unqualified geometric size (too large or too small) passes through the eddy current sensor, the sensor will output different signals.

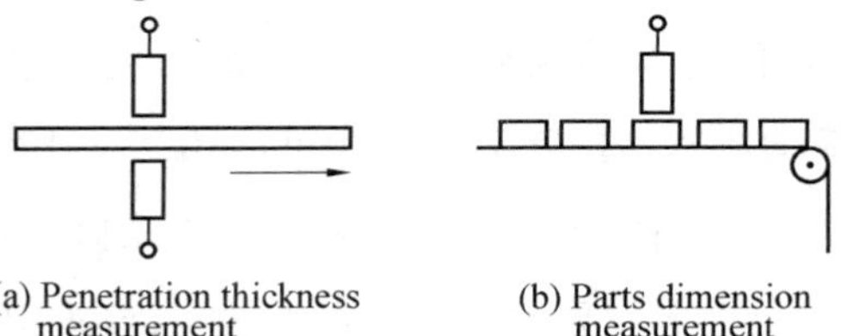

(a) Penetration thickness measurement (b) Parts dimension measurement

Figure 4.69 Eddy current sensor used for size measurement

3. Speed Measurement

The working principle of the eddy current tachometer is shown in Figure 4.70. One or more grooves or teeth are opened on the rotating body (shaft or flywheel), and an eddy current sensor is installed next to it. When the shaft rotates, the gap (periodical) change between the sensor and the shaft surface can be detected, so the output of the sensor also changes periodically. After being amplified and reshaped, it becomes a periodic pulse signal, which then the frequency counter can count and indicate the frequency value, namely the speed.

The frequency of the pulse signal is proportional to the speed of the shaft, that is

$$n = \frac{f}{z} \times 60 \tag{4.79}$$

where n —shaft speed;

f —pulse signal frequency;

z —the number of slots or teeth on the shaft.

4. Surface Crack Inspection of Metal Parts (Eddy Current Flaw Detection)

The eddy current sensor can detect defects such as surface or near surface cracks, heat treatment cracks and weld cracks on metal conductors. Eddy current flaw detection is shown in Figure 4.71. During the measurement, the sensor is close to the surface of the part, and the distance between the sensor and the measured metal part remains unchanged. When there is a crack, the resistivity and permeability of the metal will change, and the displacement of the crack will also change, so that it can change the eddy current reflection resistance and eddy current reflection inductance in the equivalent circuit of the eddy current sensor, which causes

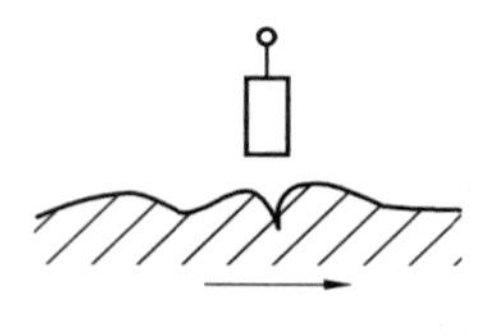

图 4.71　电涡流探伤

4.6　压电式传感器

4.6.1　压电式传感器的工作原理

压电式传感器的工作原理是利用压电晶体的压电效应。它是一种可逆型换能器,既可以将机械能转换为电能,也可以将电能转换为机械能。它被广泛用于力、压力、加速度测量,也被用于超声波发射与接收装置。当用作加速度传感器时,可测频率范围为 0.1 Hz ~20 kHz,可测振动加速度按其不同结构可达10^{-2} ~ 10^{5} ms^{-2}。当用于测力传感器时,其灵敏度可达10^{-3} N。这种传感器具有灵敏度高、固有频率高、信噪比高、结构简单、体积小、工作可靠等优点。其主要缺点是无静态输出,要求很高的输出阻抗,需要低电容低噪声电缆等。随着与其配套的后续仪器如电荷放大器等技术性能的日益提高,这种传感器应用也日益广泛。

压电效应可分为正压电效应和逆压电效应。当某些物质如石英、钛酸钡等在一定方向上受到外力作用时,不仅其几何尺寸、形状发生变化,其内部也产生极化,表面上会产生正负相反的电荷,形成电场。受力所产生的电荷量与外力的大小成正比。当作用力的方向改变时,电荷的极性也随之改变,外力去除后又重新回到原来不带电状态,这种现象称为正压电效应。

若将这些物质置于交变电场中,其几何尺寸(体积) 将发生变化,电场去除后,变形随之消失。这种因外电场作用而导致物质变形的现象称为逆压电效应或电致伸缩效应。压电式传感器大多是利用正压电效应制成的。

具有压电效应的材料称为压电材料,常见的压电材料有两类:压电单晶体,如石英、酒石酸钾钠等;多晶压电陶瓷,如钛酸钡、锆钛酸铅等。下面以石英晶体为例,说明压电效应的机理。

石英晶体的基本形状为六角形晶柱。石英晶体结构如图 4.72(a) 所示,其两端为对称的棱锥,六棱柱是它的基本组织。纵轴线 Z 称为光轴,通过六角棱线而垂直于光轴的轴线 X 称为电轴,垂直于棱面的轴线 Y 称为机械轴,如图 4.72(b) 所示。

the impedance of the coil to change, and the output signal voltage also changes accordingly.

Figure 4.70 Working principle of the eddy current tachometer

Figure 4.71 Eddy current flaw detection

4.6 Piezoelectric Sensors

4.6.1 Working Principle of Piezoelectric Sensor

The working principle of the piezoelectric sensor is to use the piezoelectric effect of the piezoelectric crystal. It is a reversible transducer, which can convert mechanical energy into electrical energy, or convert electrical energy into mechanical energy. It is widely used in force, pressure, acceleration measurement, and also used in ultrasonic transmitter and receiver devices. When used as an acceleration sensor, the piezoelectric sensor's measurable frequency range is 0.1 Hz - 20 kHz, and the measurable vibration acceleration can reach 10^{-2} - 10^{5} ms^{-2} according to its different structures. When used in a load cell, its sensitivity can reach 10^{-3} N. This kind of sensor has the advantages of high sensitivity, high natural frequency, high signal-to-noise ratio, simple structure, small size, reliable operation, and so on. The main disadvantage is that there is no static output, high output impedance is required, and it has the equivalent need of low capacitance and low noise cables, and so on. With the increasing technical performance of its supporting follow-up instruments, such as charge amplifiers, this sensor is more and more widely used.

The piezoelectric effect can be divided into the direct piezoelectric effect and the inverse piezoelectric effect. When certain substances, such as quartz and barium titanate, are subjected to external forces in a certain direction, not only the geometric size and shape will change, but also the internal polarization will generate positive and negative charges on the surface to form an electric field. The amount of charge generated by the force is proportional to the magnitude of the external force. When the direction of the force changes, the polarity of the charge also changes. After the external force is removed, it returns to the original uncharged state whose phenomenon is called the direct piezoelectric effect.

If these substances are placed in an alternating electric field, with their geometric dimensions (volume) changing, the deformation disappears after the electric field is removed. This phenomenon of material deformation due to the action of an external electric field is called the inverse piezoelectric effect or electrostrictive effect. Most piezoelectric sensors are made by using the direct piezoelectric effect.

Materials with piezoelectric effect are called piezoelectric materials. There are two types of common piezoelectric materials: piezoelectric single crystals, such as quartz, potassium sodium tartrate; polycrystalline piezoelectric ceramics, such as barium titanate, lead zirconate titanate. Let's take quartz crystal as an example to illustrate the mechanism of piezoelectric effect.

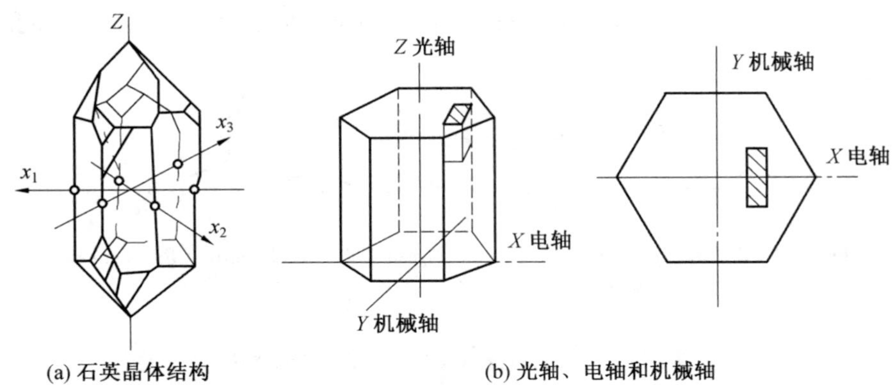

(a) 石英晶体结构　　(b) 光轴、电轴和机械轴

图 4.72　石英晶体

从晶体上沿轴线切下的薄片称为晶体切片，并使其晶面分别平行于 Z、Y、X 轴线，这个晶片在正常状态下不呈现电性。当沿 X 方向对晶片施加外力 F 时，晶片极化，沿 X 方向形成电场，其电荷分布在垂直于 X 轴的平面上，如图 4.73(a) 所示，这种现象称为纵向压电效应。当沿 Y 方向对晶片施加外力 F 时，其电荷仍在与 X 轴垂直的平面上出现，如图 4.73(b) 所示，这种现象称为横向压电效应。当沿 Z 轴对晶片施加外力时，无论外力的大小和方向如何，晶片的表面都不会极化，不产生电荷。

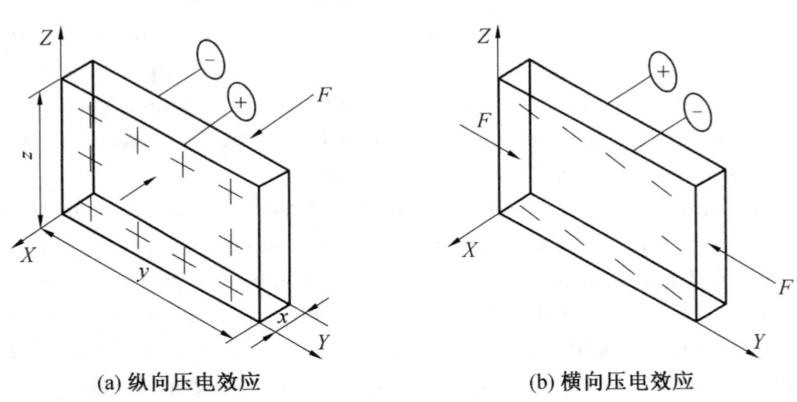

(a) 纵向压电效应　　(b) 横向压电效应

图 4.73　石英晶体受力后的极化现象

实验证明，在极板上积聚的电荷量 q(库仑) 与晶片所受的外力 F 成正比，即

$$q = d_c F \tag{4.80}$$

式中　q —— 电荷量；

d_c—— 压电常数，与材质和切片方向有关，如 $d_c = -2.31 \times 10^{-12}$ C/N，所以输出电荷极微，常需要接电荷放大器。

F—— 作用力。

The basic shape of quartz crystal is hexagonal crystal column. Quartz crystal structure is shown in Figure 4.72(a), the two ends are symmetrical pyramids, the hexagonal prism is its basic organization. The longitudinal axis Z is called the optical axis, the axis X perpendicular to the optical axis through the hexagonal ridgeline is named the electrical axis, and the axis Y perpendicular to the edge is known as the mechanical axis, as shown in Figure 4.72(b).

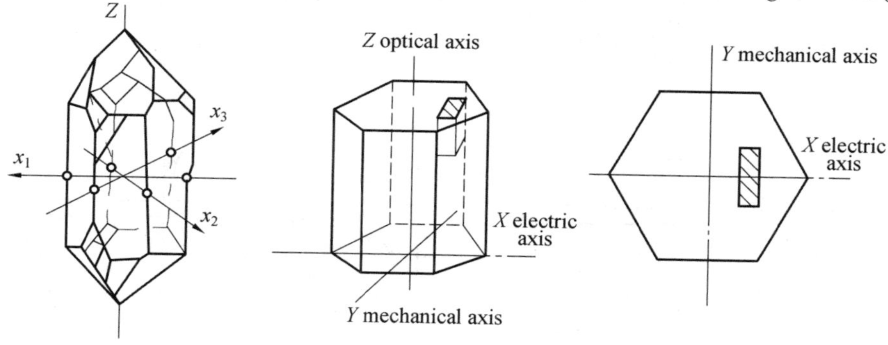

(a) Quartz crystal structure (b) Optical axis,electrical axis and mechanical axis

Figure 4.72 Quartz crystal

The slices cut from the crystal along the axis are called crystal wafers, and their crystal planes are parallel to the Z, Y, X axes. This crystal wafer does not exhibit electrical properties under normal conditions. When an external force F is applied to the wafer along the X direction, the crystal wafer is polarized and an electric field is formed along the X direction. The charge is distributed on a plane perpendicular to the X axis, as shown in Figure 4.73(a). This phenomenon is called the longitudinal piezoelectric effect. When an external force F is applied to the wafer along the Y direction, the charge still appears on the plane perpendicular to the X axis, as shown in Figure 4.73(b). This phenomenon is known as the horizontal piezoelectric effect. When an external force is applied to the wafer along the Z axis, the surface of the wafer will not be polarized and no charge will be generated regardless of the magnitude and direction of the external force.

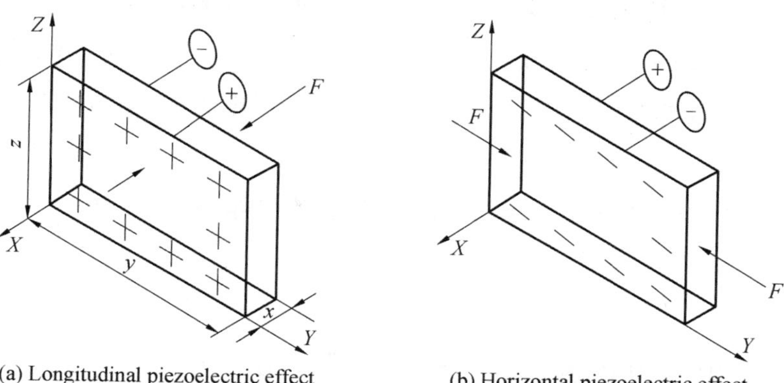

(a) Longitudinal piezoelectric effect (b) Horizontal piezoelectric effect

Figure 4.73 Polarization phenomenon of quartz crystal after being stressed

Experiments have proved that the amount of charge q (Coulomb) accumulated on the plate is proportional to the external force F on the wafer, that is

由式(4.80)可知,应用压电式传感器测力 F,实质上就是如何测量电荷量 q 的问题。

晶体切片上电荷的符号与受力方向的关系如图4.74所示。图4.74(a)所示为在 X 轴方向上受压力,图4.74(b)所示为在 X 轴方向受拉力,图4.74(c)所示为在 Y 轴方向受压力,图4.74(d)所示为在 Y 轴方向受拉力。当沿电轴 X 方向加作用力 F_X 时,则在与电轴 X 垂直的平面上产生电荷。当作用力 F_Y 沿着机械轴 Y 方向时,其电荷仍在与 X 轴垂直的平面上出现。

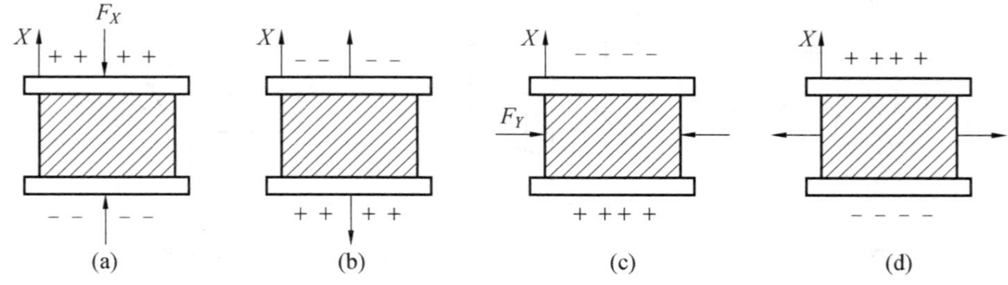

图4.74　晶体切片上电荷符号与受力方向的关系

若压电体受到多方向的力,则压电体各表面都会积聚电荷。每个表面上的电荷量不仅与作用于该面上的垂直力有关,而且还与压电体其他面上所受的力有关。

压电材料是力敏感元件,可测最终能变换为力的物理量,如力、压力、加速度等,具有结构简单、灵敏度高、响应频率宽、信噪比高、工作可靠、质量轻等优点。

4.6.2　压电式传感器的等效电路

在压电晶片的两个工作面上进行金属蒸镀,形成金属膜,构成两个电极,如图4.75(a)所示。当压电晶片受到力的作用时,便有电荷聚集在两极上,一面为正电荷,另一面为等量的负电荷。这种情况与电容器十分相似,所不同的是晶片表面上的电荷会随着时间的推移而逐渐漏掉,因为压电晶片材料的绝缘电阻(又称漏电阻)虽然很大,但毕竟不是无穷大。

压电晶片受力后,两极板上聚集电荷,中间为绝缘体,使它成为一个电容器,其电容量为

$$C_a = \varepsilon_0 \varepsilon A / \delta$$

式中　ε_0—— 真空介电常数,$\varepsilon_0 = 8.85 \times 10^{-12}$ F/m;

ε—— 压电材料的相对介电常数,石英晶体 $\varepsilon = 4.5$;

A—— 极板的面积,即压电晶片工作面的面积,m^2;

δ—— 极板间距,即晶片厚度,m。

$$q = d_c F \tag{4.80}$$

where q — the amount of charge;

d_c —the piezoelectric constant, related to the material and the wafer direction, such as $d_c = -2.31 \times 10^{-12}$ C/N, so the output charge is very small, and it is often necessary to connect a charge amplifier.

F — force.

From formula(4.80), it can be seen that applying piezoelectric sensors to measure force F is essentially a question of how to measure the amount of charge q.

The relationship between the sign of the charge on the crystal wafer and the direction of the force is shown in Figure 4.74. Figure 4.74(a) shows the pressure in the X axis direction, Figure 4.74(b) shows the tension in the X axis direction, Figure 4.74(c) shows the compressed in the Y axis direction, Figure 4.74(d) shows the tension in the Y axis direction. When a force F_X is applied in the direction of the electric axis X, charges are generated on a plane perpendicular to the electric axis X. When the force F_Y is along the mechanical axis Y, its charge still appears on the plane perpendicular to the X axis.

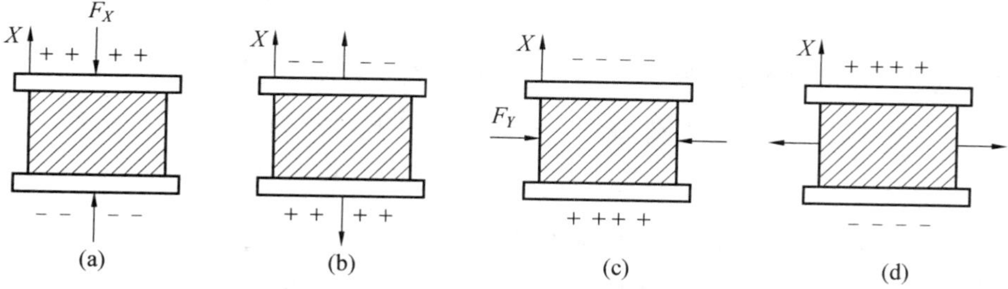

Figure 4.74 The relationship between the sign of the charge on the crystal wafer and the direction of the force

If the piezoelectric body is subjected to multi-directional forces, charges will accumulate on each surface of the piezoelectric body. The amount of charge on each surface is not only related to the vertical force acting on the surface, but also related to the force on the other surfaces of the piezoelectric body.

Piezoelectric materials are force-sensitive components that can measure physical quantitiesthat can ultimately be transformed into forces, such as force, pressure, acceleration, and have the advantages such as simple structure, high sensitivity, wide response frequency, high signal-to-noise ratio, reliable work, and light weight.

4.6.2 The Equivalent Circuit of the Piezoelectric Sensor

Metal evaporation is performed on the two working surfaces of the piezoelectric wafer to form a metal film and two electrodes will be formed, as shown in Figure 4.75(a). When the piezoelectric wafer is subjected to force, there will be charges accumulated on the two electrodes, one side is positive, and the other side is the same amount of negative charges. This situation is very similar to capacitors, the difference is that the charge on the surface of the wafer will gradually leak out over time due to the fact that the insulation resistance (also called leakage resistance) of the piezoelectric wafer material is very large, but it is not infinite after all.

After the piezoelectric chip is stressed, the charge is collected on the two plates with an

压电晶片受力后，两极板间电压（又称极板上的开路电压）e_a 为

$$e_a = \frac{q}{C_a} \tag{4.81}$$

式中　q—— 压电晶片表面上的电荷；

C_a—— 压电晶片的电容。

从信号变换角度看，压电元件相当于一个电荷发生器。从结构上看，它又是一个电容器。因此，通常将压电元件等效为一个电荷源 q 与一个电容 C_a 相并联的电路，如图 4.75(b)。

如果仅由单片压电晶片工作，则为产生足够的表面电荷，需要很大的作用力。实际的压电传感器中，往往用两片或两片以上的压电晶片串联或并联而成。由于压电材料是有极性的，因此并联、串联两种情况下的总电荷将不同，可视测试要求合理选用并联或串联。

两压电晶片串联时（图 4.75(c)），正电荷集中在上极板，负电荷集中在下极板。设总电容量为 C'，总电压为 U'，总电荷为 Q'，则 C'、U'、Q' 与单片的 C、U、Q 的关系为

$$C' = C/2, U' = 2U, Q' = Q \tag{4.82}$$

串联时，传感器的总电容量变小，输出电压变大，适用于以电压为输出信号的场合。

当两压电晶片并联时（图 4.75(d)），负电荷集中在中间极板上，正电荷集中在两侧的电极上，其关系为

$$C' = 2C, U' = U, Q' = 2Q \tag{4.83}$$

并联时，总电容量变大，输出电荷量变大，时间常数大，适用于测量缓变信号并以电荷量作为输出。

压电传感器总是在有负载的情况下工作。设 C_i 为负载的等效电容，C_c 为压电传感器与负载间连接电缆的分布电容，R_i 为负载的输入电阻，R_a 为传感器本身的漏电阻。压电传感器接负载后，等效电荷源电路中的等效电容 C 为

$$C = C_a + C_c + C_i \tag{4.84}$$

等效电阻 R_0 为

$$R_0 = \frac{R_a R_i}{R_a + R_i} \tag{4.85}$$

压电元件在外力作用下产生的电荷 q 除给等效电容 C 充电外，还将通过等效电阻 R_0 泄漏掉。根据电荷平衡方程式，可得压电元件在外力 F 作用下产生的电荷 q 为

$$q = Ce + \int i\mathrm{d}t \tag{4.86}$$

insulator in the middle, making it a capacitor with a capacitance of

$$C_a = \varepsilon_0 \varepsilon A/\delta$$

where ε_0—vacuum dielectric constant, $\varepsilon_0 = 8.85 \times 10^{-12}$ F/m;

ε —the relative permittivity of piezoelectric materials, quartz crystal $\varepsilon = 4.5$;

A —the area of the electrode plate, that is, the area of the working surface of the piezoelectric wafer, m^2;

δ —the distance between the plates, which is the thickness of the wafer, m.

After the piezoelectric chip is stressed, the voltage between the two plates (also called the open circuit voltage on the plates) e_a is

$$e_a = \frac{q}{C_a} \tag{4.81}$$

where q —the charge on the surface of the piezoelectric wafer;

C_a —the capacitance of the piezoelectric chip.

From the perspective of signal conversion, the piezoelectric element is equivalent to a charge generator. From the structural point of view, it is a capacitor. Therefore, the piezoelectric element is usually equivalent to a circuit in which a charge source q is connected in parallel with a capacitor C_a , as shown in Figure 4.75(b).

If only a single piezoelectric wafer works, a large force is required in order to generate sufficient surface charge. In actual piezoelectric sensors, two or more piezoelectric wafers are often connected in series or in parallel. Since the piezoelectric material has polarity, the total charge will be different in the two cases of parallel and series. Parallel or series can be selected reasonably according to the measurement requirements.

When two piezoelectric wafers are connected in series (Figure 4.75(c)), the positive charge is concentrated on the upper plate, and the negative charge is concentrated on the lower plate. Suppose the total capacitance is C' , the total voltage is U' , and the total charge is Q' so the relatiohnship between C', U', Q' and C, U, Q of the single chip is

$$C' = C/2 \ , \ U' = 2U \ , \ Q' = Q \tag{4.82}$$

When connected in series, the total capacitance of the sensor becomes smaller, and the outputvoltage tends to be larger, which is suitable for occasions where voltage is the output signal.

When two piezoelectric wafers are connected in parallel (Figure 4.75(d)), the negative charge is concentrated on the middle plate, and the positive charge is concentrated on the electrodes on both sides. The relationship is

$$C' = 2C \ , \ U' = U \ , \ Q' = 2Q \tag{4.83}$$

When connected in parallel, the total capacitance becomes larger, and the output charge becomes larger, and the time constant is large, which is suitable to measure the slowly changing signal and use the charge as the output.

Piezoelectric sensors always work under load. Suppose C_i is the equivalent capacitance of the load, C_c is the distributed capacitance of the cable connecting the piezoelectric sensor and the load, R_i is the input resistance of the load, and R_a is the leakage resistance of the sensor itself. After the piezoelectric sensor is connected to the load, the equivalent capacitance C in the equivalent charge source circuit is

$$C = C_a + C_c + C_i \tag{4.84}$$

The equivalent resistance R_0 is

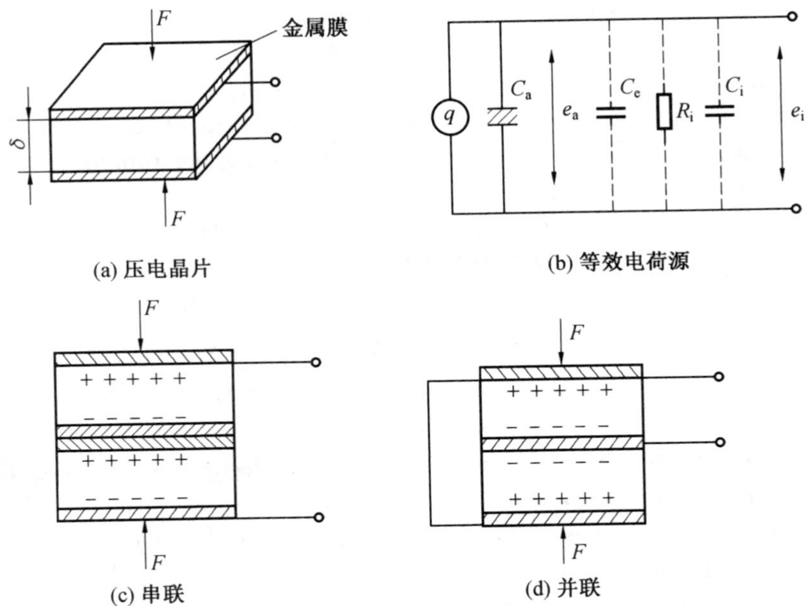

(a) 压电晶片　　(b) 等效电荷源

(c) 串联　　(d) 并联

图 4.75　压电晶片及等效电路

$$q = DF = DF_0 \sin \omega t$$

式中　F_0、ω—— 交变外力的幅值、圆频率；

e—— 接负载后压电元件的输出电压(即等效电容 C 上的电压值)，$e = R_0 i$；

i—— 泄漏电流。

式(4.86) 可写为

$$q = q_0 \sin \omega t = CR_0 i + \int i \mathrm{d}t \tag{4.87}$$

忽略过渡过程，其稳态解为

$$i = \frac{\omega q_0}{\sqrt{1 + (\omega C R_0)^2}} \sin(\omega t + \varphi) \tag{4.88}$$

$$\varphi = \arctan \frac{1}{\omega C R_0} \tag{4.89}$$

接负载后，压电元件的输出电压 e_i 为

$$e_i = R_0 i = \frac{D}{C} \frac{1}{\sqrt{1 + \left(\frac{1}{\omega C R_0}\right)^2}} F_0 \sin(\omega t + \varphi) \tag{4.90}$$

由以上分析可得以下结论。

(1) 根据压电传感器的输出电压 e_i 推算被测力 $F_0 \sin \omega t$，受到 $\frac{1}{\sqrt{1 + \left(\frac{1}{\omega C R_0}\right)^2}}$ 及 C 的

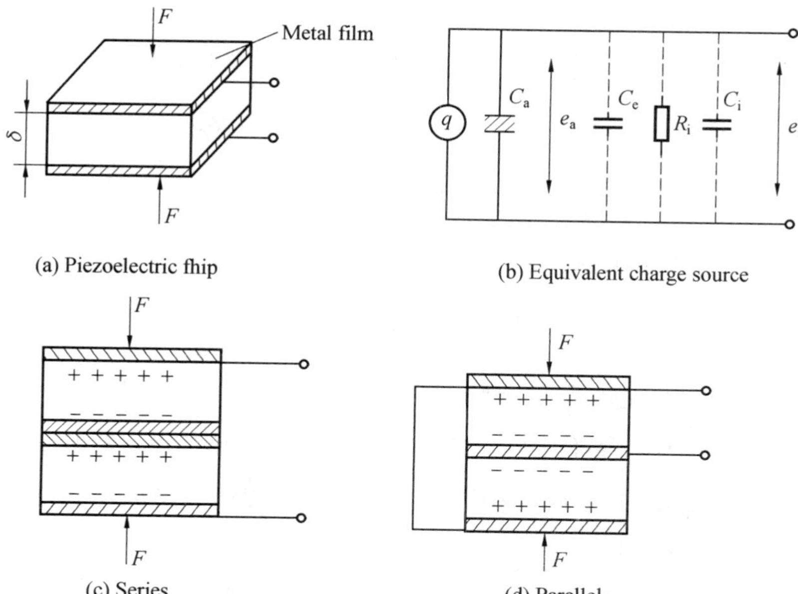

Figure 4.75 Piezoelectric chip and equivalent circuit

$$R_0 = \frac{R_a R_i}{R_a + R_i} \tag{4.85}$$

The charge q generated by the piezoelectric element under the action of external force will not only charge the equivalent capacitance C , but will also leak out through the equivalent resistance R_0. According to the charge balance equation, the charge q generated by the piezoelectric element under the action of the external force F is

$$q = Ce + \int i\mathrm{d}t \tag{4.86}$$

$$q = DF = DF_0 \sin \omega t$$

where F_0, ω —amplitude and circular frequency of alternating external force;

e —the output voltage of the piezoelectric element after the load is connected (that is, the voltage value on the equivalent capacitance C), $e = R_0 i$;

i —leakage current.

Formula (4.86) can be written as

$$q = q_0 \sin \omega t = CR_0 i + \int i\mathrm{d}t \tag{4.87}$$

Ignoring the transition process, the steady-state solution is

$$i = \frac{\omega q_0}{\sqrt{1 + (\omega C R_0)^2}} \sin(\omega t + \varphi) \tag{4.88}$$

$$\varphi = \arctan \frac{1}{\omega C R_0} \tag{4.89}$$

After the load is connected, the output voltage e_i of the piezoelectric element is

$$e_i = R_0 i = \frac{D}{C} \frac{1}{\sqrt{1 + \left(\frac{1}{\omega C R_0}\right)^2}} F_0 \sin(\omega t + \varphi) \tag{4.90}$$

影响。

(2) 当被测信号频率 ω 足够高时，压电传感器的输出电压 e_i 与频率无关，这时才可能实现不失真测试，即

$$\omega \gg \frac{1}{CR_0} \tag{4.91}$$

此时，信号频率的下限式(4.90) 可写为

$$e_i = \frac{DF_0}{C}\sin(\omega t + \varphi) \tag{4.92}$$

式(4.91) 表明，压电传感器实现不失真测试的条件与被测信号的频率 ω 及回路的时间常数 R_0C 有关。为使测量信号频率的下限范围扩大，压电式传感器的后接测量电路必须有高输入阻抗，即很高的负载输入阻抗R_i(当R_i 值很大时，在图 4.75 所示压电晶片的等效电路中可将其视为断开)。

(3) 当测量静态信号或缓变信号时，为使压电晶片上的电荷不消耗或泄漏，负载电阻 R_i 必须非常大，否则将会因电荷泄漏而产生测量误差。但R_i 值不可能无限加大，因此用压电传感器测量静态信号或缓变信号，或者作用在压电元件上的力是静态力($\omega = 0$) 时，电荷会通过放大器的输入电阻和传感器本身的泄漏电阻漏掉，这从原理上决定了压电式传感器不能测量静态量。

当压电式传感器用于动态信号的测量时，由于动态交变力的作用，因此压电晶片上的电荷可以不断补充，供给测量电路一定的电流，使测量成为可能。当被测信号频率足够高时，压电传感器的输出电压与R_i 无关。

上述三点分析表明压电传感器适用于动态信号的测量，高频响应很好(高频时输入电压与作用力的频率 ω 几乎无关)，这是压电式传感器的一个突出优点。但测量信号频率的下限受 R_0C 的影响，上限则受压电传感器固有频率的限制。

压电传感器的输出理论上应当是压电晶片表面上的电荷 q。由图 4.75(b) 可知，实际测试中往往是取等效电容 C 上的电压值作为压电传感器的输出。因此，压电式传感器就有两种输出形式：电荷和电压。相应地，其灵敏度也有电荷灵敏度和电压灵敏度两种表示方法。两种灵敏度之间的关系为

$$S_q = CS_e = (C_a + C_c + C_i)S_e \tag{4.93}$$

或

$$S_e = \frac{S_q}{C} = \frac{S_q}{C_a + C_c + C_i} \tag{4.94}$$

From the above analysis, we can get following conclusions.

(1) Calculate the measured force $F_0 \sin \omega t$ based on the output voltage e_i of the piezoelectric sensor, which is affected by $\frac{1}{\sqrt{1+\left(\frac{1}{\omega C R_0}\right)^2}}$ and C.

(2) When the frequency ω of the measured signal is high enough, the output voltage e_i of the piezoelectric sensor has nothing to do with the frequency, and then it is possible to realize the non-distortion measurement, namely

$$\omega \gg \frac{1}{CR_0} \tag{4.91}$$

At this time, the lower limit of the signal frequency formula(4.90) can be written as

$$e_i = \frac{DF_0}{C}\sin(\omega t + \varphi) \tag{4.92}$$

Formula (4.91) shows that the condition for the piezoelectric sensor to realize the non-distortion measurement is related to the frequency ω of the measured signal and the time constant R_0C of the loop. In order to expand the lower limit range of the measurement signal frequency, the piezoelectric sensor's subsequent measurement circuit must have a high input impedance, that is, a very high load input impedance R_i (when the value of R_i is large, the pressure shown in Figure 4.75 can be regarded as disconnected in the equivalent circuit of the electric chip).

(3) When measuring static signals or slowly changing signals, the load resistance R_i must be very large in order to prevent the charge on the piezoelectric chip from being consumed or leaking, otherwise measurement errors will occur due to charge leakage. However, R_i value cannot be increased indefinitely. Therefore, when a piezoelectric sensor is used to measure a static signal or a slowly changing signal, or when the force acting on the piezoelectric element is a static force ($\omega = 0$), the charge will pass through the input resistance of the amplifier and the leak resistance of sensor itself, which determines in principle that the piezoelectric sensor cannot measure the static quantity.

When the piezoelectric sensor is used for the measurement of dynamic signals, due to the dynamic alternating force, the charge on the piezoelectric wafer can be continuously supplemented to supply a certain current to the measurement circuit, making the measurement possible. When the frequency of the measured signal is high enough, the output voltage of the piezoelectric sensor has nothing to do with R_i.

The above three-point analysis shows that the piezoelectric sensor is suitable for the measurement of dynamic signals, the high frequency response is very good (the input voltage is almost independent of the frequency ω of the force at high frequency), which is a prominent advantage of the piezoelectric sensor. However, the lower limit of the measurement signal frequency is affected by R_0C, and the upper limit is limited by the natural frequency of the piezoelectric sensor.

The output of the piezoelectric sensor should theoretically be the charge q on the surface of the piezoelectric wafer. According to Figure 4.75 (b), it can be seen that in actual measurements, the voltage value on the equivalent capacitance C is often taken as the output of the piezoelectric sensor. Therefore, piezoelectric sensors have two output forms: charge and voltage. Correspondingly, its sensitivity can be expressed in charge sensitivity and voltage

式中　S_e—— 电压灵敏度;

　　　S_q—— 电荷灵敏度。

注意,压电传感器结构和材料确定之后,其电荷灵敏度 S_q 便已确定。由于等效电容 C 受电缆电容 C_c 的影响,因此电压灵敏度 S_e 会因所用电缆长度的不同而有所变化。

4.6.3　前置放大器

一般来说,压电式传感器的绝缘电阻 $R_a \geqslant 10^{10}\ \Omega$,因此传感器可近似看作开路。当传感器与测量仪器连接后,在测量回路中就应当考虑电缆电容和前置放大器的输入电容、输入电阻对传感器的影响。要求前置放大器的输入电阻尽量高,至少大于 $10^{11}\ \Omega$。这样才能减小漏电造成的电压(或电荷)损失,不致引起过大的测量误差。

压电式传感器后面的前置放大器有以下两个作用。

(1) 阻抗转换功能。在压电式传感器的输出端先接入高输入阻抗的前置放大器,将传感器高阻抗输出转换为低阻抗输出,然后才能接入通用的放大、检波等电路及显示记录仪表。

(2) 将压电式传感器输出的微弱信号放大。压电材料内阻很高,输出的信号能量很小,这就要求测量电路的输入电阻应非常大。

压电式传感器的输出可以是电压,也可以是电荷;压电式传感器可以等效为电压源或电荷源。因此,与它配套的测量电路的前置放大器也有电压型、电荷型两种形式:一种是带电阻反馈的电压放大器,其输出电压与输入电压(即传感器的输出电压)成正比;另一种是带电容反馈的电荷放大器,其输出电压与输入电荷量成正比。

1. 电压放大器

电压放大器具有很高的输入阻抗(1 000 MΩ 以上)、很低的输出阻抗(小于 100 Ω)。图 4.76 所示为压电传感器 - 连接电缆 - 电压前置放大器等效电路。放大器的输入电压(即传感器的输出电压)e_i 为

$$e_i = \frac{q}{C_a + C_e + C_i} \tag{4.95}$$

系统的输出电压为

$$e_y = A_p, e_i = \frac{qA_p}{C_a + C_c + C_i} \tag{4.96}$$

式(4.96) 表明,测量系统的输出电压对电缆电容 C_c 敏感。当电缆长度变化时,C_c 随之变化,使得放大器输入电压 e_i 变化,系统的电压灵敏度也将发生变化,这就增加了测量

sensitivity. The relationship between the two sensitivities is

$$S_q = CS_e = (C_a + C_c + C_i)S_e \tag{4.93}$$

or

$$S_e = \frac{S_q}{C} = \frac{S_q}{C_a + C_c + C_i} \tag{4.94}$$

where S_e —voltage sensitivity;

S_q —charge sensitivity.

Note that after the piezoelectric sensor structure and material are determined, its charge sensitivity S_q has been determined. Since the equivalent capacitance C is affected by the cable capacitance C_c, the voltage sensitivity S_e will vary depending on the length of the cable used.

4.6.3 Preamplifier

Generally speaking, the insulation resistance of a piezoelectric sensor is $R_a \geqslant 10^{10}\ \Omega$, so the sensor can be approximated as an open circuit. When the sensor is connected to the measuring instrument, the influence of the cable capacitance and the input capacitance and input resistance of the preamplifier on the sensor should be considered in the measurement loop. The input resistance of the preamplifier is required to be as high as possible, at least greater than $10^{11}\ \Omega$. In this way, the voltage (or charge) loss caused by leakage can be reduced, and excessive measurement errors will not be caused.

The preamplifier behind the piezoelectric sensor has the following two functions.

(1) Impedance conversion function. At the output end of the piezoelectric sensor, a high-impedance preamplifier is first connected to convert the sensor's high-impedance output into a low-impedance output, and then it can be connected to general amplifying and detecting circuits and display and recording instruments.

(2) Amplify the weak signal output by the piezoelectric sensor. Due to the high internal resistance of the piezoelectric material, the output signal energy is very small, requiring the input resistance of the measurement circuit to be very large.

The output of a piezoelectric sensor can be a voltage or a charge; the piezoelectric sensor canbe equivalent to a voltage source or a charge source. Therefore, the preamplifier of the measuring circuit that is matched with it also has the two forms voltage type and charge type: one is a voltage amplifier with resistance feedback, whose output voltage is proportional to the input voltage (that is, the output voltage of the sensor); the other is a charge amplifier with capacitive feedback, and its output voltage is proportional to the amount of input charge.

1. Voltage Amplifier

A voltage amplifier has a very high input impedance (above 1 000 MΩ), and a very low output impedance (less than 100 Ω). Figure 4.76 shows the piezoelectric sensor–connecting cable–voltage preamplifier equivalent circuit. The input voltage of the amplifier (that is, the output voltage of the sensor) e_i is

$$e_i = \frac{q}{C_a + C_e + C_i} \tag{4.95}$$

The output voltage of the system is

$$e_y = A_p, e_i = \frac{qA_p}{C_a + C_c + C_i} \tag{4.96}$$

的困难。

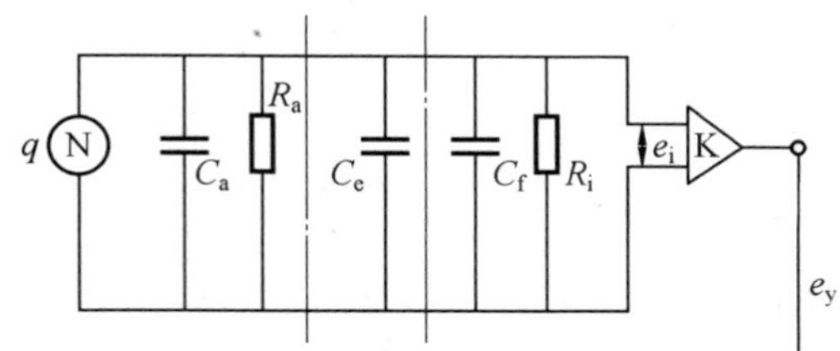

图 4.76　压电传感器 – 连接电缆 – 电压前置放大器等效电路

2. 电荷放大器

电荷放大器克服了上述电压放大器的缺点，它是一个高增益带电容反馈的运算放大器，能将高内阻的电荷源转换为低内阻的电压源，且输出电压正比于输入电荷，因此电荷放大器同样起阻抗变换的作用，其输入阻抗高达 $10^{12}\sim10^{14}\ \Omega$，输出阻抗小于 100 Ω。图 4.77 所示为压电传感器连接电荷放大器等效电路。当略去传感器的漏电阻 R_a 和电荷放大器输入电阻 R_i 的影响时，有

$$q \approx e_i(C_a + C_c + C_i) + (e_i - e_y)C_f$$
$$= e_iC + (e_i - e_y)C_f \tag{4.97}$$

式中　e_i—— 放大器输入端电压；

e_y—— 放大器输出端电压，$e_y = -Ke_i$；

K—— 电荷放大器开环放大倍数；

C_f—— 电荷放大器反馈电容。

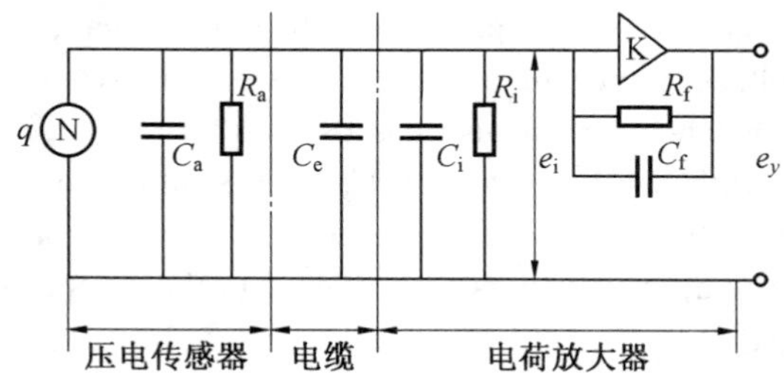

图 4.77　压电传感器连接电荷放大器等效电路

将 e_y 代入式(4.97) 中，可得到放大器输出端电压 e_y 与传感器电荷 q 的关系式为

$$e_y = \frac{-Kq}{(C + C_f) + KC_f} \tag{4.98}$$

当放大器的开环增益足够大时，式(4.59) 简化为

$$e_y \approx -\frac{q}{C_f} \tag{4.99}$$

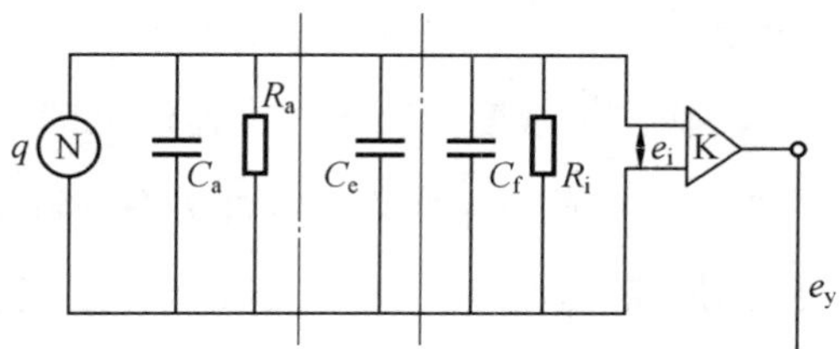

Figure 4.76　Piezoelectric sensor-connecting cable-voltage preamplifier equivalent circuit

Formula (4.96) indicates the output voltage of the measurement system sensitive to cable capacitance C_c. When the length of the cable changes, C_c changes accordingly, making the amplifier input voltage e_i change, and the voltage sensitivity of the system will also change, which increases the difficulty of measurement.

2. Charge Amplifier

The charge amplifier that overcomes the shortcomings of the above voltage amplifier is an operational amplifier with high gain and capacitive feedback, which can convert a high internal resistance charge source into a low internal resistance voltage source, and the output voltage is proportional to the input charge, so the charge amplifier also plays the role of impedance transformation, its input impedance is as high as $10^{12}-10^{14}$ Ω, and the output impedance is less than 100 Ω. Figure 4.77 shows the equivalent circuit of connecting a sensor to a charge amplifier. When the sensor's leakage resistance R_a and the input resistance R_i of the charge amplifier are omitted, there are

$$q \approx e_i(C_a + C_c + C_i) + (e_i - e_y)C_f$$
$$= e_iC + (e_i - e_y)C_f \tag{4.97}$$

where　e_i —the voltage at the input of the amplifier;
　e_y —amplifier output voltage, $e_y = -Ke_i$;
　K —charge amplifier open-loop magnification;
　C_f —feedback capacitance of charge amplifier.

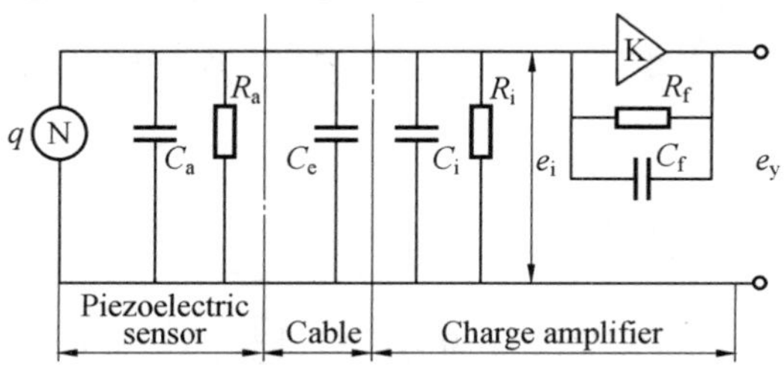

Figure 4.77　Equivalent circuit of connecting a piezoelectric sensor to a charge amplifier

Substituting e_y into formula(4.97), the relationship between the amplifier output voltage e_y and the sensor charge q can be obtained as

$$e_y = \frac{-Kq}{(C + C_f) + KC_f} \tag{4.98}$$

When the open-loop gain of the amplifier is large enough, the formula (4.59) is simplified as

$$e_y \approx -\frac{q}{C_f} \tag{4.99}$$

式(4.99)表明，在一定条件下，电荷放大器的输出电压 e_y 与传感器的电荷量 q 成正比，而与电缆的分布电容无关，输出灵敏度取决于放大器的反馈电容 C_f。因此，只要保持反馈电容的数值不变，就可得到与电荷量 q 变化呈线性关系的输出电压。同时，反馈电容 C_f 小，输出就大。因此，要达到一定的输出灵敏度，必须选择适当容量的反馈电容 C_f。采用电荷放大器时，即使连接电缆长度在 100 m 以上，其灵敏度也无明显变化，即传感器的灵敏度与电缆长度无关，这是电荷放大器的突出优点。

在电荷放大器的实际电路中，考虑到被测物理量的不同及后级放大器不致因输入信号太大而引起饱和，反馈电容 C_f 需是可调的，范围一般在 100 ~ 10 000 pF。为减小零漂，使电荷放大器工作稳定，一般在反馈电容的两端并联一个大电阻 R_f(10^8 ~ 10^{10} Ω)，其功能是提供直流反馈。

图4.76为压电式传感器以电压灵敏度表示时的等效电路；图4.77为传感器以电荷灵敏度表示时的等效电路。二者的意义是一样的，只是表示的方式不同。

与电荷放大器相比，电压放大器电路更简单、元件更少、价格更便宜、工作更可靠。使用电压放大器时，必须按传感器的出厂要求严格选择电缆长度，否则须另行标定。由于电缆长度对传感器测量精度的影响较大，因此限制了压电式传感器的应用场合。将放大器装入传感器之中组成一体化传感器，可以消除长电缆对传感器灵敏度的影响。

电荷放大器的显著优点是，放大器输出电压只与传感器的电荷量及反馈电容有关，无须考虑电缆的电容，这为远距离测试提供了很大的方便，因此电荷放大器目前使用较多。

4.7 磁电式传感器

4.7.1 磁电式传感器的工作原理

磁电式传感器是把被测物理量转换为感应电动势的一种传感器，又称感应式或电动式传感器。

根据电磁感应定律，一个匝数为 W 的线圈，当穿过该线圈的磁通量 Φ 发生变化时，线圈两端就会产生出感应电势 e，即

$$e=-W\frac{\mathrm{d}\Phi}{\mathrm{d}t} \tag{4.100}$$

负号表明感应电势的方向与磁通变化的方向相反。对于一个线圈，当穿过该线圈的磁通发生变化时，其感应电动势的大小取决于匝数和穿过线圈的磁通变化率。磁通变化率与

Formula (4.99) shows that under certain conditions, the output voltage e_y of the charge amplifier is proportional to the charge q of the sensor, and has nothing to do with the distributed capacitance of the cable. The output sensitivity depends on the amplifier feedback capacitance C_f. Therefore, as long as the value of the feedback capacitor is kept unchanged, an output voltage with a linear relationship with the change in the amount of charge q can be obtained. At the same time, if the feedback capacitor C_f is small, the output will be large. Therefore, it is necessary to select an appropriate capacity C_f in order to achieve a certain output sensitivity. When the charge amplifier is used, even if the length of the connecting cable is more than 100 m, its sensitivity does not change significantly, that is, the sensitivity of the sensor has nothing to do with the cable length, which proves a prominent advantage of the charge amplifier.

In the actual circuit of the charge amplifier, taking into account the difference of the measured physical quantity and the subsequent amplifier will not cause saturation due to the input signal being too large, thus the feedback capacitor C_f needs to be adjustable, with the general range between 100–10 000 pF. In order to reduce the zero drift, and make the charge amplifier work stable, a large resistor R_f (about 10^8–10^{10} Ω) is generally connected in parallel at both ends of the feedback capacitor, whose function is to provide DC feedback.

Figure 4.76 is the equivalent circuit of a piezoelectric sensor in terms of voltage sensitivity; Figure 4.77 is the equivalent circuit when the sensor is expressed by charge sensitivity. The meaning of the two is the same, but the way of expression is different.

Compared with the charge amplifier, the voltage amplifier has simpler circuit, fewer components, lower price, more reliable operation. When using a voltage amplifier, the cable length must be strictly selected according to the factory requirements of the sensor, otherwise it must be calibrated separately. The application of piezoelectric sensors is limited, for the cable length has a greater impact on the measurement accuracy of the sensor. Put the amplifier into the sensor to form an integrated sensor, which can eliminate the influence of long cables on the sensitivity of the sensor.

The remarkable advantage of the charge amplifier is that the output voltage of the amplifier is only related to the charge amount of the sensor and the feedback capacitance, without considering the capacitance of the cable, which provides great convenience for long-distance measurement, so the charge amplifier is currently used more.

4.7 Magnetoelectric Sensors

4.7.1 Working Principle of Magnetoelectric Sensors

Magnetoelectric sensor is a kind of sensor that can convert the measured physical quantity into induced electromotive force, also known as inductive or electric sensor.

According to the law of electromagnetic induction, a coil with W turns, when the magnetic flux Φ passing through the coil changes, an induced electric potential e will be generated at both ends of the coil, that is

$$e = -W\frac{d\Phi}{dt} \tag{4.100}$$

The negative sign indicates that the direction of the induced electric potential is opposite to the

磁场强度、磁路磁阻和线圈的运动速度有关。因此,改变其中任何一个因素,线圈的感应电动势都会改变。

根据穿过线圈的磁通发生变化方法不同,磁电式传感器可分成两大类型:动圈式(定磁通式、动磁式)与磁阻式(变磁通式、可动衔铁式)。

4.7.2 磁电式传感器的分类

1. 动圈式

图4.78(a)所示为线速度型动圈式磁电传感器工作原理。在铁芯产生的直流磁场内放置一个可动线圈,当线圈在磁场中做直线运动时,所产生的感应电动势为

$$e = NBlV\sin\theta \tag{4.101}$$

式中 B—— 磁场的磁感应强度,T;

l—— 单匝线圈有效长度,m;

N—— 线圈匝数;

V—— 线圈与磁场的相对运动速度,m/s;

θ—— 线圈运动方向与磁场方向的夹角。

此式表明,当N、B、l均为常数时,感应电动势大小与线圈运动的线速度成正比。一般常见的地震式速度计就是按此原理工作的。

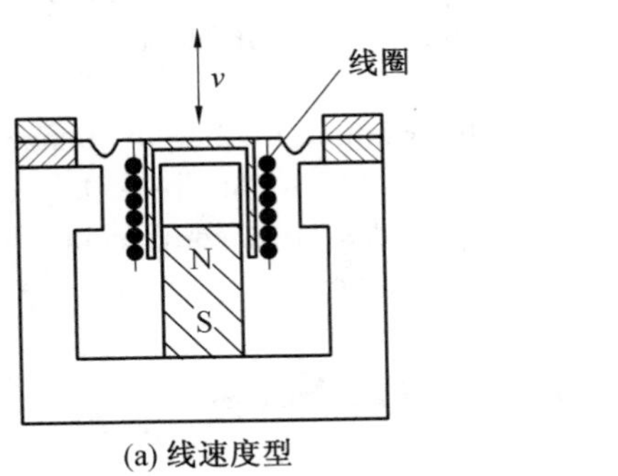

(a) 线速度型

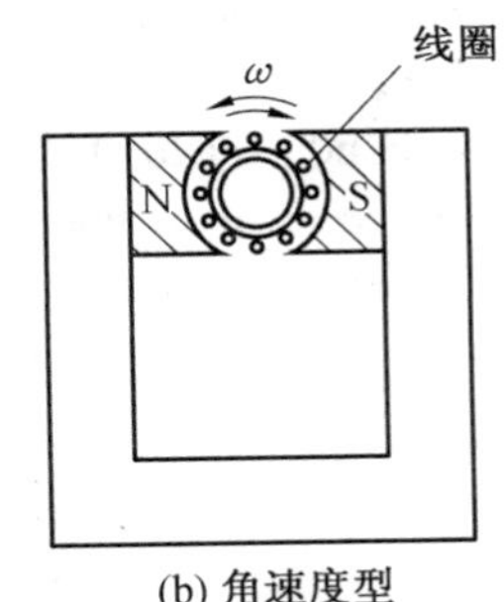

(b) 角速度型

图4.78 动圈式磁电传感器工作原理

图4.78(b)所示为角速度型动圈式磁电传感器工作原理。线圈在磁场中转动时产生的感应电动势与线圈相对磁场的角速度成正比,这种传感器用于转速测量。

将传感器中线圈产生的感应电动势与电压放大器连接时,动圈式磁电传感器等效电路如图4.79所示。图中,e是线圈的感应电势;z_0是线圈内阻;R_L是负载电阻(放大器输入电阻);C_c是电缆导线的分布电容;R_c是电缆导线的电阻。一般情况下,感应电动势经放大检波后即可推动指示仪表。如果经过微分或积分网络,则还可以得到加速度或位移。

direction of the magnetic flux change. For a coil, when the magnetic flux passing through the coil changes, the magnitude of its induced electromotive force depends on the number of turns and the rate of change of the magnetic flux passing through the coil. The rate of change of magnetic flux is related to the strength of the magnetic field, the magnetic resistance of the magnetic circuit, and the speed of the coil. Therefore, one of these factors is changed, the induced electromotive force of the coil will be changed.

According to the methods of causing changes in the megnetic flux through the coil, moving coil type (constant flux type, moving magnetic type) and magnetoresistive type (variable flux type, movable armature).

4.7.2 The Classification of Magnetoelectric Sensors

1. Moving Coil

Figure 4.78(a) shows the working principle of linear velocity moving coil magnetoelectric sensors. In the DC magnetic field generated by the iron core, a movable coil is placed, when the coil moves linearly in the magnetic field, the induced electromotive force is

$$e = NBlV\sin\theta \tag{4.101}$$

where B —magnetic induction intensity of the magnetic field, T;

l —effective length of single-turn coil, m;

N —number of coil turns;

V —the relative speed of the coil and the magnetic field, m/s;

θ —the angle between the moving direction of the coil and the direction of the magnetic field.

This formula shows that when N, B and l are all constants, the magnitude of the induced electromotive force is proportional to the linear velocity of the coil movement. The common seismic speedometer works according to this principle.

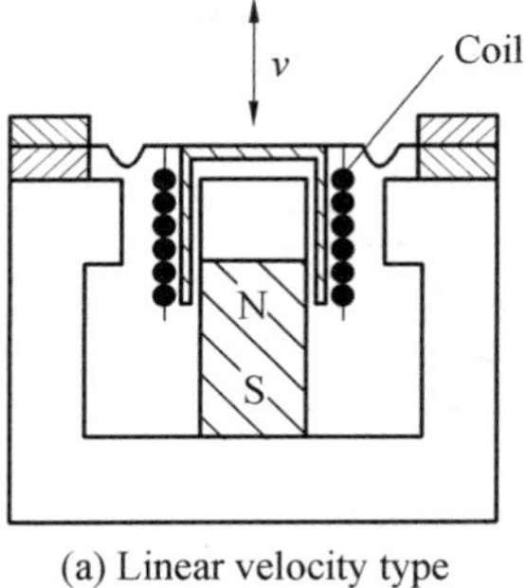

(a) Linear velocity type

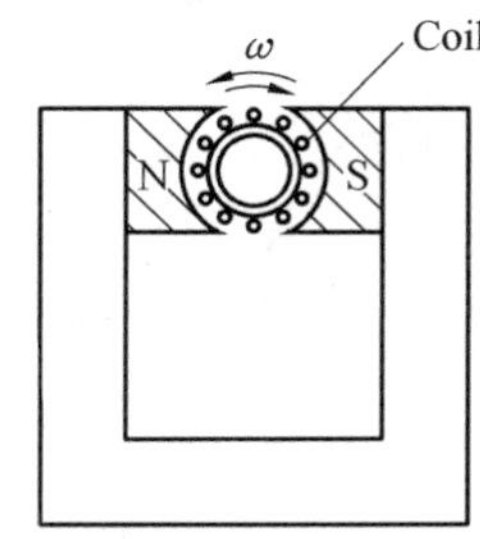

(b) Anguar velocity type

Figure 4.78 Working principle of moving coil magnetoelectric sensors

Figure 4.78(b) shows the working principle of angular velocity moving coil magnetoelectric sensors. The induced electromotive force generated during the coil rotating in a magnetic field is proportional to the angular velocity of the coil relative to the magnetic field. This kind of sensor is usually used for rotational speed measurement.

When the induced electromotive force generated by the coil in the sensor is connected to the voltage amplifier, the equivalent circuit of moving coil magnetoelectric sensors is shown in Figure 4.79. In the figure, e is the induced potential of the coil; z_0 is the internal resistance of the coil; R_L is the load resistance (amplifier input resistance); C_c is the distributed capacitance

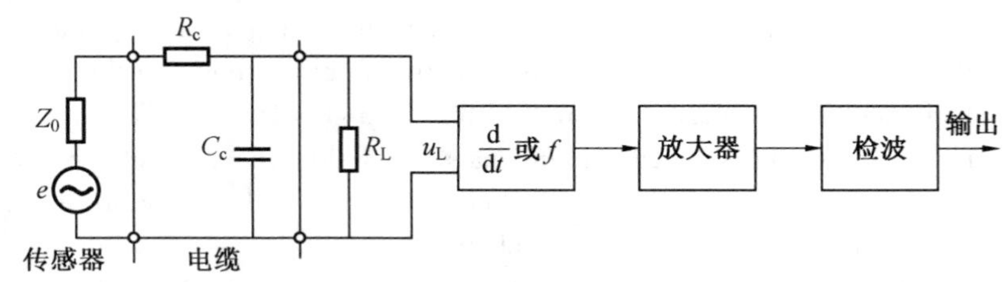

图 4.79　动圈式磁电传感器等效电路

2. 磁阻式

磁阻式传感器的线圈与磁铁彼此不做相对运动，由运动着的物体（导磁材料）来改变磁路的磁阻，引起磁力线增强或减弱，使线圈产生感应电动势。磁阻式传感器工作原理及应用实例如图 4.80 所示。这种传感器是由永久磁铁及缠绕其上的线圈组成的。例如，图 4.80（a）可测旋转体频数，当齿轮旋转时，齿的凸凹引起磁阻变化，使磁通量变化，在线圈中感应出交流电动势，其频率等于齿轮的齿数和转速的乘积。

磁阻式传感器使用简便、结构简单，在不同场合下可用来测量转速、偏心量、振动等。

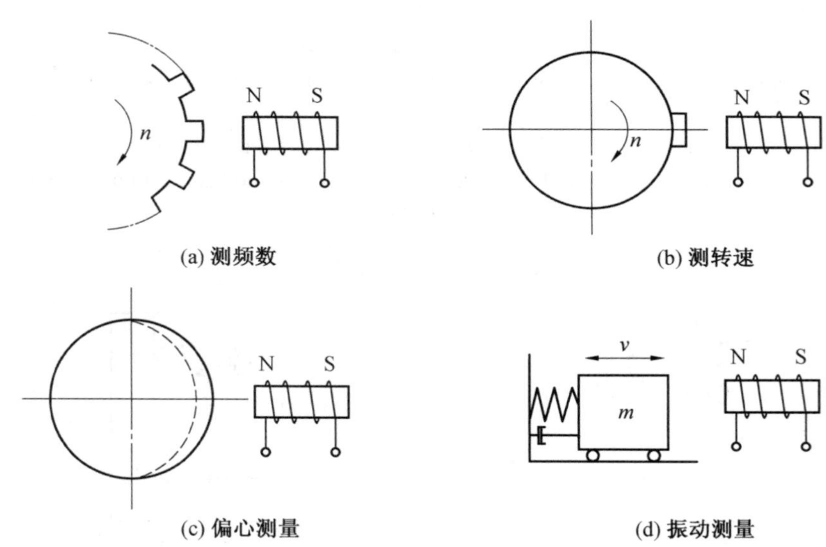

图 4.80　磁阻式传感器工作原理及应用实例

4.8　霍尔传感器

霍尔传感器是利用霍尔元件（一种半导体材料）的霍尔效应进行工作的一种传感器，它可以将被测量转换成电动势输出，实现磁电转换。

霍尔元件一般由锗（Ge）、锑化铟（InSb）、砷化铟（InAs）等半导体材料制成。在静止

of the cable wire; R_c is the resistance of the cable wire. Under normal circumstances, the induced electromotive force can push the indicating instrument after being amplified and detected. If it passes through a differential or integral network, acceleration or displacement will also be obtained.

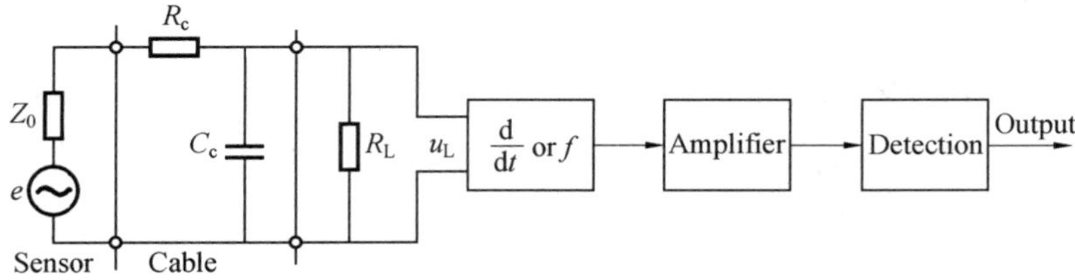

Figure 4.79 Equivalent circuit of moving coil magnetoelectric sensors

2. Magnetoresistive Type

The coil and the magnet of the magnetoresistive sensor do not move relative to each other, and the moving object (magnetic material) changes the magnetic resistance of the magnetic circuit, which will cause the magnetic lines of force to increase or decrease, making coil generate induced electromotive force. The working principle and application examples of magnetoresistive sensor are shown in Figure 4.80. This type of sensor is composed of a permanent magnet and a coil wound on it. For example, Figure 4.80(a) can measure the frequency of the rotating body. When the gear rotates, the convex and concave of the tooth will cause the magnetic resistance to change, which causes the magnetic flux to change, and the AC electromotive force is induced in the coil, whose frequency is equal to the product of the number of gear teeth and the speed.

Magnetoresistive sensors are easy to use and simple in structure, which can be used to measure speed, eccentricity, vibration, etc. in different situations.

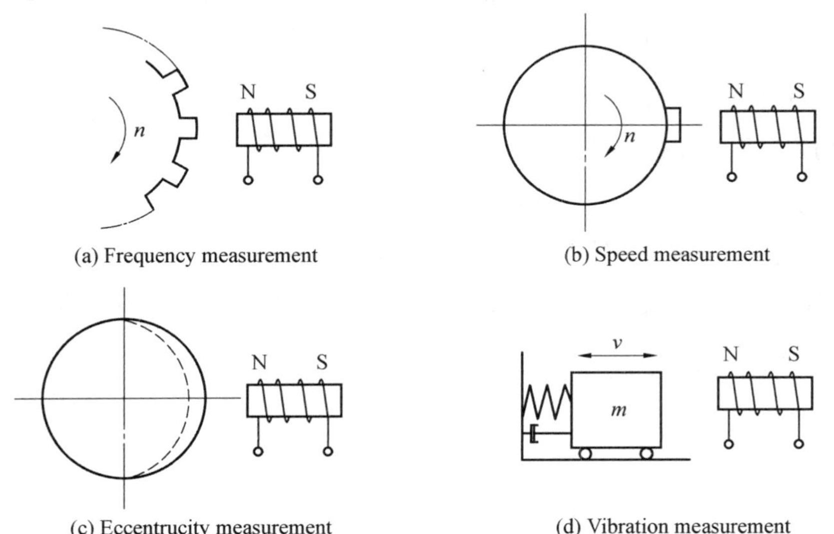

Fig. 4.80 Working principle and application examples of magnetoresistive sensor

状态下，其具有感受磁场的独特能力，并且具有结构简单、体积小、噪声低、频率范围宽、动态范围大和寿命长等特点。

4.8.1 霍尔效应

霍尔（E. H. Hall）发现：一个半导体薄片置于磁场中，当有电流流过时，在垂直于电流和磁场的方向上将产生电动势，这种现象称为霍尔效应。

霍尔效应与霍尔元件如图4.81所示。假设薄片为N型半导体，磁感应强度为B的磁场方向垂直于薄片。在薄片左右两端通以控制电流I，那么半导体中的载流子（电子）将沿着与电流I相反的方向运动。由于外磁场B的作用，因此电子受到磁场力F_L（洛仑兹力）而发生偏转，结果在半导体的后端面上积累电子带负电，而前端面缺少电子带正电，在前后端面间形成电场，该电场产生的电场力F_E阻止电子继续偏转。当$F_E = F_L$时，电子积累达到动态平衡。这时，在半导体前后两端面之间（即垂直于电流和磁场方向）建立电场，称为霍尔电场E_H，相应的电势称为霍尔电势U_H。霍尔电势U_H表示为

$$U_H = R_H IB/\delta = k_H IB$$

式中 R_H—— 霍尔系数，由半导体材料的物理性质决定；

I—— 流经霍尔元件的电流；

B—— 磁场的磁感应强度；

k_H—— 灵敏度系数，与载流材料的物理性质和几何尺寸有关，表示在单位磁感应强度和单位控制电流时的霍尔电势的大小；

δ—— 霍尔元件薄片的厚度。

如果磁场和薄片法线有α角，则有

$$U_H = k_H IB\cos\alpha$$

改变B、I、α中的任何一个参数，都会使霍尔电势发生变化。

霍尔效应的产生机理为运动电荷（载流子）受磁场中洛仑兹力作用的结果。

4.8 Hall Sensors

The Hall sensor is a kind of sensor that uses the Hall effect of its element (a kind of semiconductor material) to work, which can convert the measured into electromotive force output, and realize the magnetoelectric conversion.

Hall elements are generally made of semiconductor materials such as germanium (Ge), indium antimonide (InSb) and indium arsenide (InAs). In the static state, it has the unique ability to feel the magnetic field, and has the characteristics of simple structure, small size, low noise, wide frequency range, large dynamic range, and long life, etc.

4.8.1 Hall Effect

Hall (E. H. Hall) found that: a semiconductor chip is placed in a magnetic field, when a current flows through, an electromotive force will be generated in the direction perpendicular to the current and the magnetic field, whose phenomenon is called the Hall effect.

Hall effect and Hall element are shown in Figure 4.81. Assuming that the sheet is an N-type semiconductor, the direction of the magnetic field with a magnetic induction intensity of B is perpendicular to the sheet. Pass the control current I at the left and right ends of the sheet, then the carriers (electrons) in the semiconductor will move in the opposite direction to the current I. Due to the action of the external magnetic field B, the electrons are deflected by the magnetic field force F_L (Lorentz force). As a result, the electrons accumulate negatively on the back side of the semiconductor, while the front side lacks electrons to be positively charged, and an electric field is formed between the front and back sides. The electric field force F_E generated by this electric field prevents the electrons from continuing to deflect. When $F_E = F_L$, the accumulation of electrons reaches dynamic equilibrium. At this time, an electric field is established between the front and rear end faces of the semiconductor (that is, perpendicular to the direction of the current and magnetic field), which is called the Hall electric field E_H, and the corresponding potential is called the Hall potential U_H. Hall potential U_H can be expressed by

$$U_H = R_H IB/\delta = k_H IB$$

where R_H —hall coefficient, which is determined by the physical properties of semiconductor materials;

I —the current flowing through the Hall element;

B —the magnetic induction intensity of the magnetic field;

k_H —the sensitivity coefficient, which related to the physical properties and geometric dimensions of the current-carrying material, and represents the size of the Hall potential per unit magnetic induction intensity and unit control current;

δ —the thickness of the Hall element sheet.

If the magnetic field and the sheet normal have an angle α, then we have

$$U_H = k_H IB\cos\alpha$$

Changing any of B, I and α will make the Hall potential change.

The mechanism of the Hall effect is the result of the Lorentz force acting on the moving charge (carrier) in the magnetic field.

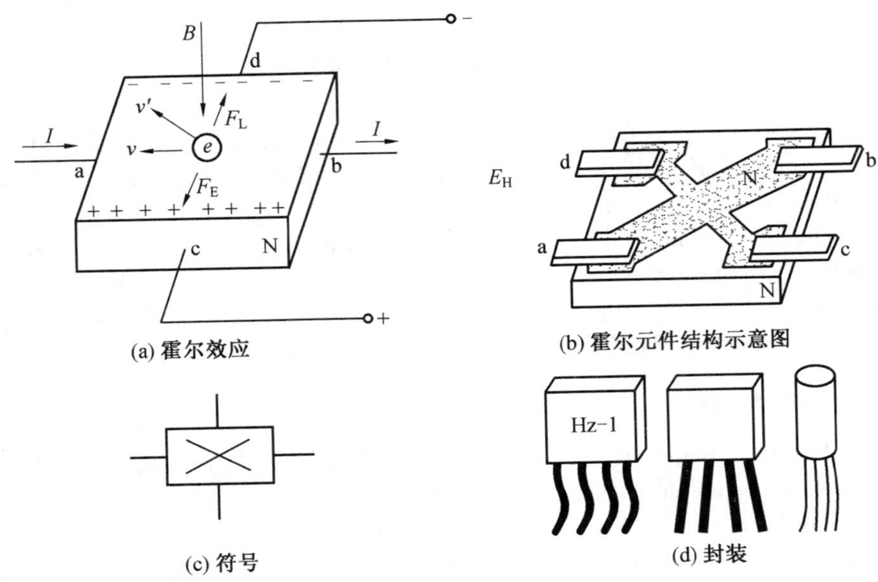

(a) 霍尔效应

(b) 霍尔元件结构示意图

(c) 符号

(d) 封装

图 4.81　霍尔效应与霍尔元件

4.8.2　霍尔元件

霍尔元件是根据霍尔效应原理制成的磁电转换元件，多采用 N 型半导体材料。霍尔元件越薄（δ 越小），k_H 就越大。因此，通常薄膜霍尔元件厚度只有 1 mm 左右。

霍尔效应产生的电压（霍尔电势）与磁场强度 B 成正比。为减小元件的输出阻抗，使其易于与外电路实现阻抗匹配，半导体霍尔元件多数都采用十字形结构，如图 4.82 所示。

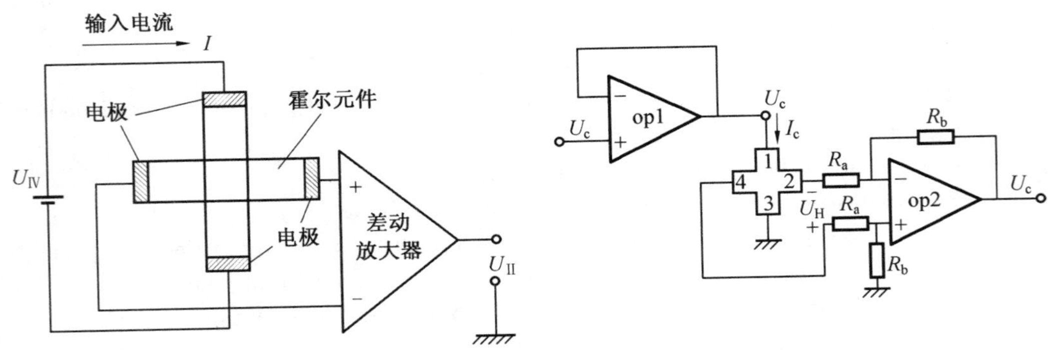

图 4.82　半导体霍尔元件的结构

霍尔元件由霍尔片、四根引线和壳体组成，如图 4.83 所示。霍尔片是一块半导体单晶薄片（一般为4 mm × 2 mm × 0.1 mm），在它的长度方向两端面上焊有 a、b 两根引线，称

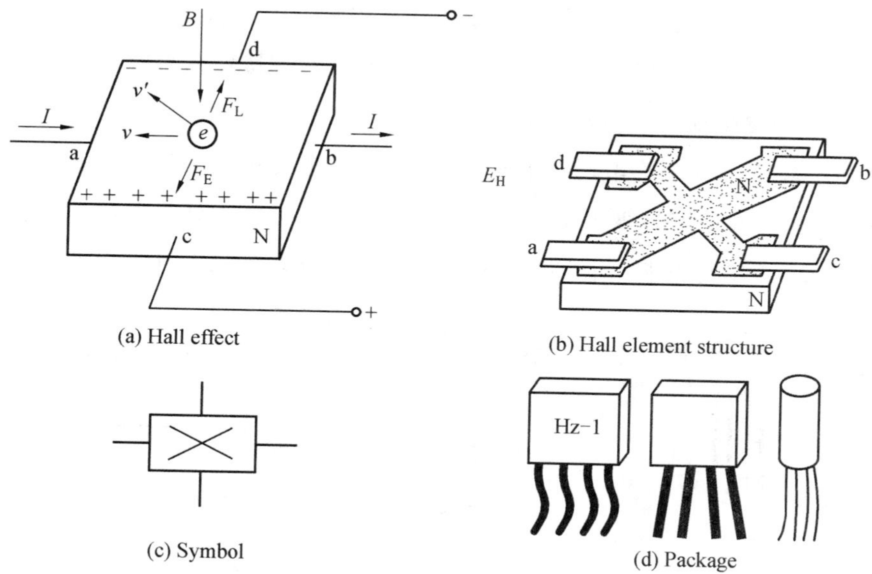

(a) Hall effect
(b) Hall element structure
(c) Symbol
(d) Package

Figure 4.81 Hall effect and Hall element

4.8.2 Hall Element

The Hall element is a magnetoelectric conversion element made according to the principle of the Hall effect, and the Hall element are mostly N-type semiconductor materials. The thinner the Hall element (the smaller the δ), the larger the k_H. Therefore, the thickness of the thin-film Hall element is usually only about 1 mm.

The voltage (Hall potential) generated by the Hall effect is proportional to the magnetic field strength B. In order to reduce the output impedance of the element and make it easy to achieve impedance matching with the external circuit, most of the semiconductor Hall elements adopt a cross-shaped structure, as shown in Figure 4.82.

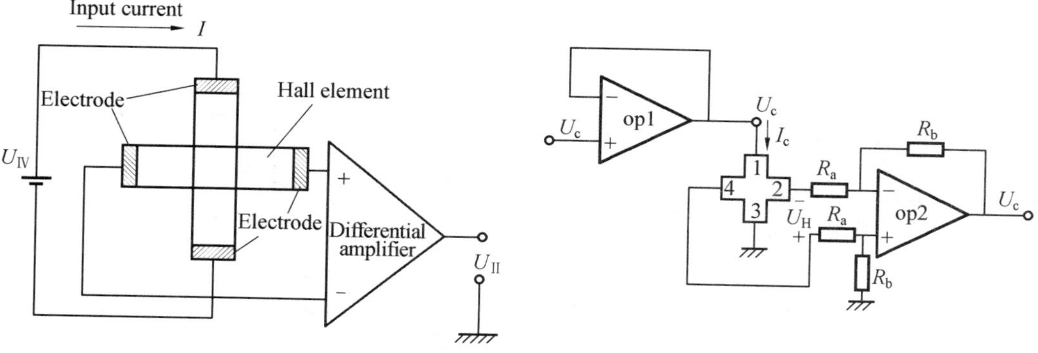

Figure 4.82 The structure of the semiconductor Hall element

The Hall element consists of a Hall plate, four leads and a housing, as shown in Figure 4.83. The Hall plate is a semiconductor single crystal sheet (usually 4 mm×2 mm×0.1 mm). Two leads a and b called control current terminal leads are welded on both ends of its length direction, of which the red lead whose welding place is called the control electrode;

为控制电流端引线,通常用红色导线,其焊接处称为控制电极;在它的另两侧端面的中间点对称地焊有 c、d 两根霍尔输出引线,通常用绿色导线,其焊接处称为霍尔电极。霍尔元件的壳体是用非导磁金属、陶瓷或环氧树脂封装的。目前,一般用锗(Ge)、硅(Si)、锑化铟(InSb)、砷化钢(InAs) 等高电阻率半导体材料制作霍尔元件。由于材料本身对弱磁场的灵敏度较低,因此在使用时要加入数特斯拉的偏置磁场,使元件处于强磁场的范围内工作,从而可以检测微弱的磁场变化。图 4.84 所示为霍尔元件的基本电路。

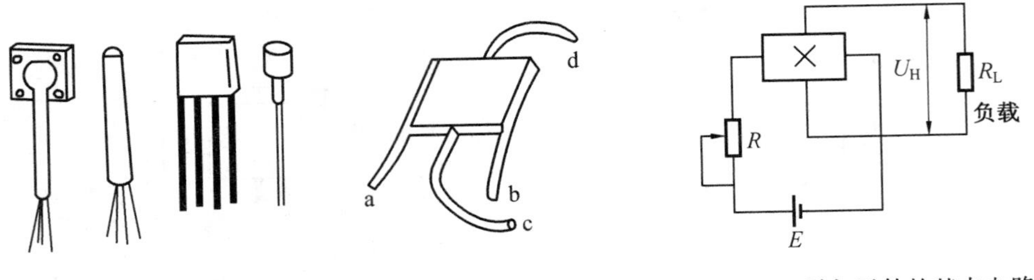

图 4.83　霍尔元件　　　图 4.84　霍尔元件的基本电路

4.8.3　应用分析

霍尔元件的应用原理是:当被测量以某种方式改变了霍尔元件在磁场中所处的位置时,作用在元件上的有效磁场强度就随之改变,所以输出的霍尔电势 U_H 就称为霍尔元件位置(即被测量) 的函数。

霍尔元件应用范围(图 4.85) 为:能转换为磁感应强度变化的参数的测量,如位移(线位移、角位移)、压力或压力差、加速度、转速、力、磁场、工件计数和钢丝绳探伤等;可以对能转换为电流变化的参数进行测量;还可以用作乘法器(电功率测量中的电流与电压的相乘等)。

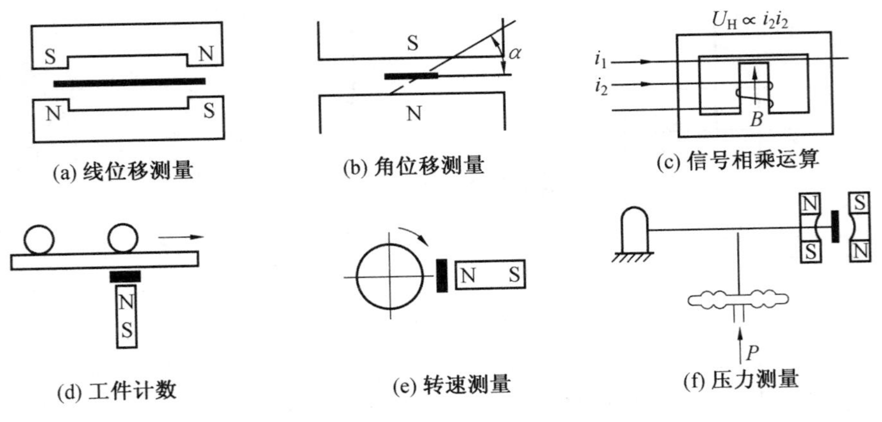

图 4.85　霍尔传感器的几种应用

of which two Hall output leads c and d are welded symmetrically at the middle point of the other two end faces, usually with green wires, and the welding place is called the Hall electrode. The housing of the Hall element is encapsulated with non-magnetic metal, ceramic, or epoxy resin. At present, high-resistivity semiconductor materials such as germanium (Ge), silicon (Si), indium antimonide (InSb), and arsenide steel (InAs) are generally used to make Hall elements. Due to the material itself being less sensitive to weak magnetic fields, a bias magnetic field of several tesla should be added to make the component work within the range of a strong magnetic field when in use, so that weak magnetic field changes can be detected. Figure 4.84 shows the basic circuit of the Hall element.

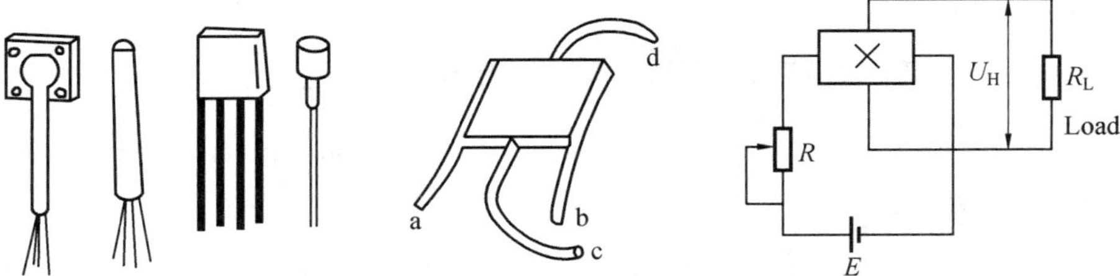

Figure 4.83 Hall element

Figure 4.84 Basic circuit of the Hall element

4.8.3 Application Analysis

The application principle of the Hall element is: when the position of the Hall element in the magnetic field is changed in some way by the measured quantity, the effective magnetic field strength acting on the element will change accordingly, so the output Hall potential U_H becomes a function of the position of the Hall element (that is, measured quantity).

Hall element application range (Figure 4.85) included that: it can be converted into the measurement of magnetic induction intensity changes, such as displacement (linear displacement, angular displacement), pressure or pressure difference, acceleration, rotation speed, force, magnetic field, workpiece count, and wire rope flaw detection; it not only can measure the parameters that can be converted into current changes; but can be used as a multiplier (multiplication of current and voltage in electric power measurement, etc).

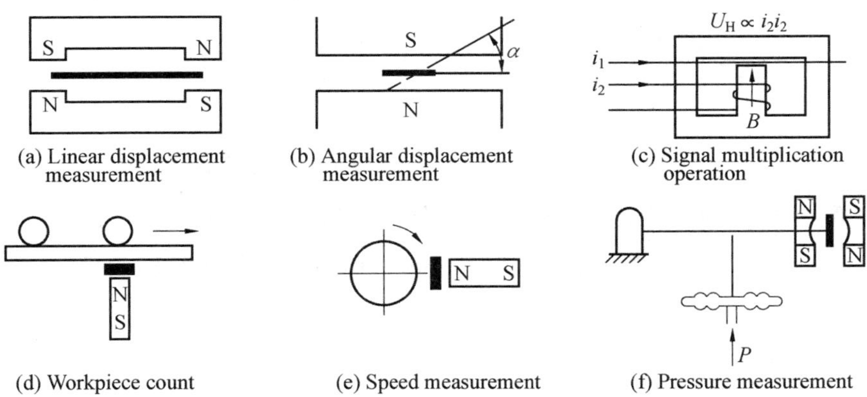

Figure 4.85 Several applications of Hall sensors

1. 霍尔效应位移传感器

图 4.86 所示为霍尔效应位移传感器工作原理。将霍尔元件置于磁场中，左半部磁场方向向上，右半部磁场方向向下。从 a 端通入电流 I，根据霍尔效应，左半部产生霍尔电势 V_{H1}，右半部产生霍尔电势 V_{H2}，其方向相反。因此，c、d 两端电势为 $V_{H1}-V_{H2}$。如果霍尔元件在初始位置时 $V_{H1}=V_{H2}$，则输出为零。当改变磁极系统与霍尔元件的相对位置时，即可得到输出电压，其大小正比于位移量。

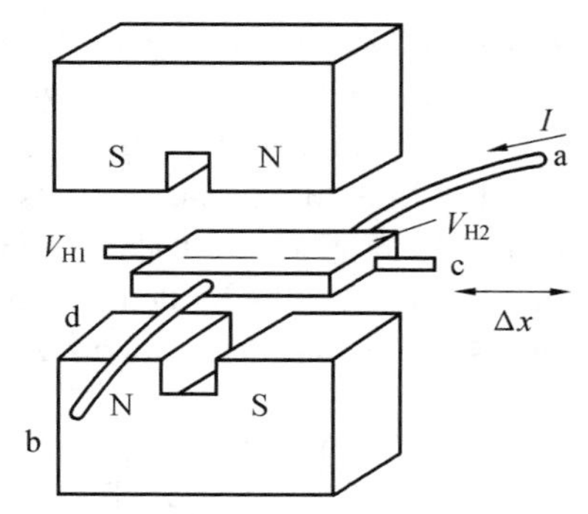

图 4.86　霍尔效应位移传感器工作原理

2. 霍尔效应电流计

图 4.87 所示为卡形电流计，被测电流所通过的导线不必切断就可穿过铁芯张开的缺口，放开扳手后铁芯闭合。它将导线电流产生的磁场引入高磁导率的磁路中，通过磁路中插入的霍尔元件对该磁场进行检测，以此测量导线上的电流。这种电流计的测量范围很宽，可以测量从直流到高频的电流，是一种不需断开电路即可直接测量电路交流电流的携带式仪表，在电气检修中应用非常方便。

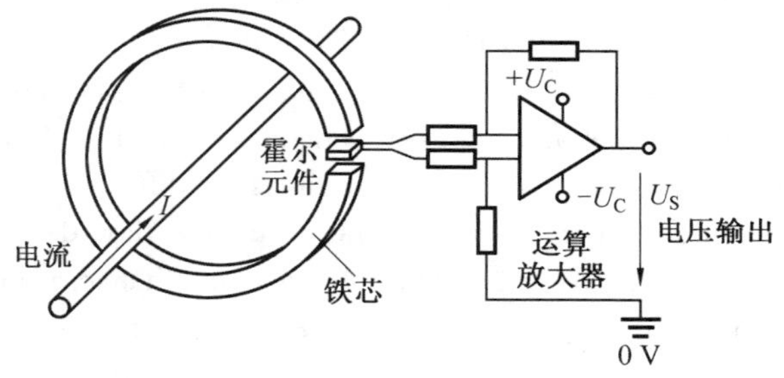

图 4.87　卡形电流计

3. 霍尔电动机

霍尔电动机是一种由检测位置的霍尔元件制成的一种无刷电动机，一个元件能够控制两组晶体管，是当今无刷电动机中使用最多的一种电动机。由于其无电刷，因此具有体积小、无噪声等优点，广泛用于盒式录音机、VTR、FDD 等需要进行转动控制的精密机械中，其结构和等效电路图分别如图 4.88(a)、(b) 所示。

4. 钢丝绳断丝检测

断丝是钢丝绳损伤的主要形式。运行中的钢丝绳一旦断裂，就会造成严重的人员伤

1. Hall Effect Displacement Sensors

Figure 4. 86 shows the working principle of a Hall-effect displacement sensor. Place the Hall element in the magnetic field, with the left half of the magnetic field pointing upwards, and the right half of the magnetic field pointing downwards. A current I is introduced from end a. According to the Hall effect, the left half produces the Hall electric potential V_{H1} , the right half produces the Hall electric potential V_{H2} , its direction is opposite. Therefore, the potential at both ends of c and d is $V_{H1} - V_{H2}$. If the Hall element is in the initial position $V_{H1} = V_{H2}$, the output is zero. When the relative position of the magnetic pole system and the Hall element is changed, the output voltage can be obtained, and its magnitude is proportional to the displacement.

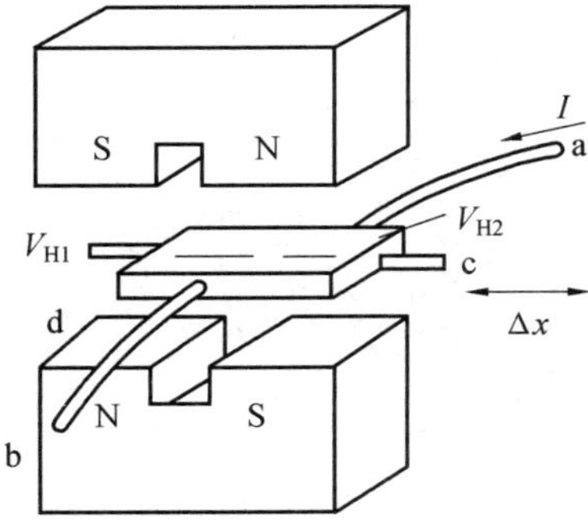

Figure 4. 86 The working principle of Hall effect displacement Sensors

2. Hall Effect Current Galvanometer

Figure 4. 87 shows a card-shaped current galvanometer. The wire through which the measured current passes can pass through the opening of the iron core without cutting off, and the iron core is closed when the wrench is released. It introduces the magnetic field generated by the wire current into the magnetic circuit with high permeability, and detects the magnetic field through the Hall element inserted in the magnetic circuit to measure the current on the wire. This kind of ammeter has a wide measuring range and can measure currents from DC to high-frequency. It is a portable instrument that can directly measure the AC current of the circuit without disconnecting the circuit, which is very convenient to apply in electrical maintenance.

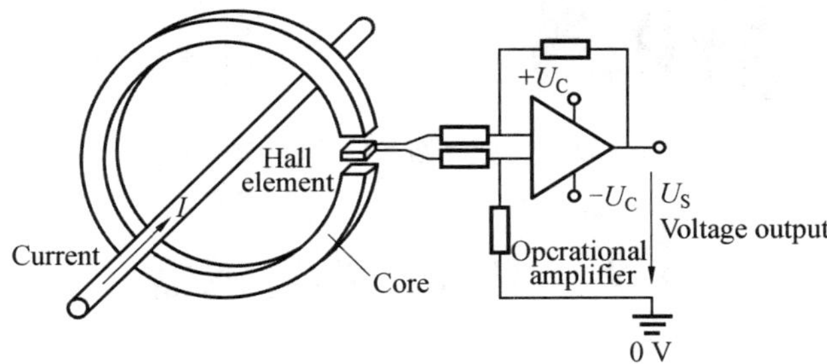

Figure 4. 87 Card-shaped current galvanometer

3. Hall Electromotor

Hall electromotor is a kind of brushless motor made of Hall element for detecting position, with one element being able to control two groups of transistors, which is the most used motor in brushless motors today. Without brushes, it has the advantages of small size and no noise, thus it is widely used in precision machinery such as cassette recorders, VTR and FDD that require rotation control, whose structure and equivalent circuit diagrams are shown in Figure 4. 88(a) and (b), respectively.

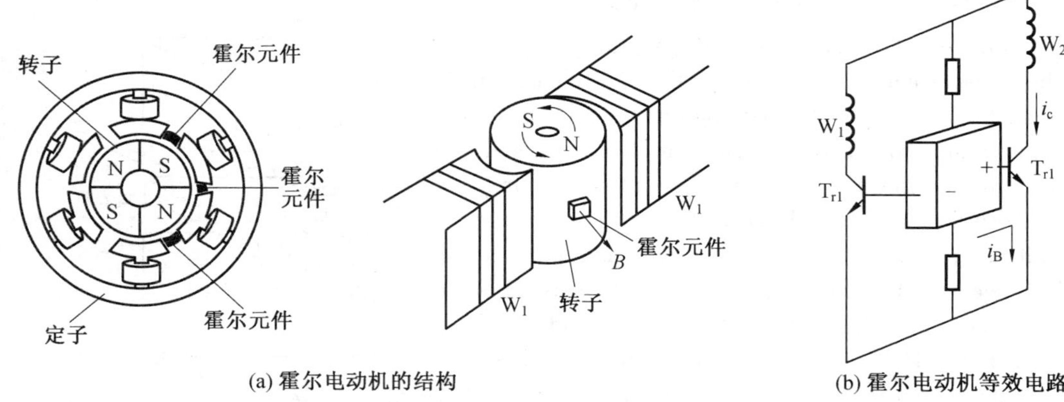

图 4.88　霍尔电动机

亡和经济损失。图 4.89 所示为钢丝绳断丝检测仪工作原理。永磁铁将对钢丝绳局部磁化,当有断丝时,在断口处会出现漏磁场。霍尔元件经过此磁场时,将其转换为一个脉动的电压信号。此信号经放大、滤波、A/D 转换后进入计算机进行分析,识别出断丝根数及位置。

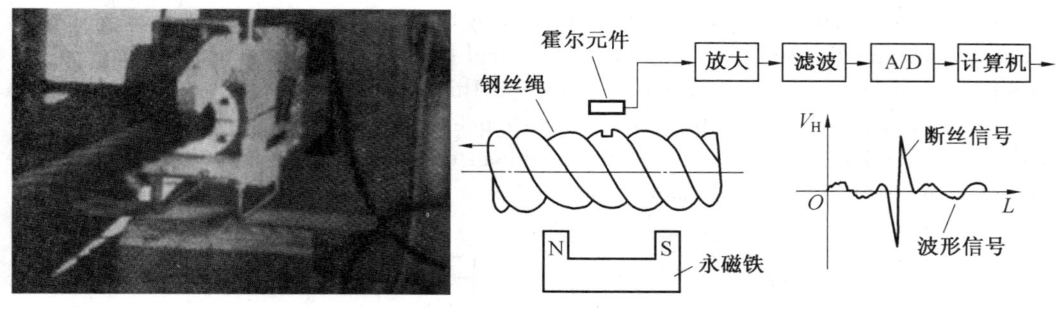

图 4.89　钢丝绳断丝检测仪工作原理

4.9　光导纤维传感器

光纤传感器与以机 - 电转换为基础检测的传感器不同,它是将被测量转换成可测的光信号,以光学测量为基础。光纤传感器具有灵敏度高、抗电磁干扰能力强、耐腐蚀、体积小、质量轻等优点,在各个领域获得了广泛应用。

4.9.1　光导纤维的结构及传光原理

1. 光导纤维的结构

光导纤维(简称光纤)是由玻璃、石英或塑料等光透射率高的电介质拉制而成的极细

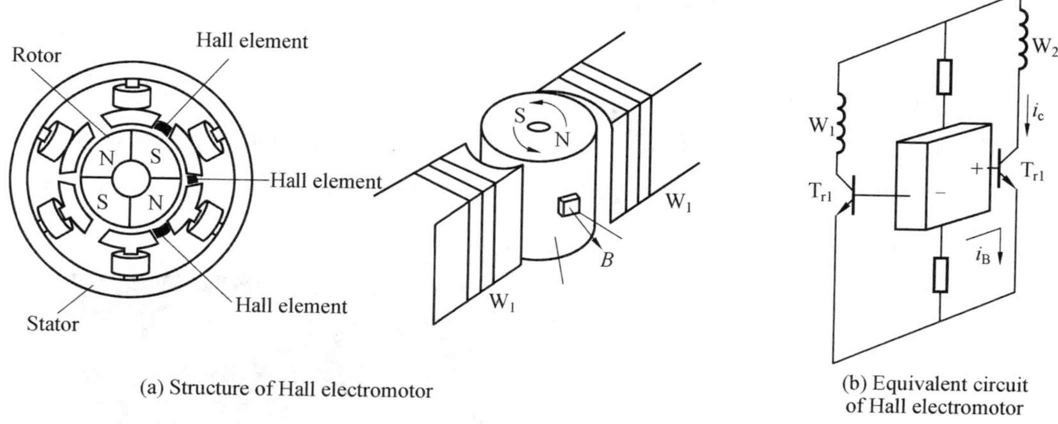

(a) Structure of Hall electromotor

(b) Equivalent circuit of Hall electromotor

Figure 4.88　Hall electromotor

4. Inspection of Broken Wire

Broken wire is the main form of wire rope damage. Once the wire rope is broken in operation, it will cause serious casualties and economic losses. Figure 4.89 shows the working principle of wire rope broken detector. The permanent magnet will locally magnetize the wire rope. When there is a broken wire, a leakage magnetic field will appear at the fracture. When the Hall element passes through this magnetic field, it will be converted into a pulsating voltage signal. After the signal is amplified, filtered and A/D converted, it enters the computer for analysis to identify the number and location of broken wires.

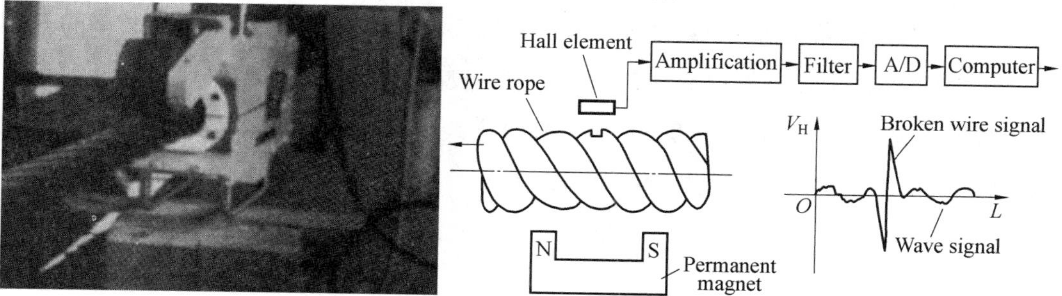

Figure 4.89　Working principle of wire rope broken detector

4.9　Optical Fiber Sensors

Different from sensors based on mechanical-electrical conversion, an optical fiber sensor, based on optical measurement, will convert the measured quantity into a measurable optical signal. Optical fiber sensors have many advantages such as high sensitivity, strong anti-electromagnetic interference, corrosion resistance, small size, light weight, which have been widely used in various fields.

纤维(直径 $\phi4 \sim 10\ \mu m$)。光导纤维一般为圆柱形结构,每一根光纤由纤芯、包层和保护层组成。纤芯位于光纤中心,纤芯外是包层,包层有一层或多层结构,总直径为 $100 \sim 200\ \mu m$。包层材料也是玻璃或塑料(一般为纯 SiO_2 中掺微量杂质),包层的折射率略低于纤芯的折射率。包层外面涂有涂料(即保护层),其作用是保护光纤不受损害,增强机械强度,保护层折射率远大于包层。这种结构能将光波限制在纤芯中传输。光纤结构及外观如图 4.90 所示。

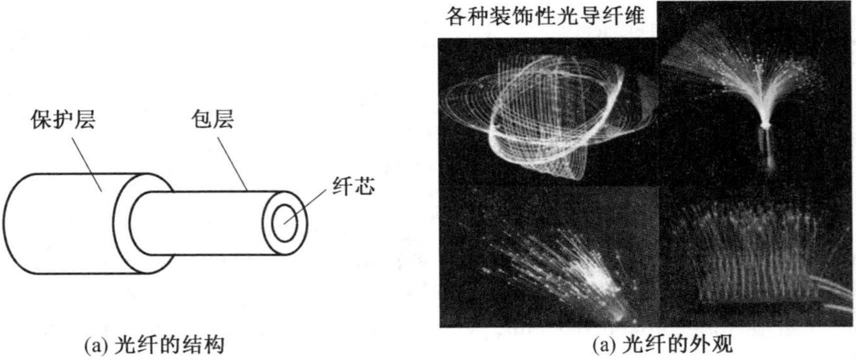

(a) 光纤的结构　　(a) 光纤的外观

图 4.90　光纤结构及外观

2. 传光原理

光的全反射现象是研究光纤传光原理的基础。在几何光学中,当光线以较小的入射角 φ($\varphi_1 < \varphi_c$,φ_c 为临界角)由光密物质(折射率较高,设为 n_1)射入光疏物质(折射率较低,设为 n_2)时,一部分光线被反射,另一部分光线折射入光疏物质。光的传输原理如图 4.91 所示。折射角 φ_2 满足折射定律,即

$$n_1 \sin \varphi_1 = n_2 \sin \varphi_2 \tag{4.102}$$

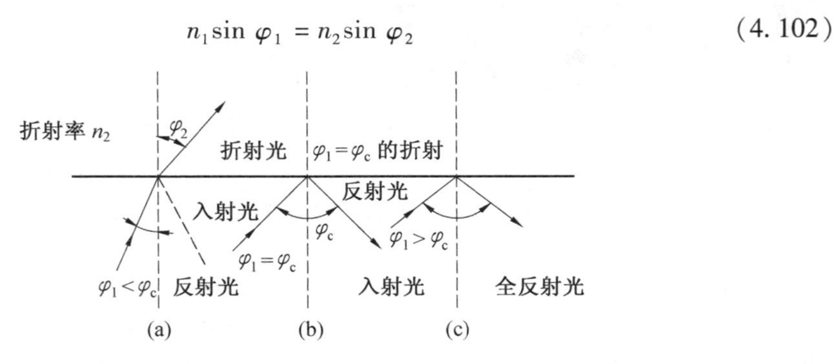

图 4.91　光的传输原理

根据能量守恒定律,反射光与折射光的能量之和等于入射光的能量。

当逐渐加大入射角 φ_1,直到 φ_c 时,折射光会沿着界面传播,此时折射角 $\varphi_2 = 90°$(图 4.91(b))。此时,入射角 $\varphi_1 = \varphi_c$,φ_c 为临界角。临界角 φ_c 为

4.9.1 The Structure and Light Transmission Principle of Optical Fiber

1. The Structure of Optical Fiber

Optical fiber (fiber for short) is an extremely thin fiber (diameter ϕ 4−10 μm) drawn from a dielectric with high light transmittance such as glass, quartz or plastic. The optical fiber is generally cylindrical, and each fiber is composed of a core, a cladding, and a protective layer. The core is located in the center of the optical fiber, and the cladding that has one or more layers with a total diameter of 100−200 μm is outside the core. The cladding material is also glass or plastic (generally pure SiO_2 doped with trace impurities), and the refractive index of the cladding is slightly lower than that of the core. The outer surface of the cladding is coated with paint (that is, protective layer), which protects the optical fiber from damaged, enhances mechanical strength, and the refractive index of the protective layer is much greater than that of the cladding. This structure can limit the transmission of light waves in the core. The structure and the appearance of the optical fiber are shown in Figure 4.90.

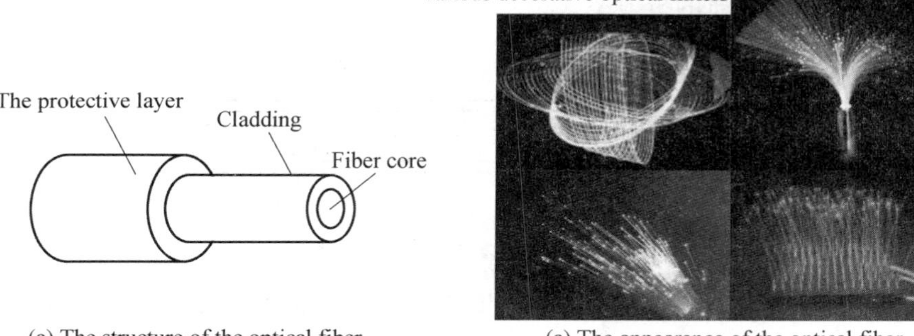

(a) The structure of the optical fiber (a) The appearance of the optical fiber

Figure 4.90 The structure and the appearance of the optical fiber

2. The Principle of Light Transmission

The phenomenon of total reflection of light is the basis for studying the principle of optical fiber transmission. In geometrical optics, when light is incident at a small incident angle φ ($\varphi_1 < \varphi_c$, φ_c is the critical angle), and the light-dense substance (high refractive index, set as n_1) enters the optically thin substance (low refractive index, set as n_2), part of the light will be reflected, and the other part of the light will be refracted into the light sparse material. The optical transmission principle is shown in Figure 4.91. The refraction angle φ_2 satisfies the law of refraction, namely

$$n_1 \sin \varphi_1 = n_2 \sin \varphi_2 \tag{4.102}$$

According to the law of conservation of energy, the sum of the energy of reflected light and refracted light is equal to the energy of incident light.

When we gradually increase the incident angle φ_1, until φ_c, the refracted light will propagate along the interface, at this time, the refraction angle $\varphi_2 = 90°$ (Figure 4.91(b)). At this time, the incident angle $\varphi_1 = \varphi_c$, φ_c is critical angle. Critical angle φ_c is determined by

$$\sin \varphi_c = \frac{n_2}{n_1} \tag{4.103}$$

When we continue to increase the incident angle φ_1 (that is, $\varphi_1 > \varphi_c$), the light no longer

$$\sin \varphi_c = \frac{n_2}{n_1} \tag{4.103}$$

当继续加大入射角 φ_1（即 $\varphi_1 > \varphi_c$）时，光不再产生折射，只有反射。这一现象就是光的全反射现象，如图 4.91（c）所示。

光纤传光原理如图 4.92 所示。当光线从光密物质（折射率较高）射向光疏物质（折射率较低），且入射角大于临界角（折射角为 90° 时的入射角，称为临界角）时，即满足关系式

$$\sin \alpha > \frac{n_2}{n_1} \tag{4.104}$$

式中　α—— 入射角；

n_1—— 光密物质的折射率；

n_2—— 光疏物质的折射率。

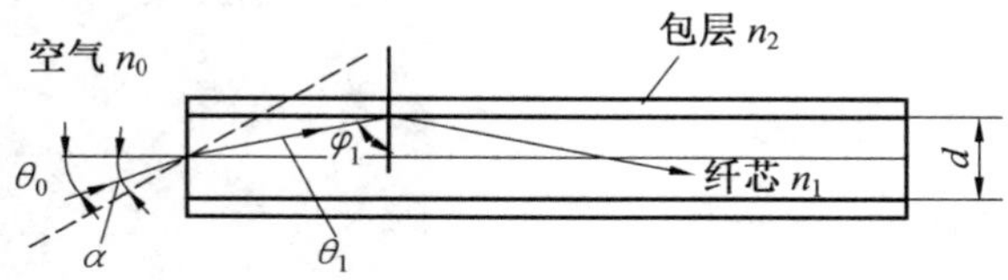

图 4.92　光纤传光原理

此时，光线将在两物质的交界面上发生全反射。根据这个原理，光纤由于其圆柱形内芯的折射率 n_1 大于包层的折射率 n_2，因此在角度为 2θ 范围内的入射光（图 4.93），除在玻璃中吸收和散射损耗的一部分外，其余大部分在界面上产生多次的全反射，以锯齿形的路线在纤芯中传播，并在光纤的末端以与入射角相等的反射角射出光纤，即光线在光芯（内层）内传播。当入射角大于全反射临界角时，光线在内外层的界面上发生全反射，光线在光纤中呈锯齿形轨迹传播。

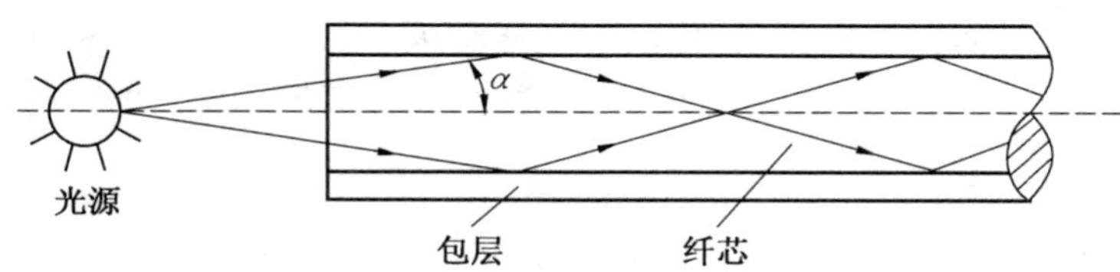

图 4.93　光纤的基本结构与传光原理

4.9.2　光纤传感器的工作原理

光纤传感器由光发送器、敏感元件、光接收器、信号处理系统及光纤等部分组成。光纤传感器应用系统如图 4.94 所示。

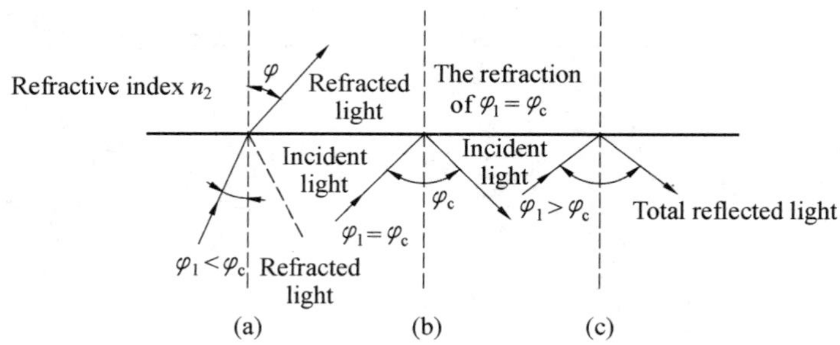

Figure 4.91 Optical transmission principle

produces refraction, only reflection. This phenomenon is the total reflection of light, as shown in Figure 4.91(c).

The optical fiber transmission principle is shown in Figure 4.92. When the light radiates from the dense material (higher refractive index) to the lighter material (lower refractive index), and the incident angle is greater than the critical angle (when refraction angle is 90°, the incident angle is called the critical angle), the relationship satisfies

$$\sin \alpha > \frac{n_2}{n_1} \tag{4.104}$$

where α — incident angle;

n_1—the refractive index of the light dense substance;

n_2—the refractive index of light-thin substances.

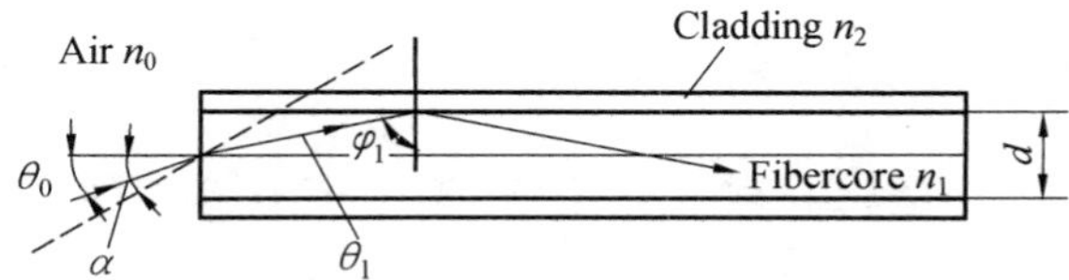

Figure 4.92 Optical fiber transmission principle

At this time, the light will be totally reflected at the interface of the two substances. According to this principle, the refractive index n_1 of the cylindrical core of the optical fiber is greater than the refractive index n_2 of the cladding, so most of the incident light (Figure 4.93) within the range of angle 2θ will produce multiple total reflections at the interface except for the part of the absorption and scattering loss in the glass, which propagates in the core in a zigzag path, and exits the fiber at the end of the fiber at a reflection angle equal to the incident angle. That is, the light propagates in the optical core (inner layer). When the incident angle is greater than the critical angle of total reflection, the light will totally be reflected on the interface between the inner and outer layers, and the light travels in a zigzag trajectory in the optical fiber.

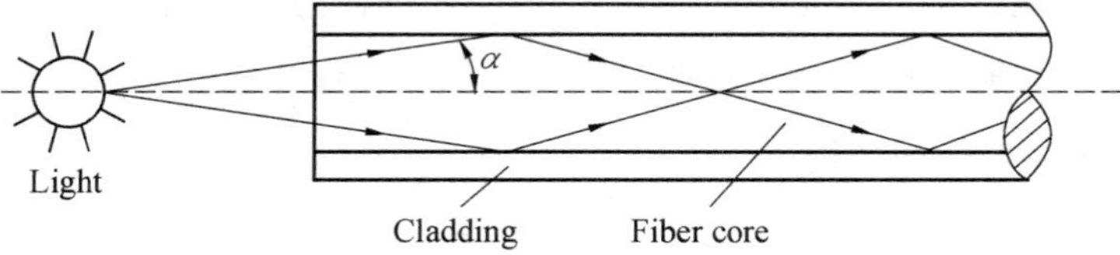

Figure 4.93 Basic structure and light transmission principle of optical fiber

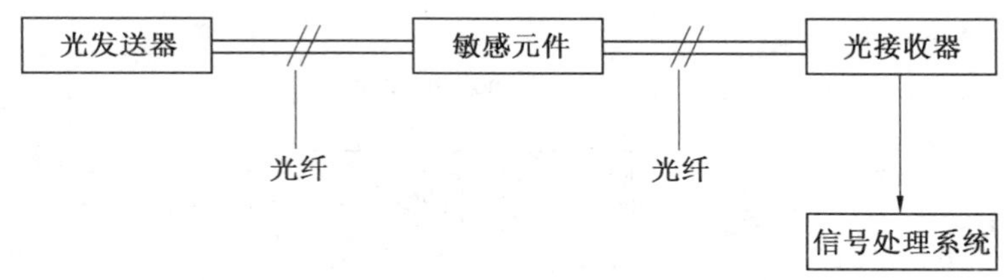

图 4.94　光纤传感器应用系统

按光纤的作用分类，光纤传感器可分为传感型（又称物性型）和传光型（又称结构型）。传感型光纤传感器利用对外界环境变化具有敏感性和检测功能的光纤，构成“传”和“感”合为一体的传感器。这里，光纤不仅起传输光的作用，而且还作为敏感元件。工作时，利用被测量（力、压力、温度等）去改变光束的一些基本参数（如光的强度、相位、偏振、频率等），这些参数的改变反映了被测量的变化。由于对光信号的检测通常使用光敏二极管等光电器件，因此光的这些参数的变化最终都要被光接收器接收并被转换成光强度及相位的变化，经信号处理后，即可得到被测的物理量。应用光纤传感器的这种特性，可以实现力、压力、温度等物理参数的测量。传光型光纤传感器的光纤仅起传输光信号的作用。

4.9.3　光纤传感器的应用

光纤传感器以其高灵敏度、抗电磁干扰、耐腐蚀、柔软、可弯曲、体积小、结构简单、与光纤传输线路相容及能够实现动态非接触测量等独特优点而受到广泛重视。光纤传感器可应用于位移、振动、转速、压力、弯曲、应变、速度、加速度、电流、磁场、电压、温度、湿度、声场、流量、浓度、pH 值等 70 多个物理量的测量，具有十分广泛的应用潜力和发展前景。

1. 半导体吸光式光纤温度传感器

如图 4.95（a）所示，在一根切断的光纤的两端面间夹有一块半导体感温薄片，这种感温薄片入射光的强度随温度而变化。当光纤一端输入恒定光强的光时，另一端接收元件所接受的光强将随被测温度的变化而变化。图 4.95（b）所示为一种双光纤差动测温光纤传感器。该结构中增加了一条参考光纤作为基准通道。两条光纤对来自同一光源的发光强度进行传输，测量光纤的发光强度随温度变化，通过在同一硅片上对称式的光探测器获得两发光强度的整值。此法可消除一定程度的干扰，提高测量精度，在 －40 ～ 400 ℃ 范围内，测温精度可达 ±0.5 ℃。

4.9.2 The Working Principle of Optical Fiber Sensors

The optical fiber sensor is composed of optical transmitter, sensing element, optical receiver, signal processing system and optical fiber. Optical fiber sensor application system is shown in Figure 4.94.

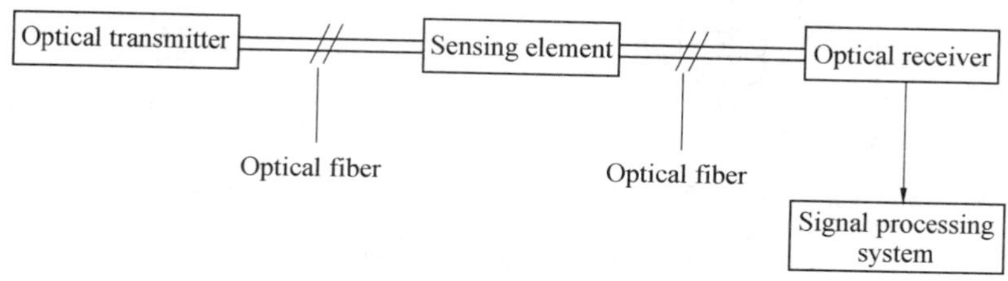

Figure 4.94 Optical fiber sensor application system

Optical fiber sensor can be divided into sensing type (also known as physical type) and light transmission type (also known as structural type) by the function. Sensing optical fiber sensors use optical fibers that are sensitive and detective to changes in the external environment to form a sensor that combines "transmission" and "sensation". Here the optical fiber not only plays the role of transmitting light, but also acts as a sensing element. When working, we can use the measured quantity (force, pressure, temperature, etc.) to change some basic parameters of the beam (such as light intensity, phase, polarization, frequency), of which changes reflect the changes in the measured quantity. Since photodiodes and other photoelectric devices are usually used for the detection of light signals, the changes in these parameters of light will eventually be received by the light receiver and converted into changes in light intensity and phase. After signal processing, the measured quantity can be obtained. The application of this characteristic of the optical fiber sensor can realize the measurement of physical parameters such as force, pressure and temperature. The optical fiber of the light-transmitting optical fiber sensor only plays the role of transmitting optical signals.

4.9.3 The Application of Optical Fiber Sensors

Optical fiber sensors are widely valued for their unique advantages such as high sensitivity, anti-electromagnetic interference, corrosion resistance, softness, flexibility, small size, simple structure, compatibility with optical fiber transmission lines, and the ability to achieve dynamic non-contact measurement. Fiber optic sensors can be applied to the measurement of more than 70 physical quantities such as displacement, vibration, rotation speed, pressure, bending, strain, speed, acceleration, current, magnetic field, voltage, temperature, humidity, sound field, flow, concentration, pH value which will generate potential in wide application and development prospects.

1. Semiconductor Absorption Optical Fiber Temperature Sensors

As shown in Figure 4.95(a), a semiconductor temperature sensing sheet is sandwiched between the two ends of a cut optical fiber. The intensity of the incident light from this temperature sensing sheet changes with temperature. When one end of the optical fiber is input with constant intensity light, the light intensity received by the receiving element at the other end will change with the change of the measured temperature. Figure 4.95(b) shows a kind of

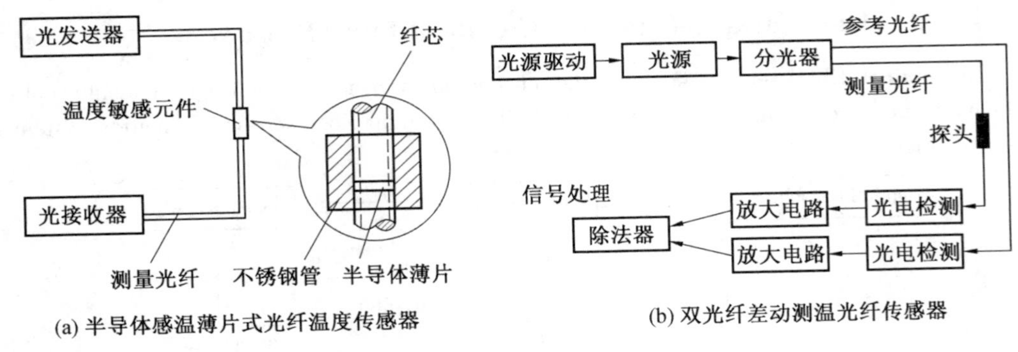

图 4.95　光纤温度传感器

2. 光纤转速传感器

图 4.96 所示为光纤转速传感器。凸块随被测转轴转动，在转到透镜组内时，将光路遮断，形成光脉冲信号，再由光电转换元件将光脉冲信号转变为电脉冲信号，经计数器处理得到转速值。

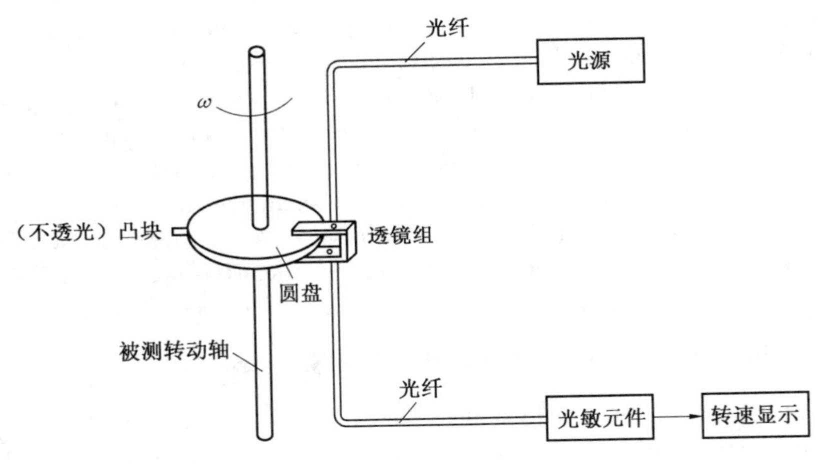

图 4.96　光纤转速传感器

3. 传光型光纤位移传感器

光纤位移传感器工作原理示意图如图 4.97 所示。当光纤探头紧贴被测件时，发射光纤中的光不能反射到接收光纤中，就不能产生光电信号。当被测表面逐渐远离光纤探头时，发射光纤照亮被测表面的面积 A 越来越大，相应的反射光锥重合面积 B_1 越来越大，因此接收光纤端面上照亮的 B_2 区也越来越大，即接收的光信号越来越强，光电流也越来越强，可以用来测量微位移。当整个接收光纤端面被全部照亮时，输出信号就达到了位移 — 输出曲线上的“光峰”点。强度变化的灵敏度比位移变化的灵敏度大得多。此时，可用于对表面状况进行光学检测。

dual-fiber differential temperature measurement fiber sensor. A reference fiber is added to the structure as a baseline channel. The two optical fibers transmit the luminous intensity from the same light source, and the luminous intensity of the optical fiber changes with temperature. The integral value of the two luminous intensities is obtained by the symmetrical photodetector on the same silicon chip. This method can eliminate a certain degree of interference and improve the measurement accuracy. In the range of $-40-400$ °C, the temperature measurement accuracy can reach ±0.5 °C.

2. Optical Fiber Speed Sensors

Figure 4.96 shows an optical fiber speed sensor. The convex block rotates with the measured rotating shaft, and when it turns into the lens group, the light path is blocked to form a light pulse signal, and then the light pulse signal is converted into an electric pulse signal by a photoelectric conversion element, and the rotation speed value is obtained by the counter processing.

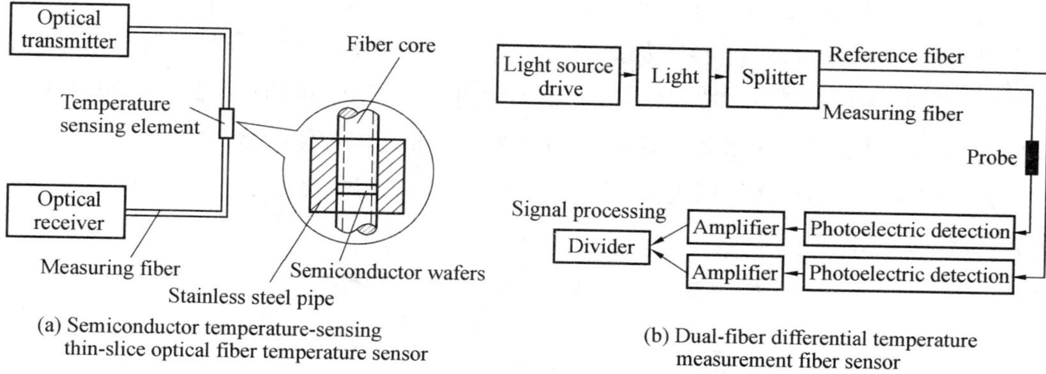

(a) Semiconductor temperature-sensing thin-slice optical fiber temperature sensor

(b) Dual-fiber differential temperature measurement fiber sensor

Figure 4.95 Optical fiber temperature sensors

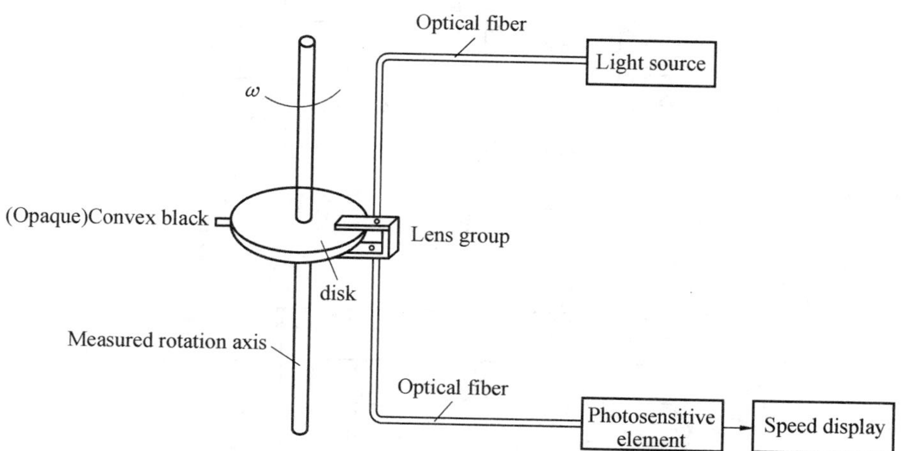

Figure 4.96 Optical fiber speed sensor

3. Light Transmission Type Optical Fiber Displacement Sensors

The working principle of the optical fiber displacement sensor is shown in Figure 4.97.

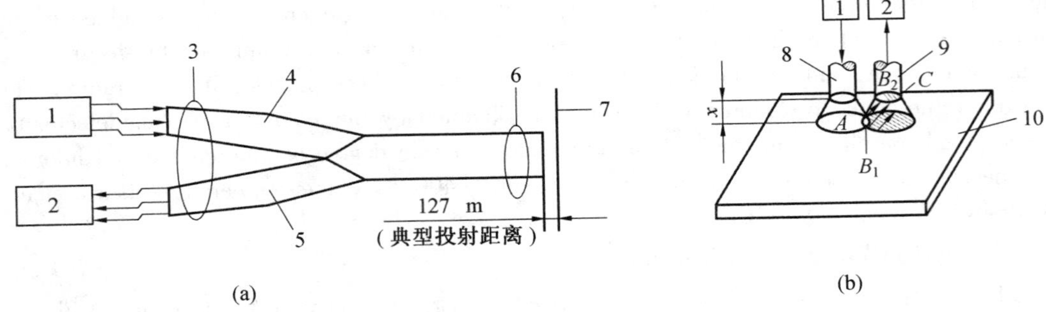

图 4.97　光纤位移传感器工作原理示意图

1— 发光器件;2— 光敏元件;3— 分叉端;4— 发射光纤束;5— 接收光纤束;6— 测量端;7— 被测体;8— 发射光纤;9— 接收光纤;10— 被测面

当被测表面继续远离时,由于被反射光照亮的B_2面积大于接收光纤截面积C,即有部分反射光没有反射进入接收光纤,因此接收到的发光强度逐渐减小,光敏检测器信号逐渐减弱。信号的减弱与探头和被测表面之间的距离平方成反比,可用于距离较远而灵敏度、线性度和精度要求不高的测量。图 4.98 所示为一种光纤液位传感器。

光纤位移传感器在光纤探头前方固定一个膜片可以用来测压了。图 4.99 所示为一种光纤测压传感器。

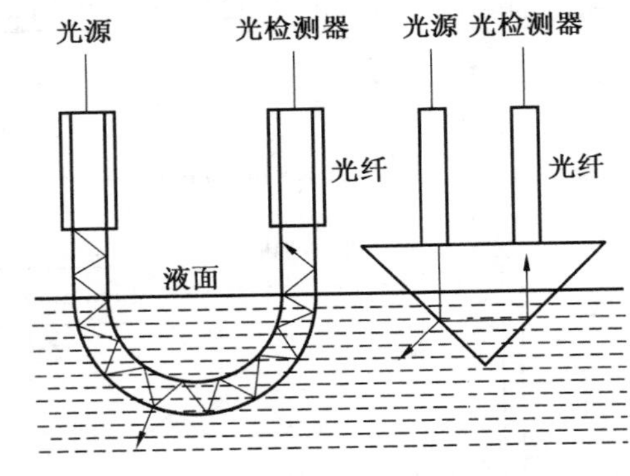

图 4.98　光纤液位传感器

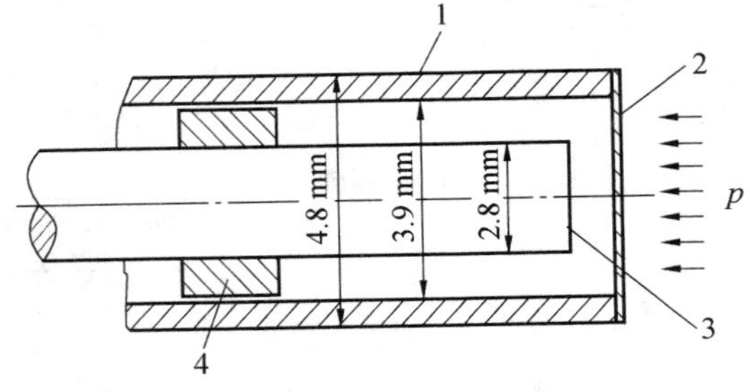

图 4.99　光纤测压传感器

1— 外套;2—0.25 mm 厚膜片;3— 光纤测端;4— 对中套管

When the optical fiber probe is attached to themeasurement piece, the light in the transmitting fiber cannot be reflected into the receiving fiber, and no photoelectric signal can be generated. When the measured surface is gradually away from the optical fiber probe, with the area A of the emitting fiber illuminating the measured surface larger, the corresponding reflected light cone overlap area B_1 will become larger and larger, so the area B_2 illuminated on the end face of the receiving optical fiber does likewise, that is, the stronger the received optical signal, the stronger the photocurrent, which can be used to measure micro-displacement. When the entire receiving fiber end face is fully illuminated, the output signal reaches the "light peak" point on the displacement–output curve. The sensitivity to changes in intensity is much greater than the sensitivity to changes in displacement. It can be used for optical detection of surface conditions at this time.

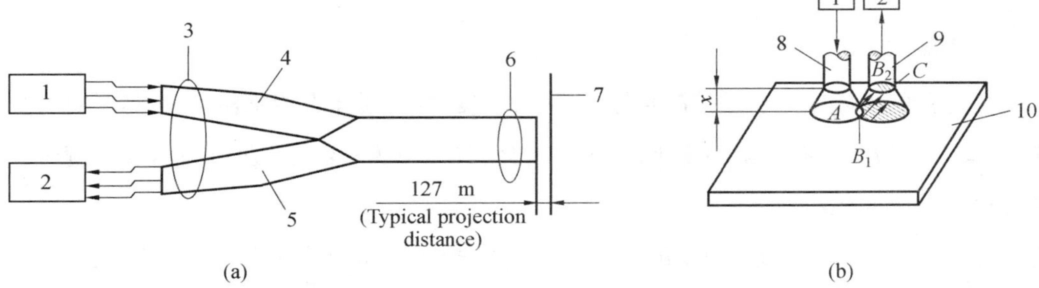

Figure 4.97 Working principle of the optical fiber displacement sensor

1—Light emitting device; 2—Photosensitive element; 3—Fork end; 4—Transmitting fiber bundle; 5—Receiving fiber bundle; 6—Measurement terminal; 7—Measured body; 8—Launching fiber; 9—Receiving fiber; 10—Measured surface

When the measured surface continues to move away, the received luminous intensity gradually decreases and the photosensitive detector signal gradually weakens, for the area B_2 illuminated by the reflected light is larger than the cross-sectional area C of the receiving fiber, that is, part of the reflected light is not reflected into the receiving fiber. The attenuation of the signal is inversely proportional to the square of the distance between the probe and the measured surface. It can be harnessed for measurements that are far away and do not require high sensitivity, linearity and accuracy. Figure 4.98 shows an optical fiber level sensor.

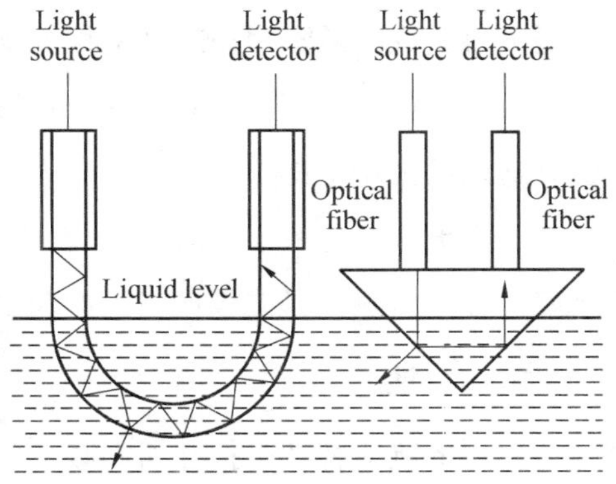

Figure 4.98 Optical fiber liquid level sensor

The optical fiber displacement sensor can be used for pressure measurement by fixing a diaphragm in front of the optical fiber probe. Figure 4.99 shows a optical fiber pressure sensor.

4.10 传感器的选用

了解传感器的结构及其原理后,如何根据测试目的和实际条件正确合理地选用传感器也是需要认真考虑的问题。

选择传感器主要考虑灵敏度、响应特性、线性范围、稳定性、精确度、测量方式等六个方面。

4.10.1 选用原则

1. 灵敏度

传感器的灵敏度一般是越高越好,因为灵敏度越高,意味着传感器所能感知的变化量越小,即只要被测量有一微小变化,传感器就会有较大的输出。但是,在确定灵敏度时,应考虑以下几个问题。

(1) 当传感器的灵敏度很高时,那些与被测信号无关的外界噪声也会同时被检测到,并通过传感器输出,从而干扰被测信号。因此,为既能使传感器检测到有用的微小信号,又能使噪声干扰小,就要求传感器的信噪比越大越好。也就是说,要求传感器本身的噪声小,而且不易从外界引进干扰噪声。

(2) 与灵敏度紧密相关的是量程范围。当传感器的线性工作范围一定时,传感器的灵敏度越高,噪声干扰越大,难以保证传感器的输入在线性区域内工作。过高的灵敏度会影响其适用的测量范围。

(3) 当被测量是一个单向量时,就要求传感器单向灵敏度越高越好,而横向灵敏度越小越好。如果被测量是二维或三维的向量,则还应要求传感器的交叉灵敏度越小越好。

2. 响应特性

传感器的响应特性是指在所测频率范围内保持不失真的测量条件。实际上,传感器的响应不可避免地有一定的延迟,只是希望延迟的时间越短越好。

3. 线性范围

传感器工作在线性区域内,是保证测量精度的基本条件。任何传感器都有一定的线性工作范围。然而,要保证传感器绝对工作在线性区域内是不容易的。在某些情况下,在许可限度内,可以选取其近似线性的区域。例如,变间隙型的电容、电感式传感器,其工作区均选在初始间隙附近。同时,必须考虑被测量的变化范围,令其非线性误差在允许限度

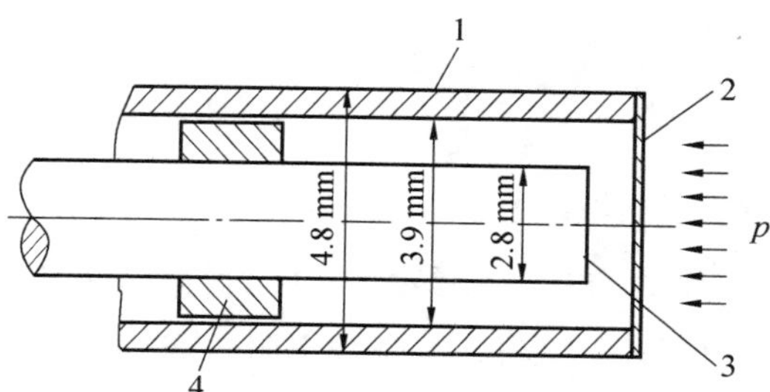

Figure 4.99 Optical fiber pressure sensor
1—Coat; 2—0.25 mm thick film; 3—Optical fiber measuring end; 4—Aligning sleeve

4.10 Sensors Selection

After comprehending the structure and principle of the sensors, how to select a sensor correctly and reasonably according to the measurement purpose and actual conditions is also a problem that needs to be seriously considered.

Selecting a sensor mainly considers six aspects such as sensitivity, response characteristics, linear range, stability, accuracy, measurement methods.

4.10.1 Selection Principle

1. Sensitivity

The sensitivity of the sensor is generally the higher the better, because the higher the sensitivity, the smaller the amount of change the sensor can perceive, that is, as long as there is a small change in the measurement, the sensor will have a larger output. However, when determining the sensitivity, the following issues should be considered.

(1) When the sensitivity of the sensor is high, those external noises that have nothing to do with the measured signal will also be detected at the same time, and output through the sensor, thereby interfering with the measured signal. Therefore, not only detect useful tiny signals, but also to make the noise interference small, the signal-to-noise ratio of the sensor is required as large as possible. In other words, the sensor itself is required to have low noise and it is not easy to introduce interference noise from the outside world.

(2) The measurement range is closely related to the sensitivity. When the linear working range of the sensor is constant, the higher the sensitivity of the sensor, the greater the noise interference, and it is difficult to ensure that the input of the sensor works in the linear region. Too high sensitivity will affect its applicable measurement range.

(3) When the measured is a one-way vector, the unidirectional sensitivity of the sensor is required as high as possible, while the smaller the lateral sensitivity, the better. If the measured is a two-dimensional or three-dimensional vector, the cross sensitivity of the sensor is as small as possible as well.

2. Response Characteristics

The response characteristic of the sensor refers to the measurement condition that remains undistorted within the measured frequency range. In fact, the response of the sensor inevitably has a certain delay, but it is hoped that the delay time is as short as possible.

以内。

4. 稳定性

稳定性是表示传感器经过长期使用以后，其输出特性不发生变化的性能。影响传感器稳定性的因素是时间与环境。为保证稳定性，在选择传感器时，一般应注意以下两点。

（1）根据环境条件选择传感器。例如，选择电阻应变式传感器时，应考虑到湿度会影响其绝缘性、湿度会产生零点漂移、长期使用会产生蠕动现象等。又如，对变极距型电容式传感器，环境湿度的影响或油剂浸入间隙会改变电容器的介质，光电传感器的感光表面有灰尘或水汽时会改变感光性质。

（2）创造或保持一个良好的环境。在要求传感器长期工作而不需要经常更换或校准的情况下，应对传感器的稳定性有严格的要求。

5. 精确度

传感器的精确度表示传感器的输出与被测量的对应程度。实际上，传感器的精确度并非越高越好，还需要考虑测量目的和经济性。因为传感器的精确度越高，其价格就越贵，所以应从实际出发来选择传感器。

6. 测量方法

传感器在实际条件下的工作方式也是选择传感器时应考虑的重要因素。例如，接触与非接触测量、破坏与非破坏性测量、在线与非在线测量等，条件不同，对测量方式的要求也不同。

除以上选用传感器时应充分考虑的一些因素外，还应尽可能兼顾结构简单、体积小、质量轻、价格低廉、易于维修、互换性好等条件。

4.10.2 传感器性能的改善

传感器的性能指标包括很多方面。要使某一传感器的各个指标都优良，不仅设计制造困难，实用时也没有很大的必要性。因此，恰当的办法是根据实际的需要来确保主要性能指标，放宽次要性能指标的要求，从而提高传感器的性价比。在选择使用传感器时，应根据实际测试目的恰当地选用能满足使用要求的产品，切忌盲目追求高指标。同时，在设计、使用传感器时，可采取下列技术措施来改善传感器的性能。

1. 平均技术

在传感器中常用的平均技术有数据平均处理和误差平均效应。误差平均效应的原理是使用 n 个传感器同时被测量，其输出是这些传感器输出的总和的平均值。若将每个传

3. Linear Range

The sensor works in the linear region, which is the basic condition to ensure the measurement accuracy. Any sensor has a certain linear working range. However, it is not easy to ensure that the sensor absolutely works in the linear region. In some cases, it is possible to select an approximate linear region within the allowable limit. For example, for variable-gap capacitive and inductive sensors, the working area is selected near the initial gap. At the same time, the measured variation range must be considered, so that its non-linear error is within the allowable limit.

4. Stability

Stability refers to the performance that the output characteristics of the sensor will not change after long-term use. Time and environment are factors that affect the stability of the sensor. In order to ensure stability, the following two points should generally be noted when selecting a sensor.

(1) Choosing a sensor according to the environmental conditions. For example, choosing a resistance strain sensor, we should consider that humanity will affect its insulation, and will cause zero point drift to cause creeping with long-term use. For another example, variable electrode distance capacitance sensors will be changed due to the influence of environmental humidity or the oil immersed in the gap. When the photosensitive surface of the photoelectric sensor has dust or moisture, it will change the photosensitive properties.

(2) To create or maintain a good environment. When the sensor is required to work for a long time without frequent replacement or calibration, there should be strict requirements on the stability of the sensor.

5. Accuracy

The accuracy of the sensor indicates the degree of correspondence between the output of the sensor and the measured. In fact, the accuracy of the sensor is not as high as possible. The purpose of measurement and economy should also be taken into consideration. Due to the fact that the higher the accuracy of the sensor, the more expensive its price, the sensor should be selected based on the actual situation.

6. Measurement Method

The way the sensor works under actual conditions is also an important factor that should be considered when selecting a sensor. For example, contact and non-contact measurement, destructive and non-destructive measurement, online and off-line measurement, etc., different conditions require different measurement methods.

In addition to the above factors that should be fully considered when selecting the sensor, the conditions of simple structure, small size, light weight, low price, easy maintenance, and good interchangeability, etc. should also be considered as much as possible.

4.10.2 Improvement of Sensor Performance

The performance index of the sensor includes many aspects. It is not only difficult to design and manufacture in an attempt to make each index of a certain sensor excellent, but it is also not necessary in practical use. Therefore, the appropriate method is to ensure the main performance indicators according to actual needs, relax the requirements of the secondary performance indicators, so as to improve the cost-effectiveness of the sensor. When choosing to use a sensor, we should appropriately select products that can meet the requirements of use according to the actual measurement purpose, and avoid blindly pursuing high indicators. At

感器可能带来的误差看作随机误差且服从正态分布,根据误差理论,总的误差将大大减小。例如,当 $n=10$ 时,误差将减小为 31.6%;当 $n=500$ 时,误差将减小为 4.5%。

数据平均处理的做法是,在同样条件下进行 n 次重复测量或 n 次采样,然后求其平均值,随机误差也将减小为原来的 $\frac{1}{\sqrt{n}}$(即 $n^{-0.5}$)。

可见,在传感器中利用平均技术不仅可使传感器误差减小,且可增大信号量,即增大传感器的灵敏度。

2. 差动技术

通常要求传感器输出 — 输入关系成线性,但实际上难以做到。如果输入量变化范围不大,而且非线性项的方次不高,则可以用切线或割线来代替实际曲线的某一段。这种方法为静态特性的线性化,但是这种方法存在很大的局限性。

差动技术是传感器中普遍采用的技术,它可显著减小温度变化、电源波动、外界干扰等对传感器精度的影响,抵消共模误差,减小非线性误差等。不少传感器(如电阻应变式、电感式、电容式等)中因采用了差动技术而提高了传感器的灵敏度。

3. 补偿与修正技术

当传感器的误差变化规律过于复杂,采取一定的技术措施后仍难满足要求,或虽可满足要求,但经济上不合算或技术过于复杂而无现实意义时,可以利用电子线路(硬件)或通过软件进行补偿与修正。

补偿与修正技术在传感器中的应用较多,尤其适用于下面两种情况:一种是针对传感器本身特性的;另一种是针对传感器的工作条件或外界环境的。

对于传感器特性,可以找出误差的变化规律,或者测出其大小和方向,采用适当的方法加以补偿或修正。

针对传感器工作条件或外界环境进行误差补偿是提高传感器精度的有效措施。不少传感器对温度敏感,温度变化引起的误差较大。为解决这个问题,可以由恒温装置来控制温度,但费用太高或使用现场不允许。较可行的方法是先找出温度对测量值影响的规律,然后在传感器内引入温度误差补偿措施,根据外界环境情况修正误差以满足要求。

4. 稳定性处理

传感器作为长期测量或反复使用的器件,其稳定性特别重要,甚至胜过精度指标,尤其是在很难或无法定期鉴定的场合,因为精度只需知道误差的规律就可以进行补偿或修正,稳定性则不然。

the same time, when designing and using the sensor, we can take the following technical measures to improve the performance of the sensor.

1. Average Technique

Commonly used averaging techniques in sensors include data averaging processing and error averaging effect. The principle of the error averaging effect is to use n sensors to be measured at the same time, and its output is the average of the sum of the outputs of these sensors. If the possible errors of each sensor are regarded as random errors and obey normal distribution, the total error will be greatly reduced according to the error theory. For example, when $n=10$, the error will be reduced to 31.6%; when $n=500$, the error will be reduced to 4.5%.

The method of data averaging is to perform n repeated measurements or n samplings under the same conditions, and then to find the average value, the random error will also be reduced by $\frac{1}{\sqrt{n}}$ (that is, $n^{-0.5}$) times the original value.

It can be seen that the use of averaging technology in the sensor can not only reduce the sensor error, but also increase the amount of signal, that is, increase the sensitivity of the sensor.

2. Differential Technology

The sensor output-input relationship is usually required to be linear, but it is difficult to achieve in practice. If the input variable range is not large, and the nonlinear term is not high, we can use tangent or secant to replace a certain segment of the actual curve. This method is the linearization of static characteristics, but this method has great limitation.

Differential technology is a technology commonly used in sensors, it can significantly reduce the impact of temperature changes, power fluctuations, and external interference, etc. on the accuracy of the sensor, offset common mode errors, and reduce nonlinear errors, etc. Many sensors (such as resistance strain type, inductive type, capacitive type) have improved the sensitivity of the sensor due to the use of differential technology.

3. Compensation and Correction Technology

The electronic circuit (hardware) or software can be used for compensation and correction when the error change law of the sensor proves too complicated, it is still difficult to meet therequirements after taking certain technical measures, or although the requirements can be met, but it is not economically cost-effective or the technology is too complex to be of practical significance.

There are many applications of compensation and correction technology in sensors, especially in the following two situations: one is for the characteristics of the sensor itself; and the other is for the working conditions of the sensor or the external environment.

For the sensor characteristics, it is a good way to find out the change rule of the error, or measure its size and direction, and use appropriate methods to compensate or correct it.

It is an effective measure to carry on the error compensation according to the working conditions of the sensor or the external environment in order to improve the accuracy of the sensor. Many sensors are sensitive to temperature, and the error caused by temperature changes is relatively large. In order to solve this problem, the temperature can be controlled by a thermostat, but the cost is too high or the site does not allow it. A more feasible method is to first find out the law of the influence of temperature on the measured value, and then introduce temperature error compensation measures in the sensor, and correct the error according to the

造成传感器性能不稳定的原因是:随着时间的推移和环境条件的变化,构成传感器的各种材料与元器件性能发生了变化。为提高传感器性能的稳定性,应对材料、元器件或传感器整体进行必要的稳定性处理,如结构材料的时效处理、冰冷处理,永磁材料的时间老化、温度老化、机械老化及交流稳磁处理,电气元件的老化与筛选等。

在使用传感器时,若测量要求较高,必要情况下还应对附加的调整元件和后续电路中的关键元器件进行老化处理和筛选。

5. 屏蔽、隔离与干扰抑制

传感器大都需要在现场工作,现场的条件往往难以充分预料,有时是极其恶劣的。传感器输入信号中除被测量外,外界各种干扰因素势必影响传感器的精度及其他性能。为减小测量误差,保证其原有性能,应设法削弱或消除外界干扰因素对传感器的影响。有两种方法:一是减少传感器对影响因素的灵敏度或者影响传感器灵敏度的因素;二是降低外界干扰因素对传感器实际作用的烈度或功率。

对电磁干扰,可采用屏蔽、隔离措施,也可用滤波方法加以抑制。但由于传感器是感受非电量的器件,因此还应考虑与被测量有关的其他影响因素,如湿度、温度、辐射、机械振动、气流等。为此,需采用相应的隔离措施(如隔热、隔振、密封等),或者在变为电量后对干扰信号进行分离或抑制,减小其影响。

6. 集成化与智能化技术

集成化、智能化的结果可扩大传感器的功能,改善其性能,从而提高其性价比。

思考题与练习题

4.1　举例说明生活和学习中用到的一些传感器,并说明各属于什么类型的传感器。

4.2　某一造纸生产线上需要测量纸张的厚度。请问应该选择什么样的传感器,为什么?

4.3　收集资料,并说明要测量某汽轮发电机转子的振动(振动幅值约为 10 mm,振动频率为 60 Hz,温度小于 120 ℃),可以选择何种传感器。

4.4　自动售/检票系统里需要哪些传感器?其作用分别是什么?

4.5　为什么说极距变化型电容传感器是非线性的?采取什么措施可改善其非线性特性?

external environment to meet the requirements.

4. Stability Processing

As a device for long-term measurement or repeated use, the stability of the sensor is particularly important, even better than the accuracy index, especially in occasions that are difficult or impossible to identify regularly, because accuracy can be compensated or corrected only by knowing the law of error, but stability is not.

The reason for the unstable performance of the sensor is: the performance of various materials and components that constitute the sensor has changed with the passage of time and changes in environmental conditions. In order to improve the stability of sensor performance, the necessary stability treatment should be carried out on materials, components, or the entire sensor, such as aging treatment and cold quenching of structural materials, time aging, temperature aging, mechanical aging and AC magnetization treatment of permanent magnet materials, aging and screening of electrical components.

When using sensors, if the measurement requirements are high, additional adjustment components and key components in the subsequent circuit should be aged and screened if necessary.

5. Shielding, Isolation and Interference Suppression

Most sensors are required to work on site with the conditions on site being often difficult to fully predict, sometimes extremely harsh. In addition to being measured in the sensor input signal, various external interference factors will inevitably affect the accuracy and other performance of the sensor. In order to reduce the measurement error and ensure its original performance, we should try to weaken or eliminate the influence of external interference factors on the sensor. There are two methods: one is to reduce the sensitivity of the sensor to influencing factors or the factors that affect the sensitivity of the sensor; the other is to reduce the intensity or power of the actual effect of external interference factors on the sensor.

For electromagnetic interference, shielding and isolation measures can be used, and filtering methods can also be used to suppress. However, since the sensor is a device that senses non-electricity, other influencing factors related to the measurement should also be considered, such as humidity, temperature, radiation, mechanical vibration, and airflow. For this reason, it is necessary to adopt corresponding isolation measures (such as heat insulation, vibration isolation, and sealing), or to separate or suppress the interference signal after it is changed into electricity to reduce its impact.

6. Integrated and Intelligent Technology

The result of integration and intelligence can expand the function of the sensor, improve its performance, thereby increasing its cost performance.

Questions and Exercises

4.1 Give examples of some sensors used in your life and study and explain the types they belong to.

4.2 A paper production line needs to measure the thickness of paper. What kind of sensors should be selected, and why?

4.3 Collect the data, and describe what kind of sensors can be selected to measure the vibration of a turbine generator rotor (vibration amplitude is about 10 mm, vibration frequency is 60 Hz, temperature is less than 120 ℃).

4.6　如果被测物体的材质是塑料，可否用电涡流式传感器测量该物体的位移？为对该物体进行位移测量，应采取什么措施？需要考虑哪些问题？

4.7　在用应变仪测量机构的应力、应变时，如何消除温度变化产生的影响？

4.8　楼梯上的电灯如何能人来就开、人走就熄？

4.9　工业生产中所用的自动报警器、恒温烘箱是如何工作的？

4.10　现要测量机床主轴的振动，请问可以选择什么类型的传感器，为什么？

4.11　金属电阻应变片与半导体应变片在工作原理上有何不同？使用时应如何进行选用？

4.12　为什么说压电式传感器只适用于动态测量而不能用于静态测量？

4.13　热电偶是如何实现温度测量的？影响热电势与温度之间关系的因素是什么？

4.14　光电效应有哪几种？与之对应的光电元件各有哪些？光纤传感器有哪些优点？

4.15　可用于实现非接触式测量的传感器有哪些？

4.16　霍尔效应的本质是什么？用霍尔元件可测哪些物理量？请举例说明。

4.17　差动式传感器的优点是什么？

4.18　哪些传感器可选作小位移传感器？

4.19　选择或购置传感器时需注意哪些事项？

4.20　某电容测微仪，其传感器的圆形极板半径为 $r = 4$ mm，工作初始极板间距离 $\delta_0 = 0.3$ mm，介质为空气。

(1) 工作时，若传感器与被测体之间距离的变化量 $\Delta\delta = \pm 1$ μm，电容的变化量为多少？

(2) 若测量电路的灵敏度为 $S_1 = 100$ mV/pF，读数仪表的灵敏度 $S_2 = 5$ 格/mV，则在 $\Delta\delta = \pm 1$ μm 时，读数仪表的指示值变化多少格？

4.21　压电式传感器的灵敏度 $S_1 = 10$ pC/MPa，连接灵敏度为 $S_2 = 0.008$ V/pC 的电荷放大器，所用的笔式记录仪的灵敏度为 $S_3 = 25$ mm/V。当压力变化 $\Delta p = 8$ MPa 时，记录笔在记录纸上的偏移量为多少？

4.22　将一只灵敏度为 0.3 mV/℃ 的热电偶与毫伏表相连，已知接线端温度(即冷端温度)为 30 ℃，毫伏表的输出为 30 mV，则热电偶热测温端的温度为多少(考虑该热电偶为线性)？

4.4 What sensors are needed in the automatic ticket sales/check-in system? What are their roles?

4.5 Why is it said that the variable gap type capacitance sensor is non-linear? What measures can be taken to improve its nonlinear characteristics?

4.6 If the material of the measured object is plastic, can an eddy current sensor be used to measure the displacement of the object? What measures should be taken in order to be able to measure the displacement of the object? What issues need to be considered?

4.7 When measuring the stress and strain of a mechanism with a strain gauge, how to eliminate the influence caused by temperature changes?

4.8 How can the lights on the stairs turn on when people come and turn off when people walk?

4.9 How do the automatic alarms and constant temperature ovens used in industrial production work?

4.10 What type of sensors can be chosen to measure the vibration of the machine tool spindle, and why?

4.11 What is the difference in working principle between metal resistance strain gauges and semiconductor strain gauges? How to select when using it?

4.12 Why is piezoelectric sensor only suitable for dynamic measurement but not static measurement?

4.13 How does thermocouple realize temperature measurement? What are the factors that affect the relationship between thermoelectric potential and temperature?

4.14 How many kinds of photoelectric effects are there? What are the corresponding optoelectronic components? What are the advantages of fiber optic sensors?

4.15 What sensors can be used to achieve non-contact measurement?

4.16 What is the nature of the Hall effect? What physical quantities can be measured with Hall elements? Please give examples.

4.17 What are the advantages of differential sensors?

4.18 Which sensors can be selected as small displacement sensors?

4.19 What matters should be paid attention to when selecting or purchasing a sensor?

4.20 A certain capacitance micrometer, the sensor's circular plate $r=4$ mm, the initial working distance $\delta_0 = 0.3$ mm between the plates, and the medium is air.

(1) What is the change in capacitance when working if the distance between the sensor and the measured body is changed by $\Delta\delta = \pm 1\ \mu m$?

(2) If the sensitivity of the measuring circuit is $S_1 = 100$ mV/pF , the sensitivity of the reading instrument is $S_2 = 5$ grid/mV , how many grids does the indicating value of the reading meter change when $\Delta\delta = \pm 1\ \mu m$?

4.21 The piezoelectric sensor's sensitivity $S_1 = 10$ pC/MPa , and it is connected to the charge amplifier with sensitivity $S_2 = 0.008$ V/pC , the sensitivity of the pen recorder used is $S_3 = 25$ mm/V . What is the offset of the pen on the recording paper when the pressure changes by $\Delta p = 8$ MPa ?

4.22 Connect a thermocouple with a sensitivity of 0.3 mV/°C to a millivoltmeter. The terminal temperature (that is, the cold junction temperature) is known to be 30 °C, and the output of the millivoltmeter is 30 mV, then what is the temperature of the thermocouple's thermal temperature measurement end (consider that the thermocouple is linear)?

第5章　信号分析与处理

【本章学习目标】

1. 掌握数字信号处理系统的基本组成；

2. 掌握模拟信号和数字信号之间的转换，了解信号数字化处理过程的问题及其解决方法；

3. 了解随机信号的特征，了解信号的相关分析方法；

4. 了解信号的功率谱分析；

5. 了解常用的现代信号处理方法。

Chapter 5　Signal Analysis and Processing

【Learning Objectives】

1. To be able to master the basic composition of digital signal processing system;

2. To be able to grasp the conversion between analog signals and digital signals, to comprehend the problems of signal digitization processing and their solutions;

3. To be able to understand the characteristics of random signals and correlation analysis method of signals;

4. To be able to know the power spectrum analysis of signals;

5. To be able to be familiar with commonly used modern signal processing methods.

信号的分析与处理可分为两大类:模拟信号处理和数字信号处理。输入、输出都是模拟信号的处理系统称为模拟信号处理系统;输入输出都是数字信号的处理系统称为数字信号处理系统。

在工程测试中,信号的频域功率谱、时域相关、幅值域概率密度函数等分析的应用广泛。然而,这些分析若用模拟仪器进行,一则难以实现,二则分析误差较大。随着计算机技术的发展,信号处理的方法已由模拟技术逐渐转向数字技术,即用计算机和适当的软件来对信号进行分析和处理。

数字信号处理具有一系列的优点,如传输时有较高的抗干扰性、易于存储和处理方便等。因此,人们努力使测试信号直接数字化或者将模拟信号及早数字化。20 世纪 70 年代以来,计算机、微电子等技术迅猛发展并逐步渗透到测试和仪器仪表技术领域。在此技术推动下,测试技术与仪器得到了迅速发展和进步,相继出现了智能仪器、总线仪器、VXI 仪器、PC 仪器、虚拟仪器及互换性虚拟仪器等微机化仪器及自动测试系统。

与计算机技术紧密结合成为当今仪器与测控技术发展的主潮流。配以相应软件和硬件的计算机将能够完成许多仪器、仪表的功能,实质上相当于一台多功能的通用测试仪器。这样的现代仪器设备的功能已不再由按钮和开关的数量来限定,而是取决于其存储器内装有软件的数量。因此,数字信号处理技术对于计算机测试系统来说尤为重要。本章将研究如何利用计算机实现测试信号的处理。

5.1 数字信号处理系统的基本组成

由于要用数字计算机进行信号处理,因此需对连续信号进行离散和数字化。数字信号处理的一般过程如图 5.1 所示。

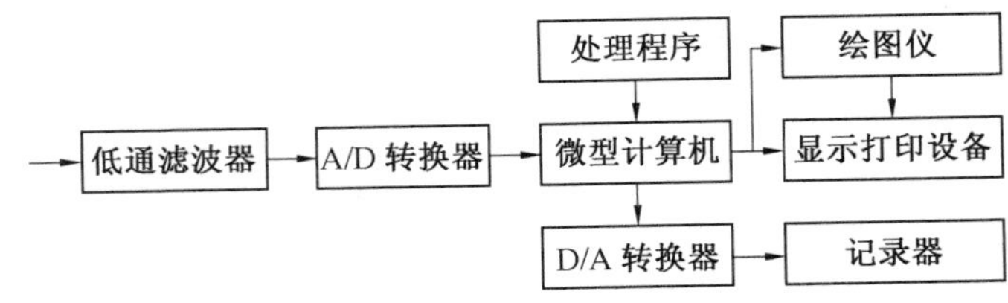

图 5.1 数字化处理的一般过程

Signal analysis and processing can be divided into two categories: analog signal processing and digital signal processing. A processing system whose input and output are both analog signals is called an analog signal processing system; a processing system whose input and output are both digital signals is called a digital signal processing system.

In engineering measurements, the analysis of signal power spectrum in frequency domain, time domain correlation, and amplitude domain probability density function is widely used. However, these analyses are carried out by analog instruments will face two problems that one is difficult to achieve, and the other is that the analysis error is large. With the development of computer technology, signal processing methods have gradually shifted from analog technology to digital technology, that is, using computers and appropriate software to analyze and process signals.

Digital signal processing has a series of advantages, such as high anti-interference performance during transmission, easy storage and convenient processing. Therefore, efforts are made to digitize the measurement signal directly or to digitize the analog signal as soon as possible. Since the 1970s, technologies such as computers and microelectronics have developed rapidly and gradually penetrated into the field of measurement and instrumentation technology. Driven by this technology, measurement technology and instruments have been rapidly developed and progressed, and intelligent instruments, bus instruments, VXI instruments, PC instruments, virtual instruments and interchangeable virtual instruments and other microcomputerized instruments and automatic measurement systems have appeared one after another.

With close integration with computer technology, it has become the main trend in the development of today's instrumentation and measurement and control technology. A computer equipped with corresponding software and hardware will be able to complete the functions of many instruments and meters, which is essentially equivalent to a multi-functional universal measurement instrument. The functions of such modern instruments are no longer limited by the number of buttons and switches, but depend on the amount of software installed in their memory. Therefore, digital signal processing technology is particularly important for computer measurement systems. This chapter will study how to use computers to process measurement signals.

5.1 Basic Composition of Digital Signal Processing System

Since a digital computer will be used for signal processing, it is necessary to discretize and digitize the continuous signal. The general process of digital signal processing is shown in Figure 5.1.

5.1.1 Signal Preprocessing

Before analyzing and processing the measured signals, it is often necessary to analyze and sort out the signals first to eliminate abnormal point data and trend item data, and then perform

5.1.1 信号的预处理

在对所测得的信号进行分析处理前，常常需首先对信号进行分析整理，消除异常点数据和趋势项数据，然后才可进行正常的处理工作。经过整理后的模拟信号，首先通过一个低通滤波器，这个滤波器有时又称抗混滤波器，它会滤掉高频干扰信号及不必要的高频分量，然后通过 A/D 转换器进行采样并转换成数字量。这个被数字化的信号在微型计算机中按照程序完成所要求的各种处理计算工作，然后根据需要得到数字形式或模拟形式的输出，如打印数据、绘出谱图或显示波形等。

5.1.2 多路模拟开关

实际的测试系统通常需要进行多参量的测量，即采集来自多个传感器的输出信号。如果每一路信号都采取独立的输入回路(信号调理、A/D 转换、采样保持)，则系统成本将比单路成倍增加，而且系统体积庞大。同时，由于模拟器件和阻容元件的参数、特性不一致，因此给系统的校准带来很大的困难。为此，通常采用多路模拟开关来实现信号测量通道的切换，将多路输入信号分时输入公用的输入回路进行测量。

目前，常采用 CMOS 场效应模拟电子开关。尽管模拟电子开关的导通电阻受电源、模拟信号电平和环境温度变化的影响会发生改变，但是与传统的机械触点式开关相比，其体积小、功耗低、容易集成、速度快且没有机械式开关抖动现象。CMOS 场效应模拟电子开关的导通电阻一般在 200 Ω 以下，关断时漏电流一般可达纳安级甚至皮安级，开关时间通常为数百纳秒。

5.1.3 A/D 转换与 D/A 转换

将模拟量转换成与其对应的数字量的过程称为模数(A/D) 转换；反之，则称为数模(D/A) 转换。实现上述过程的装置分别称为 A/D 转换器和 D/A 转换器。A/D 和 D/A 转换是数字信号处理的必需程序。通常所用的 A/D 和 D/A 转换器，其输出的数字量大多用二进制编码表示，以与计算机技术相适应。

随着大规模集成电路技术的发展，各种类型的 A/D 和 D/A 转换芯片已大量供应市场，其中大多数是采用电压数字转换方式，输入输出的模拟电压也都标准化，如单极性 0 ~ 5 V、0 ~ 10 V 或双极性 ±5 V、±10 V 等，给使用带来极大方便。

1. A/D 转换

A/D 转换过程包括三个步骤：采样、量化和编码。采样即将连续时间信号离散化。采

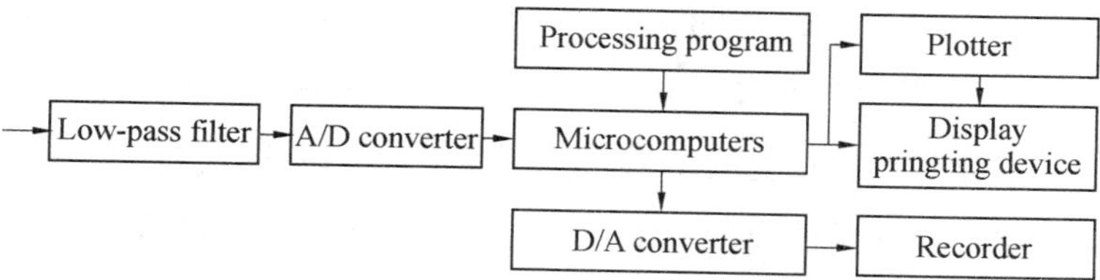

Figure 5.1　The general process of digital processing

normal processing. The processed analog signal first passes through a low-pass filter, sometimes called an anti-aliasing filter that will filter out high-frequency interference signals and unnecessary high-frequency components, and then the signal is sampled and converted into digital quantity by an A/D converter. This digitized signal is completed in a microcomputer in accordance with the program to complete various processing and calculation tasks required, and then output in digital or analog form as needed, such as printing data, drawing spectra or displaying waveforms.

5.1.2　Multi-Channel Analogue Switch

The actual measurement system usually needs to perform multi-parameter measurement, that is, collecting the output signals from multiple sensors. If each signal adopts an independent input loop (signal conditioning, A/D conversion, sample and hold), then the system cost will be multiplied compared to the single-channel, and the system is bulky. At the same time, due to the inconsistent parameters and characteristics of the analog device and the resistor-capacitor unit, it brings great difficulties to the calibration of the system. For this reason, multiple analog switches are usually used to switch the signal measurement channels, and the multiple input signals are time-shared into a common input circuit for measurement.

At present, CMOS field-effect analog electronic switches are often adopted. Although the on-resistance of the analog electronic switch is affected to change by the power supply, analog signal level, and environmental temperature changes, it is small in size, low in power consumption, easy to integrate, fast, and without mechanical switch jitter phenomenon compared with the traditional mechanical contact switch. The on-resistance of CMOS field-effect analog electronic switches is generally below 200 Ω, and the leakage current during turn-off can generally reach the nanoampere or even picoampere level, and the switching time is usually hundreds of nanoseconds.

5.1.3　A/D Conversion and D/A Conversion

The process of converting an analog quantity into its corresponding digital quantity is called analog-to-digital (A/D) conversion; and vice versa, it is called digital-to-analog (D/A) conversion. The devices implementing the above process are called A/D converters and D/A converters, respectively. A/D and D/A conversion are necessary procedures for digital signal processing. The digital output of the commonly used A/D and D/A converters is mostly expressed in binary codes to be compatible with computer technology.

With the development of large-scale integrated circuit technology, various types of A/D

样后，信号在幅值上仍然是连续取值的，必须进一步通过幅值量化转换为幅值离散的信号。若信号 $x(t)$ 可能出现的最大值为 A，令其分为 d 个间隔，则每个间隔大小为 $q = A/d$，q 称为量化当量或量化步长。量化的结果是将连续信号幅值通过舍入或截尾的方法表示为量化当量的整数倍。量化后的离散幅值需通过编码表示为二进制数字以适应数字计算机处理的需要，即

$$A = qD$$

式中　D—— 编码后的二进制数。

显然，经过上述量化和编码后得到的数字信号，其幅值必然带来误差，这种误差称为量化误差。当采用舍入量化时，最大量化误差为 $\pm 1/2$；而采用截尾量化时，最大量化误差为 $-q$。量化误差的大小一般取决于二进制编码的位数，因为它决定了幅值被分割的间隔数量 d。例如，采用 8 位二进制编码时，$d = 2^8 = 256$，即量化当量为最大可测信号幅值的 1/256。

2. D/A 转换

D/A 转换器将输入的数字量转换为模拟电压或电流信号输出，其基本要求是输出信号 A 与输入数字量 D 成正比，即

$$A = qD \tag{5.1}$$

式中　q—— 量化当量，即数字量的二进制码最低有效位所对应的模拟信号幅值。

根据二进制计数方法，一个数是由各位数码组合而成的，每位数码均有确定的权值，即

$$D = 2^{n-1}a_{n-1} + 2^{n-2}a_{n-2} + \cdots + 2^{i}a_{i} + \cdots + 2^{1}a_{1} + 2^{0}a_{0} \tag{5.2}$$

式中　a_i—— 二进制数的第 i 位，等于 0 或 1（$i = 0,1,\cdots,n-1$）。

为将数字量表示为模拟量，应将每一位代码按其权值大小转换成相应的模拟量，然后根据叠加原理将各位代码对应的模拟分量相加，其和即与数字量成正比的模拟量。这就是 D/A 转换的基本原理。

从 D/A 转换器得到的输出电压值 U 是转换指令来到时刻的瞬时值。不断转换可得到各个不同时刻的瞬时值，这些瞬时值的集合对一个信号而言在时域仍是离散的。要将其恢复为原来的时域模拟信号，还必须通过保持电路进行波形复原。

保持电路在 D/A 转换器中相当于一个模拟存储器，其作用是在转换间隔的起始时刻接收 D/A 转换输出的模拟电压脉冲，并保持到下一转换间隔的开始（零阶保持器）。

and D/A conversion chips have been supplied to the market in large quantities, most of which will adopt voltage-to-digital conversion methods, and the input and output analog voltages are also standardized, such as unipolar 0–5 V, 0–10 V, or bipolar ±5 V, ±10 V, with bringing great convenience to use.

1. A/D Conversion

The A/D conversion process includes three steps: sampling, quantization and encoding. It is recognized that sampling is to discretize a continuous time signal. After sampling, the signal is still continuous in amplitude, and must be further converted into a discrete amplitude signal through amplitude quantization. If the maximum possible value of signal $x(t)$ is A, and it is divided into d intervals, then the size of each interval is $q = A/d$, and q is called quantity equivalent or quantization step size. The result of quantization is to express the continuous signal amplitude as an integer multiple of the equivalent quantization by rounding or truncation. The quantized discrete amplitude needs to be encoded as a binary number to meet the needs of digital computer processing, namely

$$A = qD$$

where　D—the encoded binary number.

Obviously, the amplitude of the digital signal obtained after the above-mentioned quantization and encoding will inevitably bring errors, and this error is called quantization error. When using rounding quantization, the maximum quantization error is ±1/2; when using truncated quantization, the maximum quantization error is $-q$. The size of the quantization error generally depends on the number of bits of the binary code, because it determines the number of intervals d by which the amplitude is divided. For example, when 8-bit binary coding is used, $d = 2^8 = 256$, that is, the quantization equivalent is 1/256 of the maximum measurable signal amplitude.

2. D/A Conversion

The D/A converter converts the input digital quantity into an analog voltage or current signal output, whose basic requirement is that the output signal A is proportional to the input digital quantity D, that is

$$A = qD \tag{5.1}$$

where　q —quantity equivalent, that is, the amplitude of the analog signal corresponding to the least significant bit of the binary code of the digital quantity.

According to the binary counting method, a number is composed of digits, and each digit has a certain weight, namely

$$D = 2^{n-1}a_{n-1} + 2^{n-2}a_{n-2} + \cdots + 2^i a_i + \cdots + 2^1 a_1 + 2^0 a_0 \tag{5.2}$$

where　a_i —the ith bit of a binary number, equals to 0 or 1 ($i=0,1,\cdots, n-1$).

In order to express a digital quantity as an analog quantity, each bit of code should be converted into a corresponding analog quantity according to its weight, and then the analog components corresponding to each code are added according to the principle of superposition, and the sum is proportional to the digital quantity. It is the basic principle of D/A conversion.

The output voltage value U obtained from the D/A converter is the instantaneous value at the moment when the conversion command comes. Continuous conversion can get the

5.1.4 采样保持(S/H)

在对模拟信号进行A/D变换时,从启动变换到变换结束需要一定的时间,即A/D转换器的孔径时间。当输入信号频率较高时,由于孔径时间的存在,因此会造成较大的孔径误差。要防止这种误差的产生,必须在A/D转换开始时将信号电平保持不变,而在A/D换结束后又能跟踪输入信号的变化,即输入信号处于采样状态。能完成上述功能的器件称为采样保持器。由上述分析可知,采样保持器在保持阶段相当于一个"模拟信号存储器"。在A/D转换过程中,采样保持对保证A/D转换的精确度具有重要作用。

实际系统中,是否需要采样保持电路取决于模拟信号的变化频率和A/D转换时间。通常,对直流或缓变低频信号进行采样时可不用采集保持电路。

5.2 模拟信号和数字信号之间的转换

5.2.1 数字信号处理过程举例

这里介绍一种工程上较实用的导出离散傅里叶变换(Discrete Fourier Transform, DFT)的方法,来说明信号数字化的过程。以一个时域的模拟信号在计算机中的傅里叶变换获得频谱为例进行说明。它从连续时间信号的傅里叶变换出发,通过对时间域信号及频率域频谱分别进行采样而使其离散化,并在时间域进行截断使其限于有限区间,从而导出在计算机上可以实现的离散傅里叶变换。

图5.2所示为DFT的图解法推演过程。更严密的数学推导可参阅有关信号处理等的书籍。图5.2(a)所示为一连续时间函数$x(t)$,它是一个单向指数衰减函数。求该函数的傅里叶变换,记为$X(f)$。

若要求在计算机上分析,就必须首先对$x(t)$进行采样,使其离散化。采样的实质就是在时间域将$x(t)$乘以图5.2(b)所示的采样函数$\delta_0(t)$,它是周期为T的δ函数序列,即采样频率$f_s=1/T$,其傅里叶变换如图5.2(b)右边所示。时域的采样结果即图5.2(c)所示的离散函数$x(t)\delta_0(t)$。

根据卷积定理,在时间域两函数相乘对应于其频率域的卷积,即$x(t)\delta_0(t)\rightleftharpoons X(f)*\Delta_0(f)$。显然,经过采样后得到的离散函数$x(t)\delta_0(t)$的频谱$X(f)*\Delta_0(f)$与原连续函数$x(t)$的频谱$X(f)$是不同的,并有可能在频率域的$1/T$附近出现重叠现象,从而产生误差。为避免这一误差,必须满足$f_s>2f_{max}$,此处f_{max}表示原函数$x(t)$所包含的最高频率成

instantaneous value at different moments. The set of these instantaneous values is still discrete in the time domain for a signal. The waveform must be restored through the hold circuit to restore it to the original time-domain analog signal.

The hold circuit is equivalent to an analog memory in the D/A converter, whose function is to receive the analog voltage pulse output by the D/A conversion at the beginning of the conversion interval and hold it until the beginning of the next conversion interval (zero-order holder).

5.1.4 Sample and Hold (S/H)

When performing A/D conversion on an analog signal, it takes a certain amount of time from the start of the conversion to the end of the conversion, that is, the aperture time of the A/D converter. When the input signal frequency is high, a larger aperture error will be caused due to the existence of the aperture time. To prevent this kind of error, the signal level must be kept unchanged at the beginning of the A/D conversion, and the change of the input signal can be tracked after the A/D conversion ends, that is, input signal is in sampling state. The device that can complete the above-mentioned functions is called a sampling holder. It can be seen from the above analysis that the sampling holder is equivalent to an "analog signal memory" in the hold phase. In the A/D conversion process, sample and hold plays an important role in ensuring the accuracy of the A/D conversion.

In the actual system, whether a sample and hold circuit is needed depends on the changing frequency of the analog signal and the A/D conversion time. Usually, the acquisition and hold circuit is not necessary when sampling DC or slowly changing low-frequency signals.

5.2 Conversion Between Analog Signals and Digital Signals

5.2.1 Examples of Digital Signal Processing

Here is a more practical method of deriving discrete Fourier transform (DFT) in engineering to illustrate the process of signal digitization. Take the Fourier transform of a time-domain analog signal in the computer to obtain the frequency spectrum as an example. It starts from the Fourier transform of the continuous time signal, discretizes it by sampling the time domain signal and the frequency domain spectrum separately, and truncates it in the time domain to limit it to a finite interval, in order to derive the discrete Fourier transform that can be realized on the computer.

Figure 5.2 shows the diagrammatic deduction process of DFT. Readers can see books on signal processing of more rigorous mathematical derivation. Figure 5.2(a) shows a continuous time function $x(t)$, which is a one-way exponential decay function. Find the Fourier transform of this function and denote it as $X(f)$.

If we want to analyze on the computer, $x(t)$ must first be sampled and discretized. The essence of sampling is to multiply $x(t)$ by the sampling function $\delta_0(t)$ shown in Figure 5.2(b) in the time domain. It is a sequence of δ functions with a period of T , that is, the sampling fre-

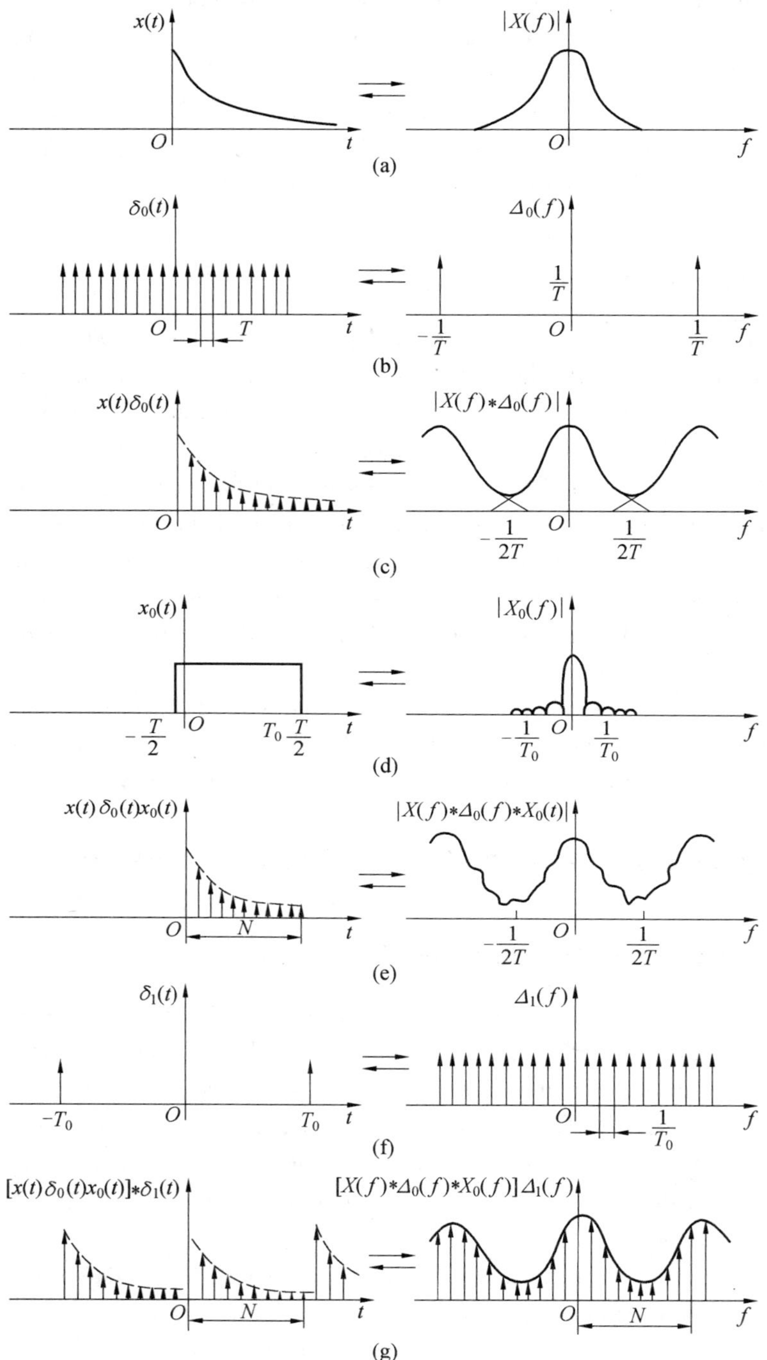

图 5.2 DFT 的图解法推演过程

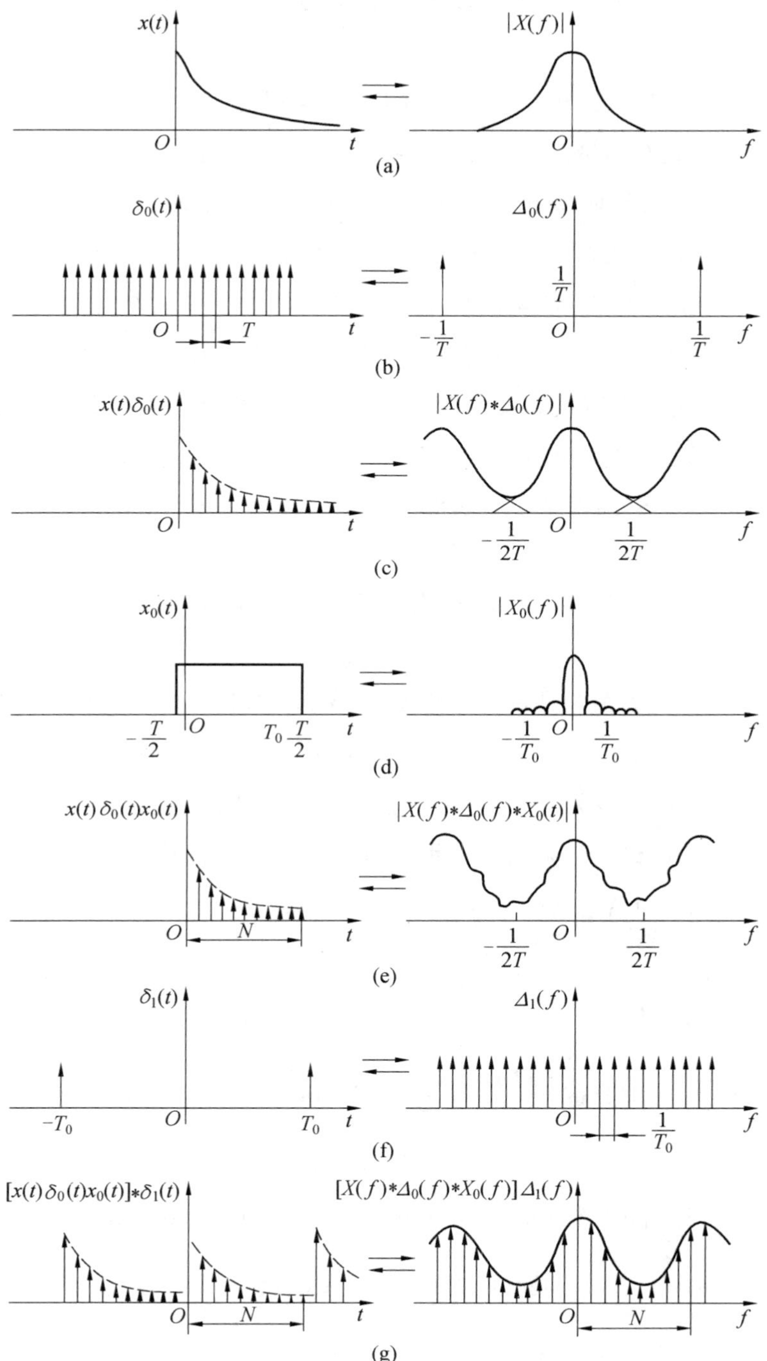

Figure 5.2　Diagrammatic deduction process of DFT

quency $f_s = 1/T$, and its Fourier transform is shown on the right side of Figure 5.2(b). The sampling result in the time domain is the discrete function $x(t)\Delta_0(t)$ shown in Figure 5.2(c).

分。这又从另一个侧面证明了采样定理:采样频率必须高于被测信号所包含最高频率的两倍。

至此,采样后的离散函数$x(t)\delta_0(t)$仍有无限个采样点,而计算机只能接收有限个点,因此要将$x(t)\delta_0(t)$进行时域截断,取出有限的N个点。这相当于用宽度为T_0的矩形窗口函数$x_0(t)$与被测信号时域相乘。图5.2(d)所示为该窗口函数$x_0(t)$及其频谱$X_0(f)$。同样,根据卷积定理,N个有限点的离散函数$x(t)\delta_0(t)x_0(t)$的频谱应等于$|X(f)*\Delta_0(f)*X_0(f)|$,即频域的卷积,如图5.2(e)所示。综上所述,由于矩形窗函数$x_0(t)$的傅里叶变换$X_0(f)$是一个抽样函数,即sinc函数,因此与它做卷积必然会出现图5.2(e)所示的旁瓣。要减少因此而带来的误差,增加截断长度T_0是有利的。

图5.2(e)所示的傅里叶变换对中,频谱函数是连续函数,这仍不是计算机可接收的。为此,还要将其离散化,即乘以频率采样函数$\Delta_1(f)$。同样,按卷积定理,频率域两函数相乘对应于时间域要做卷积,如图5.2(e)、(f)、(g)所示,此处频率采样函数$\Delta_1(f)$的采样间隔应为$1/T_0$,以保证在时间域做卷积时不会产生重叠。$f_0=1/T_0$表示频率分辨率。

这样,图5.2(g)已经成为计算机可接收的离散傅里叶变换对,它们在时间域和频率域上均周期化了。分别取一个周期的N个时间采样值与N个频率值相对应,即$T_0=NT$,从而导出了与原来连续函数$x(t)$及傅里叶变换$X(f)$相当的有限离散傅里叶变换对,分别称为离散傅里叶变换(DFT)和离散傅里叶逆变换(IDFT),即

$$
\begin{aligned}
X_\mathrm{d}(\mathrm{j}k\Delta\omega) &= X_\mathrm{d}\left(\mathrm{j}\frac{2\pi}{NT}k\right) = \sum_{n=0}^{N-1}(XnT)\mathrm{e}^{-\mathrm{j}k\Delta\omega nT} \\
&= \sum_{n=0}^{N-1}x(nT)\mathrm{e}^{-\mathrm{j}2\pi nk/N} \\
&= \sum_{n=0}^{N-1}x(nT)\cos\frac{2\pi kn}{N} - \mathrm{j}\sum_{n=0}^{N-1}x(nT)\sin\frac{2\pi kn}{N},\quad k=0,1,\cdots,N-1(\mathrm{DFT})
\end{aligned}
\tag{5.3}
$$

式中　$\Delta\omega$——与各相邻采样点间频率间隔相对应的圆频率间隔,$\Delta\omega=2\pi/(NT)$。

$$
x(nT)=\sum_{k=0}^{N-1}X_\mathrm{d}(\mathrm{j}k\Delta\omega)\mathrm{e}^{\mathrm{j}2\pi kn/N},\quad n=0,1,\cdots,N-1(\mathrm{IDFT}) \tag{5.4}
$$

式中　$x(nT)$——时间采样间隔为T(即采样频率$f_s=1/T$)的共N个($n=0,1,2,\cdots,N-1$)时间序列;

$X_\mathrm{d}(\mathrm{j}k\Delta\omega)$——频率分辨率$f_0=1/T_0=1/(NT)$中的共$N$个($k=0,1,2,\cdots,N-1$)频率序列点(即谱线)。

According to the convolution theorem, the multiplication of two functions in the time domain corresponds to the convolution in the frequency domain, that is, $x(t)\,\delta_0(t) \rightleftharpoons X(f) * \Delta_0(f)$. Obviously, the frequency spectrum $X(f) * \Delta_0(f)$ of the discrete function $x(t)\delta_0(t)$ obtained after sampling is different from the frequency spectrum $X(f)$ of the original continuous function $x(t)$, and an overlap phenomenon may occur near $1/T$ of the frequency domain, thereby causing errors. In order to avoid this error, $f_s > 2f_{max}$ must be satisfied, where f_{max} represents the highest frequency component contained in the original function $x(t)$. This proves the sampling theorem from another side: the sampling frequency must be higher than twice the highest frequency contained in the measured signal.

At this point, the sampled discrete function $x(t)\delta_0(t)$ still has an infinite number of sampling points, and the computer can only receive a finite number of points, so $x(t)\delta_0(t)$ must be truncated in the time domain to take out a limited number of N points. This is equivalent to using a rectangular window function $x_0(t)$ with a width of T_0 to multiply the time domain of the signal under measurement. Figure 5.2(d) shows the window function $x_0(t)$ and its spectrum $X_0(f)$. Also, according to the convolution theorem, the frequency spectrum of the discrete function $x(t)\delta_0(t)x_0(t)$ with N finite points should be equal to $|X(f) * \Delta_0(f) * X_0(f)|$, that is, the convolution in the frequency domain, as shown in Figure 5.2(e). As mentioned earlier, since the Fourier transform $X_0(f)$ of the rectangular window function $x_0(t)$ is a sampling function, that is, a sinc function, convolution with it will inevitably cause sidelobe as shown in Figure 5.2(e). To reduce the error caused by this, it is advantageous to increase the cut-off length T_0.

In the Fourier transform pair in Figure 5.2(e), the spectral function is a continuous function, which is still not acceptable by the computer. For this reason, it must be discretized, that is, multiplied by the frequency sampling function $\Delta_1(f)$. Similarly, according to the convolution theorem, the multiplication of two functions in the frequency domain corresponds to the convolution in the time domain. As shown in Figure 5.2(e), (f), (g), the sampling interval of the frequency sampling function $\Delta_1(f)$ should be $1/T_0$ to ensure that there is no overlap when doing convolution in the time domain. $f_0 = 1/T_0$ represents frequency resolution.

In this way, Figure 5.2(g) has become a discrete Fourier transform pair that the computer can receive. They are both periodic in the time domain and frequency domain. We should take a period of N time sampling values corresponding to N frequency values, namely $T_0 = NT$, thereby deriving a finite discrete Fourier transform pair equivalent to the original continuous function $x(t)$ and Fourier transform $X(f)$, respectively called discrete Fourier transform (DFT) and inverse discrete Fourier transform (IDFT), namely

$$
\begin{aligned}
X_d(jk\Delta\omega) = X_d\left(j\frac{2\pi}{NT}k\right) &= \sum_{n=0}^{N-1}(XnT)\,e^{-jk\Delta\omega nT} \\
&= \sum_{n=0}^{N-1}x(nT)\,e^{-j2\pi nk/N} \\
&= \sum_{n=0}^{N-1}x(nT)\cos\frac{2\pi kn}{N} - j\sum_{n=0}^{N-1}x(nT)\sin\frac{2\pi kn}{N},\quad k=0,1,\cdots,N-1\,(\text{DFT})
\end{aligned}
\tag{5.3}
$$

实用中,为提高谱线的频率分辨率,在原信号记录的末端填补一些零,相当于人为地增加了时域截断长度 T_0,从而减小了谱线间隔 f_0,也就是在原来频谱形式不变的情况下变更了谱线的位置。这样,原来看不到的频谱分量就有可能看到了,即谱线变密了。

5.2.2 信号数字化处理过程中的几个主要问题

1. 采样及采样定理

在对连续信号进行离散时,首先遇到的问题就是采样问题。下面着重讨论如何确定采样间隔。

(1) 信号的采样。对信号进行采样即进行 A/D 转换,其相当于用一个开关电路对模拟信号进行处理。设开关为 K,令它每隔 Δt 时间完成一次开关动作,并设其接通时间非常短。当模拟信号 $x(t)$ 加到该电路的输入端时,信号的输出是离散的,其间隔为 Δt,幅值为每一次接通时信号 $x(t)$ 的瞬时值,写成 $x(n\Delta t)$,对这样 $x(t)$ 所对应的一系列的离散值 $x(n\Delta t)$,通常称为离散数据。再进一步量化和编码,就得到数字信号数据。

从数学上来考查采样过程,可以设 $x(t)$ 为原始模拟时间信号,$\delta_0(t)$ 为采样信号,这样 $x_s(t)$ 就可看成是 $x(t)$ 与脉冲序列 $\delta_0(t)$ 的乘积,采样过程说明如图 5.3 所示,有

$$x_s(t) = x(t)\sum_{n=-\infty}^{\infty}\delta(t-nT) \tag{5.5}$$

式中 T—— 采样周期或采样间隔。

由于式(5.5) 中 $x(t)$ 只在 $t = nT$ 时才有定义,因此式(5.5) 可进一步写成

$$x_s(t) = \sum_{n=-\infty}^{\infty}x(nT)\delta(t-nT) \tag{5.6}$$

(2) 采样定理。我们希望采样后离散信号的外包络线应与模拟信号一致,并能够将离散信号还原成原始的模拟信号而不发生失真。这方面主要取决于采样间隔 T。如果 T 过大,则会产生失真;如果 T 过小,则数据的计算量将会增大,影响计算效率。因此,需对 T 值进行合适的选取。

由式(5.6) 可知,采样数据信号 $x_s(t)$ 等于 $x(t)$ 与脉冲序列 $\delta_0(t)$ 的乘积。

由卷积定理可知

$$x(t)\delta_0(t) \rightleftharpoons X(f) * \Delta_0(f) \tag{5.7}$$

式中 $X(f)$——$x(t)$ 的傅里叶变换;

$\Delta_0(f)$——$\delta_0(t)$ 的傅里叶变换。

假如 $x(t)$ 为图 5.3(a) 所示的时域函数,其傅里叶变换 $X(f)$ 如图 5.3(e) 所示,脉冲

where　$\Delta\omega$ —circular frequency interval corresponding to the frequency interval between adjacent sampling points, $\Delta\omega = 2\pi/(NT)$.

$$x(nT) = \sum_{k=0}^{N-1} X_{\mathrm{d}}(\mathrm{j}k\Delta\omega)\mathrm{e}^{\mathrm{j}2\pi kn/N}, \quad n = 0,1,\cdots,N-1(\mathrm{IDFT}) \tag{5.4}$$

where　$x(nT)$ — a total of n ($n=0,1,2,\cdots,N-1$) time series with time sampling interval T (that is, sampling frequency $f_s = 1/T$);

$X_{\mathrm{d}}(\mathrm{j}k\Delta\omega)$ —a total of N ($k=0,1,2,\cdots,N-1$) frequency sequence points (that is, spectral lines) in the frequency resolution $f_0 = 1/T_0 = 1/(NT)$.

In practice, in order to improve the frequency resolution of the spectral line, some zeros are filled at the end of the original signal record, which is equivalent to artificially increasing the time-domain truncation length T_0, thereby reducing the spectral line interval f_0 , that is, the position of the spectral line is changed while the original form of the spectrum is unchanged. In this way, the previously invisible spectral components may be seen, that is, the spectral lines become denser.

5.2.2　Several Main Problems in the Process of Signal Digitization

1. Sampling and Sampling Theorem

When discretizing a continuous signal, we will encounter the first problem, the sampling problem. How to determine the sampling interval is emphatically discussed below.

(1) Signal sampling. Signal sampling means A/D conversion, which is equivalent to using a switch circuit to process the analog signal. Set the switch as K, make it complete a switching action every time Δt , with turn-on time being set very short. When the analog signal $x(t)$ is added to the input terminal of the circuit, the output of the signal is discrete, the interval is Δt, and the amplitude is the instantaneous value of the signal $x(t)$ each time it is turned on, written as $x(n\Delta t)$, for such a series of discrete values $x(n\Delta t)$ corresponding to $x(t)$, it is usually called discrete data. After further quantization and coding, the digital signal data will be obtained.

Observing the sampling process mathematically, we can set $x(t)$ as the original analog time signal and $\delta_0(t)$ as the sampled signal, so that $x_s(t)$ can be regarded as the product of $x(t)$ and the pulse sequence $\delta_0(t)$, the description of the sampling process is shown in Figure 5.3, we have

$$x_s(t) = x(t)\sum_{n=-\infty}^{\infty}\delta(t-nT) \tag{5.5}$$

where　T — sampling period or sampling interval.

Since $x(t)$ in formula (5.5) is only defined when $t = nT$, formula (5.5) can be further written as

$$x_s(t) = \sum_{n=-\infty}^{\infty} x(nT)\delta(t-nT) \tag{5.6}$$

(2) Sampling theorem. We hope that the outer envelope of the discrete signal after sampling should be consistent with the analog signal, and the discrete signal can be restored to the original analog signal without distortion. This aspect mainly depends on the sampling interval T. If T is too large, distortion will occur; and if T is too small, the amount of data cal-

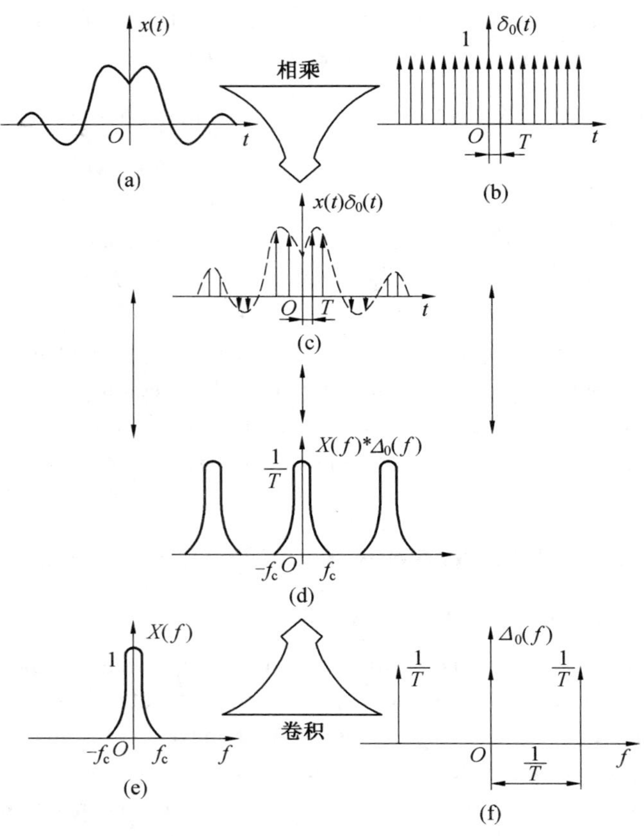

图 5.3　采样过程说明

序列$\delta_0(t)$ 如图5.3(b) 所示，则其傅里叶变换如图5.3(f) 所示。根据图5.3可以看出，采样信号 $x_s(t)$ 的频谱中包含了原信号 $x(t)$ 的频谱 $X(f)$ 。如果用一个低通滤波器把 $X(f)$ 从 $X_s(f)$ 中选出，则就可以通过$X(f)$ 还原出$x(t)$。同时，为使$X_s(f)$ 的各波形不发生重叠现象且能不失真地恢复原信号，要求f_s 的取值满足$f_s - f_c \geq f_c$，所以可以得到

$$f_s \geq 2f_c \tag{5.8}$$

式中　f_c—— 被测信号中所包含的最高频率，称为信号的截止频率；

f_s—— 采样频率。

式(5.8) 说明，若要不失真地恢复原信号，采样频率至少应为被测信号中所包含的最高频率的 2 倍，这就是著名的采样定理。

在不考虑相位的理想情况下，低通滤波器的截止频率只要选为f_c，即可得还原的信号。但考虑到信号的相位和滤波器的特性，f_s 一般应该选择大于 $2f_c$，工程中常取$f_s = 5f_c$以上。

culation will increase, which will affect the calculation and efficiency. Therefore, the value of T needs to be selected appropriately.

From formula (5.6), we can see that the sampled data signal $x_s(t)$ is equal to the product of $x(t)$ and the pulse sequence $\delta_0(t)$.

It can be known from the convergence theorem that

$$x(t)\,\delta_0(t) \rightleftharpoons X(f) * \Delta_0(f) \tag{5.7}$$

where　$X(f)$ —Fourier transform of $x(t)$;

$\Delta_0(f)$ —Fourier transform of $\delta_0(t)$.

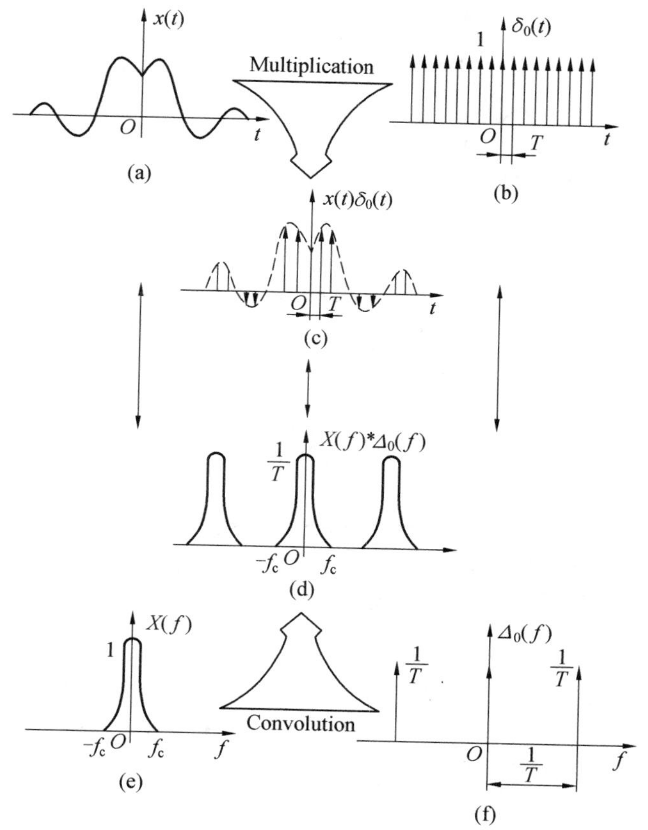

Figure 5.3　Description of the sampling process

If $x(t)$ is the time domain function shown in Figure 5.3(a), its Fourier transform $X(f)$ is shown in Figure 5.3(e), and the pulse sequence $\delta_0(t)$ is shown in Figure 5.3(b), then its Fourier transform is shown in Figure 5.3(f). According to Figure 5.3, it can be seen that the frequency spectrum of the sampled signal $x_s(t)$ contains the frequency spectrum $X(f)$ of the original signal $x(t)$. If we use a low-pass filter to select $X_s(f)$ from $X(f)$, then $x(t)$ can be restored through $X(f)$. At the same time, in order to prevent the overlapping of the waveforms of $X_s(f)$ and restore the original signal without distortion, the value of f_s is required to satisfy $f_s - f_c \geq f_c$, so we get

$$f_s \geq 2f_c \tag{5.8}$$

where　f_c —the highest frequency contained in the measured signal, which is called the cut-off

2. 频率混叠效应

当采样间隔T取得过大，即采样频率f_s过低，使$f_s < 2f_c$时，将会产生信号$x(t)$的高频分量与其低频分量发生重叠，这种现象称为频率混叠。频率混叠效应如图5.4所示，当采样间隔T取得太大时，频域$\Delta_0(f)$的间隔变小，它与$X(f)$的卷积就产生了相互的重叠，即频率混叠效应。可见，一旦发生了混叠效应，则采用何种低通滤波器都难以无失真地恢复频谱。

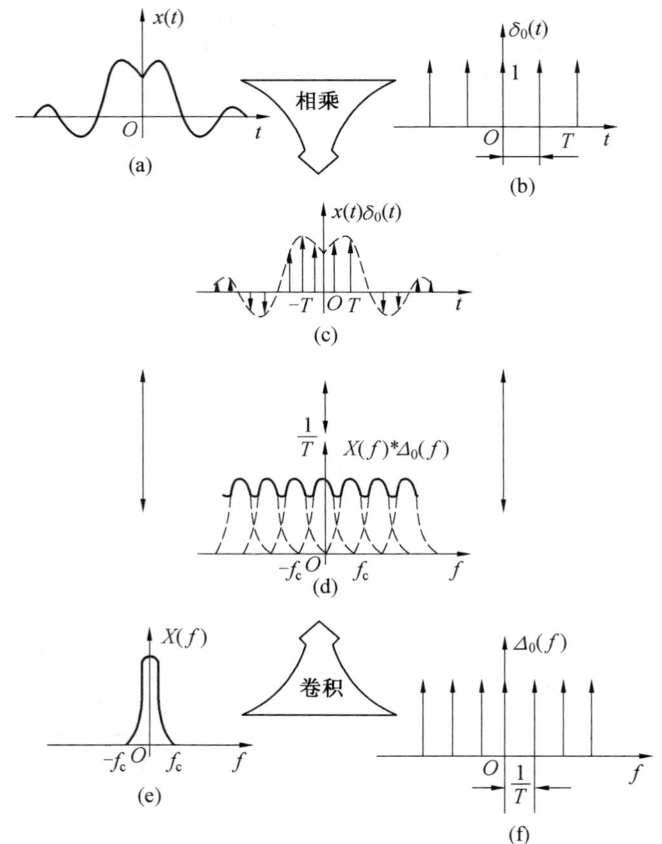

图5.4　频率混叠效应

减小混叠效应的办法可采用以下两个。

(1) 选取足够大的采样频率f_s。若信号中的最高频率为f_m，可选取$f_c = (1.5 \sim 2.5)f_m$，则$f_s = (3 \sim 5)f_c$。

(2) 采样前对信号进行低通滤波，即让信号$x(t)$通过一个低通滤波器，衰减掉高频分量，然后根据滤波后信号的最高频率选取采样频率f_s。

frequency of the signal;

f_s —sampling frequency.

Formula(5.8) shows that to restore the original signal without distortion, the sampling frequency should be at least twice the highest frequency contained in the measured signal, which is the famous sampling theorem.

In the ideal case without considering the phase, the restored signal can be obtained as long as the cut-off frequency of the low-pass filter is selected as f_c. However, considering the phase of the signal and the characteristics of the filter, f_s should generally be selected to be greater than $2f_c$, and in engineering, it is often selected above $f_s = 5f_c$.

2. Frequency Aliasing Effect

When the sampling interval T is too large, that is, the sampling frequency f_s is too low, when $f_s < 2f_c$, the high-frequency component of signal $x(t)$ will overlap with its low-frequency component. This phenomenon is called frequency aliasing. Frequency aliasing effect is shown in Figure 5.4, when the sampling interval T is too large, the interval of the frequency domain $\Delta_0(f)$ becomes smaller, and the convolution of it and $X(f)$ produces mutual overlap, that is, frequency aliasing effect. It can be seen that once the aliasing effect occurs, it is difficult to restore the spectrum without distortion by using any low-pass filter.

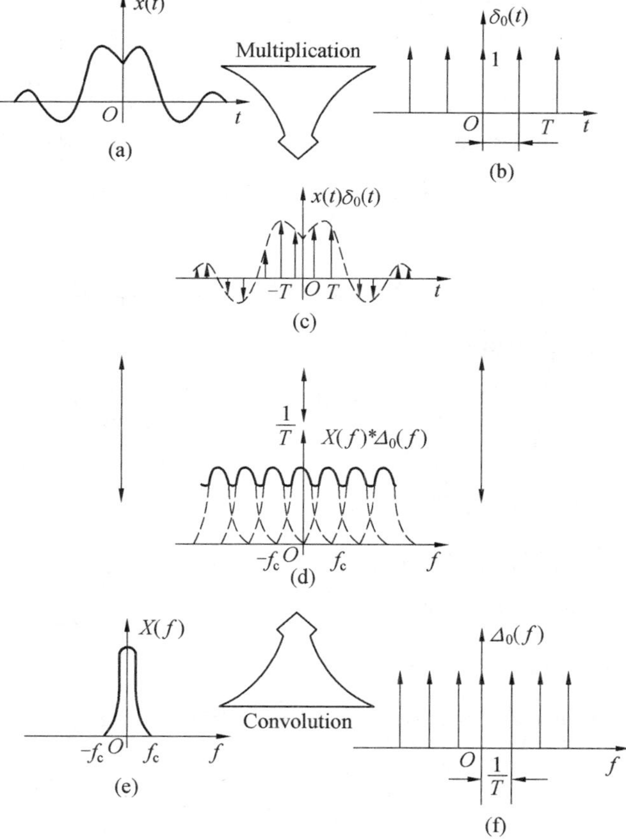

Figure 5.4　Frequency aliasing effect

3. 量化和量化误差

采样所得的离散信号的电压幅值若用二进制数码组来表示,就会使离散信号变成数字信号,这一过程称为量化。量化是从一组有限个离散电平中取一个来近似代表采样点的信号实际幅值电平。这些离散电平称为量化电平,每个量化电平对应一个二进制数码。

A/D 转换器的位数是一定的。一个 b 位(又称数据字长)的二进制数共有 $L=2^b$ 个数码。如果 A/D 转换器允许的动态工作范围为 D(如 ±5 V 或 0 ~ 10 V),则两相邻量化电平之间之差 Δx 为

$$\Delta x = D/2^{b-1} \tag{5.9}$$

其中,采用 2^{b-1} 而不用 2^b,是因为实际上字长的第一位用作符号位。

当离散信号采样值 $x(n)$ 的电平落在两个相邻量化电平之间时,就要舍入到相近的一个量化电平上。该量化电平与信号实际电平之间的差值称为量化误差 $\varepsilon(n)$。量化误差的最大值为 $\pm(\Delta x/2)$,可认为量化误差在 $(-\Delta x/2, +\Delta x/2)$ 区间各点出现的概率是相等的,其概率密度为 $1/\Delta x$,均值为零,其均方值 σ_ε^2 为 $\Delta x^2/12$,误差的标准差 σ_ε 为 $0.29\Delta x$。实际上,与信号获取和处理的其他误差相比,量化误差通常是不大的。

量化误差 $\varepsilon(n)$ 将形成叠加在信号采样值 $x(n)$ 上的随机噪声。假设字长 $b=8$,峰值电平为 $2^{8-1}\Delta x = 128\Delta x$。这样,峰值电平与 σ_ε 之比为 $128\Delta x/0.29\Delta x \approx 450$,即约近于 26 dB。

A/D 转换器位数选择应视信号的具体情况和量化的精度要求而定。然而,应考虑位数增多后成本显著增加、转换速率下降的影响。

为讨论简便,今后假设各采样点的量化电平就是信号的实际电平,即假设 A/D 转换器的位数为无限多,量化误差等于零。

4. 截断、泄漏和窗函数

由于实际只能对有限长的信号进行处理,因此必须截断过长的时间信号历程。截断就是将信号乘以时域的有限宽矩形窗函数。“窗”的含义是指透过窗口能够“看见”“外景”(信号的一部分),将时窗以外的信号视为零。

从采样后的信号中截取一段,就相当于在时域中用矩形窗函数 $w(t)$ 乘采样后信号。经这些处理后,其时、频域的相应关系为

$$x_a(t)\,p_g(t)\,w(t) \rightleftharpoons X_a(j\omega) * P_g(j\omega) * W(j\omega) \tag{5.10}$$

一般信号记录中,常以某时刻作为起点截取一段信号,这实际上就是采用单边时窗,

The following two methods can be used to reduce the aliasing effect.

(1) Select a sufficiently large sampling frequency f_s. If the highest frequency in the signal is f_m, $f_c = (1.5 - 2.5)f_m$ can be selected, then $f_s = (3 - 5)f_c$.

(2) We need to low-pass filter the signal before sampling, that is, let signal $x(t)$ pass through a low-pass filter to attenuate high-frequency components, and then select the sampling frequency f_s according to the highest frequency of the filtered signal.

3. Quantization and Quantization Error

If the voltage amplitude of the discrete-time signal obtained by sampling is represented by a binary code group, the discrete signal will become a digital signal, whose process is called quantification. Quantization is to take a level from a set of finite discrete levels which can be used to approximate the actual voltage amplitude at that point of the sampled signal. These discrete levels are called quantization levels, and each quantization level corresponds to a binary number.

The number of bits of the A/D converter is fixed. A binary number with b bits (also known as data word length) has $L = 2^b$ digits in total. If the allowable dynamic working range of the A/D converter is D (such as ±5 V or 0 – 10 V), then the difference Δx between two adjacent quantization levels is

$$\Delta x = D/2^{b-1} \tag{5.9}$$

Among them, 2^{b-1} is used instead of 2^b, because the first bit of the word length is actually used as the sign bit.

When the level of the discrete signal sample value $x(n)$ falls between two adjacent quantization levels, it must be rounded to a similar quantization level. The difference between the quantization level and the actual level of the signal is called the quantization error $\varepsilon(n)$. The maximum value of the quantization error is $\pm(\Delta x/2)$. It can be considered that the probability of the quantization error at each point in the interval $(-\Delta x/2, +\Delta x/2)$ is equal, its probability density is $1/\Delta x$, the mean value is zero, its mean square value σ_ε^2 is $\Delta x^2/12$, and the standard deviation of the error σ_ε is $0.29\Delta x$. In fact, the quantization error is usually not large compared with other errors in signal acquisition and processing.

The quantization error $\varepsilon(n)$ will form random noise superimposed on the signal sample value $x(n)$. Assuming the word length $b = 8$, the peak level is equal to $2^{8-1}\Delta x = 128\Delta x$. In this way, the ratio of the peak level to σ_ε is $128\Delta x/0.29\Delta x \approx 450$, which is approximately 26 dB.

The selection of the number of bits of the A/D converter should depend on the specific conditions of the signal and the accuracy requirements of the quantization. However, it should be considered that the cost increases significantly and the conversion rate decreases after the number of bits increases.

In order to simplify the discussion, assume that the quantization level of each sampling point is the actual level of the signal, that is, assume that the number of bits of the A/D converter is infinite, and the quantization error is equal to zero.

4. Truncation, Leakage and Window Functions

Since the actual signal can only be processed with a limited length, the excessively long time signal history must be cut off. The truncation is to multiply the signal by a finite-width rec-

这时的矩形窗函数为

$$w(t)=\begin{cases}1, & 0\leqslant t\leqslant T\\ 0, & \text{其他}\end{cases} \tag{5.11}$$

由于 $W(\mathrm{j}\omega)$ 是一个无限带宽的 sinc 函数,因此即使原时域信号是带限信号,在截断后也必然成为无限带宽信号。这种信号的能量在频率轴分布扩展的现象称为泄漏。同时,由于截断后信号带宽变宽,因此无论采样频率多高,信号总是不可避免地出现混叠,故信号截断必然导致出现一些误差。

为减小或抑制泄漏,提出了各种不同形式的窗函数来对时域信号进行加权处理,以改变时域截断处的不连续状况。例如,选择的窗函数应力求其频谱的主瓣宽度变窄些、旁瓣幅度变小些。窄的主瓣可以提高分辨能力,小的旁瓣可以减小泄漏。这样,窗函数的优劣大致可以从三个方面来评价:最大旁瓣峰值与主瓣峰值之比、最大旁瓣10倍频程衰减率和主瓣宽度。

5. 频率分辨率、整周期截断

频率采样间隔 Δf 也是频率分辨率的指标。此间隔越小,频率分辨率越高,被"挡住"的频率成分越少。在利用离散傅里叶变换将有限时间序列变换成相应的频谱序列的情况下,Δf 和分析的时间信号长度 T 的关系是

$$\Delta f=f_s/N=1/T \tag{5.12}$$

这种关系是DFT算法固有的特征。这种关系往往加剧了频率分辨率与计算工作量的矛盾。

另外,在分析简谐信号的场合下,需要了解某特定频率 f_0 的谱值,希望DFT谱线落在 f_0 上。单纯减小 Δf 并不一定会使谱线落在频率 f_0 上。从DFT的原理来看,谱线落在 f_0 处的条件是 $f_0/\Delta f$ 等于一个整数。考虑到 Δf 是分析时长 T 的倒数,简谐信号的周期 T_0 是其频率 f_0 的倒数,因此只有截取的信号长度 T 正好等于信号周期的整数倍时,才可能使分析谱线落在简谐信号的频率上,从而获得准确的频谱。显然,这个结论适用于所有周期信号。因此,对周期信号实行整周期截断是获得准确频谱的先决条件。从概念来说,DFT的效果相当于将时窗内信号向外周期延拓。若事先按整周期截断信号,则延拓后的信号将与原信号完全重合,无任何畸变;反之,延拓后将在 $t=kT$ 交接处出现间断点,波形和频谱都发生畸变,其中 k 为某个整数。

6. 频域采样、时域周期延拓和栅栏效应

经过时域采样和截断后,其频谱在频域是连续的。如果要用数字描述频谱,就意味着

tangular window function in the time domain. The meaning of "window" means that we can "see" the "external scene" (part of the signal) through the window, and the signal outside the time window is regarded as zero.

Taking a segment from the sampled signal is equivalent to multiplying the sampled signal with the rectangular window function $w(t)$ in the time domain. After these treatments, the corresponding relationship between time and frequency domain is

$$x_a(t)\, p_g(t) w(t) \rightleftharpoons X_a(j\omega) * P_g(j\omega) * W(j\omega) \tag{5.10}$$

In general signal recording, a segment of signal is often intercepted at a certain moment as the starting point In fact, a single-sided time window is used, and the rectangular window function at this time is

$$w(t) = \begin{cases} 1, & 0 \leqslant t \leqslant T \\ 0, & \text{other} \end{cases} \tag{5.11}$$

Since $W(j\omega)$ is a sinc function with unlimited bandwidth, even if the original time-domain signal is a band-limited signal, it will inevitably become an infinite bandwidth signal after truncation. The phenomenon that the energy of this signal is distributed on the frequency axis is called leakage. Meanwhile, since the signal bandwidth becomes wider after truncation, the signal will inevitably be aliased no matter how high the sampling frequency is, so signal truncation will inevitably lead to some errors.

In order to reduce or suppress leakage, various forms of window functions are proposed to weight the time-domain signal to change the discontinuity at the time-domain truncation. For example, the selected window function should narrow the main lobe width and the side lobe amplitude of the frequency spectrum. A narrow main lobe can improve the resolution, and a small side lobe can reduce leakage. In this way, the pros and cons of the window function can be roughly evaluated from three aspects: the ratio of the maximum side lobe peak to the main lobe peak, the maximum side lobe 10-octave attenuation rate and the main lobe width.

5. Frequency Resolution, Whole Cycle Truncation

The frequency sampling interval Δf is also an indicator of frequency resolution. The smaller the interval, the higher the frequency resolution, and the fewer frequency components "blocked". In the case of using discrete Fourier transform to transform a finite time sequence into a corresponding spectrum sequence, the relationship between Δf and the analyzed time signal length T is

$$\Delta f = f_s / N = 1/T \tag{5.12}$$

This relationship is an inherent feature of the DFT algorithm. This relationship tends to exacerbate the contradiction between frequency resolution and computational workload.

In addition, in the case of analyzing harmonic signals, it is necessary to know the spectrum value of a certain frequency f_0 , and it is hoped that the DFT spectrum line falls on f_0 . Simply reducing Δf does not necessarily cause the spectrum to fall on the frequency f_0 . From the principle of DFT, the condition for the spectrum line to fall at f_0 is that $f_0/\Delta f$ is equal to an integer. Considering that Δf is the reciprocal of the analysis duration T , and the period T_0 of the harmonic signal is the reciprocal of its frequency f_0 , so only when the intercepted signal length T is exactly equal to an integer multiple of the signal period, it is possible to make the analysis spectrum fall on the harmonic signal in order to obtain an accurate frequency spectrum.

首先必须使频率离散化,实行频域采样。频域采样与时域采样相似,在频域中用脉冲序列乘以信号频谱函数。这一过程在时域中相当于将信号与周期脉冲序列做卷积,其结果是将时域信号平移至各脉冲坐标位置重新构图,从而相当于在时域中将窗内的信号波形在窗外进行周期延拓。因此,频率离散化无疑已将时域信号"改造"成周期信号。总之,经过时域采样、截断、频域采样之后的信号是一个周期信号,与原信号是不一样的。

对一函数实行采样的实质就是"摘取"采样点上对应的函数值。其效果犹如透过栅栏的缝隙观看外景,只有落在缝隙前的少数景象被看到,其余景象都被栅栏挡住,视为零,这种现象称为栅栏效应。无论是时域采样还是频域采样,都有相应的栅栏效应。只不过时域采样若满足采样定理要求,栅栏效应不会有什么影响。而频域采样的栅栏效应则影响颇大,"挡住"或丢失的频率成分有可能是重要的或具有特征的成分,以致整个处理失去意义。

7. 常用的窗函数

下面介绍几种常用窗函数各自的特点。

(1) 矩形窗。

矩形窗的定义及其频谱已在第 1 章讨论过。矩形窗是使用最多的窗,在信号处理时,凡是将信号截断、分块都相当于对信号加了矩形窗。矩形窗的主瓣高为 T,宽为 $2/T$,第一旁瓣幅值为 −13 dB,相当于主瓣高的 20%,旁瓣衰弱为 20 dB/10 倍频程。与其他窗相比,矩形窗主瓣最窄,旁瓣则较高,泄漏较大。在需要获得精确频谱主峰的所在频率,而对幅值精度要求不高的场合,可选用矩形窗。

(2) 三角窗。

三角窗(图 5.5) 可表示为

$$w(t)=\begin{cases}1-\dfrac{2}{T}|t|, & |t|\leqslant \dfrac{T}{2}\\[2ex] 0, & |t|>\dfrac{T}{2}\end{cases} \tag{5.13}$$

其频谱为

$$W(f)=\frac{T}{2}\left(\frac{\sin\dfrac{\pi fT}{2}}{\dfrac{\pi fT}{2}}\right)^2=\frac{T}{2}\operatorname{sinc}^2\left(\frac{\pi fT}{2}\right) \tag{5.14}$$

Obviously, this conclusion applies universally to all periodic signals. Therefore, the truncation of the entire period of the periodic signal is a prerequisite for obtaining an accurate spectrum. Conceptually, the effect of DFT proves equivalent to extending the signal in the time window to the outer period. If the signal is truncated according to the entire period in advance, the signal after extension will completely coincide with the original signal without any distortion. On the contrary, a discontinuity will appear at the junction of $t = kT$, and both the waveform and the frequency spectrum will be distorted after prolongation, where k is an integer.

6. Frequency Domain Sampling, Time Domain Period Extension and Fence Effect

After sampling and truncation in the time domain, the frequency spectrum is continuous in the frequency domain. Describing the frequency spectrum digitally means that we must first discretize the frequency and implement frequency domain sampling. Frequency domain sampling is similar to time domain sampling, in which the pulse sequence is multiplied by the signal spectrum function in the frequency domain. In the time domain, this process is equivalent to convolving the signal with the periodic pulse sequence, of which the result is to translate the time domain signal to the coordinate position of each pulse to recompose the image. This is equivalent to periodical extension of the signal waveform inside the window outside the window in the time domain. Therefore, the frequency discretization has undoubtedly "transformed" the time domain signal into a periodic signal. In short, the signal after time domain sampling, truncation, and frequency domain sampling is a periodic signal, which is different from the original signal.

The essence of sampling a function is to "extract" the corresponding function value at the sampling point. The effect is like viewing the exterior scene through a gap in a fence. Only a few scenes falling in front of the gap are seen, and the rest are blocked by the fence and regarded as zero, whose phenomenon is called the fence effect. Whether it is time domain sampling or frequency domain sampling, there is a corresponding fence effect. It's just that the fence effect will not have any impact if time domain sampling meets the requirements of the sampling theorem. The fence effect of frequency domain sampling has a great influence. The frequency components that are "blocked" or lost may be important or characteristic components, so that the entire process will lose its meaning.

7. Commonly Used Window Functions

Here are the characteristics of several commonly used window functions.

(1) Rectangular window.

The definition of the rectangular window and its frequency spectrum have been discussed in Chapter 1. The rectangular window is the most popular. In signal processing, cutting off and dividing the signal into blocks is equivalent to adding a rectangular window to the signal. The main lobe height of the rectangular window is T, the width is $2/T$, the first side lobe amplitude is -13 dB, which is equivalent to 20% of the main lobe height, and the side lobe attenuation is 20 dB/10 octave. Compared with other windows, the main lobe of the rectangular window is the narrowest, the side lobe is higher, and the leakage is larger. A rectangular window can be selected when the frequency of the main peak of the accurate spectrum needs to be obtained, but the amplitude accuracy is not high.

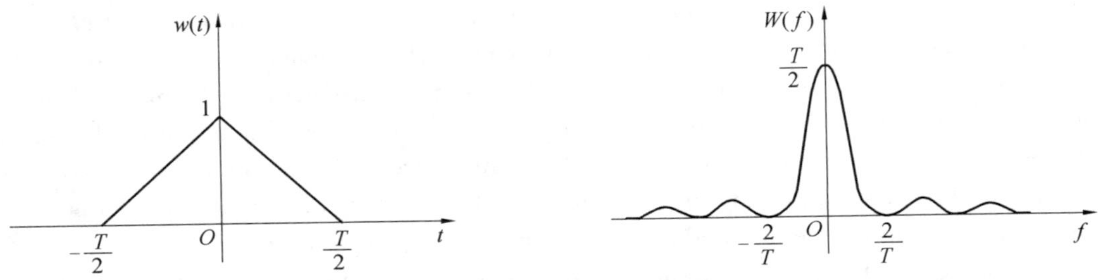

图 5.5　三角窗函数及其幅频谱

三角窗与矩形窗相比，其主瓣宽度约为矩形窗的2倍，但旁瓣低并且不会出现负值。

(3) 汉宁窗。

汉宁窗(图5.6)又称余弦窗，其定义为

$$w(t)=\begin{cases}\frac{1}{2}+\frac{1}{2}\cos\frac{2\pi t}{T}, & |t|>\frac{T}{2}\\ 0, & |t|\leqslant\frac{T}{2}\end{cases} \tag{5.15}$$

其频谱为

$$W(f)=\frac{1}{2}W_{R}(f)+\frac{1}{4}\left[W_{R}\left(f+\frac{1}{T}\right)+W_{R}\left(f-\frac{1}{T}\right)\right] \tag{5.16}$$

式中

$$W_{R}(f)=T\mathrm{sinc}(\pi fT)$$

汉宁窗的主瓣高为$T/2$，为矩形窗的1/2；宽为$4/T$，为矩形窗主瓣宽的2倍；第一旁瓣幅值为 - 32 dB，约为主瓣高的2.4%；旁瓣衰减率为60 dB/10倍频程。相比之下，汉宁窗的旁瓣明显降低，具有抑制泄漏的作用，但主瓣较宽，致使频率分辨能力较差。

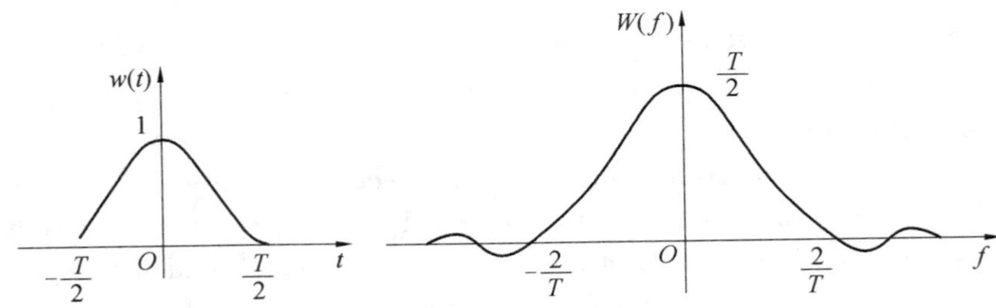

图 5.6　汉宁窗函数及其幅频谱图

在截断随机信号或非整周期截断周期函数时，DFT的周期延拓功能会在信号中出现间断点，造成新的泄漏。为平滑或削弱截取信号的两端，减小泄漏，宜加汉宁窗。

(2) Triangular window.

Triangular window (Figure 5.5) can be expressed as

$$w(t) = \begin{cases} 1 - \dfrac{2}{T}|t|, & |t| \leqslant \dfrac{T}{2} \\ 0, & |t| > \dfrac{T}{2} \end{cases} \tag{5.13}$$

Its frequency spectrum is

$$W(f) = \frac{T}{2}\left(\frac{\sin\dfrac{\pi f T}{2}}{\dfrac{\pi f T}{2}}\right)^2 = \frac{T}{2}\mathrm{sinc}^2\left(\frac{\pi f T}{2}\right) \tag{5.14}$$

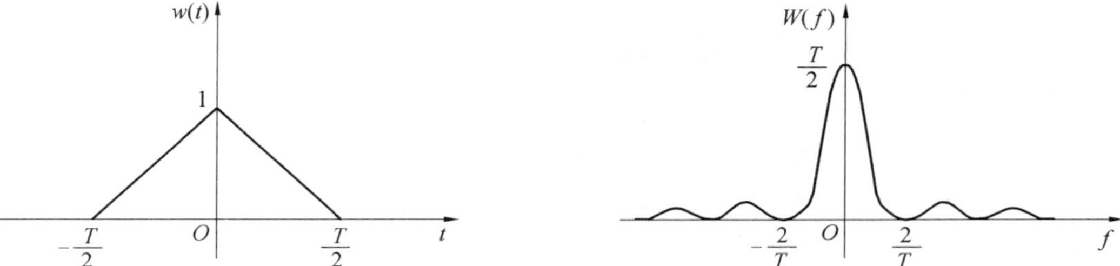

Figure 5.5　Triangular window function and its amplitude spectrum

Comparing the triangular window with the rectangular window, the width of the main lobe is about twice that of the rectangular window, but the side lobe is low and does not appear negative value.

(3) Hanning window.

Hanning window (Figure 5.6) is also called cosine window, which is defined as

$$w(t) = \begin{cases} \dfrac{1}{2} + \dfrac{1}{2}\cos\dfrac{2\pi t}{T}, & |t| > \dfrac{T}{2} \\ 0, & |t| \leqslant \dfrac{T}{2} \end{cases} \tag{5.15}$$

Its frequency spectrum is

$$W(f) = \frac{1}{2}W_R(f) + \frac{1}{4}\left[W_R\left(f + \frac{1}{T}\right) + W_R\left(f - \frac{1}{T}\right)\right] \tag{5.16}$$

In the formula

$$W_R(f) = T\mathrm{sinc}(\pi f T)$$

The height of the main lobe of the Hanning window is $T/2$, which is 1/2 of the rectangular window; the width is $4/T$, which is twice the width of the main lobe of the rectangular window; the amplitude of the first side lobe is −32 dB, which is about 2.4% of the height of the main lobe; the sidelobe attenuation rate is 60 dB/10 octave. In contrast, the side lobe of the Hanning window is significantly reduced, which has the effect of suppressing leakage, but the main lobe is wider, resulting in poor frequency resolution.

When truncating a random signal or truncating a periodic function with a non-integral period, the period extension function of DFT will cause discontinuities in the signal, causing new leakage. In order to smooth or weaken both ends of the intercepted signal and reduce leakage, a Hanning window should be added.

(4) 指数窗。

在测量系统的脉冲响应时,由于信号随时间而衰减,而许多噪声和误差的影响却是定值,因此开始部分信噪比较好,随着响应信号的衰减,信噪比变坏。如果加汉宁窗,则会把最重要的开始一段信号大大削弱,这样得到的频谱的可靠程度就受到影响。因此,对脉冲响应这类信号不宜加汉宁窗、三角窗等对称型的窗函数。

指数窗(图 5.7) 定义为

$$w(t)=\begin{cases}e^{-at}, & |t|\geqslant 0\\ 0, & |t|<0\end{cases} \tag{5.17}$$

式中 a—— 衰减系数,$a>0$,其频谱为

$$W(f)=\frac{1}{\sqrt{a^2+(2\pi f)^2}} \tag{5.18}$$

指数窗的特点是无旁瓣,但主瓣很宽,其频率分辨能力低。

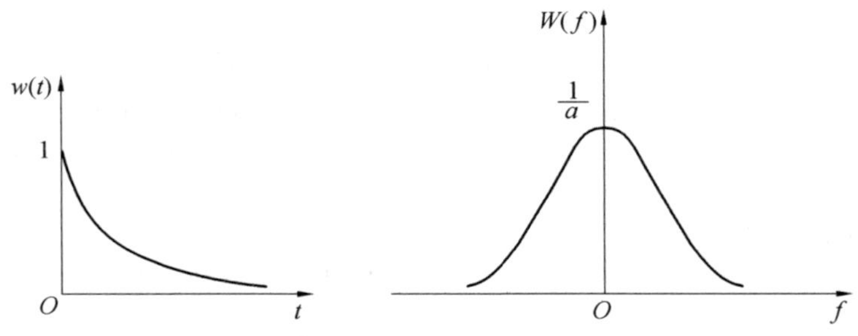

图 5.7 指数窗及其幅频谱

如果把脉冲响应信号加上指数窗,并适当选择衰减系数 a,就可以较显著地衰减信噪比差的后一部分的信号,起到抑制噪声的作用,从而所得到的频谱曲线就会平滑些。

5.3 随机信号

5.3.1 随机过程的基本概念

事物的变化过程可以广泛地分为两类:一类是变化过程具有确定的形式,变化过程带有必然性,可用一个(或者几个) 确定的时间 t 的函数来描述,称为确定性过程;另一类是变化过程没有确定的形式,没有必然的规律性,无法用一个(或者几个) 确定的时间 t 的函数来描述。独立的、重复的多次观测所得到的结果是各不相同的,无法事先预测,每一次观测的结果都是众多可能结果中的一个。一个很典型的例子是热噪声电压,电子元器件

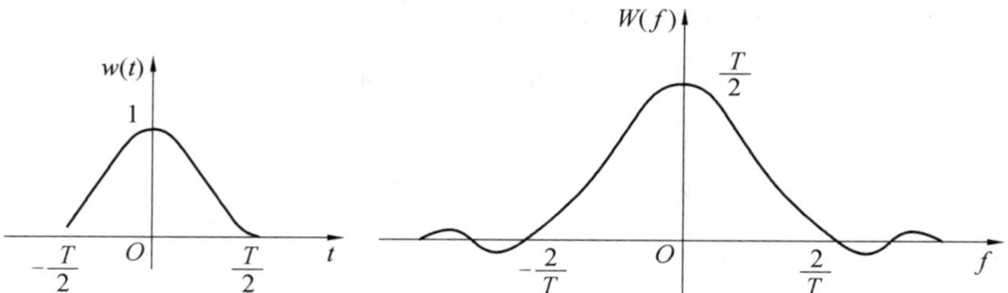

Figure 5.6　Hanning window function and its amplitude spectrum

(4) Index window.

When measuring the impulse response of the system, since the signal decays with time, and the influence of many noises and errors is a fixed value, the signal-to-noise ratio is better at the beginning, and the signal-to-noise ratio becomes worse as the response signal attenuates. If the Hanning window is added, the most important first segment of the signal will be greatly attenuated, which the reliability of the obtained spectrum will be affected. Therefore, it is not appropriate to add symmetrical window functions such as Hanning window and triangular window to such signals as impulse response.

The index window (Figure 5.7) is defined as

$$w(t) = \begin{cases} e^{-at}, & |t| \geqslant 0 \\ 0, & |t| < 0 \end{cases} \tag{5.17}$$

where　a — attenuation coefficient, $a > 0$, its frequency spectrum is

$$W(f) = \frac{1}{\sqrt{a^2 + (2\pi f)^2}} \tag{5.18}$$

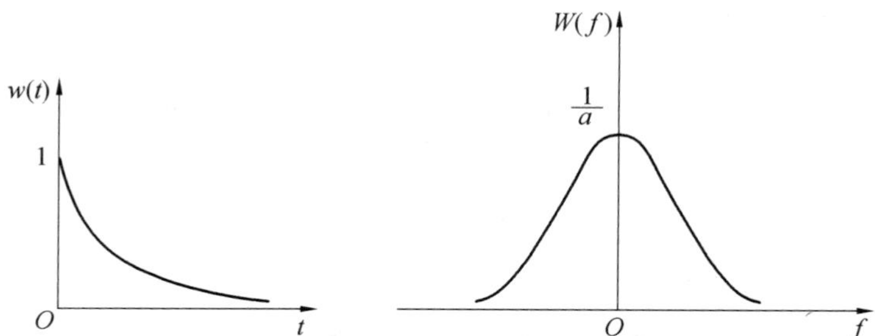

Figure 5.7　Exponential window and its amplitude spectrum

The index window is characterized by no side lobes, but the main lobe is very wide, and its frequency resolution ability is low.

If the impulse response signal is added to the index window with appropriately selecting the attenuation coefficient a, the latter part of the signal with poor signal-to-noise ratio can be attenuated significantly, and the effect of suppressing noise can be achieved, so that the resulting spectrum curve will be smoother.

因内部微观粒子(如电子)的随机热骚动而引起的端电压称为热噪声电压,它在任意确定时刻的值都是随机变量,由于热骚动的随机性,因此在相同条件下每一次观测都将产生不同的电压——时间函数 $u_i(t)$(图 5.8)。这族曲线可以想象为在相同条件下同时测量无限多个"相同"电阻的端电压的结果。这里可以用图 5.8 所示的一族电压—时间函数来描述该电阻热噪声电压的变化过程。

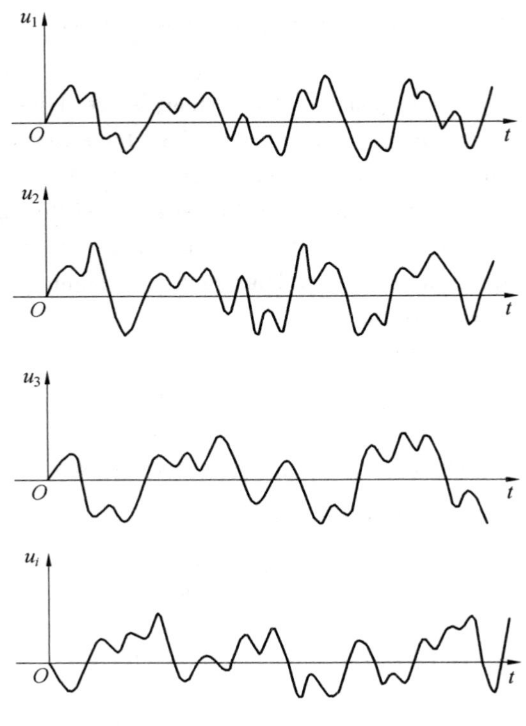

图 5.8　热噪声电压

在数学上,随机过程的概念如下。

设 E 是随机试验,即对不确定性过程做无数次物理实现,e 是每一次观测的结果,S 是 e 的集合,$S=\{e\}$,称为样本空间。对于每一个属于样本空间的 e,总可以依某种规则建立一个包含 e 和一些(待定的)常系数的时间的函数,即

$$X(e,a_1,a_2,\cdots,a_n,t) \tag{5.19}$$

与 e 相对应,对于所有的 e 来说,就得到一族时间的函数,称这族(不计其数)时间的函数为随机过程,常省去式(5.19)中的 e 和常系数,简记为 $X(t)$,它是一种集合的概念。族中的任何一个函数都称为该随机过程的样本函数,表示对随机过程的一次物理实现,用小写字母表示,记为 $x_i(t)(i=1,2,\cdots)$。此处应注意到,$X(t)$ 是一些时间函数的集合,函数中的 e 是随机变量,故 $X(t)$ 是一族时间 t 的非确定性函数。

5.3 Random Signals

5.3.1 The Basic Concept of Stochastic Process

The change process of things can be broadly divided into two categories. One is that the change process proves inevitable with having a definite form, and can be described by a function (or several) definite time t, which is called a deterministic process; the other type is that the change process has no inevitable regularity without definite form, and cannot be described by one (or several) definite time t functions. The results obtained by independent and repeated observations are different. It is impossible to predict in advance, and the result of each observation is one of many possible results. A very typical example is thermal noise voltage that is the terminal voltage of electronic components caused by random thermal disturbance of internal microscopic particles (such as electrons), of which value at any certain time is a random variable, and each observation under the same conditions will produce a different voltage—a time function $u_i(t)$ due to the randomness of thermal disturbance (Figure 5.8). This family of curves can be imagined as the result of simultaneously measuring the terminal voltages of an infinite number of "same" resistors under the same conditions. Here can use a family of voltage-time functions as shown in Figure 5.8 to describe the change process of the thermal noise voltage of the resistor.

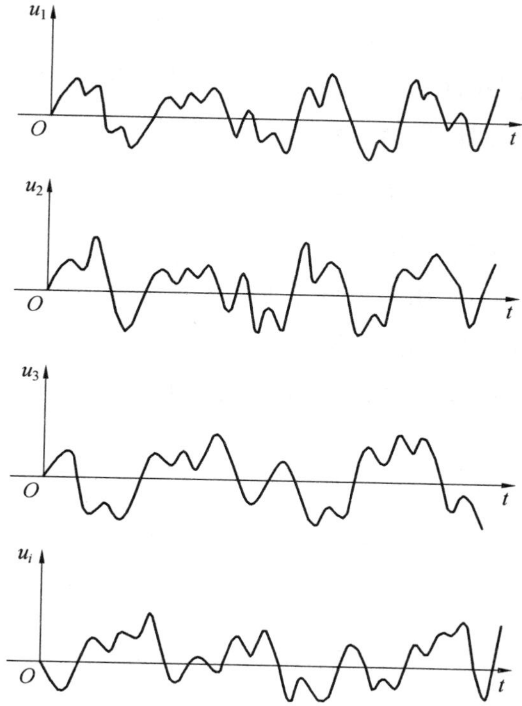

Figure 5.8　Thermal noise voltage

例如，随机地确定时间原点，观察质点 M 做匀速圆周运动时半径 OM 与横轴的夹角 θ（图 5.9（a）），可以发现 θ 随机地分布在 0 ~ 2π 内。对于每一个具体的 θ_i，可以确定一个包含 θ_i 的时间的函数使之与 θ_i 相对应，有

$$x_i(t) = A_m\cos(\omega t + \theta_i) \tag{5.20}$$

式中 A_m—— M 点做圆周运动的半径，为常系数，$A_m = OM$；

ω—— M 点做圆周运动的角速度，为常系数。

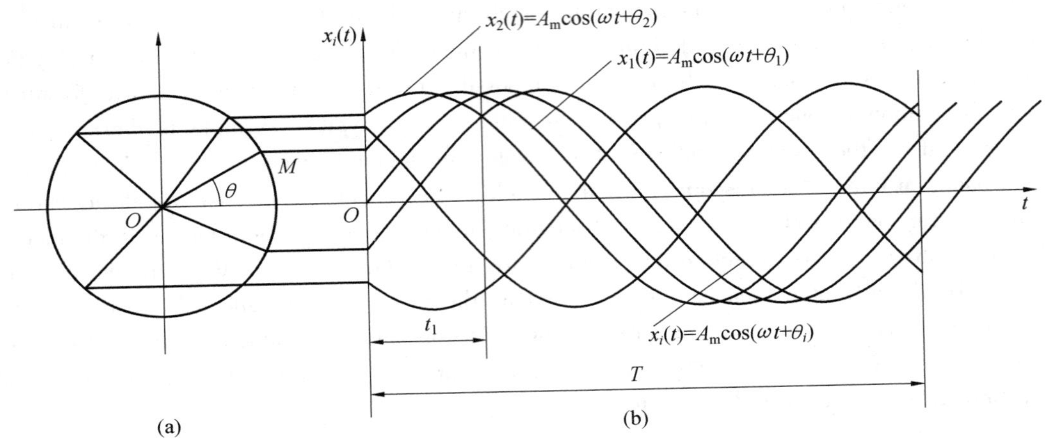

图 5.9　随机相位余弦波

每一次观测，$x_i(t)$ 都会因 θ_i 的不同而不同，事先无法预测。经过多次观测，对于所有的 $\theta_i \in \theta$ 来说，就得到一族时间 t 的函数，即

$$X(t) = A_m\cos(\omega t + \theta) \tag{5.21}$$

式中 θ—— 分布于 0 ~ 2π 的随机变量。

称此族时间 t 的函数 $X(t)$ 为随机过程。假如 θ 是均匀分布于 0 ~ 2π 的随机变量，则称 $X(t) = A_m\cos(\omega t + \theta)$ 为随机相位余弦波（图 5.9（b））。

对于上述的随机相位余弦波，若在每次观测中都特别注意 $t = t_1$ 时刻的函数值，则会发现它们各不相同。

第一次观测，有

$$x_1(t_1) = A_m\cos(\omega t_1 + \theta_1)$$

第二次观测，有

$$x_2(t_1) = A_m\cos(\omega t_1 + \theta_2)$$

第 n 次观测，有

$$x_n(t_1) = A_m\cos(\omega t_1 + \theta_n)$$

In mathematics, the concept of stochastic process is as follows.

Let E be a random experiment, that is, countless physical realizations of the uncertainty process. e is the result of each observation, and S is the set of e, $S=\{e\}$, which is called the sample space. For each sample space e, we can always build a time function containing e and some (undetermined) constant coefficients according to a certain rule, that is

$$X(e,a_1,a_2,\cdots,a_n,t) \tag{5.19}$$

Corresponding to e, for all e, a family of time functions can be obtained. We call this family of (uncountable) time functions a random process, and e and constant coefficients in formula (5.19) are often omitted, abbreviated as $X(t)$, it is a concept of set. Any function in the family is called the sample function of the random process, which represents a physical realization of the random process. It is represented by lowercase letters, and marked as $x_i(t)$ $(i=1,2,\cdots)$. It should be noted here that $X(t)$ is a collection of some time functions, e in the function is a random variable, so $X(t)$ is a non-deterministic function of a family of time t.

For example, to determine the time origin randomly, observe the angle θ between the radius OM and the horizontal axis when the mass point M moves in a uniform circular motion (Figure 5.9(a)). We will find that θ is randomly distributed between $0-2\pi$. For each specific θ_i, we can determine a function of time including θ_i to correspond to θ_i, that is

$$x_i(t)=A_m\cos(\omega t+\theta_i) \tag{5.20}$$

where　A_m —the radius of circular motion at point M, which is a constant coefficient, $A_m=OM$;

ω — the angular velocity of circular motion at point M, which is a constant coefficient.

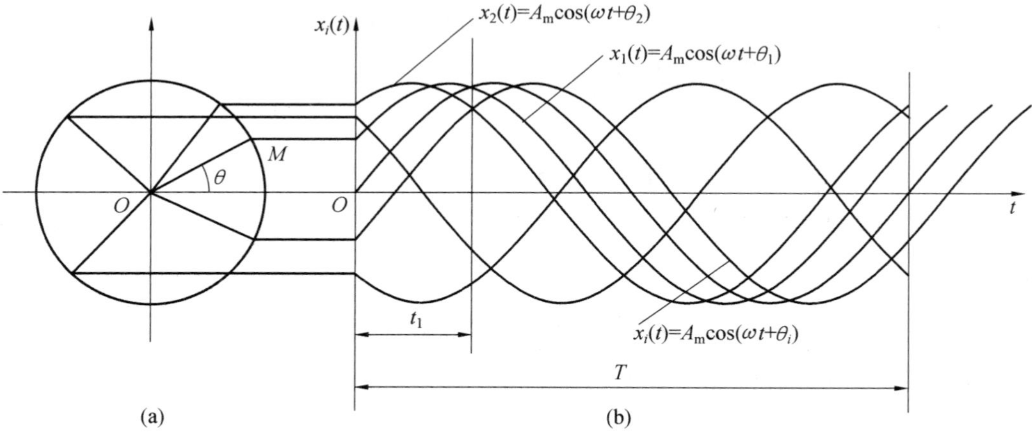

Figure 5.9　Random phase cosine wave

For each observation, $x_i(t)$ will be different depending on θ_i , which cannot be predicted in advance. After multiple observations, for all $\theta_i\in\theta$, a set of functions of time t is obtained, that is

$$X(t)=A_m\cos(\omega t+\theta) \tag{5.21}$$

where　θ —random variables distributed among $0-2\pi$.

We call the function $X(t)$ of this family of time t a random process. If θ is a random variable uniformly distributed between $0-2\pi$, then $X(t)=A_m\cos(\omega t+\theta)$ is called a random

它们不像确定性过程那样是一个确定的数值,而是变动在区间$[-A_m, A_m]$的随机变量,则有

$$X(t_1)=\{x_i(t_1)\}, \quad t_1 \in T$$

根据这种意义,随机过程还有另一种形式的定义:对于每一个确定的时刻t_1,$X(t_1)$都是随机变量,而不是一个确定的数值,则称$X(t)$为随机过程,或者说随机过程$X(t)$是依赖于时间t的一族随机变量。

以上两种定义在本质上是一样的。无论哪一种定义形式,都指出了随机过程的两个根本特点:第一,它是依赖于时间t的一族(而不是一个确定的)函数;第二,在确定时刻,它是一族随机变量(而不是一个确定值)。

研究随机过程与研究确定性过程相比,在方法上有一定的差异。确定性过程有必然的规律性,其研究目的在于发现必然规律。例如,自由落体下落的距离与时间的关系、还与哪些因素有关、在t_1时刻下落的距离究竟多少等。而随机过程本来就没有必然的变化规律,那么研究随机过程的目的当然就不在于揭示其必然规律,而在于发现它的统计规律,即隐藏在随机现象中的规律性。其中,包括发现随机过程$X(t)$中的随机变量e在它的取值范围内的分布规律和包含e的时间函数$X(t)$到底是按照什么规则来建立的,以及该时间函数中那些常系数到底是多少。只有将这些问题弄清楚,才可以说对随机过程的规律性有了一个清楚的了解。因此,研究随机过程的基本思路可以概括为以下几点。

(1) 用统计方法寻找随机过程中所包含的随机变量e的分布规律,该规律被称为随机变量的统计特性。

(2) $X(t)$到底是按照什么规则来建立的时间函数。

(3) 在(1)已知后,用数学分析的方法计算该随机过程与统计特性相对应的各种数字特征。

(4) 根据各数字特征与$X(t)$中各种系数的关系确定各个常系数的数值。

按照这种思路,将随机过程,至少是工程中常见的平稳随机过程的时间函数$X(e, a, a_2, \cdots, a_n, t)$的具体规则弄清楚,将式中随机变量$e$的统计特性及各个系数$a_1, a_2, \cdots, a_n$弄清楚,就可以说随机过程$X(t)$的规律性最后搞清楚了。

5.3.2 随机过程的统计特性及数字特征

随机过程在任一时刻的状态是随机变量,而不是一个确定值,所以人们只能用如在某一确定时刻t_1,$X(t_1)$介于某一范围内的概率等来描述随机过程的统计特性。

例如,某随机过程$X(t)$在t_1时刻的所有样本函数中$x(t_1) \leqslant x$的概率记为分布函数

phase cosine wave (Figure 5.9(b)).

For the aforementioned random phase cosine wave, if we pay special attention to the function value at time $t = t_1$ in each observation, we will find that they are different.

First observation, we have

$$x_1(t_1) = A_m\cos(\omega t_1 + \theta_1)$$

Second observation, we have

$$x_2(t_1) = A_m\cos(\omega t_1 + \theta_2)$$

The nth observation, we have

$$x_n(t_1) = A_m\cos(\omega t_1 + \theta_n)$$

They are not a deterministic value like a deterministic process, but random variables that vary in the interval $[-A_m, A_m]$, so we have

$$X(t_1) = \{x_i(t_1)\}, \quad t_1 \in T$$

According to this meaning, the stochastic process has another form of definition: for every certain moment t_1, $X(t_1)$ is a random variable, rather than a certain value, then $X(t)$ is called a random process, or a random process $X(t)$ is a family of random variables that depend on time t.

The above two definitions are essentially the same. No matter which form of definition, they point out two fundamental characteristics of stochastic processes: first, it is a family (rather than a definite) function that depends on time t; sencond, at a certain moment it is a family of random variable (rather than a certain value).

When we study random processes, there is a certain difference in method compared with studying deterministic process. The deterministic process has inevitable regularities, whose research purpose is to discover inevitable regularities. For example, the relationship between the distance of a free fall and time, other involved factors, the falling distance at time t_1, etc. and the random process does not have an inevitable law of change, so the purpose of studying a random process is of course not to reveal its inevitable law, but to discover its statistical law, that is, regularity hidden in random phenomena. These include discovering the distribution law of the random variable e in the random process $X(t)$ within its value range and by what kind of rules the time function $X(t)$ containing e is established, and what the constant coefficients in the time function are. Only by clarifying these problems can we say that we have a clear understanding of the regularity of random processes. Therefore, the basic ideas for studying random processes can be summarized as follows.

(1) Statistical methods can be used to find the distribution law of the random variable e contained in the random process. This law is called the statistical characteristics of random variables.

(2) To find by what rules did $X(t)$ establish the time function.

(3) Mathematical analysis is used to calculate various numerical characteristics corresponding to the statistical characteristics of the random process after (1) is known.

(4) The value of each constant coefficient can be determined according to the relationship between each numerical feature and various coefficients in $X(t)$.

According to this idea, it can be said that the regularity of random process $X(t)$ is finally figured out before we makes it clear that the random process is at least the time function $X(e, a, a_2, \cdots, a_n, t)$ of the stationary random process common in engineering, clarifies the statistical

$$F_1(x,t_1) = P\{X(t_1) \leqslant x\} \tag{5.22}$$

式中　$P\{X(t_1) \leqslant x\}$——t_1 时刻随机过程的状态 $X(t_1) \leqslant x$ 的概率。

显然，在 t_1 确定后，$F_1(x,t_1)$ 仅是 x 的函数。当已知 $F_1(x,t_1)$ 时，$X(t_1)$ 介于区间 $(x, x+\Delta x]$ 内的概率就应该为

$$\begin{aligned} p\{x < X(t_1) \leqslant x + \Delta x\} &= p\{X(t_1) \leqslant (x+\Delta x)\} - p\{X(t_1) \leqslant x\} \\ &= F_1(x+\Delta x, t_1) - F_1(x,t_1) \end{aligned} \tag{5.23}$$

当 $\Delta x \to 0$ 时，若下列极限存在，则定义该极限为随机过程 $X(t)$ 在 t_1 时刻的概率密度函数，记为 $f_1(x,t_1)$，则有

$$\begin{aligned} f_1(x,t_1) &= \lim_{\Delta x \to 0} \frac{P\{x < X(t_1) \leqslant x + \Delta x\}}{\Delta x} \\ &= \lim_{\Delta x \to 0} \frac{F_1(x+\Delta x, t_1) - F(x,t_1)}{\Delta x} \\ &= \frac{\mathrm{d}F_1(x,t_1)}{\mathrm{d}x} = F_1'(x,t_1) \end{aligned} \tag{5.24}$$

若 $f_1(x,t_1)$ 在 x 点处连续，则 $f_1(x,t_1)$ 与 $F_1(x,t_1)$ 的关系还可以表示为

$$F_1(x,t_1) = \int_{-\infty}^{x} f_1(x,t_1)\,\mathrm{d}x \tag{5.25}$$

当在大量试验的基础上用统计的方法获得了随机过程在各个时刻的概率密度或分布函数之后，则在任何一个时刻 t_i，随机过程的状态 $X(t_i)$ 介于指定区间 $(x_1, x_2]$ $(x_2 > x_1)$ 的概率就可以很容易地求出来，即

$$P\{x_1 < X(t_i) \leqslant x_2\} = \int_{x_1}^{x_2} f_1(x,t_i)\,\mathrm{d}x = F_1(x_2,t_i) - F_1(x_1,t_i)$$

这样就把随机过程在任何一个孤立时刻的统计规律描述清楚了。正是因为 $f_1(x,t_1)$ 与 $F_1(x,t_1)$ 仅是描述随机过程各孤立时刻的统计特性，所以它们称为一维概率密度和一维分布函数，这两个函数的脚标为“1”。一维分布函数和一维概率密度对随机信号的描述是孤立的。

假如在观察随机过程时，不仅关注 t_1 时刻的状态 $X(t_1)$，还关注 t_2 时刻的状态 $X(t_2)$，那么同样可以得到，在 t_1 时刻 $X(t_1)$ 介于 $(x_{11}, x_{12}]$，随后在 t_2 时刻，$t_2 = t_1 + \tau$，$X(t_2)$ 又介于区间 $(x_{21}, x_{22}]$ 内的概率，即

$$\begin{aligned} & P\{x_{11} < X(t_1) \leqslant x_{12}, t_1, x_{21} < X(t_2) \leqslant x_{22}, t_2\} \\ &= \int_{x_{11}}^{x_{12}} \int_{x_{21}}^{x_{22}} f_2(x_1, x_2, t_1, t_2)\,\mathrm{d}x_1 \mathrm{d}x_2 \\ &= F_2(x_{12}, x_{22}, t_1, t_2) - F_2(x_{11}, x_{21}, t_1, t_2) \end{aligned}$$

characteristics of the random variable e in the formula and the various coefficients $a_1, a_2, \cdots, a_n$.

5.3.2 The Statistical Characteristics and Numerical Characteristics of Random Processes

The state of a random process at any time is a random variable, not a definite value, so people have to use such as the probability that $X(t_1)$ is within a certain range at a certain time t to describe the statistical characteristics of the random process.

For example, the probability of $x(t_1) \leqslant x$ in all sample functions of a random process $X(t)$ at time t_1 is recorded as a distribution function

$$F_1(x,t_1) = P\{X(t_1) \leqslant x\} \tag{5.22}$$

where　$P\{X(t_1) \leqslant x\}$ —probability of state $X(t_1) \leqslant x$ in random process at time t_1.

Obviously, after t_1 is determined, $F_1(x,t_1)$ is only a function of x. When $F_1(x,t_1)$ is known, the probability that $X(t_1)$ is within the interval $(x, x+\Delta x]$ should be

$$\begin{aligned} p\{x < X(t_1) \leqslant x + \Delta x\} &= p\{X(t_1) \leqslant (x+\Delta x)\} - p\{X(t_1) \leqslant x\} \\ &= F_1(x+\Delta x, t_1) - F_1(x,t_1) \end{aligned} \tag{5.23}$$

When $\Delta x \rightarrow 0$, if the following limit exists, define the limit as the probability density function of random process $X(t)$ at time t_1, denoted as $f_1(x,t_1)$, so we have

$$\begin{aligned} f_1(x,t_1) &= \lim_{\Delta x \to 0} \frac{P\{x < X(t_1) \leqslant x + \Delta x\}}{\Delta x} \\ &= \lim_{\Delta x \to 0} \frac{F_1(x+\Delta x, t_1) - F(x,t_1)}{\Delta x} \\ &= \frac{\mathrm{d}F_1(x,t_1)}{\mathrm{d}x} = F_1'(x,t_1) \end{aligned} \tag{5.24}$$

If $f_1(x,t_1)$ is continuous at point x, the relationship between $f_1(x,t_1)$ and $F_1(x,t_1)$ can also be expressed as

$$F_1(x,t_1) = \int_{-\infty}^{x} f_1(x,t_1)\,\mathrm{d}x \tag{5.25}$$

After a large number of experiments are used to obtain the probability density or distribution function of the random process at each time by statistical methods, the probability that the state of the random process $X(t_i)$ is within the specified interval $(x_1, x_2]$ $(x_2 > x_1)$ at any time t_i can be easily calculated come out, that is

$$P\{x_1 < X(t_i) \leqslant x_2\} = \int_{x_1}^{x_2} f_1(x,t_i)\,\mathrm{d}x = F_1(x_2,t_i) - F_1(x_1,t_i)$$

In this way, the statistical law of the random process at any isolated moment is described clearly. It is precisely because $f_1(x,t_1)$ and $F_1(x,t_1)$ only describe the statistical characteristics of each isolated moment of the random process, they are called one-dimensional probability density and one-dimensional distribution function, the footer of these two functions is "1". The description of random signals by one-dimensional distribution functions and one-dimensional probability density is isolated.

If, when we observe a random process, not only focus on the state $X(t_1)$ at time t_1, but also focus on the state $X(t_2)$ at time t_2, then it can also be obtained that at time t_1, $X(t_1)$ is between $(x_{11}, x_{12}]$, and then at time t_2, $t_2 = t_1 + \tau$, and $X(t_2)$ are between $(x_{21}, x_{22}]$, that is

式中　$f_2(x_1,x_2,t_1,t_2)$—— 随机变量 $X(t)$ 的二维概率密度；

$F_2(x_1,x_2,t_1,t_2)$——$X(t)$ 的二维分布函数。

考虑不同时刻 $x(t)$ 的走势如图 5.10 所示。该概率表现的是像图 5.10(b) 所示的那样，$X(t)$ 在 t_1 时刻穿越区间 $(x_{11},x_{12}]$，随后又在 t_2 时刻穿越区间 $(x_{21},x_{22}]$ 的概率，是考虑了两个不同时刻 $X(t)$ 的"走势" 的一种概率。显然，对随机过程的刻画比仅考虑某一时刻的"走势" 要全面一些。以此推论，若像图 5.10(c) 所示那样考虑更多时刻 $X(t)$ 依次穿行指定区间的概率，那么不就等于 $X(t)$ 穿行于由不同时刻的区间所构成的"走廊" 的概率吗？这种概率比一维、二维概率对随机过程的刻画要更全面得多。当然，相应地就需要用 n 维概率密度和 n 维分布函数，有

$$f_n(x_1,x_2,\cdots,x_n,t_1,t_2,\cdots,t_n)$$

$$F_n(x_1,x_2,\cdots,x_n,t_1,t_2,\cdots,t_n)$$

且

$$F_n(x_1,x_2,\cdots,x_n,t_1,t_2,\cdots,t_n)=\int_{-\infty}^{x_1}\int_{-\infty}^{x_2}\cdots\int_{-\infty}^{x_n}f_n(x_1,x_2,\cdots,x_n,t_1,t_2,\cdots,t_n)\,\mathrm{d}x_n\cdots\mathrm{d}x_2\mathrm{d}x_1$$

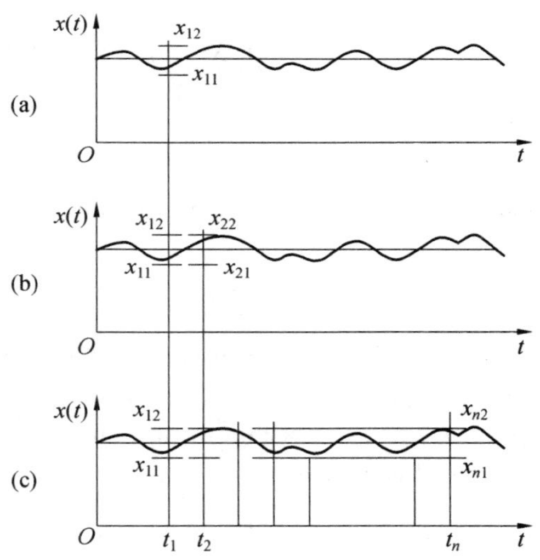

图 5.10　考虑不同时刻 $x(t)$ 的走势

当 $n\to+\infty$ 时，对随机过程的刻画就可以足够精确。概率密度及分布函数称为随机过程的统计特性，特别是多维概率密度或分布函数能精确地刻画随机过程的统计规律，但在实际使用中很不方便，甚至无法用它们来解决实际应用问题，因此引入随机过程的基本数字特征。数字特征是概率密度的导出量，它既能刻画随机过程的重要特征，又便于运算和实际测量，常用的数字特征是均值 $\mu_X(t)$ 和自相关函数 $R_{XX}(t_1,t_2)$，它们分别与一维和

$$P\{x_{11} < X(t_1) \leqslant x_{12}, t_1, x_{21} < X(t_2) \leqslant x_{22}, t_2\}$$
$$= \int_{x_{11}}^{x_{12}} \int_{x_{21}}^{x_{22}} f_2(x_1, x_2, t_1, t_2) \mathrm{d}x_1 \mathrm{d}x_2$$
$$= F_2(x_{12}, x_{22}, t_1, t_2) - F_2(x_{11}, x_{21}, t_1, t_2)$$

where $f_2(x_1, x_2, t_1, t_2)$ — the two-dimensional probability density of random variable $X(t)$;
$F_2(x_1, x_2, t_1, t_2)$— the two-dimensional distribution function of $X(t)$.

Consider the trend of $x(t)$ at different times is shown in Figure 5.10. This probability represents the probability that $X(t)$ crosses the interval $(x_{11}, x_{12}]$ at time t_1, and then crosses the interval $(x_{21}, x_{22}]$ at time t_2, as shown in Figure 5.10(b), it is a probability that considers the "trend" of $X(t)$ at two different times. Obviously, the description of the random process is more comprehensive than just considering the "trend" at a certain moment. Based on this inference, if we consider the probability that more times $X(t)$ will walk through the designated interval in turn, as shown in Figure 5.10(c), isn't it equal to the probability that $X(t)$ will walk through the "corridor" formed by intervals at different times? This probability is much more comprehensive than one-dimensional and two-dimensional probabilities in describing random processes. Of course, the corresponding n-dimensional probability density and n-dimensional distribution function are required, we have

$$f_n(x_1, x_2, \cdots, x_n, t_1, t_2, \cdots, t_n)$$
$$F_n(x_1, x_2, \cdots, x_n, t_1, t_2, \cdots, t_n)$$

and

$$F_n(x_1, x_2, \cdots, x_n, t_1, t_2, \cdots, t_n) = \int_{-\infty}^{x_1} \int_{-\infty}^{x_2} \cdots \int_{-\infty}^{x_n} f_n(x_1, x_2, \cdots, x_n, t_1, t_2, \cdots, t_n) \mathrm{d}x_n \cdots \mathrm{d}x_2 \mathrm{d}x_1$$

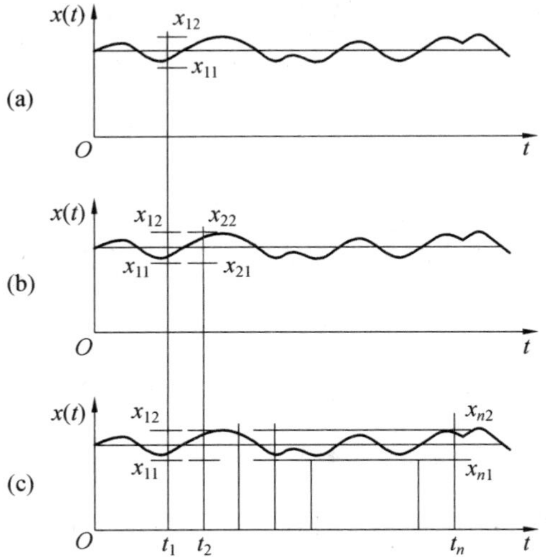

Figure 5.10 Consider the trend of $x(t)$ at different times

When $n \to +\infty$, the description of the random process can be accurate enough. Probability density and distribution functions are called the statistical characteristics of random processes. They, especially multi-dimensional probability density or distribution functions, can accurately describe the statistical laws of random processes, but they are very inconvenient in actual use, and they cannot even be used to solve practical application problems, thus the basic

二维概率密度相对应。

均值 $\mu_X(t)$ 一般与 t 有关。在指定时刻 t_1，随机过程 $X(t)$ 的均值记为 $\mu_X(t_1)$，其定义为

$$\mu_X(t_1)=E[X(t_1)]=\int_{-\infty}^{\infty}xf_1(x,t_1)\,\mathrm{d}x \tag{5.26}$$

式中 $f_1(x,t_1)$—— 随机过程 $X(t)$ 的一维概率密度。

实际上，它是随机过程 $X(t)$ 的所有样本函数在时刻 t_1 的函数值的数学期望，称为集合平均。

自相关函数 $R_X(t_1,t_2)$ 的定义为

$$\begin{aligned}R_{XX}(t_1,t_2)&=E[X(t_1)X(t_2)]\\&=\int_{-\infty}^{\infty}\int_{-\infty}^{\infty}x_1x_2f_2(x_1,x_2,t_1,t_2)\,\mathrm{d}x_1\mathrm{d}x_2\end{aligned} \tag{5.27}$$

式中 $X(t_1)$、$X(t_2)$—— 随机过程 $X(t)$ 在任意两时刻 t_1,t_2 的状态(图 5.11)；

$f_2(x_1,x_2,t_1,t_2)$—— 随机过程 $X(t)$ 的二维概率密度。

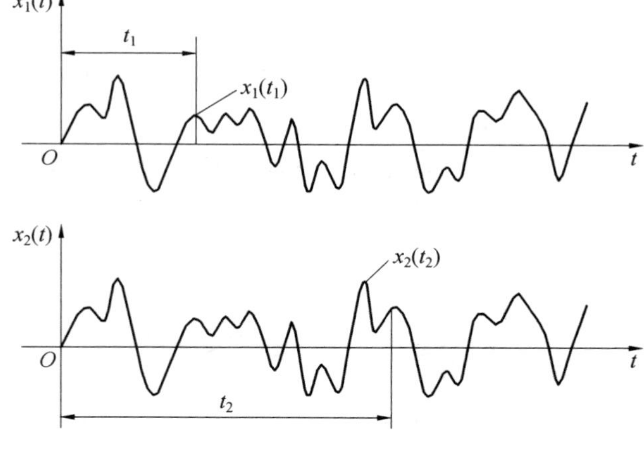

图 5.11 随机信号在不同时刻的状态

自相关函数又称二阶原点混合矩，在不致引起混淆的情况下常简记为 $R_X(t_1,t_2)$。

均值 $\mu_X(t)$ 表示了随机过程在各个不同时刻的摆动中心。自相关函数刻画了随机过程自身在两个不同时刻状态之间的依从关系。

涉及多维概率密度的数字特征一般就不再引申，因为在工程测试领域内只涉及随机信号的分析及处理问题，对概率密度或分布函数的研究仅停留在一维及二维范围内，对数字特征的研究仅集中在均值和自相关函数两方面，这对工程测试中的信号分析和处理来说已经足够了。因此，在概率数学中形成一个以研究随机过程的均值和自相关函数为主

numerical characteristics of random processes should be introduced. Digital feature is the derived quantity of probability density, which can not only describe the important features of random process, but also facilitate calculation and actual measurement. Commonly used are mean $\mu_X(t)$ and autocorrelation function $R_{XX}(t_1, t_2)$. They correspond to one-dimensional and two-dimensional probability densities respectively.

The mean value $\mu_X(t)$ is generally related to t. At the specified time t_1, the mean value of the random process $X(t)$ is denoted as $\mu_X(t_1)$, which is defined as

$$\mu_X(t_1) = E[X(t_1)] = \int_{-\infty}^{\infty} x f_1(x, t_1)\,\mathrm{d}x \tag{5.26}$$

where　$f_1(x, t_1)$ —one-dimensional probability density of random process $X(t)$.

In fact, it is the mathematical expectation of the function value of all the sample functions of the random process $X(t)$ at time t_1, which is called the set average.

The autocorrelation function $R_X(t_1, t_2)$ is defined as

$$\begin{aligned} R_{XX}(t_1, t_2) &= E[X(t_1)X(t_2)] \\ &= \int_{-\infty}^{\infty}\int_{-\infty}^{\infty} x_1 x_2 f_2(x_1, x_2, t_1, t_2)\,\mathrm{d}x_1\mathrm{d}x_2 \end{aligned} \tag{5.27}$$

where　$X(t_1), X(t_2)$ —the state of t_1 and t_2 in random process $X(t)$ at any two moments (Figure 5.11);

$f_2(x_1, x_2, t_1, t_2)$ —the two-dimensional probability density of random process $X(t)$.

The autocorrelation function is also called the second-order origin mixing moment, and is often abbreviated as $R_X(t_1, t_2)$ when it does not cause confusion.

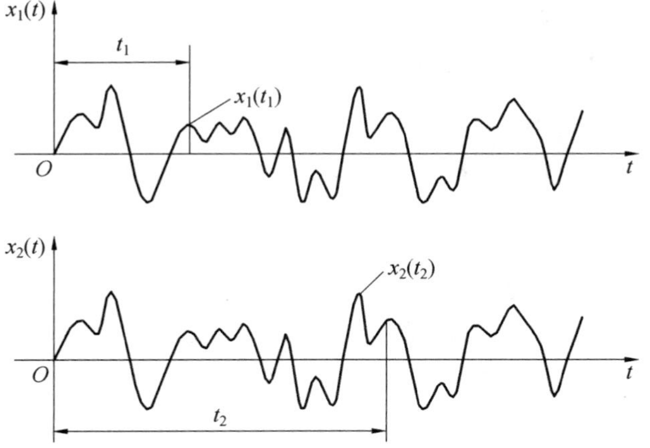

Figure 5.11　The state of random signals at different moments

The mean value $\mu_X(t)$ represents the swing center of the random process at different times. The autocorrelation function describes the dependence of the stochastic process itself between the states at two different moments.

Digital features involving multi-dimensional probability density are generally no longer extended, because in the field of engineering measurement, only random signal analysis and processing problems are involved, and the research on probability density or distribution function only stays in the range of one and two dimensions. The study of digital features only focuses on the mean value and autocorrelation function, which is sufficient for signal analysis

要内容的分支,称为相关理论,它就是信号分析的基础理论。

5.3.3 随机过程的分类

随机过程本质性的分类是按其分布函数或概率密度的不同特点来分类的,不做详述,仅指明这种分类中有一种过程称为平稳随机过程,其余几种统称为非平稳随机过程。平稳随机过程的特点是过程的统计特性不随时间原点的选择而变化,严格地说就是它的 n 维分布函数或概率密度不随其 n 个时间参数 $t_1,t_2,\cdots,t_n$ 的原点的选择而变化。如果在研究某一随机过程时,其前后的环境和主要条件都不随时间变化,则一般就可以认为它是平稳随机过程,但要严格地按 n 维分布函数或概率密度的特点来证明它的平稳性并不是容易的。因此,常根据真实工程问题的实际需要,将平稳随机过程定义中所涉及的 n 维分布函数或概率密度限定在一维、二维范围之内,即仅涉及与一维、二维概率密度有关的数字特征(即均值和自相关函数)。将 n 维降低到一、二维后所定义的平稳随机过程称为宽平稳随机过程或广义平稳过程,以区别涉及 n 维统计特性的严格平稳随机过程。本书今后所讨论的平稳过程都是指宽平稳随机过程,其具体定义如下。

给定随机过程 $X(t)$,如果其均值 $E[X(t)]$ 等于常数,且自相关函数 $R_X(t_1,t_2)$ 仅是时差 $t_2-t_1=\tau$ 的函数,它们都与时间 t 的原点的选择无关,或索性说与时间 t 无关,则当 $X(t)$ 是功率有限信号时,就称 $X(t)$ 为宽平稳随机过程,或简称平稳过程。这样,平稳随机过程的自相关函数就可以简写为 $R_X(\tau)$。

平稳过程的均值 $E[X(t)]$ 等于常数,即 $t=t_1$ 时刻随机变量 $X(t_1)$ 和 $t=t_2$ 时刻随机变量 $X(t_2)$ 的均值都等于同一常数,该常数不因 t_1 和 t_2 的变化而变化。这一特点为根据样本函数曲线粗略判断过程的平稳性提供了必要的依据。若过程是平稳的,则其必要条件体现在样本记录曲线上为:各曲线都在同一条水平线上下随机波动,正如飞机完成爬升的过渡过程以后,按预定高度飞行时,由于大气湍流的影响,因此实际飞行高度 $H(t)$ 应在控制高度 h 上下随机波动。这时,$H(t)$ 就可以看成平稳过程,h 就是它的均值,是一个不变的常数。图 5.12 所示为非平衡随机过程,其均值 $\mu_X(t)$ 不是常数,据此就可以将 $X(t)$ 排除在平稳过程之外。

在计算均值和自相关函数时,要分别针对随机过程在某一时刻状态 $X(t_1)$ 及某两个时刻状态 $X(t_1)$、$X(t_2)$,用计算数学期望的办法来进行,这种方法称为集合平均。无疑,使用集合平均的首要条件是拥有大量以至无穷多个样本函数,这是相当困难甚至是办不到的。所幸的是,平稳过程中有这样一类过程,按集合平均所计算的数字特征和按时间平均所计算的数字特征是相等的,这一类过程称为各态历经随机过程。对这类随机过程,就

and processing in engineering measurement. Therefore, a branch of probabilistic mathematics that focuses on studying the mean value and autocorrelation function of random processes is formed, called correlation theory, which is the basic theory of signal analysis.

5.3.3 The Classification of Random Processes

The essential classification of random processes is classified according to the different characteristics of its distribution function or probability density. It will not be described in detail, but only to indicate that one of the processes in this classification is called a stationary random process, and the rest are collectively called non-stationary random process. The characteristic of a stationary random process is that the statistical characteristics of the process do not change with the choice of the time origin. Strictly speaking, its n -dimensional distribution function or probability density does not change with the choice of the origin of the nth time parameters t_1, $t_2, \cdots, t_n$. If the environment and main conditions before and after a certain random process do not change with time when studying it, it can generally be considered as a stationary random process, but it is not easy to prove its stationarity strictly according to the characteristics of n -dimensional distribution function or probability density. Therefore, n -dimensional distribution function or probability density involved in the definition of a stationary random process is often limited to the one-dimensional and two-dimensional range according to the actual needs of real engineering problems, that is, only digital features related to one-dimensional and two-dimensional probability densities (that is, mean and autocorrelation functions) are involved. After reducing n dimensions to one or two dimensions, the defined stationary random process is called broad stationary random process or generalized stationary process to distinguish strictly stationary random processes involving n -dimensional statistical properties. The stationary processes discussed in this book will all refer to wide stationary random processes. Its specific definition is as follows.

Given a random process $X(t)$, if its mean value $E[X(t)]$ is equal to a constant, and the autocorrelation function $R_X(t_1, t_2)$ is only a function of the time difference $t_2 - t_1 = \tau$, they are not related to the choice of the origin of time t , or simply not related to time t , then when $X(t)$ is a signal with limited power, $X(t)$ is called a wide stationary random process, or simply a stationary process. In this way, the autocorrelation function of a stationary random process can be abbreviated as $R_X(\tau)$.

The mean value $E[X(t)]$ of a stationary process is equal to a constant, that is, the mean value of the random variable $X(t_1)$ at time $t = t_1$ and the mean value of the random variable $X(t_2)$ at time $t = t_2$ are both equal to the same constant, and the constant does not change with the change of t_1 and t_2. This feature provides a necessary basis for roughly judging the smoothness of the process based on the sample function curve. If the process is stable, the necessary condition is reflected in the sample record curve as follows: each curve fluctuates randomly on the same horizontal line, just as an airplane completes after the transition process of climb, when flying at a predetermined altitude, due to the influence of atmospheric turbulence, the actual flying altitude $H(t)$ should fluctuate randomly up and down the control altitude h . At this time, $H(t)$ can be regarded as a stable process, and h is its average value, which is a constant. The non-stationary random process is shown in Figure 5.12, the mean value $\mu_X(t)$ is not a constant, so $X(t)$ can be excluded from the stationary process.

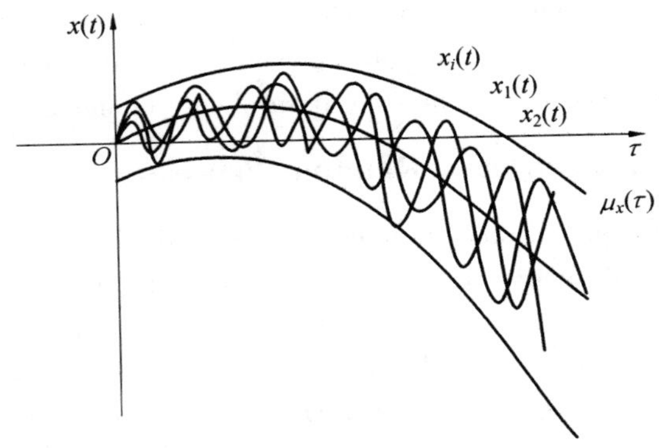

图 5.12 非平衡随机过程

可以根据一次实验所得到的样本函数 $x(t)$ 按时间平均来计算它的各种数字特征，而不必要大量的样本函数，这样在解决实际问题的时候就可以减少大量的工作量，甚至可以使原来办不到的事情成为可能。

例如，上文所提到的随机相位余弦波（或正弦波），即

$$X(t)=A_{\mathrm{m}}\cos(2\pi f_0 t+\theta)$$

在 $t=t_1$ 时刻，其均值为

$$\begin{aligned}E[X(t_1)]&=E[A_{\mathrm{m}}\cos(2\pi f_0 t_1+\theta)]\\&=\int_0^{2\pi}A_{\mathrm{m}}\cos(2\pi f_0 t_1+\theta)f(\theta)\,\mathrm{d}\theta\\&=\int_0^{2\pi}A_{\mathrm{m}}\cos(2\pi f_0 t_1+\theta)\frac{1}{2\pi}\mathrm{d}\theta=0\end{aligned}$$

式中 $f(\theta)$——随机变量 θ 在区间 $[0,2\pi]$ 内均匀分布的概率密度，有

$$f(\theta)=\begin{cases}\dfrac{1}{2\pi}, & 0<\theta<2\pi\\ 0, & 其他\end{cases}$$

其自相关函数为

$$\begin{aligned}R_X(t_1,t_2)&=E[X(t_1)X(t_2)]\\&=E[A_{\mathrm{m}}\cos(2\pi f_0 t_1+\theta)A_{\mathrm{m}}\cos(2\pi f_0 t_2+\theta)]\\&=\int_0^{2\pi}A_{\mathrm{m}}\cos(2\pi f_0 t_1+\theta)A_{\mathrm{m}}\cos(2\pi f_0 t_2+\theta)f(\theta)\,\mathrm{d}\theta\\&=A_{\mathrm{m}}^2\int_0^{2\pi}\frac{1}{2}\{\cos 2\pi f_0(t_1-t_2)+\cos[2\pi f_0(t_1+t_2)+2\theta]\}\frac{1}{2\pi}\mathrm{d}\theta\\&=\frac{A_{\mathrm{m}}^2}{2}\cos 2\pi f_0\tau=\sigma_X^2\cos 2\pi f_0\tau\end{aligned}\tag{5.28}$$

When calculating the mean value and autocorrelation function, it is necessary to calculate the mathematical expectation for the random process at a certain moment state $X(t_1)$ and a certain two moment states $X(t_1), X(t_2)$ respectively. This method is called ensemble averaging. Undoubtedly, the first condition for using it is to have a large or even infinite number of sample functions, which is quite difficult or even impossible. Fortunately, there is a kind of process in the stationary process that the digital feature calculated by the set average and the digital feature calculated by the time average are equal, which is called the random process of each state. For this kind of random process, it is possible to calculate its various numerical characteristics according to the sample function $x(t)$ obtained in an experiment on a time average, instead of a large number of sample functions, so that a large amount of work can be reduced when solving practical problems, and even things that could not be done before become possible.

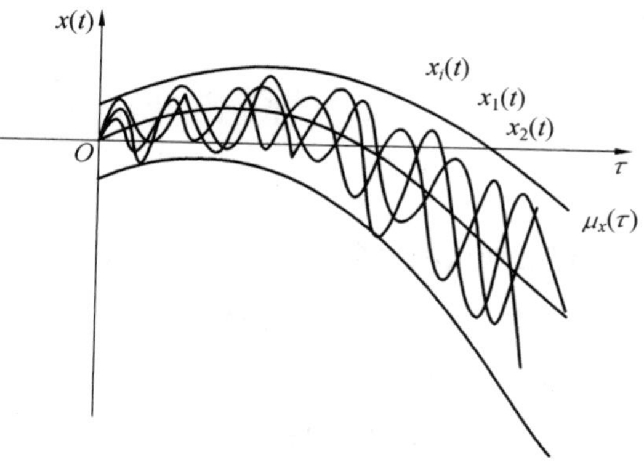

Figure 5.12　Non-stationary random process

For example, the random phase cosine wave (or sine wave) mentioned above, that is

$$X(t) = A_m\cos(2\pi f_0 t + \theta)$$

Its mean value at time $t = t_1$ is

$$\begin{aligned} E[X(t_1)] &= E[A_m\cos(2\pi f_0 t_1 + \theta)] \\ &= \int_0^{2\pi} A_m\cos(2\pi f_0 t_1 + \theta) f(\theta)\,d\theta \\ &= \int_0^{2\pi} A_m\cos(2\pi f_0 t_1 + \theta)\frac{1}{2\pi}d\theta \\ &= 0 \end{aligned}$$

where　$f(\theta)$ — the probability density of the random variable θ uniformly distributed in the interval $[0, 2\pi]$, we have

$$f(\theta) = \begin{cases} \dfrac{1}{2\pi}, & 0 < \theta < 2\pi \\ 0, & \text{other} \end{cases}$$

Its autocorrelation function is

$$\begin{aligned} R_X(t_1, t_2) &= E[X(t_1)X(t_2)] \\ &= E[A_m\cos(2\pi f_0 t_1 + \theta)A_m\cos(2\pi f_0 t_2 + \theta)] \\ &= \int_0^{2\pi} A_m\cos(2\pi f_0 t_1 + \theta)A_m\cos(2\pi f_0 t_2 + \theta) f(\theta)\,d\theta \\ &= A_m^2\int_0^{2\pi}\frac{1}{2}\{\cos 2\pi f_0(t_1 - t_2) + \cos[2\pi f_0(t_1 + t_2) + 2\theta]\}\frac{1}{2\pi}d\theta \\ &= \frac{A_m^2}{2}\cos 2\pi f_0\tau = \sigma_X^2\cos 2\pi f_0\tau \end{aligned} \tag{5.28}$$

where

式中

$$\tau = t_2 - t_1$$

$$\sigma_X^2 = \frac{A_m^2}{2}$$

可见，随机相位余弦波的均值与时间 t 的原点选择无关，恒为零，而自相关函数是 τ 的单变量函数，因此随机相位余弦波是平稳的随机过程。同理，可以证明随机相位正弦波也是平稳随机过程。

若根据某一样本函数 $x_i(t) = A_m\cos(2\pi f_0 t + \theta_i)$，按时间平均来计算时，则均值 μ_X 为

$$\mu_X = \langle x_i(t) \rangle = \lim_{T\to\infty} \frac{1}{T}\int_0^T A_m\cos(2\pi f_0 t + \theta_i)\,\mathrm{d}t = 0$$

式中　T—— 观测时间；

$\langle x_i(t) \rangle$—— 按上述极限取时间平均值时的数学符号。

自相关函数为

$$\begin{aligned} R_X(t_1, t_2) &= \langle x_i(t_1)x_i(t_2) \rangle \\ &= \lim_{T\to\infty} \frac{1}{T}\int_0^T A_m\cos(2\pi f_0 t_1 + \theta_i)A_m(2\pi f_0 t_2 + \theta_i)\,\mathrm{d}t \\ &= \lim_{T\to\infty} \frac{1}{T}A_m^2\int_0^T \frac{1}{2}\{\cos 2\pi f_0(t_1 - t_2) + \cos[2\pi f_0(t_1 + t_2) + 2\theta_i]\}\,\mathrm{d}t \\ &= \frac{A_m^2}{2}\cos 2\pi f_0\tau = \sigma_X^2\cos 2\pi f_0\tau \end{aligned}$$

通过两种平均方法所计算的随机相位余弦波的均值及自相关函数是相等的。可见，随机相位余弦波不仅是平稳随机过程，还是各态历经随机过程。读者可以用同样的方法自行证明随机相位正弦波。

在概率论中，有关平稳随机过程各态历经性的概念如下。

设 $X(t)$ 是平稳随机过程，如果

$$\langle x_i(t) \rangle = E[X(t)] = \mu_X$$

对所有的样本函数都成立，则称 $X(t)$ 的均值具有各态历经性。

如果

$$\langle x_i(t)\, x_i(t+\tau) \rangle = E[X(t)X(t+\tau)] = R_X(\tau)$$

对所有的样本函数都成立，则称过程 $X(t)$ 的自相函数具有各态历经性。

如果 $X(t)$ 的均值和自相关函数都具有各态历经性，则称 $X(t)$ 是各态历经的，或 $X(t)$ 是(宽) 各态历经过程。

$$\tau = t_2 - t_1$$

$$\sigma_X^2 = \frac{A_{\mathrm{m}}^2}{2}$$

It can be seen that the mean value of the random phase cosine wave has nothing to do with the origin of timet, it is always zero, and the autocorrelation function is a univariate function of τ , so the random phase cosine wave is a stable random process, and similarly, it can be proved that the random phase sine wave is also a stationary random process.

If according to a sample function $x_i(t) = A_{\mathrm{m}}\cos(2\pi f_0 t + \theta_i)$, when calculating by time average, the mean value μ_X is

$$\mu_X = \langle x_i(t) \rangle = \lim_{T \to \infty} \frac{1}{T} \int_0^T A_{\mathrm{m}}\cos(2\pi f_0 t + \theta_i)\,\mathrm{d}t = 0$$

where　T —observation time;

$\langle x_i(t) \rangle$ — the mathematical symbol when the time average is taken according to the above limit.

The autocorrelation function is

$$\begin{aligned} R_X(t_1, t_2) &= \langle x_i(t_1) x_i(t_2) \rangle \\ &= \lim_{T \to \infty} \frac{1}{T} \int_0^T A_{\mathrm{m}}\cos(2\pi f_0 t_1 + \theta_i) A_{\mathrm{m}}(2\pi f_0 t_2 + \theta_i)\,\mathrm{d}t \\ &= \lim_{T \to \infty} \frac{1}{T} A_{\mathrm{m}}^2 \int_0^T \frac{1}{2} \{\cos 2\pi f_0(t_1 - t_2) + \cos[2\pi f_0(t_1 + t_2) + 2\theta_i]\}\,\mathrm{d}t \\ &= \frac{A_{\mathrm{m}}^2}{2}\cos 2\pi f_0 \tau = \sigma_X^2 \cos 2\pi f_0 \tau \end{aligned}$$

The mean value and autocorrelation function of the random phase cosine wave calculated by the two averaging methods are equal. It can be seen that the random phase cosine wave is not only a stable random process, but also a random process in each state. Readers can prove the random phase sine wave by yourself in the same way.

In probability theory, the concepts of ergodicity of stationary random processes are as follows.

Let $X(t)$ be a stationary random process, if

$$\langle x_i(t) \rangle = E[X(t)] = \mu_X$$

for all the sample functions are true, the mean value of $X(t)$ is said to have ergonomics.

If

$$\langle x_i(t)\, x_i(t + \tau) \rangle = E[X(t)X(t + \tau)] = R_X(\tau)$$

for all sample functions, it is said that the self-phase function of process $X(t)$ has ergonomics.

If both the mean value and autocorrelation function of $X(t)$ have the ergodicity of each state, then it is said that $X(t)$ is the ergodic process of each state, or $X(t)$ is the (wide) ergodic process of each state.

A stable process is a very important and basic type of random process. Many of the processes encountered in the engineering field can be considered as stable processes, and the conditions required for the ergonomics of various states are relatively loose, and most random processes can be satisfied. Although we are not sure about certain processes, we can deal with them according to the process of each state first, and finally see whether the results are consistent with the actual situation. If they do not match, then modify the original assumptions

平稳过程是很重要很基本的一类随机过程,工程领域中所遇到的过程很多都可以认为是平稳过程,且各态历经性所要求的条件也比较宽松,大多数随机过程都能满足。虽然对某些过程不能肯定,但可以先按各态历经过程来处理,最后看其结果与实际情况是否符合,如果不符,则再修改原来的假设,另做处理。这样做,一方面是根据工程上遇到的大多数随机过程确实都是各态历经过程这一个基本事实;另一方面也是因为要验证某过程具有各态历经性确实十分困难。

今后,本书所涉及的随机过程仅局限在各态历经过程范畴之内。

5.3.4 随机信号的主要特征参数

1. 一维概率密度$f_1(x)$

设随机信号 $X(t)$ 的一个样本函数为 $x(t)$(图 5.13),$x(t)$ 落在$(x,x+\Delta x]$ 区间内的时间为

$$T_x = \Delta t_1 + \Delta t_2 + \cdots + \Delta t_n = \sum_{i=1}^{n} \Delta t_i$$

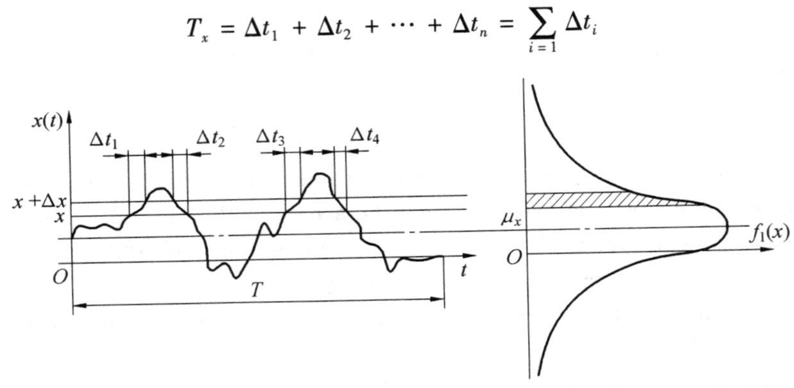

图 5.13　概率密度函数的计算

当样本函数的记录时间 T 趋于无穷大时,比值 T_x/T 的极限就是幅值 $x(t)$ 落于$(x,x+\Delta x]$ 区间的概率,记为

$$P[x < x(t) \leqslant x + \Delta x] = \lim_{T\to\infty} \frac{T_x}{T} \tag{5.29}$$

根据定义,随机信号的一维概率密度为

$$f_1(x) = \lim_{\Delta x\to 0} \frac{P[x < x(t) \leqslant x + \Delta x]}{\Delta x} \tag{5.30}$$

不同的随机信号有不同的概率密度,其密度曲线的图形各不一样,可以借此来识别信号的类型,这就是在此介绍一维概率密度的主要目的。

图 5.14 所示为常见的四种随机信号(假设这些信号的均值为零)的概率密度图形。

and deal with them separately. This is done on the one hand based on the basic fact that most random processes encountered in engineering are indeed all-state ergodic processes; and on the other hand, it is indeed very difficult to verify that a certain process has all-state ergonomics.

In the future, the random processes involved in this book are limited to the category of ergonomic processes.

5.3.4 The Main Characteristic Parameters of Random Signals

1. One-Dimensional Probability Density $f_1(x)$

Suppose a sample function of the random signal $X(t)$ is $x(t)$ (Figure 5.13), the time that $x(t)$ falls in the $(x, x+\Delta x]$ interval is

$$T_x = \Delta t_1 + \Delta t_2 + \cdots + \Delta t_n = \sum_{i=1}^{n} \Delta t_i$$

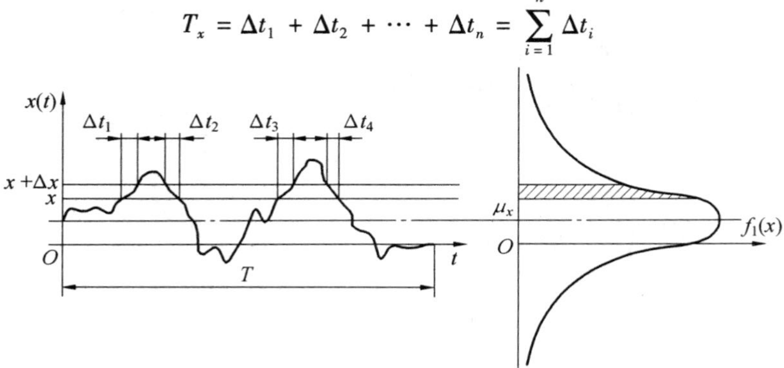

Figure 5.13　Calculation of probability density function

When the recording time T of the sample function tends to infinity, the limit of the ratio T_x/T is the probability that the amplitude $x(t)$ falls in the interval $(x, x+\Delta x]$, denoted as

$$P[x < x(t) \leqslant x + \Delta x] = \lim_{T \to \infty} \frac{T_x}{T} \tag{5.29}$$

According to the definition, the one-dimensional probability density of a random signal is

$$f_1(x) = \lim_{\Delta x \to 0} \frac{P[x < x(t) \leqslant x + \Delta x]}{\Delta x} \tag{5.30}$$

Different random signals have different probability densities, and the graphs of their density curves are different, which can be used to identify the type of signal, and this is the main purpose of introducing one-dimensional probability density here.

Figure 5.14 shows the probability density graph of four common random signals (assuming the mean value of these signals is zero). Without knowing what kind of distribution the data of the processed random signal obeys, the method of drawing a histogram to estimate the probability density can be used.

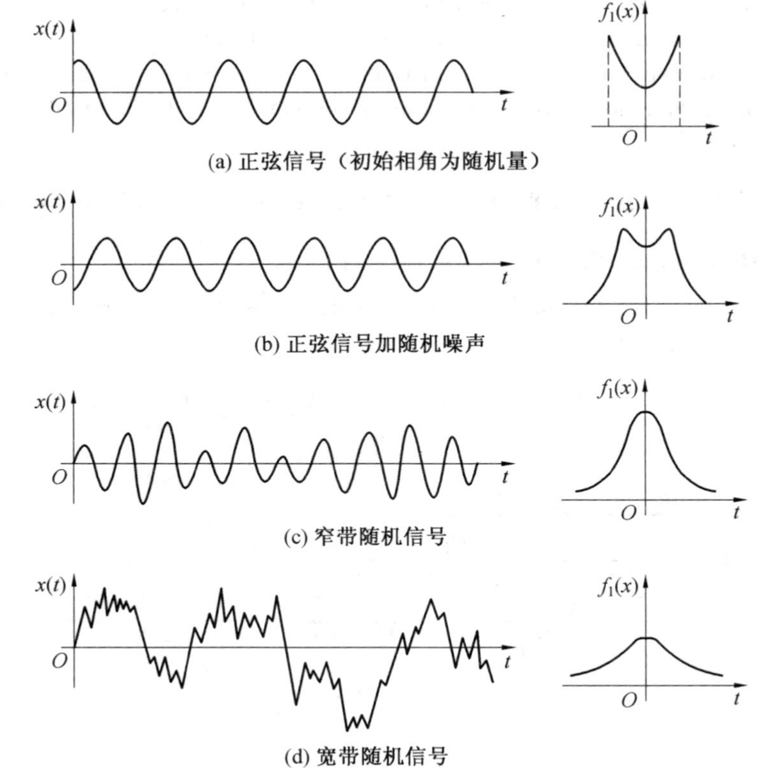

图 5.14　常见的四种随机信号的概率密度图形

当暂时不知道所处理的随机信号的数据服从何种分布时，可以用绘制直方图的方法来估计概率密度。

2. 主要数字特征

随机过程的主要数字特征是均值和自相关函数，它们分别与一维和二维概率密度相对应。对于平稳随机过程，除以上两个数字特征外，在实际应用中还有自功率谱密度函数均方值和方差，它们各从不同的侧面刻画出随机信号的统计特征，分述如下。

设 $x(t)$ 是各态历经平稳随机过程的一个样本函数，用时间平均来计算。

(1) 均值 μ_X。

$$\mu_X = E[X(t)] = \langle x(t) \rangle = \lim_{T\to\infty} \frac{1}{T}\int_0^T x(t)\,\mathrm{d}t \tag{5.31}$$

式中　T—— 观测时间；

μ_X—— 随机信号的常值分量，或波动中心。

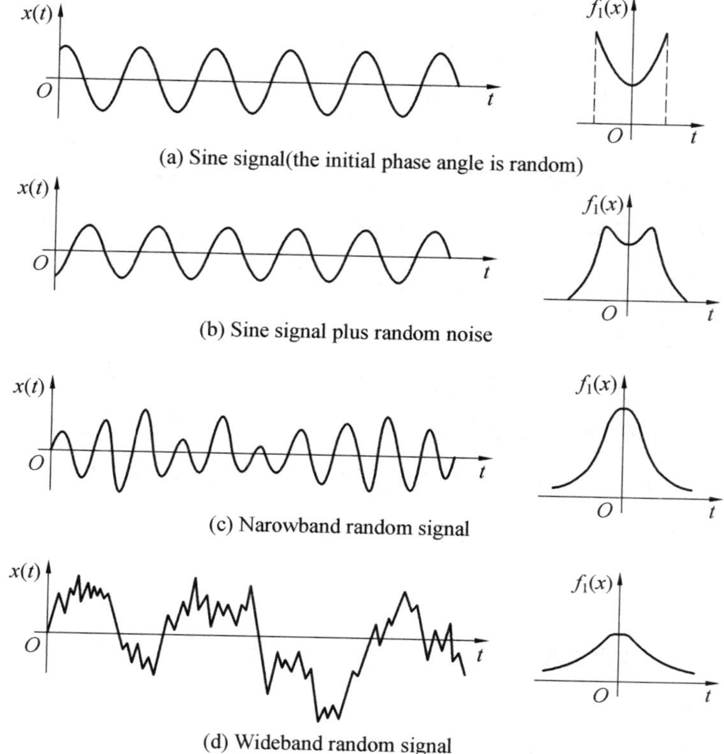

(a) Sine signal(the initial phase angle is random)

(b) Sine signal plus random noise

(c) Narowband random signal

(d) Wideband random signal

Figure 5. 14　Probability density graph of four common random signals

2. Main Digital Features

The main digital features of random processes are the mean value and the autocorrelation function, which correspond to one-dimensional and two-dimensional probability densities, respectively. For a stationary random process, there are also the mean square value and variance of the self-power spectral density function in practical applications in addition to the above two digital features. They portray the statistical characteristics of random signals from different sides. The breakdown is as follows.

Let $x(t)$ be a sample function of each state undergoing a stationary random process, and use the time average to calculate.

(1) Mean value μ_X .

$$\mu_X = E[X(t)] = \langle x(t) \rangle = \lim_{T \to \infty} \frac{1}{T} \int_0^T x(t)\,dt \tag{5.31}$$

where　T —observation time;

μ_X —the constant component of a random signal, or the center of fluctuation.

(2) Autocorrelation function $R_X(\tau)$.

$$\begin{aligned} R_X(\tau) &= E[X(t)X(t+\tau)] = \langle x(t)x(t+\tau) \rangle \\ &= \lim_{T \to \infty} \frac{1}{T} \int_0^T x(t)x(t+\tau)\,dt \end{aligned} \tag{5.32}$$

(2) 自相关函数 $R_X(\tau)$。

$$R_X(\tau) = E[X(t)X(t+\tau)] = \langle x(t)x(t+\tau) \rangle$$
$$= \lim_{T\to\infty} \frac{1}{T}\int_0^T x(t)x(t+\tau)\mathrm{d}t \tag{5.32}$$

$R_X(\tau)$ 表现了随机信号在两个不同时刻的状态的依从关系。

以上两个数字特征是最基本的。对平稳随机过程而言,μ_X 是常数,$R_X(\tau)$ 是 τ 的单变量函数。因此,根据 $R_X(\tau)$ 还可以引申出以下数字特征。

(3) 自功率谱密度函数 $S_X(f)$。

$R_X(\tau)$ 是时移 τ 的确定性函数,是平稳随机信号的动态数字特征。根据动态信号的基本属性可知,它能够分解为许多简谐波,或可以视为许多简谐波的叠加,即它有确定的频谱,因此定义 $R_X(\tau)$ 的频谱或频谱密度函数为平稳随机信号的自功率谱。若 $R_X(\tau)$ 满足傅里叶变换条件,则它的傅里叶变换

$$S_X(f) = \int_{-\infty}^{\infty} R_X(\tau)\mathrm{e}^{-\mathrm{j}2\pi f\tau}\mathrm{d}\tau \tag{5.33}$$

就是随机信号 $X(t)$ 的自功率谱密度函数,简称自功率谱或自谱,它与自相关函数是一个傅立叶变换对,即

$$R_X(\tau) = \int_{-\infty}^{\infty} S_X(f)\mathrm{e}^{\mathrm{j}2\pi f\tau}\mathrm{d}f \tag{5.34}$$

(4) 均方值 ψ_X^2。

当 $\tau=0$ 时,$R_X(\tau)=R_X(0)=E[X^2(t)]$,称 $E[X^2(t)]$ 为平稳随机过程的均方值,记为 ψ_X^2。当 $X(t)$ 或 $R_X(\tau)$ 具有各态历经性时,有

$$\psi_X^2 = \langle x^2(t) \rangle = \lim_{T\to\infty} \frac{1}{T}\int_0^T x^2(t)\mathrm{d}t \tag{5.35}$$

根据此式,当不考虑信号的实际量纲时,ψ_X^2 就是该随机信号的平均功率。其正平方根就是该信号的有效值,记为 ψ_X。

(5) 方差 σ_X^2。

当均值 $\mu_X \neq 0$ 时,$X(t)$ 可写成 $X(t)=[X(t)-\mu_X]+\mu_X$,其中 $[X(t)-\mu_X]$ 是随机过程的交流分量,有

$$\psi_X^2 = E[X^2(t)] = E\{[X(t)-\mu_X+\mu_X]^2\}$$
$$= E\{[X(t)-\mu_X]^2 + 2[X(t)-\mu_X]\mu_X + \mu_X^2\}$$
$$= E\{[X(t)-\mu_X]^2\} + \mu_X^2$$

称 $E\{[X(t)-\mu_X]^2\}$ 为平稳随机过程的方差,记为 σ_X^2,故有

$R_X(\tau)$ shows the dependence of the state of the random signal at two different moments.

The above two digital features are the most basic. For a stationary random process, μ_X is a constant, and $R_X(\tau)$ is a univariate function of τ. Therefore, according to $R_X(\tau)$, the following digital features can also be derived.

(3) Self-power spectral density function $S_X(f)$.

$R_X(\tau)$ is the deterministic function of time-shift τ. It is the dynamic digital characteristic of a stationary random signal. According to the basic properties of the dynamic signal, it can be decomposed into many simple harmonics, or it can be regarded as the superposition of many simple harmonics, which means that there is a certain frequency spectrum, so the frequency spectrum or spectral density function of $R_X(\tau)$ is defined as the self-power spectrum of a stationary random signal. If $R_X(\tau)$ satisfies the Fourier transform condition, then its Fourier transform

$$S_X(f) = \int_{-\infty}^{\infty} R_X(\tau)\, e^{-j2\pi f\tau} d\tau \tag{5.33}$$

is the self-power spectral density function of the random signal $X(t)$, referred to as self-power spectrum or self-spectrum. It is a Fourier transform pair with the autocorrelation function, that is

$$R_X(\tau) = \int_{-\infty}^{\infty} S_X(f)\, e^{j2\pi f\tau} df \tag{5.34}$$

(4) Mean square value ψ_X^2.

When $\tau = 0$, $R_X(\tau) = R_X(0) = E[X^2(t)]$, and $E[X^2(t)]$ is called the mean square value of a stationary random process, denoted as ψ_X^2. When $X(t)$ or $R_X(\tau)$ has ergonomics, we have

$$\psi_X^2 = \langle x^2(t) \rangle = \lim_{T\to\infty} \frac{1}{T}\int_0^T x^2(t)\, dt \tag{5.35}$$

According to this formula, ψ_X^2 is the average power of the random signal when the actual dimension of the signal is not considered. Its positive square root is the effective value of the signal, denoted as ψ_X.

(5) Variance σ_X^2.

When the mean value is $\mu_X \neq 0$, $X(t)$ can be written as $X(t) = [X(t) - \mu_X] + \mu_X$, where $[X(t) - \mu_X]$ is the AC component of the random process, so we have

$$\begin{aligned}\psi_X^2 &= E[X^2(t)] = E\{[X(t) - \mu_X + \mu_X]^2\} \\ &= E\{[X(t) - \mu_X]^2 + 2[X(t) - \mu_X]\mu_X + \mu_X^2\} \\ &= E\{[X(t) - \mu_X]^2\} + \mu_X^2\end{aligned}$$

$E\{[X(t) - \mu_X]^2\}$ is called variance of a stationary random process, denoted as σ_X^2. So there is

$$\psi_X^2 = \sigma_X^2 + \mu_X^2 \tag{5.36}$$

When the random signal has ergonomics, we have

$$\sigma_X^2 = E\{[X(t) - \mu_X]^2\} = \langle [x(t) - \mu_X]^2 \rangle = \lim_{T\to\infty} \frac{1}{T}\int_0^T [x(t) - \mu_X]^2 dt \tag{5.37}$$

According to this formula, σ_X^2 is the average power of the AC component of the random signal in each state. Its positive square root is called the standard deviation, denoted as σ_X,

$$\psi_X^2 = \sigma_X^2 + \mu_X^2 \tag{5.36}$$

当随机信号具有各态历经性时，有

$$\sigma_X^2 = E\{[X(t) - \mu_X]^2\} = \langle [x(t) - \mu_X]^2 \rangle = \lim_{T\to\infty} \frac{1}{T}\int_0^T [x(t) - \mu_X]^2 \mathrm{d}t \tag{5.37}$$

根据此式可知，σ_X^2 就是各态历经随机信号交流分量的平均功率，它的正平方根称为标准差，记为 σ_X，即信号交流分量的有效值。

式(5.36)表明各态历经随机信号的平均功率是交流分量的平均功率与直流分量的功率之和。信号的有效值 ψ_X 与交流分量的有效值 σ_X 的关系表示在图 5.15 所示的直角三角形中。

若令

$$\frac{R_X(\tau) - \mu_X^2}{\sigma_X^2} = \rho_X(\tau)$$

则称 $\rho_X(\tau)$ 为随机信号 $X(t)$ 的自相关系数，则有

$$R_X(\tau) = \rho_X(\tau)\sigma_X^2 + \mu_X^2 \tag{5.38}$$

若 $\tau = 0$，则 $R_X(0) = \psi_X^2 = \sigma_X^2 + \mu_X^2$。可见，当 $\tau = 0$ 时，$\rho_X(\tau) = 1$。

不考虑 $X(t)$ 的实际量纲，将样本函数 $x(t)$ 设想为电流(或电压)，令它通过 1 Ω 的纯电阻时，ψ_x^2 和 σ_X^2 可以理解为信号的平均功率。若电流(或电压)信号不是流过一个纯电阻，而是流过图 5.15 所示的 RLC 串联电路，并假设电流信号为随机相位余弦电流 $X(t) = X_m\cos(\omega t + \theta)$，其中一样本函数为

$$x(t) = X_m\cos(\omega t + \theta)$$

则 RLC 串联电路的阻抗为

$$Z = |Z|\angle\varphi$$

其中

$$\varphi = \arctan\frac{Z_L - Z_C}{R}$$

$$|Z| = \sqrt{R^2 + (Z_L - Z_C)^2}$$

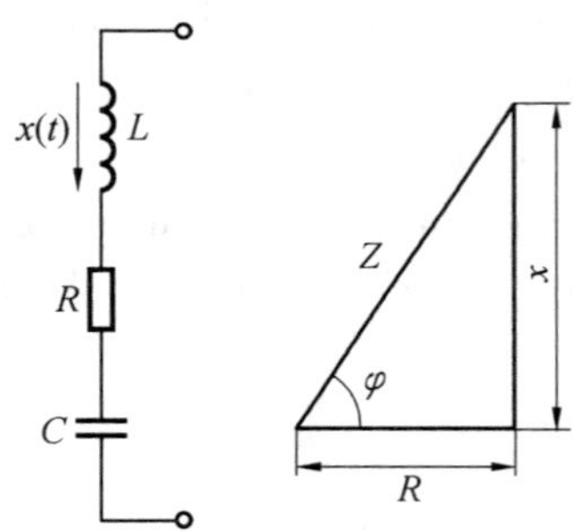

图 5.15　RLC 串联电路

设在既定频率 ω 下，$|Z| = 1$ Ω，则电路的端电压为

$$X_m\cos(\omega t + \theta + \varphi) = x(t + \tau)$$

式中　φ—— 电压与电流之间的初相差；

τ—— 电压与电流之间的时差，或电压相对于电流波形的时移，$\tau = \dfrac{\varphi}{\omega}$。

which is the effective value of the AC component of the signal.

Formula (5.36) shows that the average power of the random signal in each state is the sum of the average power of the AC component and the power of the DC component. The relationship between the effective value ψ_X of the signal and the effective value σ_X of the AC component is shown in the right triangle in Figure 5.15.

If we let

$$\frac{R_X(\tau) - \mu_X^2}{\sigma_X^2} = \rho_X(\tau)$$

then $\rho_X(\tau)$ is called the autocorrelation coefficient of random signal $X(t)$, then we have

$$R_X(\tau) = \rho_X(\tau)\sigma_X^2 + \mu_X^2 \tag{5.38}$$

If $\tau = 0$, then $R_X(0) = \psi_X^2 = \sigma_X^2 + \mu_X^2$. We can see that when $\tau = 0$, $\rho_X(\tau) = 1$.

The sample function $x(t)$ is assumed to be a current (or voltage) without considering the actual dimension of $X(t)$, and when it is passed through a 1 Ω pure resistance, ψ_x^2 and σ_X^2 can be understood as the average power of the signal. If the current (or voltage) signal does not flow through a pure resistor, it flows through the RLC series circuit shown in Figure 5.15, and suppose that the current signal is a random phase cosine current $X(t) = X_m\cos(\omega t + \theta)$, where the same function is

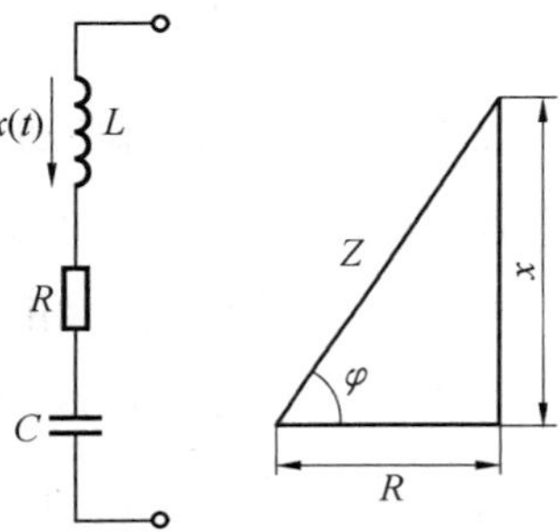

Figure 5.15　RLC series circuit

$$x(t) = X_m\cos(\omega t + \theta)$$

Then the impedance of the series circuit RLC is

$$Z = |Z|\angle\varphi$$

In it

$$\varphi = \arctan\frac{Z_L - Z_C}{R}$$

$$|Z| = \sqrt{R^2 + (Z_L - Z_C)^2}$$

Set at a given frequency ω, $|Z| = 1$ Ω, then the terminal voltage of the circuit is

$$X_m\cos(\omega t + \theta + \varphi) = x(t + \tau)$$

where　φ — the initial phase difference between voltage and current;

　　τ — the time difference between voltage and current, or the time shift of voltage with respect to the current waveform, $\tau = \dfrac{\varphi}{\omega}$.

The instantaneous power of the circuit at this time is

$$p(t) = x(t)x(t + \tau)$$

Therefore, the average power of the signal during the observation time T is

$$\begin{aligned} P &= \frac{1}{T}\int_0^T p(t)\,dt = \frac{1}{T}\int_0^T x(t)x(t+\tau)\,dt \\ &= \frac{1}{T}\int_0^T \frac{X_m^2}{2}[\cos\varphi + \cos(2\omega_t + 2\theta + \varphi)]\,dt \\ &= X^2\cos\varphi = X^2\cos\omega\tau = \sigma_X^2\cos\omega\tau \end{aligned} \tag{5.39}$$

这时，电路的瞬时功率为

$$p(t)=x(t)x(t+\tau)$$

因此，在观察时间 T 内，信号的平均功率为

$$P=\frac{1}{T}\int_0^T p(t)\,\mathrm{d}t=\frac{1}{T}\int_0^T x(t)x(t+\tau)\,\mathrm{d}t$$

$$=\frac{1}{T}\int_0^T\frac{X_{\mathrm{m}}^2}{2}[\cos\varphi+\cos(2\omega t+2\theta+\varphi)]\,\mathrm{d}t$$

$$=X^2\cos\varphi=X^2\cos\omega\tau=\sigma_X^2\cos\omega\tau \tag{5.39}$$

式中　σ_X——余弦电流、电压的有效值，$\sigma_X=X=\dfrac{X_{\mathrm{m}}}{\sqrt{2}}$。

此平均功率又称有功功率，是真正消耗在电阻上的平均功率。它不再等于电流与电压有效值的乘积，而要考虑功率因数 $\cos\varphi$。其原因是此时非电阻性电路的端电压与电流已不再同步，两波形虽然相同，但已产生了时移 τ。

将式(5.32)，式(5.39)相比，会发现它们具有相同的形式。可见，在不考虑信号实际量纲的前提下，平稳随机信号的自相关函数 $R_X(\tau)$ 就是电工学中所说的非电阻性电路中的有功功率，自相关系数 $\rho_X(\tau)$ 就是电路的功率因数，σ_X 就是把随机信号想象为电流(或电压)时电流(或电压)的交流分量的有效值，μ_X 就是电流(或电压)的直流分量。当 $\mu_X\neq 0$ 时，有

$$R_X(\tau)=\sigma_X^2\rho_X(\tau)+\mu_X^2 \tag{5.40}$$

该式表明信号的有功功率是其交流部分的有功功率和直流部分功率之和。随机信号的有功功率如图 5.16 所示。

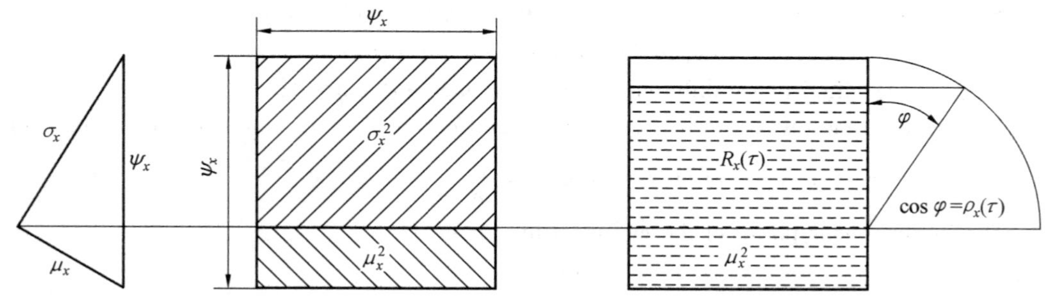

图 5.16　随机信号的有功功率

对平稳随机过程而言，σ_X 和 μ_X 都是常数。如果 $R_X(\tau)$ 表现了随机过程在两个不同时刻的状态的依从关系，那么显然是借助了自相关系数 $\rho_X(\tau)$。

至此，面对平稳随机信号和它许许多多不可预测、随机出现的瞬时值，终于找到了三

where　σ_X —the effective value of cosine current and voltage, $\sigma_X = X = \frac{X_m}{\sqrt{2}}$.

This average power, also called active power, is the average power that is actually consumed in the resistance. It is no longer equal to the product of the effective value of the current and the voltage, but the power factor cos φ should be considered. The reason is that the terminal voltage and current of the non-resistive circuit at this time are no longer synchronized. Although the two waveforms are the same, the time shift τ has been produced.

Comparing formula(5. 32) with formula(5. 39), we will find that they have the same form. Without considering the actual dimension of the signal, it can be seen that the autocorrelation function $R_X(\tau)$ of a stationary random signal is the active power in the non-resistive circuit in electrical engineering, the autocorrelation coefficient $\rho_X(\tau)$ is the power factor of the circuit, σ_X is the effective value of the AC component of the current (or voltage) when a random signal is imagined as a current (or voltage), and μ_X is the DC component of current (or voltage). When $\mu_X \neq 0$, we have

$$R_X(\tau) = \sigma_X^2 \rho_X(\tau) + \mu_X^2 \tag{5.40}$$

This formula shows that the active power of a signal is the sum of the active power of the AC part and the power of the DC part. Active power of random signal is shown in Figure 5. 16.

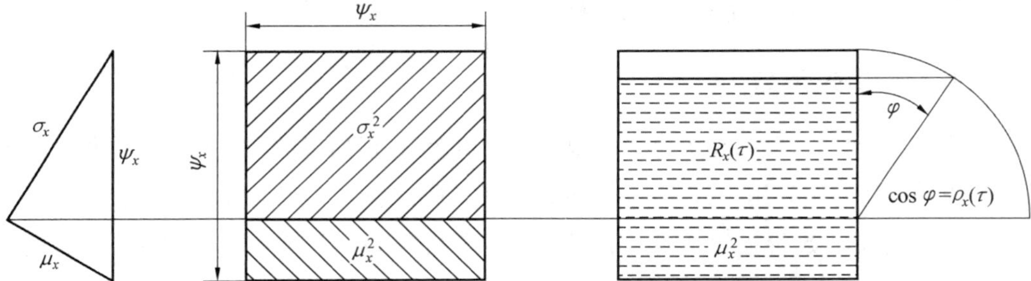

Figure 5. 16　Active power of random signal

For a stationary random process, σ_X and μ_X are always constants. If $R_X(\tau)$ represents the state dependence of the random process at two different moments, then it is obviously with the help of the autocorrelation coefficient $\rho_X(\tau)$.

So far, when facing a stationary random signal and its many unpredictable, random instantaneous values, three constants finally can be founded, that is, μ_X , σ_X^2 , ψ_X^2 , and a function $R_X(\tau)$ (or $\rho_X(\tau)$). They are inevitable in countless accidents. No matter how many times the random process is repeated and how many sample functions there are, these main digital features of a certain kind of stable random process are stable. It is inherent to the random process and is the main mark that distinguishes the random process from other random processes. They portray the statistical characteristics of random signals, just like the statistical characteristics of random processes. In the previous article, it was defined that the random process is established according to a certain rule. A time function containing a random variable e , in fact, the process of finding the mean value and autocorrelation function is the process to use mathematical analysis methods to find those certain constants in the time function. Only after obtaining these constants based on $E[X(t)]$ and $R_X(\tau)$ under the condition of mastering

个常数，即μ_X、σ_X^2、ψ_X^2，还有一个函数$R_X(\tau)$（或$\rho_X(\tau)$），它们就是寓于无数偶然中的必然。无论随机过程重复多少次，有多少个样本函数，某一种平稳随机过程的这些主要数字特征都是稳定不变的。它是该随机过程所固有的，是区别该随机过程与其他随机过程的主要标志。它们与随机过程的统计特性一样，刻画了随机信号的统计特征。前文曾经定义随机过程是依据某种规则建立的。一个包含随机变量e的时间函数，实际上求均值与自相关函数的过程就是利用数学分析的方法寻找该时间函数中那些确定的常量的过程。只有在掌握随机变量的统计规律的基础上依据$E[X(t)]$和$R_X(\tau)$求出这些常数以后，才算弄清随机信号是依什么样的规则建立的时间函数，最后才算弄清它的规律。因此，讨论均值与自相关函数对研究随机过程是至关重要的。

5.4 信号的相关分析及应用

在测试技术领域中，分析两个随机信号之间的关系或是分析一个信号在一定时移前后之间的关系都需要相关分析。因此，相关是一个非常重要的概念。

5.4.1 相关系数

在信号分析与处理中，相关是一个很重要的概念，这里所说的相关是指线性相关。如果两个变量之间的关系用函数$y = ax + b$来描述，则y与x呈线性关系。若它们的关系既不像上面的线性函数所描述的那样严格，又不像图5.17(c)所示那样毫无关系，而像图5.17(b)那样，坐标面内表示两个变量x和y的点大致都落在一条斜线的两侧，即x、y的关系介于a、c这两种极端情况之间的情况，用(线性)“相关”一词来概括。xy相关就等于说x、y之间保持某种程度的线性关系，至于线性相关的程度有多大，要用相关系数来度量。

图5.17所示为两个随机变量的相关性。图5.17(a)中变量x和y有精确的线性关系。图5.17(b)中变量x和y没有确定的关系，但从总体来看，具有某种程度的线性关系说明它们之间有着相关关系。图5.17(c)中各数据点分布很散，说明变量x和y之间是无关的。

评价变量x和y之间的相关程度常用相关系数ρ_{xy}表示，即

$$\rho_{xy} = \frac{\sigma_{xy}}{\sigma_x \sigma_y} = \frac{E[(x - \mu_x)(y - \mu_y)]}{\sqrt{E[(x - \mu_x)^2]E[(y - \mu_y)^2]}} \tag{5.41}$$

式中　E—— 数学期望值；

σ_{xy}—— 随机变量x、y的协方差；

the statistical law of random variables, can we figure out the time function established by the random signal according to what rules, and finally figure out its law. Therefore, it is very important to discuss the mean value and autocorrelation function for the study of random processes.

5.4 Correlation Analysis and Application of Signal

In the field of measurement technology, the correlation analysis can be used at the analysis of the relationship between two random signals or between a signal before and after a certain time shift. Therefore, correlation is a very important concept.

5.4.1 Correlation Coefficient

Correlation is a very important concept in signal analysis and processing. Correlation mentioned here refers to linear correlation. If the relationship between two variables is described by the function $y = ax + b$, then y and x are linear. If their relationship is neither as strict as the linear function described above, nor is it irrelevant as shown in Figure 5. 17(c), but like Figure 5. 17(b), the points of the two variables x and y in the coordinate plane are roughly on both sides of a diagonal line, that is, the relationship between x and y is between the two extreme cases of a and c. We use the term (linear) "correlation" to summarize the situation between them. To say that x and y are correlated is equivalent to saying that x and y maintain a certain degree of linear relationship. As for the degree of linear correlation, the correlation coefficient should be used to measure.

Figure 5. 17 shows the correlation of two random variables. The variables x and y in Figure 5.17(a) have an exact linear relationship. The variables x and y in Figure 5. 17(b) have no definite relationship, but from a general point of view, there is a certain degree of linear relationship indicating that there is a correlation between them. The distribution of the data points in Figure 5. 17(c) is very scattered, indicating that the variables x and y are irrelevant.

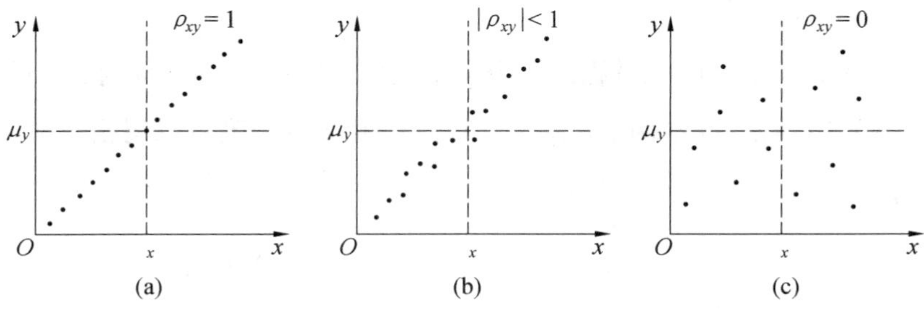

Figure 5. 17　The correlation of two random variables

The correlation degree between the evaluation variables x and y is usually expressed by the correlation coefficient ρ_{xy} , namely

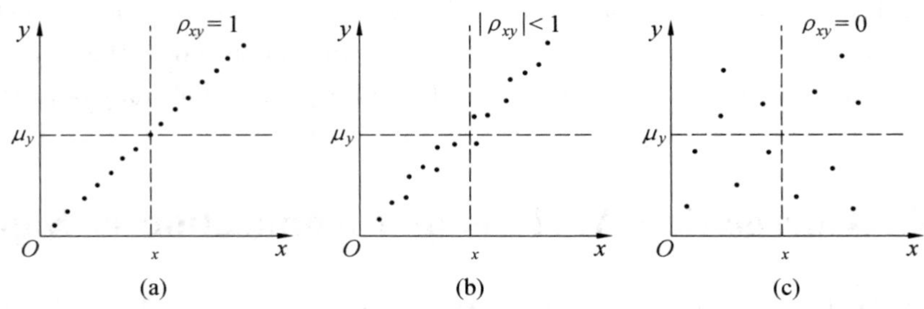

图 5.17　两个随机变量的相关性

μ_x—— 随机变量 x 的均值,$\mu_x = E(x)$;

μ_y—— 随机变量 y 的均值,$\mu_y = E(y)$。

利用柯西 - 许瓦兹不等式

$$E[(x-\mu_x)(y-\mu_y)]^2 \leqslant E[(x-\mu_x)^2]E[(y-\mu_y)^2] \tag{5.42}$$

可知$|\rho_{xy}|\leqslant 1$。ρ_{xy} 是一个无量纲的量,是描述随机变量 X、Y 之间的关系的数字特征。当 $\rho_{xy}=\pm 1$ 时,说明两变量 x、y 是理想的线性关系,只是当 $\rho_{xy}=-1$ 时,直线的斜率为负值。当 $\rho_{xy}=0$ 时,表示两变量 x、y 之间完全无关。

关于相关的概念,应该注意以下几点:独立和相关是不同的概念,X、Y 相互独立时,它们必定不相关,但 X、Y 不相关时,它们却不一定独立。X、Y 不相关是指 X、Y 之间不存在任何程度的线性关系,可能存在其他关系,如 $Y=kX^2$ 等关系。另外,独立是根据两随机变量客观存在的因果关系在事先所做出的判断,一般不需要计算。而相关是必须经过计算的,相关的前提是 X、Y 不相互独立。

5.4.2　自相关函数分析

1. 自相关函数的定义

假设 $x(t)$ 是各态历经随机过程的一个样本函数,$x(t+\tau)$ 是 $x(t)$ 时移 τ 后的样本,自相关函数如图5.18 所示,样本函数 $\rho_{x(t)x(t+\tau)}$ 和 $x(t)$ 相关的程度可以用相关系数 $x(t)$ 来表示。如果把 $\rho_{x(t)x(t+\tau)}$ 简化为 $\rho_x(\tau)$,则有

$$\rho_x(\tau)=\frac{\lim\limits_{T\to\infty}\dfrac{1}{T}\int_0^T[x(t)-\mu_x][x(t+\tau)-\mu_x]\mathrm{d}t}{\sigma_x^2}$$

$$=\frac{\lim\limits_{T\to\infty}\dfrac{1}{T}\int_0^T x(t)x(t+\tau)\mathrm{d}t-\mu_x^2}{\sigma_x^2}$$

定义自相关函数 $R_x(\tau)$ 为

$$\rho_{xy} = \frac{\sigma_{xy}}{\sigma_x \sigma_y} = \frac{E[(x-\mu_x)(y-\mu_y)]}{\sqrt{E[(x-\mu_x)^2]E[(y-\mu_y)^2]}} \tag{5.41}$$

where　E —mathematical expectation value;

σ_{xy} —the covariance of random variables x and y ;

μ_x —mean value of random variable x, $\mu_x = E(x)$;

μ_y —the mean value of the random variable y, $\mu_y = E(y)$.

Use Cauchy–Schwarz inequality

$$E[(x-\mu_x)(y-\mu_y)]^2 \leqslant E[(x-\mu_x)^2]E[(y-\mu_y)^2] \tag{5.42}$$

we get $|\rho_{xy}| \leqslant 1$. ρ_{xy} is a dimensionless quantity, is a digital feature that describes the relationship between random variables X and Y. When $\rho_{xy} = \pm 1$, it means that the two variables x and y have an ideal linear relationship, but when $\rho_{xy} = -1$, the slope of the straight line is negative. When $\rho_{xy} = 0$, it means that the two variables x and y are completely unrelated.

Regarding related concepts, the following points should be noted: independence and correlation are different concepts: when X and Y are independent of each other, they must not be related, but when X and Y are not related, they are not necessarily independent. The fact that X and Y are not related means that there is no linear relationship to any degree between X and Y, and there may be other relationships, such as $Y = kX^2$ and other relationships. In addition, independence is a judgment made in advance based on the causal relationship between two random variables, which generally does not need to be calculated. However, the correlation must be calculated, and the premise of correlation is that X and Y are not independent of each other.

5.4.2 Autocorrelation Function Analysis

1. Definition of Autocorrelation Function

Assuming that $x(t)$ is a sample function of each state undergoing a random process, $x(t+\tau)$ is the sample of $x(t)$ time-shifted by τ, autocorrelation function is shown in Figure 5.18, the degree of correlation between the sample function $\rho_{x(t)x(t+\tau)}$ and $x(t)$ can be expressed by the correlation coefficient $x(t)$. If $\rho_{x(t)x(t+\tau)}$ is simplified to $\rho_x(\tau)$, then there is

$$\rho_x(\tau) = \frac{\lim_{T\to\infty}\frac{1}{T}\int_0^T [x(t)-\mu_x][x(t+\tau)-\mu_x]\mathrm{d}t}{\sigma_x^2} = \frac{\lim_{T\to\infty}\frac{1}{T}\int_0^T x(t)x(t+\tau)\mathrm{d}t - \mu_x^2}{\sigma_x^2}$$

Define the autocorrelation function $R_x(\tau)$ as

$$R_x(\tau) = \lim_{T\to\infty}\frac{1}{T}\int_0^T x(t)x(t+\tau)\mathrm{d}t \tag{5.43}$$

we have

$$\rho_x(\tau) = \frac{R_x(\tau) - \mu_x^2}{\sigma_x^2} \tag{5.44}$$

The nature of the signal is different, and the expression of the autocorrelation function is

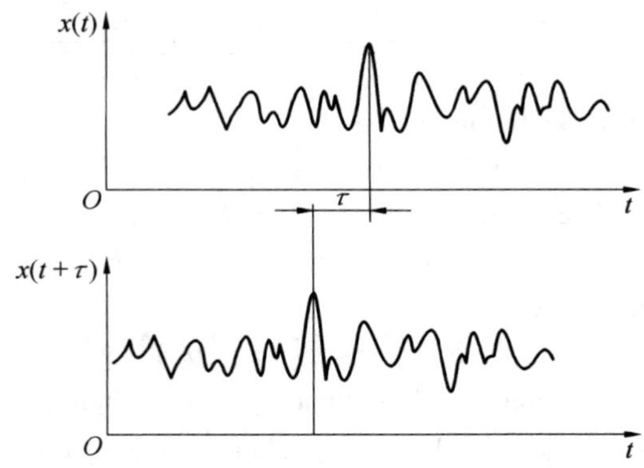

图 5.18　自相关函数

$$R_x(\tau)=\lim_{T\to\infty}\frac{1}{T}\int_0^T x(t)x(t+\tau)\mathrm{d}t \tag{5.43}$$

则有

$$\rho_x(\tau)=\frac{R_x(\tau)-\mu_x^2}{\sigma_x^2} \tag{5.44}$$

信号的性质不同,自相关函数的表达式是不同的。

周期信号自相关函数的表达式为

$$R_x(\tau)=\frac{1}{T_0}\int_0^{T_0} x(t)x(t+\tau)\mathrm{d}t \tag{5.45}$$

式中　T_0——信号的周期。

非周期信号自相关函数的表达式为

$$R_x(\tau)=\int_{-\infty}^{\infty} x(t)x(t+\tau)\mathrm{d}t \tag{5.46}$$

从以上几个公式可以看出,$\rho_x(\tau)$ 和 $R_x(\tau)$ 均与 τ 有关,并且 $\rho_x(\tau)$ 与 $R_x(\tau)$ 呈线性关系。

2. 自相关函数的性质

(1) 自相关函数是偶函数,即

$$R_X(\tau)=R_X(-\tau)$$

因为

$$R_X(-\tau)=E[X(t)X(t-\tau)]$$

故设 $t=\lambda+\tau$,则 $t-\tau=\lambda$,代入上式,得

$$R_X(-\tau)=E[X(\lambda+\tau)X(\lambda)]=R_X(\tau)$$

一般情况下,$X(t)$ 是实函数,故 $R_X(\tau)$ 是实偶函数。

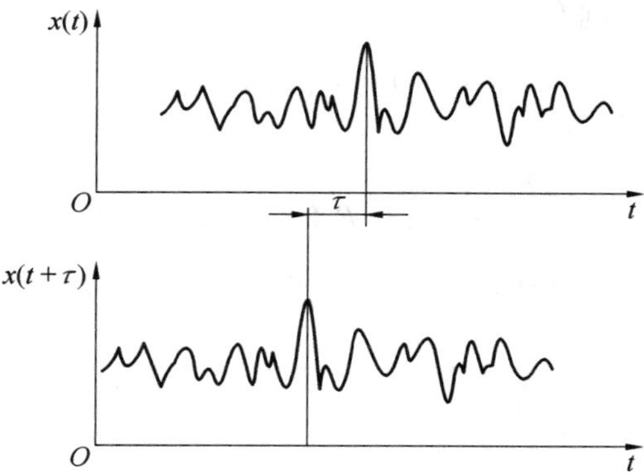

Figure 5.18　Autocorrelation function

different.

The expression of the autocorrelation function of periodic signal is

$$R_x(\tau) = \frac{1}{T_0}\int_0^{T_0} x(t)x(t+\tau)\,\mathrm{d}t \tag{5.45}$$

where　T_0 —the period of the signal.

The expression of the autocorrelation function of nonperiodic signal is

$$R_x(\tau) = \int_{-\infty}^{\infty} x(t)x(t+\tau)\,\mathrm{d}t \tag{5.46}$$

It can be seen from the above formulas that both $\rho_x(\tau)$ and $R_x(\tau)$ are related to τ, and $\rho_x(\tau)$ and $R_x(\tau)$ have a linear relationship.

2. The Nature of the Autocorrelation Function

(1) Autocorrelation function is an even function, that is

$$R_X(\tau) = R_X(-\tau)$$

Because

$$R_X(-\tau) = E[X(t)X(t-\tau)]$$

so suppose $t = \lambda + \tau$, then $t - \tau = \lambda$, substituting into the above formula, we get

$$R_X(-\tau) = E[X(\lambda+\tau)X(\lambda)] = R_X(\tau)$$

In general, $X(t)$ is a real function, so $R_X(\tau)$ is a real even function.

(2) According to formula (5.44), there is

$$R_x(\tau) = \rho_x(\tau)\sigma_x^2 + \mu_x^2$$

Because $|\rho_x(\tau)| \leqslant 1$, so

$$\mu_x^2 - \sigma_x^2 \leqslant R_x(\tau) \leqslant \mu_x^2 + \sigma_x^2$$

(3) The value of τ is different, so is the value of $R_x(\tau)$. When $\tau = 0$, the value of $R_x(\tau)$ is the largest, and it equals to the mean square value ψ_x^2 of the signal, that is

$$R_x(0) = \lim_{T\to\infty}\frac{1}{T}\int_0^T x(t)x(t)\,\mathrm{d}t = \psi_x^2 = \mu_x^2 + \sigma_x^2$$

(4) When $\tau \to \infty$, there is no internal connection between $x(t)$ and $x(t+\tau)$, and they have nothing to do with each other, so $\underset{l\to\infty}{\rho_x(\tau)} \to 0$, $\underset{l\to\infty}{R_x(\tau)} \to \mu_x^2$.

(5) The autocorrelation function of the periodic function is still a periodic function, and

(2) 根据式(5.44),则有

$$R_x(\tau)=\rho_x(\tau)\sigma_x^2+\mu_x^2$$

因为$|\rho_x(\tau)|\leqslant 1$,所以

$$\mu_x^2-\sigma_x^2\leqslant R_x(\tau)\leqslant\mu_x^2+\sigma_x^2$$

(3)τ 值不同,$R_x(\tau)$ 也不同。当 $\tau=0$ 时,$R_x(\tau)$ 的值最大,并等于信号的均方值 ψ_x^2,即

$$R_x(0)=\lim_{T\to\infty}\frac{1}{T}\int_0^T x(t)x(t)\mathrm{d}t=\psi_x^2=\mu_x^2+\sigma_x^2$$

(4) 当 $\tau\to\infty$ 时,$x(t)$ 和 $x(t+\tau)$ 之间不存在内在联系,彼此无关,故$\rho_x(\tau)\underset{l\to\infty}{\to}0$,$R_x(\tau)\underset{l\to\infty}{\to}\mu_x^2$。

(5) 周期函数的自相关函数仍为周期函数,且二者的频率相同,但是丢失了原信号的相位信息。

图 5.19 所示为自相关函数的曲线及其性质。

例 5.1 求正弦函数 $x(t)=A\sin(\omega t+\varphi)$ 的自相关函数。

解 正弦函数为周期函数,根据式(5.30)得

$$R_x(\tau)=\frac{1}{T_0}\int_0^{T_0}x(t)x(t+\tau)\mathrm{d}t$$
$$=\frac{1}{T_0}\int_0^{T_0}A^2\sin(\omega t+\varphi)\sin[\omega(t+\tau)+\varphi]\mathrm{d}t$$

式中 T_0—— 正弦函数的周期,$T_0=\frac{2\pi}{\omega}$。

令 $\omega t+\varphi=\theta$,则 $\mathrm{d}t=\mathrm{d}\theta/\omega$,则有

$$R_x(\tau)=\frac{A^2}{2\pi}\int_0^{2\pi}\sin\theta\sin(\theta+\omega\tau)\mathrm{d}\theta=\frac{A^2}{2}\cos\omega\tau$$

从例 5.1 中可以看出,正弦函数的自相关函数是一个余弦函数,在 $\tau=0$ 时具有最大值。它保留了原信号的幅值和频率信息,但是丢失了原正弦信号中的初始相位信息。

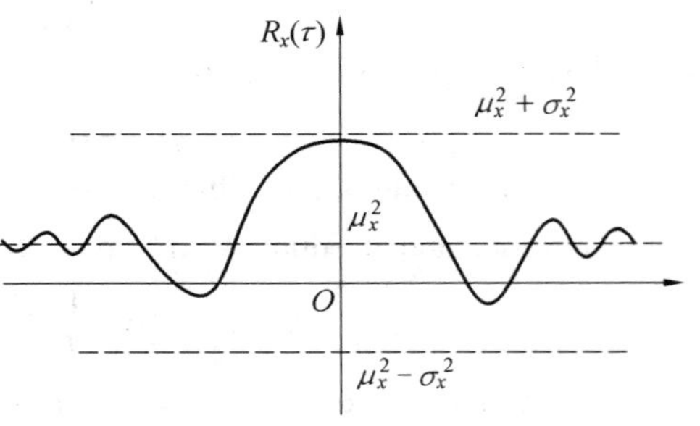

图 5.19 自相关函数的曲线及其性质

从图 5.13 所示的四种常见的随机信号的自相关函数 $R_x(\tau)$ 图中可以看出,只要信号中含有周期成分,其自相关函数 $R_x(\tau)$ 在 τ 很大时都不

the frequencies of the two are the same, but the phase information of the original signal is lost.

Figure 5.19 shows the curve of the autocorrelation function and its properties.

Example 5.1 Find the autocorrelation function of the sine function $x(t) = A\sin(\omega t + \varphi)$.

Solution The sine function is a periodic function, according to formula (5.30), we get

$$R_x(\tau) = \frac{1}{T_0}\int_0^{T_0} x(t)x(t+\tau)\,\mathrm{d}t$$

$$= \frac{1}{T_0}\int_0^{T_0} A^2\sin(\omega t+\varphi)\sin[\omega(t+\tau)+\varphi]\,\mathrm{d}t$$

where　T_0 —the period of the signal, $T_0 = \frac{2\pi}{\omega}$.

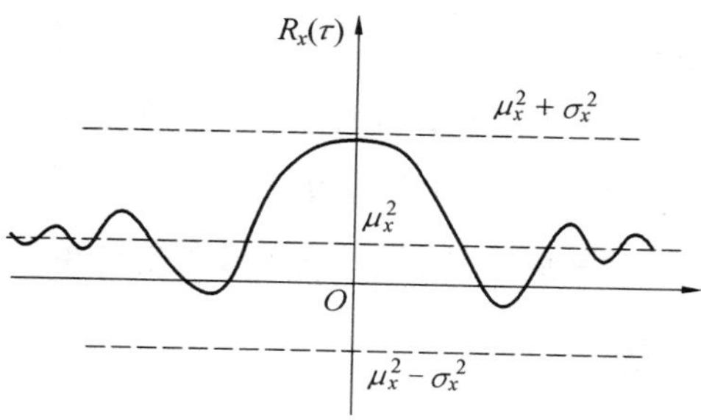

Figure 5.19　The curve of the autocorrelation function and its properties

Let $\omega t + \varphi = \theta$, then $\mathrm{d}t = \mathrm{d}\theta/\omega$, then we have

$$R_x(\tau) = \frac{A^2}{2\pi}\int_0^{2\pi}\sin\theta\sin(\theta+\omega\tau)\,\mathrm{d}\theta = \frac{A^2}{2}\cos\omega\tau$$

It can be seen from Example 5.1 that the autocorrelation function of the sine function is a cosine function, which has a maximum value at $\tau = 0$. It retains the amplitude and frequency information of the original signal, but loses the initial phase information in the original sinusoidal signal.

From the autocorrelation function $R_x(\tau)$ diagram of the four common random signals shown in Figure 5.13, it can be seen that as long as the signal contains periodic components, the autocorrelation function $R_x(\tau)$ will not attenuate even when τ is very large, and there will be obvious periodicity. For random signals without periodic components, when τ is slightly larger, the autocorrelation function $R_x(\tau)$ will approach zero. The autocorrelation function of narrowband random signals (such as noise) has slower attenuation characteristics, while the autocorrelation function of broadband random signals (such as noise) quickly decays to zero.

Autocorrelation analysis is of great significance in engineering applications. Figure 5.20 shows the surface roughness and autocorrelation function. The diamond contact converts the unevenness of the workpiece surface into a time domain signal through an inductive sensor (Figure 5.20(a)), and then obtains an autocorrelation graph through correlation analysis (Figure 5.20(b)). It can be seen that this is a waveform mixed with periodic signals in a random signal. The random signal has a large correlation at the origin, and decreases with the increase of the value of τ. After that, it shows periodicity, which indicates that the surface roughness is caused. The reason contains a certain cyclical factor, so we can further analyze

衰减,并且有明显的周期性。不含有周期成分的随机信号,在 τ 稍大时,自相关函数 $R_x(\tau)$ 将趋近于零。窄带随机信号(如噪声)的自相关函数有较慢的衰减特性,而宽带随机信号(如噪声)的自相关函数很快衰减到零。

自相关分析在工程应用中有着重要的意义。图 5.20 所示为表面粗糙度与自相关函数。金刚石触头将工件表面的凸凹不平度通过电感式传感器转换为时间域信号(图 5.20(a)),再经过相关分析得到自相关图形(图 5.20(b))。可以看出,这是一种随机信号中混杂着周期信号的波形。随机信号在原点处有较大相关性,随 τ 值增大而减小,此后呈现出周期性,这表明造成表面粗糙度的原因中包含了某种周期因素,从而可进一步分析其原因。例如,沿工件轴向,可能是走刀运动的周期性变化;沿工件切向,则可能是主轴回转振动的周期性变化等。又如,在分析汽车车座位置的振动信号时,利用自相关分析来检测该信号是否含有某种周期成分(如发动机工作所产生的周期振动信号),从而可进一步改进座位的结构来消除这种周期性影响,达到改善舒适度的目的。

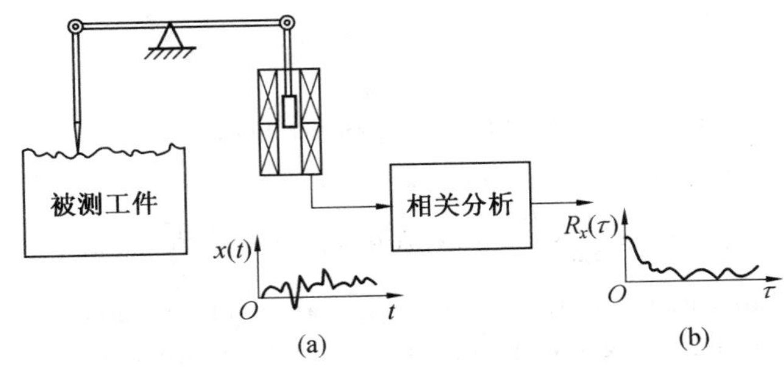

图 5.20　表面粗糙度与自相关函数

在实际工程测试中,经常会碰到确定性信号受到随机噪声干扰的情况。这些信号都是许多简谐波叠加起来的,确定性信号受随机噪声的干扰,实际上就是许多正(余)弦信号与随机噪声相叠加。根据上面的分析可知,如此叠加起来的信号虽然不是平稳随机信号,但其自相关函数却具有各态历经性,这样就可以根据该过程的一个样本函数来分析或计算它的自相关函数,极大地简化了计算。

综上,不同的随机信号其自相关函数是各不相同的,自相关函数是区别信号类型的重要手段。

(1) 那些单纯由频率相连续的成分(随机相位或确定相位简谐波)叠加而成的随机噪声或确定信号,其自相关函数在时移内无周期性,但都有收敛性。例如,白噪声、宽带、窄带随机噪声以及一切由连续频率成分所构成的确定性信号,这些信号的自谱也是连续的。自谱越宽广,即信号的频率成分越丰富,自相关函数衰减得越快。例如,白噪声的

the reason. For example, along the axis of the workpiece, it may be the periodic change of the cutting motion; along the workpiece tangential direction, it may be due to the periodic change of the spindle rotation vibration. Another example is the use of autocorrelation analysis to detect whether the signal contains a certain periodic component (such as the periodic vibration signal produced by the engine) when analyzing the vibration signal of the car seat position, so that the structure of the seat can be further improved to eliminate this periodic impact to achieve the purpose of improving comfort.

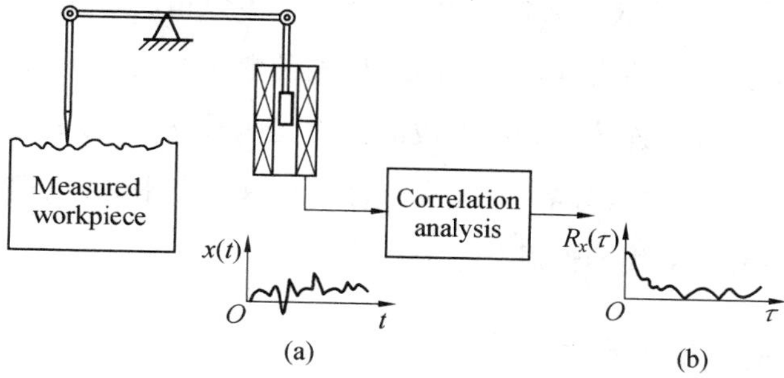

Figure 5.20　Surface roughness and autocorrelation function

In actual engineering measurements, it is often encountered that the deterministic signal is interfered by random noise. These signals are superimposed by many simple harmonics. The deterministic signal is interfered by random noise due to the fact that many sine (cosine) signals are superimposed with random noise. According to the above analysis, although the signal superimposed in this way is not a stationary random signal, its autocorrelation function has ergonomics, so that it can be analyzed or calculated according to a sample function of the process, which greatly simplified the calculation.

In summary, the autocorrelation function of different random signals is different, and the autocorrelation function is an important means to distinguish signal types.

(1) Those components that are purely continuous in frequency (random phase or definite phase harmonics) superimposed random noise or definite signal whose autocorrelation function has no periodicity in time shift, but all have convergence. Such as white noise, wideband, narrowband random noise, and all deterministic signals composed of continuous frequency components. The self-spectrum of these signals is also continuous. The wider the autospectrum, that is, the richer the frequency components of the signal, and the faster the autocorrelation function decays. For example, $R_X(\tau)$ of white noise is only a pulse, which is fleeting. Broadband has richer frequency components than narrowband random noise, so its attenuation is faster than narrowband.

(2) Those random noises or deterministic signals superimposed by random phases or definite phases and harmonics, which are simply discrete by frequency phase, their autocorrelation function has no attenuation in the time-shift domain, and sometimes has periodicity. Their self-spectrum is discrete, or expressed as a continuous form in isolated pulse form.

(3) When the signal contains both continuous components and discrete components, in most cases, periodic signals are interfered by random noise or random noise contains periodic

$R_X(\tau)$ 仅是一个脉冲，稍纵即逝。宽带比窄带随机噪声所含得频率成分丰富，故其衰减要快于窄带。

(2) 那些单纯由频率相离散的随机相位或确定相位简谐波叠加起来的随机噪声或确定信号，它们的自相关函数在时移域内无衰减性，有时还有周期性，它们的自谱是离散的，或以孤立的脉冲形式表示为连续形式。

(3) 当信号即含有连续成分又含有离散成分时，多数情况下是周期信号受随机噪声干扰，或随机噪声中含有周期成分。它们的自相关函数是两类不同成分的信号的自相关函数的叠加。当 τ 增大时，连续成分所形成的自相关函数逐渐衰减，以至消逝，最终所显示的自相关函数曲线是离散成分的，它无衰减性。从这一点来看，自相关函数像一个过滤器，通过它可以将信号中的周期性成分过滤出来。

5.4.3　互相关函数分析

1. 互相关函数的定义

假设 $x(t)$ 和 $y(t)$ 是各态历经随机过程的两个样本函数，它们互相关函数的定义 $R_{xy}(\tau)$ 为

$$R_{xy}(\tau)=\lim_{T\to\infty}\frac{1}{T}\int_0^T x(t)y(t+\tau)\mathrm{d}t \tag{5.47}$$

时移为 τ 的两信号 $x(t)$ 和 $y(t)$ 的互相关系数为

$$\rho_{xy}(\tau)=\frac{\lim\limits_{T\to\infty}\frac{1}{T}\int_0^T[x(t)-\mu_x][y(t+\tau)-\mu_y]\mathrm{d}t}{\sigma_x\sigma_y}$$

$$=\frac{\lim\limits_{T\to\infty}\frac{1}{T}\int_0^T x(t)y(t+\tau)\mathrm{d}t-\mu_x\mu_y}{\sigma_x\sigma_y}=\frac{R_{xy}(\tau)-\mu_x\mu_y}{\sigma_x\sigma_y} \tag{5.48}$$

2. 互相关函数的性质

(1) 互相关函数 $R_{xy}(\tau)$ 是可正、可负的实函数。

(2) 根据式(5.48) 可得

$$R_{xy}(\tau)=\mu_x\mu_y+\rho_{zy}(\tau)\sigma_x\sigma_y$$

因为 $|p_{xy}(\tau)|\leqslant 1$，所以互相关函数的范围为

$$\mu_x\mu_y-\sigma_x\sigma_y\leqslant R_{xy}(\tau)\leqslant\mu_x\mu_y+\sigma_x\sigma_y$$

(3) 互相关函数既不是偶函数，也不是奇函数，即 $R_{xy}(\tau)$ 一般不等于 $R_{xy}(-\tau)$，但满足 $R_{xy}(\tau)=R_{xy}(-\tau)$，因为所讨论的随机过程是平稳的，在 t 和 $t-\tau$ 时刻从样本函数计算

components. Their autocorrelation function is the superposition of the autocorrelation functions of two types of signals with different components. When τ increases, the autocorrelation function formed by the continuous component gradually attenuates, and even disappears. The autocorrelation function curve finally displayed is a discrete component, and it has no attenuation. From this point of view, the autocorrelation function is like a filter, through which the periodic components in the signal can be filtered.

5.4.3 Cross-Correlation Function Analysis

1. Definition of Cross-Correlation Function

Assuming that $x(t)$ and $y(t)$ are two sample functions of each state undergoing a random process, the definition of their cross-correlation function $R_{xy}(\tau)$ is

$$R_{xy}(\tau) = \lim_{T\to\infty} \frac{1}{T}\int_0^T x(t)y(t+\tau)\,\mathrm{d}t \tag{5.47}$$

The correlation coefficient of the two signals $x(t)$ and $y(t)$ with a time shift of τ is

$$\rho_{xy}(\tau) = \frac{\lim\limits_{T\to\infty} \frac{1}{T}\int_0^T [x(t)-\mu_x][y(t+\tau)-\mu_y]\,\mathrm{d}t}{\sigma_x\sigma_y} = \frac{\lim\limits_{T\to\infty} \frac{1}{T}\int_0^T x(t)y(t+\tau)\,\mathrm{d}t - \mu_x\mu_y}{\sigma_x\sigma_y} = \frac{R_{xy}(\tau)-\mu_x\mu_y}{\sigma_x\sigma_y} \tag{5.48}$$

2. The Nature of the Cross-Correlation Function

(1) The cross-correlation function $R_{xy}(\tau)$ is a real function that can be positive or negative.

(2) According to formula (5.48), we get

$$R_{xy}(\tau) = \mu_x\mu_y + \rho_{zy}(\tau)\sigma_x\sigma_y$$

Because of $|p_{xy}(\tau)| \leqslant 1$, the range of the cross-correlation function is

$$\mu_x\mu_y - \sigma_x\sigma_y \leqslant R_{xy}(\tau) \leqslant \mu_x\mu_y + \sigma_x\sigma_y$$

(3) The cross-correlation function is neither an even function nor an odd function, that is, $R_{xy}(\tau)$ is generally not equal to $R_{xy}(-\tau)$. But it satisfies $R_{xy}(\tau) = R_{xy}(-\tau)$, because the random process in question is stationary, and the cross-correlation function calculated from the sample function at time t and $t-\tau$ is the same, that is

$$R_{xy}(\tau) = \lim_{T\to\infty} \frac{1}{T}\int_0^T x(t)y(t+\tau)\,\mathrm{d}t = \lim_{T\to\infty} \frac{1}{T}\int_0^T x(t+\tau)y(t)\,\mathrm{d}t$$
$$= \lim_{T\to\infty} \frac{1}{T}\int_0^T y(t)x(t+\tau)\,\mathrm{d}t = R_{yx}(-\tau)$$

(4) The maximum value of the cross-correlation function $R_{xy}(\tau)$ is not at $\tau = 0$, but deviates from the origin τ_0. The time shift τ_0 reflects the lag time between $x(t)$ and $y(t)$. Figure 5.21 shows the curve of the cross-correlation function and its properties.

(5) Two statistically independent random signals, when the mean value is zero, $R_{xy}(\tau) = 0$.

(6) The cross-correlation function of two signals with the same period is still a period

的互相关函数是一致的,即

$$R_{xy}(\tau) = \lim_{\tau\to\infty}\frac{1}{T}\int_0^{\tau} x(t)y(t+\tau)\,\mathrm{d}t = \lim_{\tau\to\infty}\frac{1}{T}\int_0^{\tau} x(t+\tau)y(t)\,\mathrm{d}t$$

$$= \lim_{T\to\infty}\frac{1}{T}\int_0^{T} y(t)x(t+\tau)\,\mathrm{d}t = R_{yx}(-\tau)$$

(4) 互相关函数 $R_{xy}(\tau)$ 的最大值不在 $\tau = 0$ 处,而在偏离原点 τ_0 处。时移 τ_0 反映 $x(t)$ 和 $y(t)$ 之间的滞后时间。图 5.21 所示为互相关函数的曲线及其性质。

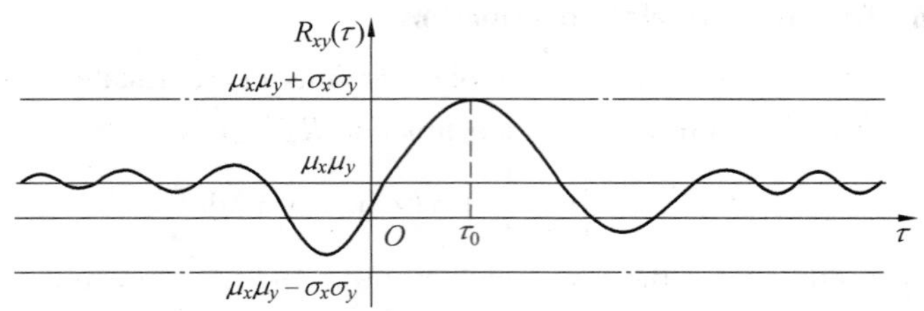

图 5.21　互相关函数的曲线及其性质

(5) 两个统计独立的随机信号,当均值为零时,$R_{xy}(\tau) = 0$。

(6) 两个同周期信号的互相关函数仍然是同频率的周期信号,但保留了原信号的相位信息。两个非同频率的周期信号互不相关,即“同频相关,不同频不相关”。

例 5.2　设周期信号 $x(t)$ 和 $y(t)$ 分别为

$$x(t) = A\sin(\omega t + \theta)$$

$$y(t) = B\sin(\omega t + \theta - \varphi)$$

试求两个周期信号的互相关函数 $R_{xy}(\tau)$。

解　由于 $x(t)$ 和 $y(t)$ 是周期信号,因此可以用一个共同周期内的平均值替代整个时间历程的平均值,有

$$R_{xy}(\tau) = \lim_{T\to\infty}\frac{1}{T}\int_0^{T} x(t)y(t+\tau)\,\mathrm{d}t$$

$$= \frac{1}{T_0}\int_0^{T}[A\sin(\omega t+\theta)]B\sin[(t+\tau)+\theta-\varphi]\,\mathrm{d}t$$

$$= \frac{1}{2}AB\cos(\omega t - \varphi)$$

由例 5.2 的结果可知,两个均值为零且具有相同频率的周期信号,其互相关函数保留了这两个信号的频率 ω、对应的幅值 A 和 B,以及相位差 φ 的信息。

3. 互相关技术的工程应用

在测试技术中,互相关技术得到了广泛的应用。利用互相关函数可以测量系统的延

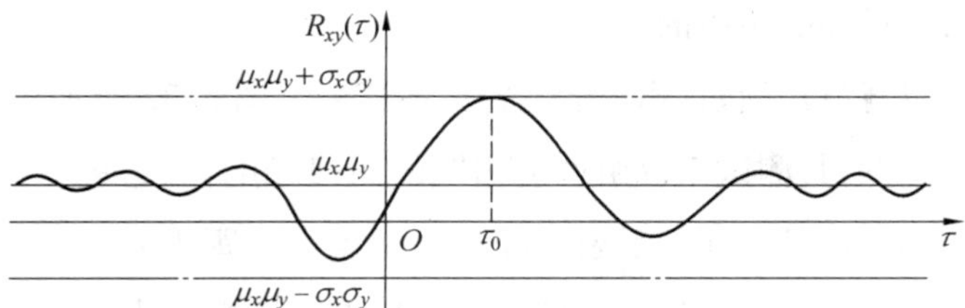

Figure 5.21 The curve and properties of the cross-correlation function

signal with the same frequency, but the phase information of the original signal is retained. Two periodic signals with different frequencies are not related to each other, that is, "the same frequency is related, but the different frequencies are not related".

Example 5.2 Let the periodic signals $x(t)$ and $y(t)$ respectively be

$$x(t) = A\sin(\omega t + \theta)$$
$$y(t) = B\sin(\omega t + \theta - \varphi)$$

Try to find the cross-correlation function $R_{xy}(\tau)$ of two periodic signals.

Solution Since $x(t)$ and $y(t)$ are periodic signals, the average value in a common period can be used to replace the average value of the entire time history, we have

$$\begin{aligned} R_{xy}(\tau) &= \lim_{T\to\infty}\frac{1}{T}\int_0^T x(t)y(t+\tau)\mathrm{d}t \\ &= \frac{1}{T_0}\int_0^T [A\sin(\omega t+\theta)]B\sin[(t+\tau)+\theta-\varphi]\mathrm{d}t \\ &= \frac{1}{2}AB\cos(\omega t-\varphi) \end{aligned}$$

From the results of Example 5.2, it can be seen that the cross-correlation function of two periodic signals with zero mean and the same frequency retains the information of the frequency ω , the corresponding amplitude A and B , and the phase difference φ of the two signals.

3. Engineering Application of Cross-Correlation Technology

In the measurement technology, the cross-correlation technology has been widely used. The cross-correlation function can be used to measure the delay of the system, and can also identify and extract the signals that are confused in the noise.

(1) Related speed and distance measurement. For example, measuring the speed of a moving object. Figure 5.22 shows a schematic diagram of non-contact measurement of the speed of a hot-rolled steel strip. The test system consists of two sets of photocells, lenses, adjustable delayers and correlators with the same performance. When the reflected light from the surface of the moving hot-rolled steel strip is focused on two photocells with a distance ofd through a lens, the reflected light is converted into an electrical signal by the photocell, delayed by an adjustable delayer, and then processed. When the adjustable delay is equal to the time τ required for a certain point on the steel strip to pass between two measuring points, the cross-correlation function proves the maximum value. The measured speed of the steel strip is $v = d/\tau_{\mathrm{m}}$.

(2) Use correlation analysis for fault diagnosis. The leakage point k in Figure 5.23 is the sound source that propagates sound to both sides. Place sensors 1 and 2 on the pipelines on

时,也可识别、提取混淆在噪声中的信号等。

(1)相关测速和测距。例如,测量运动物体的速度。图5.22所示为非接触测量热轧钢带运动速度的示意图,其测试系统由性能相同的两组光电池、透镜、可调延时器和相关器组成。当运动的热轧钢带表面的反射光经透镜聚焦在相距为d的两个光电池上时,反射光通过光电池转换为电信号,经可调延时器延时,再进行相关处理。当可调延时等于钢带上某点在两个测点之间经过所需的时间τ时,互相关函数为最大值,所测钢带的运动速度为$v = d/\tau_m$。

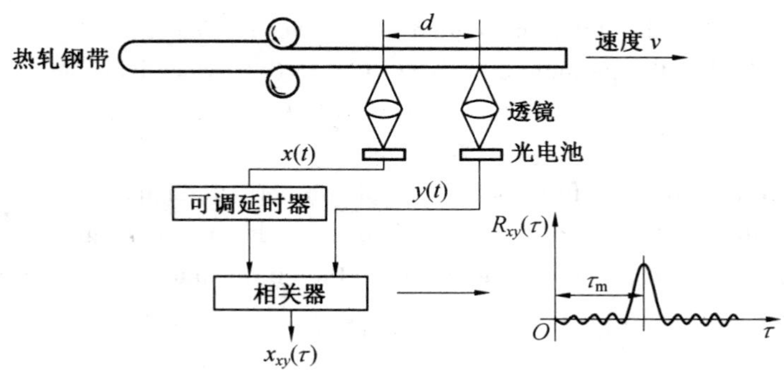

图5.22　非接触测量热轧钢带运动速度的示意图

(2)利用相关分析进行故障诊断。图5.23中漏损处k为向两侧传播声响的声源。在两侧管道上分别放置传感器1和2。因为放传感器的两点距漏损处不等远,所以漏油的声响传至两传感器就有时差τ_m,在互相关函数图$\tau = \tau_m$处,$R_{x_1x_2}(\tau)$有最大值。由τ_m可确定漏损处的位置。设两传感器的中点至漏损处的距离为s,声音通过管道的传播速度为v,则有

$$s = \frac{1}{2}v\tau_m$$

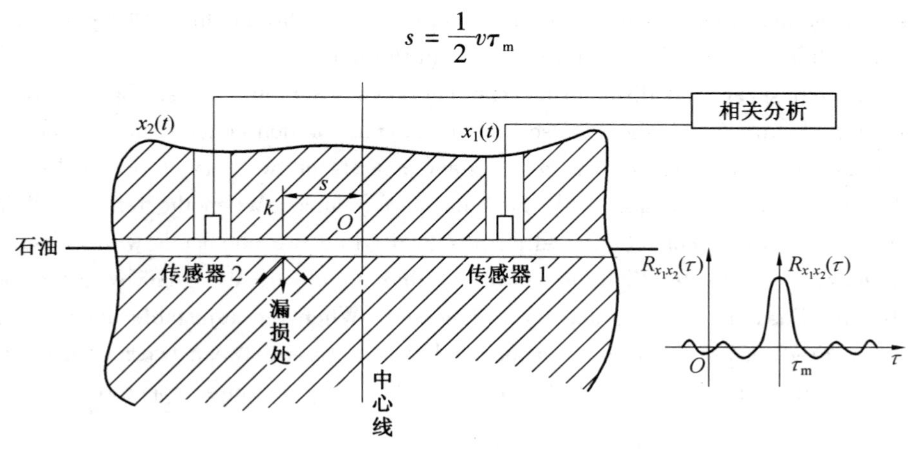

图5.23　利用相关分析找出油管漏损处

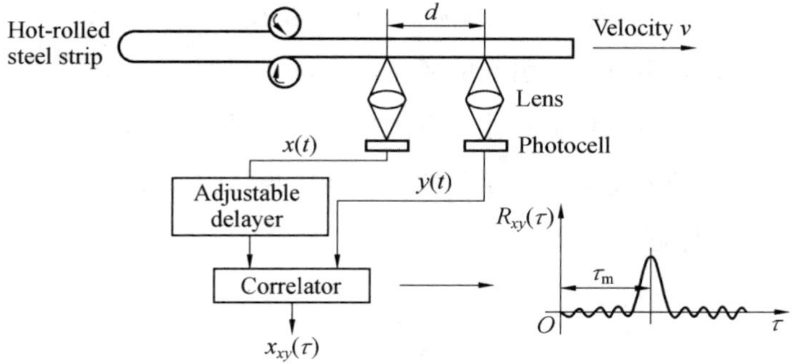

Figure 5.22 A schematic diagram of non-contact measurement of the speed of a hot-rolled steel strip

both sides. Because the two points where the sensors are placed are not equidistant from the leakage point, the sound of oil leakage will have a time difference τ_m to the two sensors. At the cross-correlation function diagram $\tau = \tau_m$, $R_{x_1x_2}(\tau)$ has a maximum value. From τ_m , the location of the leakage can be determined. Suppose the distance from the midpoint of the two sensors to the leakage point is s, and the propagation speed of sound through the pipe is v , then

$$s = \frac{1}{2}v\tau_m$$

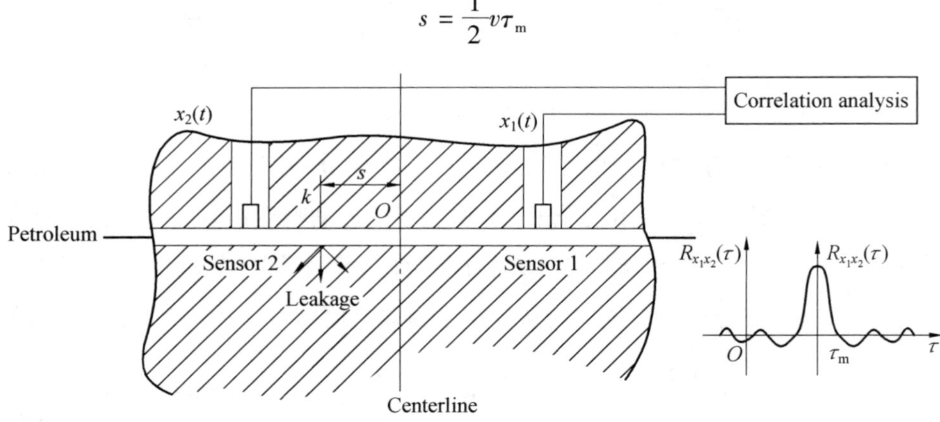

Figure 5.23 Use correlation analysis to find out where the oil pipe is leaking

For example, vehicle vibration fault detection shown in Figure 5.24 is a diagnostics performed without disassembly to vehicle seat vibration by using cross-correlation function. It is necessary to arrange acceleration sensors on the engine, the driver's seat, and the rear axle to measure whether the vibration is caused by the engine or the rear axle and then the signals obtained by the sensors are amplified and analyzed. It can be seen from the cross-correlation function that the cross-correlation between the rear axle and the driver's seat is greater than that of the engine and the driver's seat. Therefore, the vibration of the car seat is mainly caused by the vibration of the rear axle.

(3) Extract a specific frequency component from the signal mixed with periodic components for the excitation measurement of a linear system. Figure 5.25 shows the measurement system block diagram of the machine tool vibration measurement. The measured

例如,图 5.24 所示的车辆振动故障检测是利用互相关函数对汽车座位的振动进行不解体诊断,要测出振动是由发动机引起的还是由后桥引起的,可在发动机、驾驶人座位、后桥上布置加速度传感器,然后将传感器获取的信号放大并进行相关分析。通过互相关函数可以看出,后桥与驾驶人座位的互相关性比发动机与驾驶人座位的大。因此,汽车座位的振动主要是后桥的振动引起的。

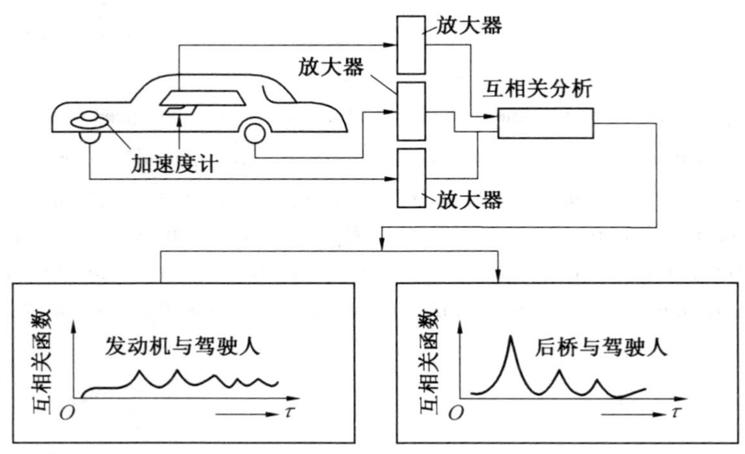

图 5.24　车辆振动故障检测

(3) 在混有周期成分的信号中提取特定的频率成分,用于线性系统的激振试验。图 5.25 所示为机床激振试验测试系统框图,所测得的振动响应信号中常常会含有大量的噪声干扰。根据线性系统频率保持特性,只有与激振频率相同的频率成分才可能是由激振引起的响应,其他成分均是干扰。为从噪声中提取有用信号,只需将激振信号和所测得

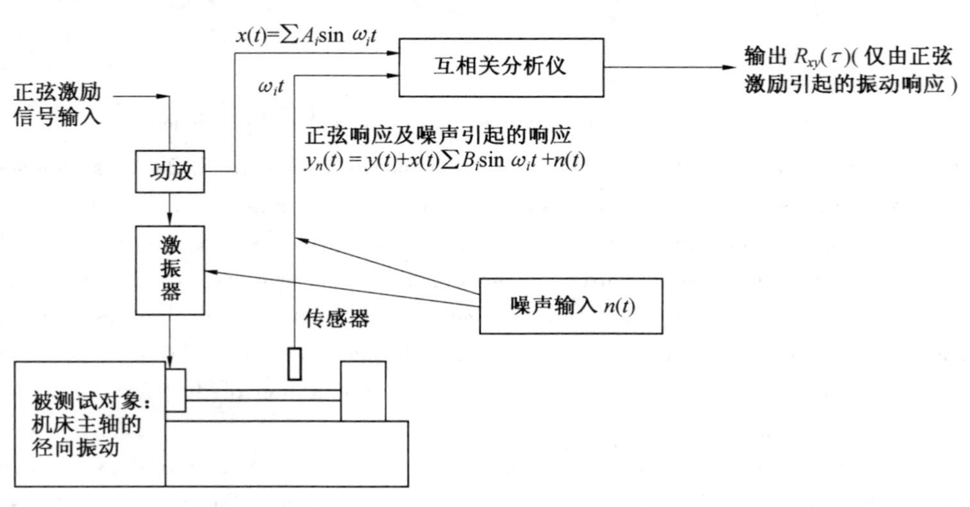

图 5.25　机床激振试验测试系统框图

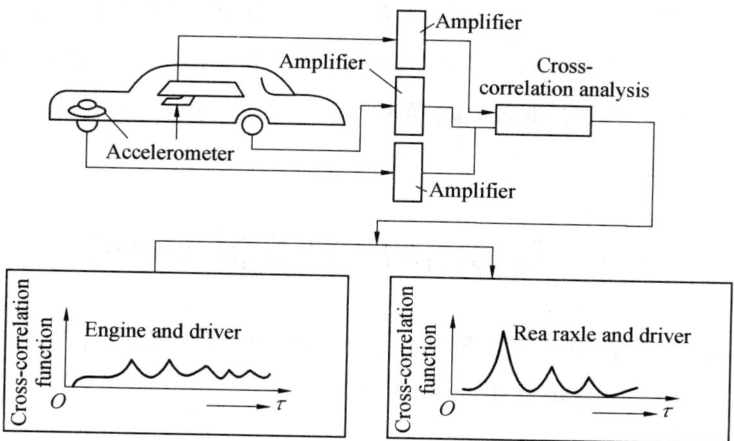

Figure 5.24　Vehicle vibration fault detection

vibration response signal often contains a lot of noise interference. According to the frequency retention characteristics of the linear system, only the frequency component with the same excitation frequency can be the response caused by the excitation, and other components are interference. In order to extract useful signals from noise, only the cross-correlation analysis of the excitation signal and the measured response signal is required to obtain the response amplitude and phase difference caused by the excitation, thereby eliminating the influence of noise interference. If the excitation frequency is changed, the frequency response function of the system formed by the corresponding signal transmission channel can be obtained. This method of applying the principle of correlation analysis to eliminate noise interference and extract useful information is called correlation filter.

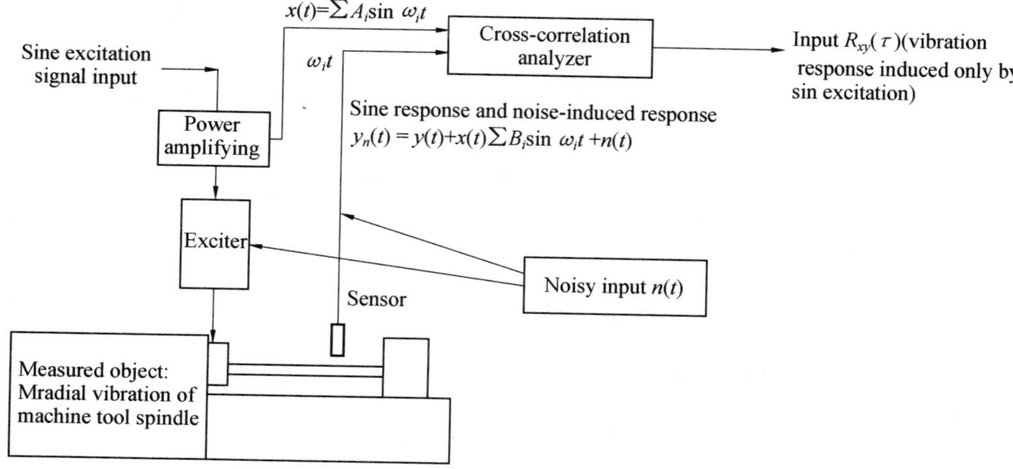

Figure 5.25　The measurement system block diagram of machine tool vibration measurement

的响应信号进行互相关分析，就可以得到由激振引起的响应幅值和相位差，从而消除噪声干扰的影响。如果改变激振频率，就可以求得相应信号传输通道构成的系统的频率响应函数。这种应用相关分析原理来消除噪声干扰、提取有用信息处理的方法称为相关滤波。

5.5 信号的功率谱分析及应用

相关分析从时域为在噪声背景下提取有用信息提供了途径，功率谱分析则从频域为研究平稳随机过程提供了重要方法。

5.5.1 帕塞瓦尔(Paseval)定理

帕塞瓦尔定理：在时域中信号的总能量等于在频域中信号的总能量，即

$$\int_{-\infty}^{\infty} x^2(t)\,\mathrm{d}t = \int_{-\infty}^{\infty} |X(f)|^2 \mathrm{d}f \tag{5.49}$$

式(5.49)又称能量等式。该定理可以用傅里叶变换的卷积来证明。

假设

$$x_1(t) \Leftrightarrow X_1(f),\ x_2(t) \Leftrightarrow X_2(f)$$

根据频域卷积定理，则有

$$x_1(t)x_2(t) \Leftrightarrow X_1(f) * X_2(f)$$

$$\begin{aligned}\int_{-\infty}^{\infty} x_1(t)x_2(t)\mathrm{e}^{-\mathrm{j}2\pi f_0 t}\mathrm{d}t &= \int_{-\infty}^{\infty}\int_{-\infty}^{\infty}[X_1(f)\mathrm{e}^{-\mathrm{j}2\pi f\tau}\mathrm{d}f]x_2(t)\mathrm{e}^{-\mathrm{j}2\pi f_0\tau}\mathrm{d}t \\ &= \int_{-\infty}^{\infty}\int_{-\infty}^{\infty}[X_1(f)\mathrm{d}f]x_2(t)\mathrm{e}^{-\mathrm{j}2\pi(f_0-f)\tau}\mathrm{d}t \\ &= \int_{-x}^{\infty} X_1(f)x_2(f_0-f)\mathrm{d}f\end{aligned}$$

令 $f_0=0$，则

$$\int_{-\infty}^{\infty} x_1(t)x_2(t)\mathrm{d}t = \int_{-\infty}^{\infty} X_1(f)X_2(-f)\mathrm{d}f$$

又令 $x_1(t)=x_2(t)=x(t)$，可得

$$\int_{-\infty}^{\infty} x^2(t)\mathrm{d}t = \int_{-\infty}^{\infty} X(f)X(-f)\mathrm{d}f$$

$x(t)$ 为实函数，则 $X(-f)=X^*(f)$，为 $X(f)$ 的共轭函数，所以

$$\int_{-\infty}^{\infty} x^2(t)\mathrm{d}t = \int_{-x}^{\infty} X(f)X^*(f)\mathrm{d}f = \int_{-\infty}^{\infty} |X(f)|^2\mathrm{d}f$$

$|X(f)|^2$ 称为能量谱，它是沿频率轴的能量分布密度。这样，在整个时间轴上信号的

5.5 Signal Power Spectrum Analysis and Application

Correlation analysis provides a way to extract useful information from the background of noise in the time domain, and power spectrum analysis provides an important method for studying stationary random processes in the frequency domain.

5.5.1 Paseval's Theorem

Paseval's theorem: the total energy of the signal in the time domain is equal to the total energy of the signal in the frequency domain, namely

$$\int_{-\infty}^{\infty} x^2(t)\,\mathrm{d}t = \int_{-\infty}^{\infty} |X(f)|^2 \mathrm{d}f \tag{5.49}$$

Formula (5.49) is also called the energy equation. The theorem can be proved by Fourier transform convolution.

Assuming

$$x_1(t) \Leftrightarrow X_1(f),\ x_2(t) \Leftrightarrow X_2(f)$$

According to the frequency domain convolution theorem, then we have

$$x_1(t)x_2(t) \Leftrightarrow X_1(f) * X_2(f)$$

$$\begin{aligned}\int_{-\infty}^{\infty} x_1(t)x_2(t)\mathrm{e}^{-\mathrm{j}2\pi f_0 t}\mathrm{d}t &= \int_{-\infty}^{\infty}\int_{-\infty}^{\infty}[X_1(f)\mathrm{e}^{-\mathrm{j}2\pi f\tau}\mathrm{d}f]x_2(t)\mathrm{e}^{-\mathrm{j}2\pi f_0\tau}\mathrm{d}t \\ &= \int_{-\infty}^{\infty}\int_{-\infty}^{\infty}[X_1(f)\mathrm{d}f]x_2(t)\mathrm{e}^{-\mathrm{j}2\pi(f_0-f)\tau}\mathrm{d}t \\ &= \int_{-x}^{\infty} X_1(f)x_2(f_0-f)\mathrm{d}f\end{aligned}$$

Let $f_0 = 0$, then

$$\int_{-\infty}^{\infty} x_1(t)x_2(t)\mathrm{d}t = \int_{-\infty}^{\infty} X_1(f)X_2(-f)\mathrm{d}f$$

And let $x_1(t) = x_2(t) = x(t)$, we get

$$\int_{-\infty}^{\infty} x^2(t)\mathrm{d}t = \int_{-\infty}^{\infty} X(f)X(-f)\mathrm{d}f$$

$x(t)$ is a real function, then $X(-f) = X^*(f)$, it is the conjugate function of $X(f)$, so

$$\int_{-\infty}^{\infty} x^2(t)\mathrm{d}t = \int_{-x}^{\infty} X(f)X^*(f)\mathrm{d}f = \int_{-\infty}^{\infty} |X(f)|^2\mathrm{d}f$$

$|X(f)|^2$ is called the energy spectrum, which is the energy distribution density along the frequency axis. In this way, the average power of the signal on the entire time axis can be calculated as

$$P_{\mathrm{av}} = \lim_{T\to\infty}\frac{1}{T}\int_0^T x^2(t)\mathrm{d}t = \int_{-\infty}^{\infty}\lim_{T\to\infty}\frac{1}{T}|X(f)|^2\mathrm{d}f \tag{5.50}$$

Formula (5.50) is another expression of Paseval's theorem.

5.5.2 Power Spectral Density Function

1. Self-Power Spectral Density Function Definition and Physical Meaning

Assuming that the mean value of the stationary random signal $x(t)$ is zero and does not

平均功率可计算为

$$P_{\mathrm{av}} = \lim_{T\to\infty}\frac{1}{T}\int_0^T x^2(t)\,\mathrm{d}t = \int_{-\infty}^{\infty}\lim_{T\to\infty}\frac{1}{T}\,|X(f)|^2\mathrm{d}f \tag{5.50}$$

式(5.50)为帕塞瓦尔定理的另一种表达形式。

5.5.2　功率谱密度函数

1. 自功率谱密度函数定义及物理意义

设平稳随机信号 $x(t)$ 的均值为零且不含周期成分，则其自相关函数 $R_x(\tau)$ 在 $\tau\to\infty$ 时有 $R_x(\tau\to\infty)=0$，则该自相关函数满足傅里叶变换的条件 $\int_{-\infty}^{\infty}|R_x(\tau)|\mathrm{d}\tau<\infty$。于是，存在 $R_x(\tau)$ 的傅里叶变换对，即

$$S_x(f) = \int_{-\infty}^{\infty}R_x(\tau)\,\mathrm{e}^{-\mathrm{j}2\pi f\tau}\,\mathrm{d}\tau \tag{5.51}$$

其逆变换为

$$R_x(\tau) = \int_{-\infty}^{\infty}S_x(f)\,\mathrm{e}^{\mathrm{j}2\pi f\tau}\,\mathrm{d}f \tag{5.52}$$

$S_x(f)$ 称为 $x(t)$ 的自功率谱密度函数，简称自谱（或自功率谱）。$R_x(\tau)$ 是对信号的时域分析，$S_x(f)$ 是对信号的频域分析，它们包含的信息是完全相同的。

$R_x(\tau)$ 为实偶函数，则 $S_x(f)$ 也为实偶函数，$S_x(f)$ 是在 $(-\infty,\infty)$ 频率范围内的自功率谱，所以称为双边自谱。由于 $S_x(f)$ 为实偶函数，而在实际应用中频率不能为负值，因此用在 $(0,\infty)$ 频率范围内的单边自谱 $G_x(f)$ 表示信号的全部功率谱，即 $G_x(f)=2S_x(f)$。单边自谱和双边自谱如图 5.26 所示。

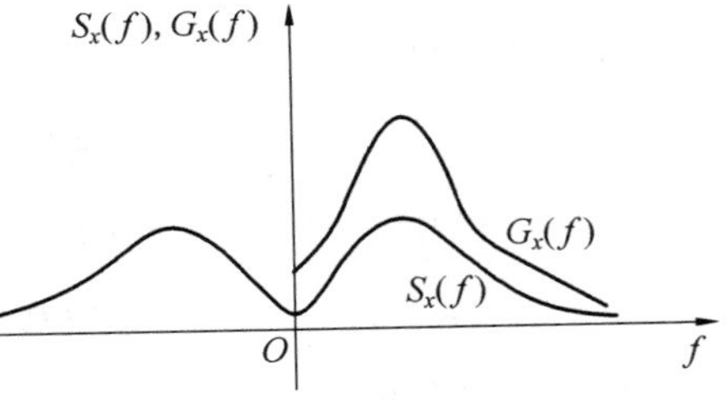

图 5.26　单边自谱和双边自谱

当 $\tau=0$ 时，根据自相关函数 $R_x(\tau)$ 和自功率谱密度函数 $S_x(f)$ 的定义，有

$$R_x(0) = \lim_{T\to\infty}\frac{1}{T}\int_0^T x^2(t)\,\mathrm{d}t = \lim_{T\to\infty}\int_0^T\frac{x^2(t)}{T}\mathrm{d}t = \int_{-\infty}^{\infty}S_x(f)\,\mathrm{d}f \tag{5.53}$$

从物理意义上讲，$x(t)^2$ 看作信号的能量，$x(t)^2/T$ 看作信号的功率，则 $\lim\limits_{T\to\infty}\int_0^T\frac{x^2(t)}{T}\mathrm{d}t$ 为信号 $x(t)$ 的总功率。由式(5.53)可以看出，$S_x(f)$ 曲线下与频率轴包围的总面积和 $\frac{x^2(t)}{T}$ 曲线下的总面积相等，所以 $S_x(f)$ 曲线下和频率轴包围的总面积就是信号的总功率，

contain periodic components, then its autocorrelation function $R_x(\tau)$ has $R_x(\tau \to \infty) = 0$ at the condition of $\tau \to \infty$, then the autocorrelation function satisfies the Fourier transform condition $\int_{-\infty}^{\infty} |R_x(\tau)| \mathrm{d}\tau < \infty$. So there is a Fourier transform pair of $R_x(\tau)$, that is

$$S_x(f) = \int_{-\infty}^{\infty} R_x(\tau) \mathrm{e}^{-\mathrm{j}2\pi f \tau} \mathrm{d}\tau \tag{5.51}$$

Its inverse transformation is

$$R_x(\tau) = \int_{-\infty}^{\infty} S_x(f) \mathrm{e}^{\mathrm{j}2\pi f \tau} \mathrm{d}f \tag{5.52}$$

$S_x(f)$ is called the power spectrum density function of $x(t)$, referred to as the self spectrum (or self power spectrum). $R_x(\tau)$ is the time domain analysis of the signal, $S_x(f)$ is the frequency domain analysis of the signal, they contain exactly the same information.

$R_x(\tau)$ is a real-even function, then $S_x(f)$ is also a real-even function, and $S_x(f)$ is the self-power spectrum in $(-\infty, \infty)$ frequency range, so it is called a bilateral self-spectrum. Since $S_x(f)$ is a real-even function, and the frequency cannot be negative in practical applications, the unilateral autospectrum $G_x(f)$ in the frequency range of $(0, \infty)$ represents the entire power spectrum of the signal, that is, $G_x(f) = 2S_x(f)$. One-sided self-spectrum and two-sided self-spectrum is shown in Figure 5.26.

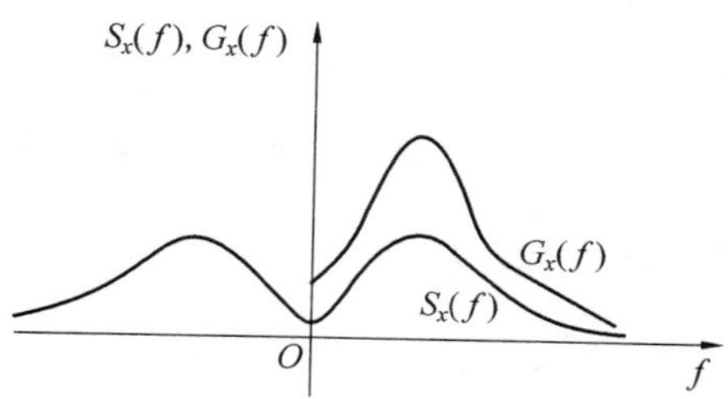

Figure 5.26　One-sided self-spectrum and two-sided self-spectrum

When $\tau = 0$, according to the definition of autocorrelation function $R_x(\tau)$ and self power spectral density function $S_x(f)$, there is

$$R_x(0) = \lim_{T\to\infty} \frac{1}{T} \int_0^T x^2(t)\,\mathrm{d}t = \lim_{T\to\infty} \int_0^T \frac{x^2(t)}{T} \mathrm{d}t = \int_{-\infty}^{\infty} S_x(f)\,\mathrm{d}f \tag{5.53}$$

In a physical sense, $x^2(t)$ is regarded as the energy of the signal, $x^2(t)/T$ is regarded as the power of the signal, and $\lim_{T\to\infty} \int_0^T \frac{x^2(t)}{T} \mathrm{d}t$ is the total power of the signal $x(t)$. It can be seen from formula(5.53) that the total area under $S_x(f)$ curve and the frequency axis is equal to the total area under $\frac{x^2(t)}{T}$ curve, so the total area under $S_x(f)$ curve and the frequency axis is the total power of the signal, the size of $S_x(f)$ represents the distribution of the self-power spectral density function along the frequency axis.

According to formula (5.50) and formula (5.53), we can get

$$S_x(f) = \lim_{T\to\infty} \frac{1}{T} |X(f)|^2 \tag{5.54}$$

Formula (5.54) reflects the relationship between the self-power spectral density function

$S_x(f)$ 的大小表示自功率谱密度函数沿频率轴的分布。

根据式(5.50) 和式(5.53) 可以得出

$$S_x(f) = \lim_{T\to\infty} \frac{1}{T} |X(f)|^2 \tag{5.54}$$

式(5.54) 反映了自功率谱密度函数和幅值谱之间的关系。利用这一关系,就可以对时域信号直接做傅里叶变换来计算其功率谱。可见,自谱 $S_x(f)$ 反映信号的频域结构,这与幅值谱 $|X(f)|$一致,但自谱所反映的是信号幅值的二次方,因此其频域结构特征更为明显。幅值谱与自功率谱如图 5.27 所示。

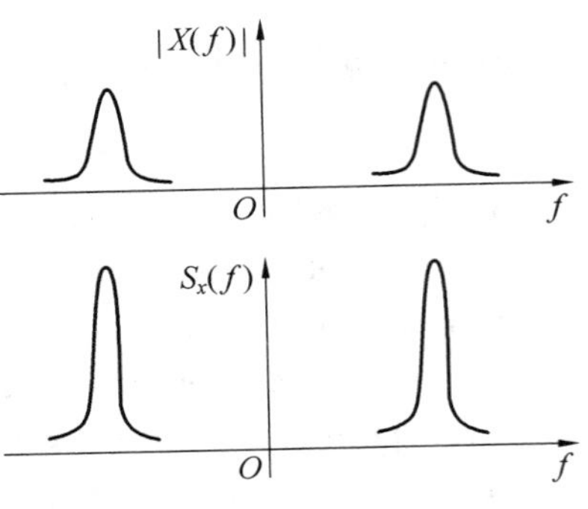

图 5.27　幅值谱与自功率谱

2. 自功率谱密度的估计

在实际测试中,信号的自功率谱密度只能在有限长度的时间区域内近似估计。根据自功率谱密度函数的定义,信号自功率谱密度估计应当先根据原始信号计算出其相关函数,然后对自相关函数做傅里叶变换。在实际自功率谱密度估计时,往往采用更为方便可行的方法。根据式(5.54) 初步估计自功率谱密度,以 $\tilde{S}_x(f)$、$\tilde{G}_x(f)$ 分别表示双边谱和单边谱,即

$$\tilde{S}_x(f) = \frac{1}{T} |X(f)|^2 \tag{5.55}$$

$$\tilde{G}_x(f) = \frac{2}{T} |X(f)|^2 \tag{5.56}$$

对于数字信号,初步估计自功率谱密度为

$$\tilde{S}_x(k) = \frac{1}{N} |X(k)|^2 \tag{5.57}$$

$$\tilde{G}_x(k) = \frac{2}{N} |X(k)|^2 \tag{5.58}$$

这是对离散随机序列信号 $x(n)$ 进行快速傅里叶变换(FFT),取其模的二次方,再乘以 $1/N$ 或 $2/N$,得到自功率谱密度的初步估计。由于该变换具有周期函数的性质,因此这种自功率谱密度的估计方法称为周期图法。它是一种最常见、常用的自功率谱密度的估计算法。

3. 互功率谱密度函数

互相关函数 $R_{xy}(\tau)$ 在 $\tau\to\infty$ 时,有 $R_{xy}(\tau\to\infty)=0$,则该互相关函数满足傅里叶变换

and the amplitude spectrum. Using this relationship, the time domain signal can be directly Fourier transformed to calculate its power spectrum. It can be seen that the self-spectrum $S_x(f)$ reflects the frequency domain structure of the signal, which is consistent with the amplitude spectrum $|X(f)|$, but the self-spectrum reflects the second power of the signal amplitude, so its frequency domain structure characteristics are more obvious. Amplitude spectrum and self-power spectrum are shown in Figure 5.27.

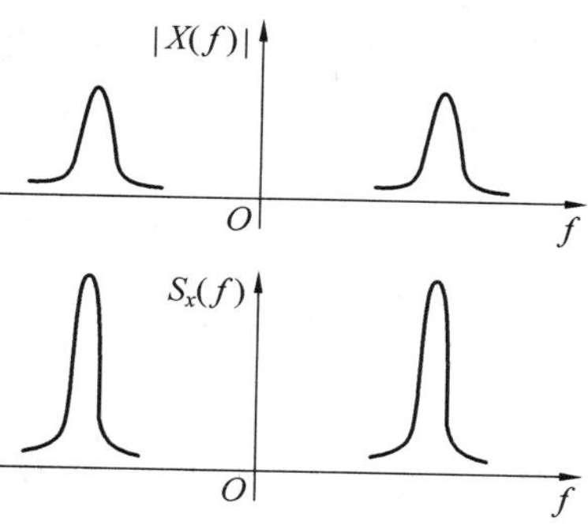

Figure 5.27 Amplitude spectrum and self-power spectrum

2. Estimation of Self-Power Spectral Density

In actual measurement, the self-power spectral density of the signal can only be approximated within a limited time area. According to the definition of self-power spectral density function, signal self-power spectral density estimation should first calculate its correlation function based on the original signal, and then perform Fourier transform on the self-correlation function. In actual self-power spectral density estimation, more convenient and feasible methods are often adopted. According to formula (5.54), the self-power spectral density can preliminarily be estimated to denote the two-sided spectrum and the one-sided spectrum with $\tilde{S}_x(f)$, $\tilde{G}_x(f)$ respectively, namely

$$\tilde{S}_x(f) = \frac{1}{T}|X(f)|^2 \tag{5.55}$$

$$\tilde{G}_x(f) = \frac{2}{T}|X(f)|^2 \tag{5.56}$$

For digital signals, the initial estimation of the self-power spectral density is

$$\tilde{S}_x(k) = \frac{1}{N}|X(k)|^2 \tag{5.57}$$

$$\tilde{G}_x(k) = \frac{2}{N}|X(k)|^2 \tag{5.58}$$

This is to perform fast Fourier transform (FFT) on the discrete random sequence signal $x(n)$, take the quadratic of its module, and then multiply it by $1/N$ or $2/N$ to get a preliminary estimate of the self-power spectral density. Because the transformation has the property of a periodic function, this method of estimating the self-power spectral density is called the periodogram method. It is the most common and commonly used self-power spectral density estimation algorithm.

3. Cross-Power Spectral Density Function

When the cross-correlation function $R_{xy}(\tau)$ is at $\tau \to \infty$, there is $R_{xy}(\tau \to \infty) = 0$, then the cross-correlation function satisfies the Fourier transform condition $\int_{-\infty}^{\infty} |R_{xy}(\tau)| d\tau < \infty$. So, there is a Fourier transform pair of $R_{xy}(\tau)$, that is

$$S_{xy}(f) = \int_{-\infty}^{\infty} R_{xy}(\tau) e^{-j2\pi f\tau} dt \tag{5.59}$$

的条件$\int_{-\infty}^{\infty} |R_{xy}(\tau)| \mathrm{d}\tau < \infty$。于是，存在 $R_{xy}(\tau)$ 的傅里叶变换对，即

$$S_{xy}(f) = \int_{-\infty}^{\infty} R_{xy}(\tau) \mathrm{e}^{-\mathrm{j}2\pi f\tau} \mathrm{d}t \tag{5.59}$$

其逆变换为

$$R_{xy}(\tau) = \int_{-\infty}^{+\infty} S_{xy}(f) \mathrm{e}^{\mathrm{j}2\pi f\tau} \mathrm{d}f \tag{5.60}$$

式中 $S_{xy}(f)$——信号 $x(t)$ 和 $y(t)$ 的互功率谱密度函数，简称互谱。

互相关函数 $R_{xy}(\tau)$ 并非偶函数，因此 $S_{xy}(f)$ 具有虚、实两部分。同样，$S_{xy}(f)$ 保留了 $R_{xy}(\tau)$ 中的全部信息。

5.5.3 功率谱的应用

1. 功率谱密度与幅值谱及系统的频率响应函数的关系

图5.28所示线性系统的输出 $y(t)$ 等于其输入 $x(t)$ 和系统的脉冲响应函数 $h(t)$ 的卷积，即

$$y(t) = x(t) * h(t)$$

根据卷积定理，则上式在频域中为

$$Y(f) = X(f)H(f)$$

式中 $H(f)$——系统的频率响应函数，反映了系统的传递特性；

$Y(f)$、$X(f)$、$H(f)$——f 的复函数。

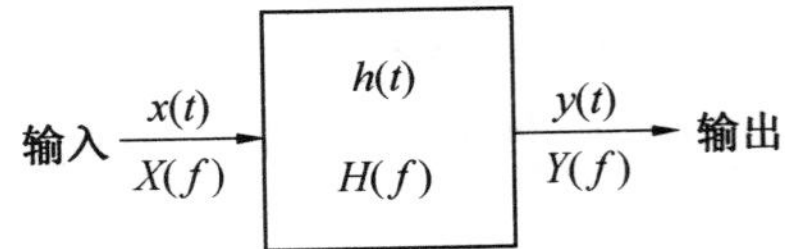

图 5.28 理想输入、输出系统

如果 $X(f)$ 表示为 $X(f) = X_{\mathrm{R}}(f) + \mathrm{j}X_{\mathrm{I}}(f)$，则 $X(f)$ 的共轭值为

$$X^*(f) = X_{\mathrm{R}}(f) - \mathrm{j}X_{\mathrm{I}}(f)$$

则

$$X(f)X^*(f) = X_{\mathrm{R}}^2(f) + X_{\mathrm{I}}^2(f) = |X(f)|^2$$

通过自谱可求得 $H(f)$ 为

$$H(f)H^*(f) = \frac{Y(f)}{X(f)} \frac{Y^*(f)}{X^*(f)} = \frac{S_y(f)}{S_x(f)} = |H(f)|^2 \tag{5.61}$$

通过输入、输出自谱分析，就能得出系统的幅频特性，但丢失了相位信息。变换式(5.61)，则得到输入、输出自谱与系统的频率响应函数的关系为

$$S_y(f) = |H(f)|^2 S_x(f) \tag{5.62}$$

也可用自谱和互谱求得 $H(f)$，即

Its inverse transformation is

$$R_{xy}(\tau) = \int_{-\infty}^{\infty} S_{xy}(f)\, e^{j2\pi f\tau} df \tag{5.60}$$

where　$S_{xy}(f)$ —called the cross power spectrum density function of signals $x(t)$ and $y(t)$, referred to as cross spectrum.

The cross-correlation function $R_{xy}(\tau)$ is not an even function, so $S_{xy}(f)$ has imaginary part and real part. Similarly, $S_{xy}(f)$ retains all the information in $R_{xy}(\tau)$.

5.5.3 The Application of Power Spectrum

1. The Relationship Between the Power Spectral Density and the Amplitude Spectrum and the Frequency Response Function of the System

The output $y(t)$ of the linear system shown in Figure 5.28 is equal to the convolution of its input $x(t)$ and the impulse response function $h(t)$ of the system, namely

$$y(t) = x(t) * h(t)$$

Figure 5.28　Ideal input and output system

According to the convolution theorem, the above formula in the frequency domain is

$$Y(f) = X(f)H(f)$$

where　$H(f)$ —the frequency response function of the system, reflecing the transfer characteristics of the system;

$Y(f)$, $X(f)$, $H(f)$ —the complex function of f.

If $X(f)$ is denoted as $X(f) = X_R(f) + jX_I(f)$, then the conjugate value of $X(f)$ is

$$X^*(f) = X_R(f) - jX_I(f)$$

We get

$$X(f)X^*(f) = X_R^2(f) + X_I^2(f) = |X(f)|^2$$

Through self-spectrum, $H(f)$ can be obtained as

$$H(f)H^*(f) = \frac{Y(f)}{X(f)}\frac{Y^*(f)}{X^*(f)} = \frac{S_y(f)}{S_x(f)} = |H(f)|^2 \tag{5.61}$$

Through the input and output autospectrum analysis, the amplitude-frequency characteristics of the system can be obtained, but the phase information will be lost. Transform formula (5.61), then the relationship between the input and output self-spectrum and the frequency response function of the system is

$$S_y(f) = |H(f)|^2 S_x(f) \tag{5.62}$$

$H(f)$ can also be obtained by self-spectrum and cross-spectrum, namely

$$H(f) = \frac{Y(f)}{X(f)}\frac{X^*(f)}{X^*(f)} = \frac{S_{xy}(f)}{S_x(f)} = \frac{G_{xy}(f)}{G_x(f)} \tag{5.63}$$

Formula (5.63) shows that the frequency response function of the system can be obtained by the ratio of the input and output cross-spectrum and the input self-spectrum. Since $S_{xy}(f)$

$$H(f)=\frac{Y(f)}{X(f)}\frac{X^*(f)}{X^*(f)}=\frac{S_{xy}(f)}{S_x(f)}=\frac{G_{xy}(f)}{G_x(f)} \tag{5.63}$$

式(5.63)说明，系统的频率响应函数可以通过输入、输出互谱与输入自谱之比得出。由于 $S_{xy}(f)$ 包含频率和相位信息，因此 $H(f)$ 含幅频和相频信息。变换式(5.63)，则有

$$S_{xy}(f)=H(f)S_x(f) \tag{5.64}$$

2. 利用互谱排除噪声影响

通常一个测试系统往往受到内部噪声和外部噪声的干扰，从而输出也会带入噪声干扰。但由于输入信号与噪声无关，因此它们的互相关函数为零。这一点说明，在利用自谱和互谱求系统频率函数时不会受到影响。

受外界干扰的系统如图5.29所示，一个测试系统输入信号为 $x(t)$，受到外界各种工作噪声干扰，$n_1(t)$ 为从输入端侵入的噪声，$n_2(t)$ 为从中间环节侵入的噪声，$n_3(t)$ 为从输出端侵入的噪声。该系统的输出 $y(t)$ 为

$$y(t)=x'(t)+n_1'(t)+n_2'(t)+n_3'(t) \tag{5.65}$$

式中　$x'(t)$、$n_1'(t)$、$n_2'(t)$、$n_3'(t)$——系统对 $x(t)$、$n_1(t)$、$n_2(t)$、$n_3(t)$ 的响应。

输入 $x(t)$ 与输出 $y(t)$ 的互相关函数为

$$R_{xy}(\tau)=R_{xx'}(\tau)+R_{xn_1'}(\tau)+R_{xn_2'}(\tau)+R_{xn_3'}(\tau) \tag{5.66}$$

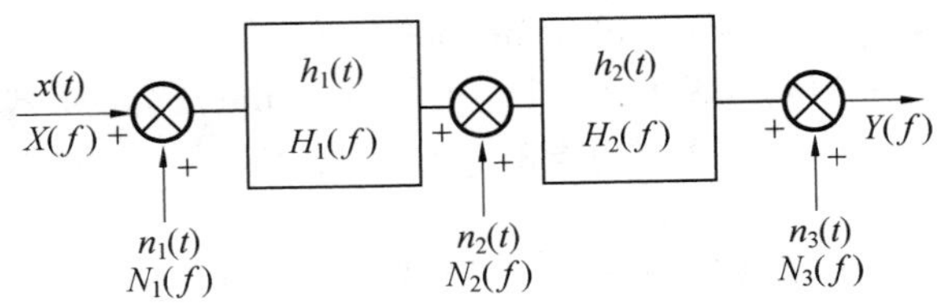

图5.29　受外界干扰的系统

由于输入 $x(t)$ 与噪声 $n_1(t)$、$n_2(t)$、$n_3(t)$ 是独立无关的，因此互相关函数 $R_{xn_1'}(t)$、$R_{xn_2'}(t)$ 和 $R_{xn_3'}(t)$ 均为零，所以

$$R_{xy}(\tau)=R_{xx'}(\tau) \tag{5.67}$$

因此，尽管输出 $y(t)$ 中含有大量噪声，但输入和输出的互相关函数 $R_{xy}(\tau)$ 仍与无干扰时一样，仍然是激励 $x(t)$ 和它的响应 $y(t)$ 的互相关函数 $R_{xx'}(\tau)$。因此，可以说线性系统输入和输出的互相关函数对外界噪声具有“免疫力”，$S_{xy}(f)=S_{xy}(\tau)$ 也具有免疫力，故

$$S_{xy}(f)=S_{xx'}(f)=H(f)S_x(f) \tag{5.68}$$

式中　$H(f)$——所研究系统的频率响应函数，$H(f)=H_1(f)H_2(f)$。

contains frequency and phase information, $H(f)$ contains amplitude-frequency and phase-frequency information. Transform formula(5.63), then we have

$$S_{xy}(f) = H(f)S_x(f) \tag{5.64}$$

2. Using Cross-Spectrum to Eliminate Noise Influence

Usually a measurement system is often interfered by internal noise and external noise, so the output will also bring noise interference, while their cross-correlation function is zero because the input signal has nothing to do with noise. This point shows that it will not be affected when using self-spectrum and cross-spectrum to find the system frequency function.

System affected by external interference is shown in Figure 5.29, the input signal of amea-surement system is $x(t)$, which is interfered by various external operating noises, $n_1(t)$ is the noise invaded from the input end, $n_2(t)$ is the noise invaded from the intermediate link, and $n_3(t)$ is the noise invaded from the output end. The output $y(t)$ of the system is

$$y(t) = x'(t) + n_1'(t) + n_2'(t) + n_3'(t) \tag{5.65}$$

where　$x'(t), n_1'(t), n_2'(t), n_3'(t)$ —the system's response to $x(t), n_1(t), n_2(t), n_3(t)$.

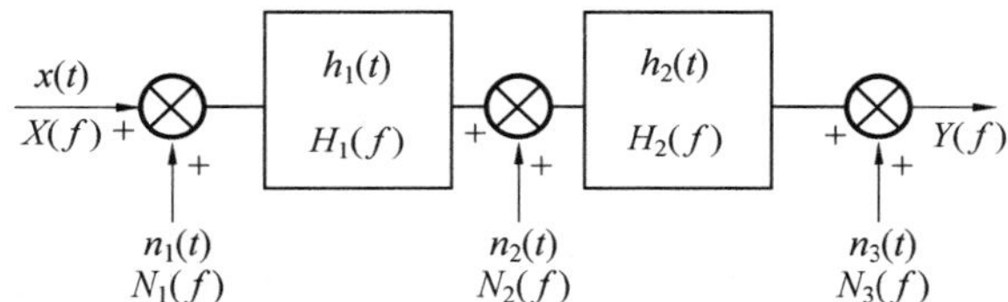

Figure 5.29　System affected by external interference

The cross-correlation function of input $x(t)$ and output $y(t)$ is

$$R_{xy}(\tau) = R_{xx'}(\tau) + R_{xn_1'}(\tau) + R_{xn_2'}(\tau) + R_{xn_3'}(\tau) \tag{5.66}$$

Since input $x(t)$ and noise $n_1(t), n_2(t), n_3(t)$ are independent and unrelated, the cross-correlation functions $R_{xn_1'}(t)$、$R_{xn_2'}(t)$ and $R_{xn_3'}(t)$ are both zero, so

$$R_{xy}(\tau) = R_{xx'}(\tau) \tag{5.67}$$

Therefore, although the output $y(t)$ contains a lot of noise, the cross-correlation function $R_{xy}(\tau)$ of the input and output is still the same as without interference. It is still the cross-correlation function $R_{xx'}(\tau)$ between the excitation $x(t)$ and its response $y(t)$. Therefore, it can be said that the cross-correlation function of the input and output of the linear system has "immunity" against external noise, and $S_{xy}(f) = S_{xy}(\tau)$ also has immunity, thus

$$S_{xy}(f) = S_{xx'}(f) = H(f)S_x(f) \tag{5.68}$$

where　$H(f)$ —the frequency response function of the studied system, $H(f) = H_1(f)H_2(f)$.

It can be seen that the use of cross-spectrum analysis can eliminate the influence of noise, which is the outstanding advantage of this analysis method. It should be noted, however, that when using formula(5.68) to find the $H(f)$ of a linear system, although the cross-spectrum $S_{xy}(f)$ is not affected by noise, the auto-spectrum $S_x(f)$ of the input signal still cannot exclude the influence of the measurement noise at the input end, thus forming error of measurement.

In order to test the dynamic characteristics of the system, sometimes adding to the running system with a specific known disturbance—input $n(t)$.

由此可见,利用互谱分析可排除噪声的影响,这是这种分析方法的突出优点。然而应当注意到,利用式(5.68) 求线性系统的 $H(f)$ 时,尽管其中的互谱 $S_{xy}(f)$ 可不受噪声的影响,但是输入信号的自谱 $S_x(f)$ 仍然无法排除输入端测量噪声的影响,从而形成测量的误差。

为测试系统的动特性,有时故意给正在运行的系统加特定的已知扰动 —— 输入 $n(t)$。

3. 功率谱在设备诊断中的应用

图 5.30 所示为汽车变速器上加速度信号功率谱图。图 5.30(a) 所示是变速器正常工作谱图,图 5.30(b) 所示是机器运行不正常时的谱图。可以看到,图 5.30(b) 比图 5.30(a) 增加了 9.2 Hz 和 18.4 Hz 两个谱峰,这两个频率为设备故障的诊断提供了依据。

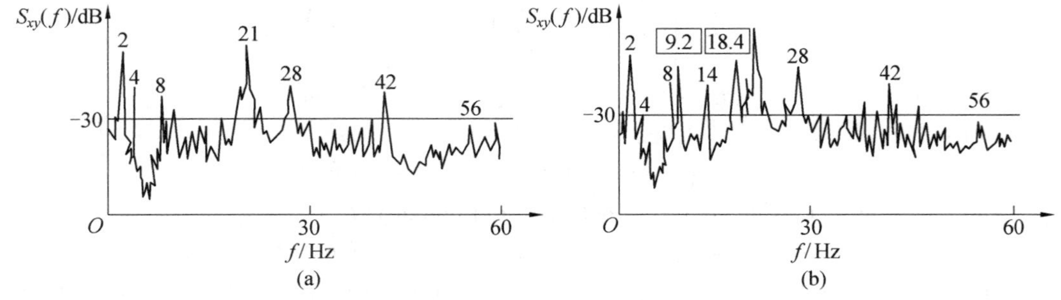

图 5.30　加速度信号功率谱图

4. 瀑布图

机器在增速或降速过程中,对不同转速时的振动信号进行等间隔采样,并进行功率谱分析。将各转速下的功率谱组合在一起成为一个转速功率谱三维图,又称瀑布图。图 5.31 所示为柴油机振动信号的瀑布图。图中在转速为 1 480 r/min 的三次频率上和 1 990 r/min 的六次频率上谱峰较高,即在这两个转速上产生两种阶次的共振,这就可以定出危险旋转速度,进而找到共振根源。

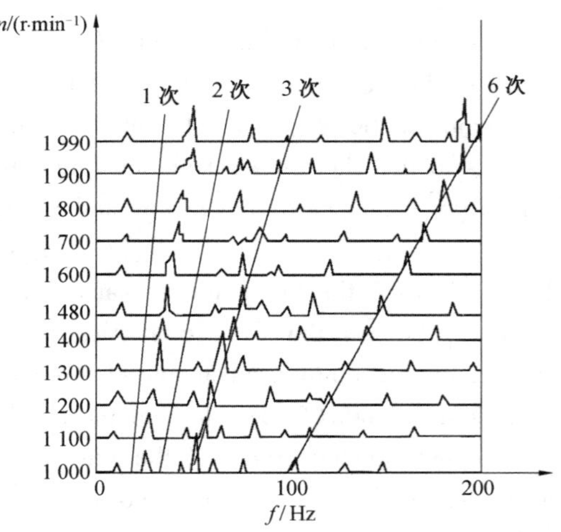

图 5.31　柴油机振动信号的瀑布图

3. The Application of Power Spectrum in Equipment Diagnosis

Figure 5.30 shows the power spectrum of the acceleration signal on the automobile transmission. Figure 5.30(a) shows the normal working spectrum of the transmission, and Figure 5.30(b) shows the spectrum when the machine is running abnormally. It can be seen that Figure 5.30(b) has two additional peaks of 9.2 Hz and 18.4 Hz compared to Figure 5.30(a). These two frequencies provide a basis for the diagnosis of equipment faults.

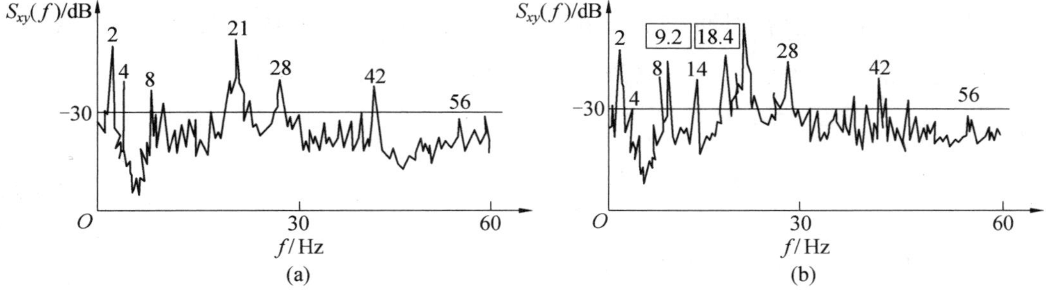

Figure 5.30 The power spectrum of the acceleration signal on the automobile transmission

4. Waterfall Chart

In the process of speed increasing or slowing down, the machine samples the vibration signals at different speeds at equal intervals and conducts power spectrum analysis. The power spectrum at each speed is combined into a three-dimensional graph of speed power spectrum, also known as waterfall chart. Figure 5.31 shows the waterfall diagram of the diesel engine vibration signal. In the figure, the spectral peaks are relatively higher at the third frequency of 1 480 r/min and the sixth frequency of 1 990 r/min, that is, two orders of resonance are generated at these two speeds, which can determine the dangerous rotation speed, and then find the source of resonance.

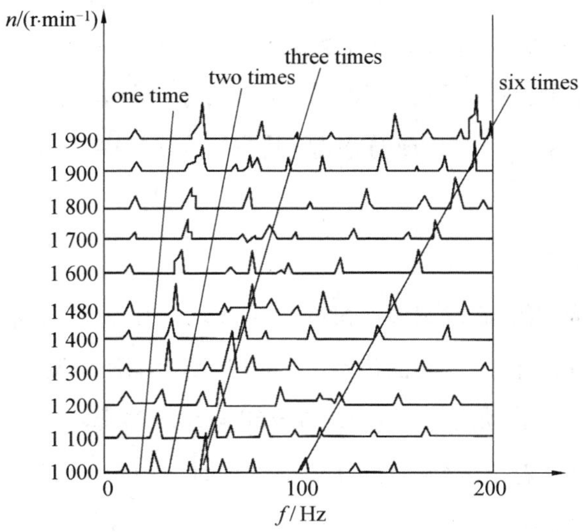

Figure 5.31 The waterfall diagram of the diesel engine vibration signal

5.5.4 相干函数

1. 相干函数定义

相干函数是用来评价测试系统的输入信号与输出信号之间因果关系的函数，即通过相干函数判别系统中输出信号的功率谱有多少是所测输入信号引起的响应。在线测试中，系统的输出信号包含大量的噪声干扰，是一种混合物。其中，哪些成分是输入信号引起的，哪些是噪声引起的，有时是必须分辨清楚的。通常，相干函数用$\gamma_{xy}^2(f)$ 表示，其定义为

$$\gamma_{xy}^2(f) = \frac{|S_{xy}(f)|^2}{S_x(f)\,S_y(f)},\quad 0 \leqslant \gamma_{xy}^2(f) \leqslant 1 \tag{5.69}$$

式中　$S_{xy}(f)$—— 系统输入与输出的互谱；

$S_x(f)$、$S_y(f)$—— 分别为系统输入、输出的自谱。

(1) 如果相干函数为0，则表示输出信号与输入信号不相干。

(2) 当相干函数为1时，表示输出信号与输入信号完全相干。若系统为线性系统，则根据式(5.62)和式(5.64)可得

$$\begin{aligned}\gamma_{xy}^2(f) &= \frac{|S_{xy}(f)|^2}{S_x(f)S_y(f)} = \frac{|H(f)S_x(f)|^2}{S_x(f)S_y(f)} \\ &= \frac{S_y(f)S_x(f)}{S_x(f)S_y(f)} = 1\end{aligned} \tag{5.70}$$

式(5.70)表明，对于线性系统，输出完全是由输入引起的响应。

(3) 若相干函数在0 ~ 1内，则可能测试中有外界噪声干扰，或输出$y(t)$是输入$x(t)$和其他输入的综合输出，或联系$x(t)$和$y(t)$的线性系统是非线性的。这说明输出中频率为f的成分不完全由输入的同频成分引起，其中掺杂有噪声的干扰。干扰越强，$\gamma_{xy}^2(f)$越低。

2. 相干分析的应用

图5.32所示为船用柴油机润滑油泵的油压脉动与压油管振动的相干分析。润滑油泵转速为$n = 781$ r/min，油泵齿轮的齿数为$z = 14$，测得油压脉动信号$x(t)$和压油管振动信号$y(t)$。压油管压力脉动的基频为$f_0 = nz/60 = 182.24$ Hz。

由图5.32(c)可以看出，当$f = f_0 = 182.24$ Hz时，$\gamma_{xy}^2(f) = 0.9$；当$f = 2f_0 \approx 361.12$ Hz时，$\gamma_{xy}^2(f) = 0.37$；当$f = 3f_0 \approx 546.54$ Hz时，$\gamma_{xy}^2(f) = 0.8$；当$f = 4f_0 \approx 722.24$ Hz时，$\gamma_{xy}^2(f) = 0.75$。齿轮引起的各次谐频对应的相干函数值都较大，而其他频率对应的相干函数值很

5.5.4 The Coherence Function

1. The Definition of Coherence Function

The coherence function is a function used to evaluate the cause and effect relationship between the input signal and the output signal of the measurement system, that is, the coherence function is used to determine how much of the power spectrum of the output signal in the system is the response caused by the measured input signal. In online measurements, the output signals of the system contain a lot of noise interference that belongs to a kind of mixture. Sometimes, it is necessary for us to clearly distinguish which components are caused by the input signal and which are caused by noise. Usually, the coherence function is represented by $\gamma_{xy}^2(f)$, which is defined as

$$\gamma_{xy}^2(f) = \frac{|S_{xy}(f)|^2}{S_x(f)\, S_y(f)},\quad 0 \leqslant \gamma_{xy}^2(f) \leqslant 1 \tag{5.69}$$

where $S_{xy}(f)$ —the cross spectrum of system input and output;

$S_x(f)$, $S_y(f)$ —the self-spectrum of system input and output respectively.

(1) If the coherence function is 0, it means that the output signal is not coherent with the input signal.

(2) When the coherence function is 1, it means that the output signal is completely coherent with the input signal. If the system is a linear system, according to formula (5.62) and formula (5.64), we can get

$$\begin{aligned}\gamma_{xy}^2(f) &= \frac{|S_{xy}(f)|^2}{S_x(f)S_y(f)} = \frac{|H(f)S_x(f)|^2}{S_x(f)S_y(f)} \\ &= \frac{S_y(f)S_x(f)}{S_x(f)S_y(f)} = 1\end{aligned} \tag{5.70}$$

Formula (5.70) shows that the output is completely the response caused by the input for a linear system.

(3) If the coherence function is between 0–1, there may be external noise interference in themeasurement, or output $y(t)$ is the integrated output of input $x(t)$ and other inputs, or the linear system connecting $x(t)$ and $y(t)$ is non-linear. It shows that the component of frequency f in the output is not completely caused by the same frequency component of the input, which is doped with noise interference. The stronger the interference is, the lower $\gamma_{xy}^2(f)$ is.

2. The Application of Coherence Analysis

Figure 5.32 shows the coherence analysis of oil pressure pulsation and pressure oil pipe vibration of marine diesel engine lubricating oil pump. The speed of the lubricating oil pump is $n = 781$ r/min , the number of teeth of the oil pump gear is $z = 14$, the oil pressure pulsation signal $x(t)$ and the oil pressure pipe vibration signal $y(t)$ are measured. The fundamental frequency of the pressure pulsation of the oil pressure pipe is $f_0 = nz/60 = 182.24$ Hz .

It can be seen from Figure 5.32(c) that when $f = f_0 = 182.24$ Hz , $\gamma_{xy}^2(f) = 0.9$; when $f = 2f_0 \approx 361.12$ Hz , $\gamma_{xy}^2(f) = 0.37$; when $f = 3f_0 \approx 546.54$ Hz , $\gamma_{xy}^2(f) = 0.8$; when $f = 4f_0 \approx 722.24$ Hz , $\gamma_{xy}^2(f) = 0.75$. The coherence function value corresponding to each

小。由此可见，油管的振动主要是由油压脉动引起的。从图5.32(a)、(b)所示的$x(t)$和$y(t)$的自谱图中也明显可见油压脉动的影响。

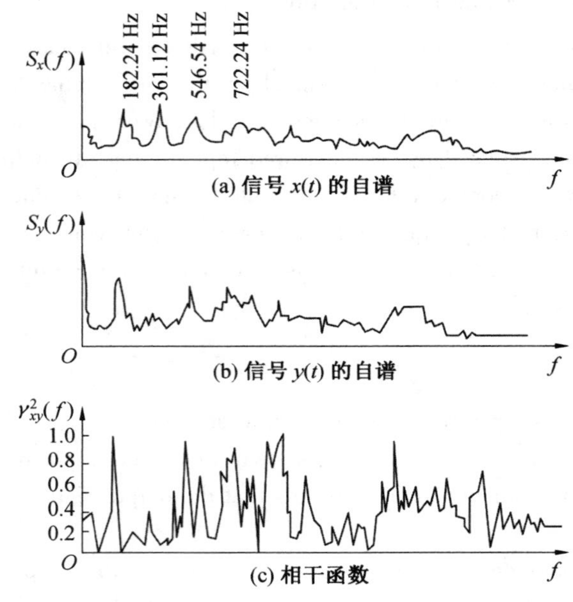

图 5.32　船用柴油机润滑油泵的油压脉动与压油管振动的相干分析

5.5.5　倒频谱分析及应用

倒频谱分析又称二次频谱分析，是近代信号处理科学的一项新技术，是检测复杂谱图中周期分量的有效工具，在语音分析、回声剔除、振动和噪声源识别、设备故障振动等方面均有成功的应用。

1. 倒频谱的数学描述

已知时域信号$x(t)$经过傅里叶变换后，可得到频域函数$X(f)$或功率谱密度函数$S_x(f)$。对功率谱密度函数取对数后，再对其进行傅里叶变换并取二次方，则可以得到倒频谱函数，其数学表达式为

$$C_p(q) = |F\{\lg S_x(f)\}|^2 \tag{5.71}$$

式中　$C_p(q)$—— 功率倒频谱，或称对数功率谱的功率谱。

工程上常用的是式(5.71)的开方形式，即

$$C_o(q) = \sqrt{C_p(q)} = |F\{\lg S_x(f)\}| \tag{5.72}$$

$C_o(q)$称为幅值倒频谱，简称倒频谱。

harmonic frequency caused by the gear is relatively large, while the value corresponding to other frequencies is very small. It can be seen that the vibration of the oil pipe is mainly caused by the oil pressure pulsation. From the autospectra of $x(t)$ and $y(t)$ shown in Figure 5.32 (a) and (b), the influence of oil pressure pulsation can also be clearly seen.

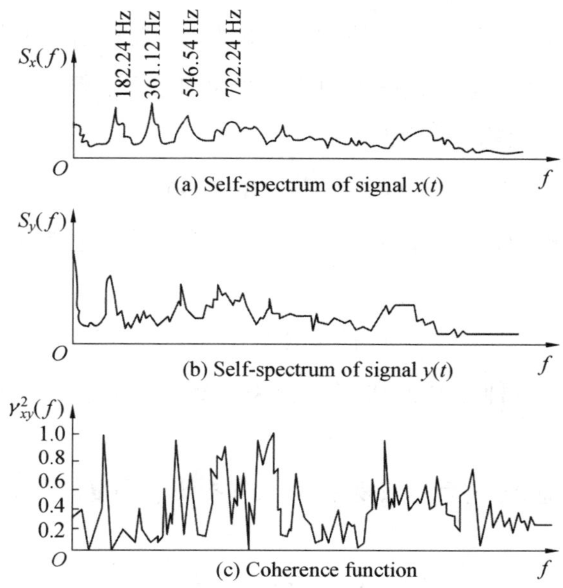

Figure 5.32　Coherence analysis of oil pressure pulsation and pressure oil pipe vibration of marine diesel engine lubricating oil pump

5.5.5　Cepstrum Analysis and Application

Cepstrum analysis, also known as secondary spectrum analysis, is a new technology in modern signal processing science and an effective tool for detecting periodic components in complex spectrograms. It has successful applications in speech analysis, echo cancellation, vibration and noise source identification, and equipment failure vibration, etc.

1. Mathematical Description of Cepstrum

Knowing that after the time domain signal $x(t)$ is Fourier transformed, the frequency domain function $X(f)$ or the power spectral density function $S_x(f)$ can be obtained. After taking the logarithm of the power spectral density function, the Fourier transform is performed on it and the square is taken, and then the cepstral function can be obtained. Its mathematical expression is

$$C_p(q) = |F\{\lg S_x(f)\}|^2 \tag{5.71}$$

where　$C_p(q)$ — power cepstrum, or power spectrum of logarithmic power spectrum.

The square root of formula (5.71) is commonly used in engineering, namely

$$C_o(q) = \sqrt{C_p(q)} = |F\{\lg S_x(f)\}| \tag{5.72}$$

where　$C_o(q)$ — the amplitude cepstrum, or cepstrum for short.

2. 倒频谱自变量 q 的物理意义

自变量 q 称为倒频率,它具有与自相关函数 $R_x(\tau)$ 中的自变量 τ 相同的时间量纲,一般取 ms 或 s。因为倒频谱是傅里叶变换,积分变量是频率 f 而不是时间 τ,故倒频谱 $C_o(q)$ 的自变量 q 具有时间的量纲。q 值大的称为高倒频率,表示谱图上的快速波动和密集谐频;q 值小的称为低倒频率,表示谱图上的缓慢波动和散离谐频。

为使 q 的定义更加明确,还可以定义

$$C_y(q) = F^{-1} | \{ \lg S_y(f) \} | \tag{5.73}$$

即倒频谱定义为信号的双边功率谱对数加权,再取其傅里叶逆变换,联系信号的自相关函数为

$$R(\tau) = F^{-1} \{ S_y(f) \} \tag{5.74}$$

由上述内容看出,这种定义方法与自相关函数很相近,变量 q 与 τ 在量纲上完全相同。

为反映出相位信息,分离后能恢复原信号,又提出一种复倒频谱的运算方法。若信号 $x(t)$ 的傅里叶变换为

$$X(f) = X_R(f) + jX_I(f) \tag{5.75}$$

则 $x(t)$ 的倒频谱为

$$C_o(q) = F^{-1} \{ \lg X(f) \} \tag{5.76}$$

显然,它保留了相位的信息。

倒频谱与相关函数的不同之处只是对数加权,其目的是使再变换以后的信号能量集中,扩大动态分析的频谱范围和提高再变换的精度,还可以解卷积(褶积)成分,易于对原信号进行分离和识别。

3. 倒频谱的应用

对于高速大型旋转机械,其旋转状况是复杂的,尤其当设备出现不对中,轴承或齿轮出现缺陷、油膜涡动、摩擦、陷流及质量不对称等现象时,则振动更为复杂,用一般频谱分析方法已经难于辨识(识别反映缺陷的频率分量),而用倒频谱则会增强识别能力。

例如,一对工作中的齿轮,在实测得到的振动或噪声信号中包含着一定数量的周期分量。如果齿轮产生缺陷,则其振动或噪声信号还将大量增加谐波分量及所谓的边带频率成分。

边带频率的定义:设在旋转机械中有两个频率 ω_1 与 ω_2 存在,在这两个频率的激励下,机械振动的响应呈现出周期性脉冲的拍,也就是呈现其振幅以差频($\omega_2 - \omega_1$)(假设

2. The Physical Meaning of the Cepstrum Independent Variable q

The independent variable q is called the reciprocal frequency, which has the same time dimension as the independent variable τ in the autocorrelation function $R_x(\tau)$, and generally takes ms or s. Because the cepstrum is a Fourier transform, the integral variable is the frequency f instead of time τ, so the independent variable q of the cepstrum $C_o(q)$ has the dimension of time. A large value of q is called a high cepstrum, which represents fast fluctuations and dense harmonic frequencies on the spectrogram; a small value of q is called a low cepstrum, which represents slow fluctuations and scattered harmonic frequencies on the spectrogram.

In order to make the definition of q more clear, we can also define

$$C_y(q) = F^{-1} | \{ \lg S_y(f) \} | \tag{5.73}$$

That is, the cepstrum is defined as the logarithmic weighting of the bilateral power spectrum of the signal, and then the inverse Fourier transform is taken, and the autocorrelation function of the contact signal is

$$R(\tau) = F^{-1} \{ S_y(f) \} \tag{5.74}$$

It can be seen from the above content that this definition method is very similar to the autocorrelation function, and the variable q and τ are exactly the same in dimension.

In order to reflect the phase information and restore the original signal after separation, a complex cepstrum calculation method is proposed. If the Fourier transform of signal $x(t)$ is

$$X(f) = X_R(f) + jX_I(f) \tag{5.75}$$

then the cepstrum of $x(t)$ is

$$C_o(q) = F^{-1} \{ \lg X(f) \} \tag{5.76}$$

Obviously, it retains the phase information.

The difference between the cepstrum and the correlation function is only logarithmic weighting. The purpose is to concentrate the signal energy after re-transformation, expand the spectrum range of dynamic analysis and improve the accuracy of re-transformation. It can also deconvolute (convolution) components, which is easy to separate and identify the original signal.

3. The Application of Cepstrum

For high-speed and large-scale rotating machinery, the rotation conditions are complex, especially when the equipment has misalignment, bearing or gear defects, oil film vortex, friction, trapping, and mass asymmetry, the vibration is more complicated. The spectrum analysis method has been difficult to identify (identify the frequency components reflecting the defect), and the use of cepstrum will enhance the identification ability.

For example, a pair of working gears contains a certain number of periodic components in the measured vibration or noise signal. If the gear is defective, its vibration or noise signal will also greatly increase the harmonic components and the so-called sideband frequency components.

Definition of sideband frequency: suppose that there are two frequencies ω_1 and ω_2 in a rotating machine. Under the excitation of these two frequencies, the response of the mechanical vibration presents a periodic pulse beat, that is, a signal whose amplitude is amplitude-

$\omega_2/\omega_1 > 1$）进行幅度调制的信号，从而形成拍的波形，这种调幅信号是自然产生的。例如，调幅波起源于齿轮啮合频率（齿数 × 轴转数）ω_0 的正弦载波，其幅值受齿轮偏心影响成为随时间而变化的某一函数 $S_m(t)$，于是有

$$y(t) = S_m(t)\sin(\omega_0 t + \varphi) \tag{5.77}$$

假设齿轮轴转动频率为 ω_m，则式(5.77)可写成

$$y(t) = A(1 + m\cos\omega_m t)\sin(\omega_0 t + \varphi) \tag{5.78}$$

其图形如图5.33(a)所示，看起来像一周期函数，实际上它并非一个周期函数，除非 ω_0 与 ω_m 成整倍数关系，在实际应用中，这种情况并不多见。根据三角半角关系，式(5.78)可写成

$$y(t) = A\sin n(\omega_0 t + \varphi) + \frac{mA}{2}\sin[(\omega_0 + \omega_m)t + \varphi] + \frac{mA}{2}\sin[(\omega_0 - \omega_m)t + \varphi] \tag{5.79}$$

从式(5.79)不难看出，它是 ω_0、$\omega_0 + \omega_m$ 与 $\omega_0 - \omega_m$ 三个不同的正弦波之和，具有图5.33(b)所示的频谱图。这里，$\omega_0 - \omega_m$ 与 $\omega_0 + \omega_m$ 分别称为差频与和频，统称为边带频率。

实际上，如果齿轮缺陷严重或有多种故障存在，以致许多机械中经常出现不对准、松动、非线性刚度，或者出现拍波截断等，则边带频率将大幅度增加。

在一个频谱图上出现过多的差频是难以识别的，而倒频谱图则有利于识别。减速器的频谱图与倒频谱图如图5.34所示。

图5.34(a)所示为一个减速器的频谱图，图5.34(b)是图5.34(a)的倒频谱图。从倒谱图上清楚地看出，有两个主要频率分量：117.6 Hz(8.5 ms)及48.8 Hz(20.5 ms)。

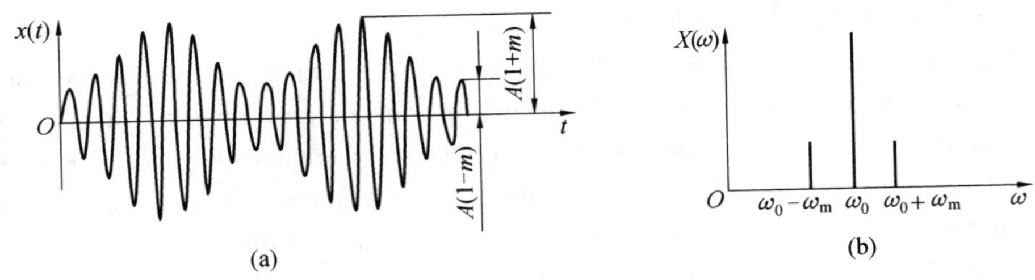

图5.33　齿轮啮合中拍波现象

modulated at the difference frequency ($\omega_2 - \omega_1$) (assuming $\omega_2/\omega_1 > 1$) to form a beat waveform. This amplitude-modulated signal is naturally generated. For example, the amplitude modulation wave originates from the sine carrier wave of gear meshing frequency (number of teeth × number of shaft revolutions) ω_0 , and its amplitude becomes a certain function $S_m(t)$ that changes with time due to the influence of gear eccentricity, so we have

$$y(t) = S_m(t)\sin(\omega_0 t + \varphi) \tag{5.77}$$

Assuming that the rotation frequency of the gear shaft is ω_m , then formula (5.77) can be written as

$$y(t) = A(1 + m\cos\omega_m t)\sin(\omega_0 t + \varphi) \tag{5.78}$$

The graph shown in Figure 5.33 (a) looks like a periodic function, but in fact it is not a periodic function, unless ω_0 and ω_m are in an integral multiple relationship, which is rare in practical applications. According to the triangle half-angle relationship, formula (5.78) can be written as

$$y(t) = A\sin n(\omega_0 t + \varphi) + \frac{mA}{2}\sin[(\omega_0 + \omega_m)t + \varphi] + \frac{mA}{2}\sin[(\omega_0 - \omega_m)t + \varphi] \tag{5.79}$$

It is not difficult to see from formula(5.79) that it is the sum of three different sine waves of ω_0, $\omega_0 + \omega_m$ and $\omega_0 - \omega_m$, and has the frequency spectrum shown in Figure 5.33(b). Here, $\omega_0 - \omega_m$ and $\omega_0 + \omega_m$ are called difference frequency and sum frequency respectively, and are commonly called sideband frequencies.

In fact, if the gear defect is serious or there are multiple faults, so that misalignment, looseness, nonlinear stiffness, or beat wave truncation occurs frequently in many machines, the sideband frequency will be greatly increased.

It is difficult to identify too many difference frequencies on a spectrogram, while a cepstrum graph is helpful for identification. Spectrogram and Cepstrum of the reducer are shown in Figure 5.34.

Figure 5.34(a) shows the frequency spectrum of a reducer, and Figure 5.34(b) is the cepstrum chart of Figure 5.34(a). It is clear from the cepstrum chart that there are two main frequency components; 117.6 Hz (8.5 ms) and 48.8 Hz (20.5 ms).

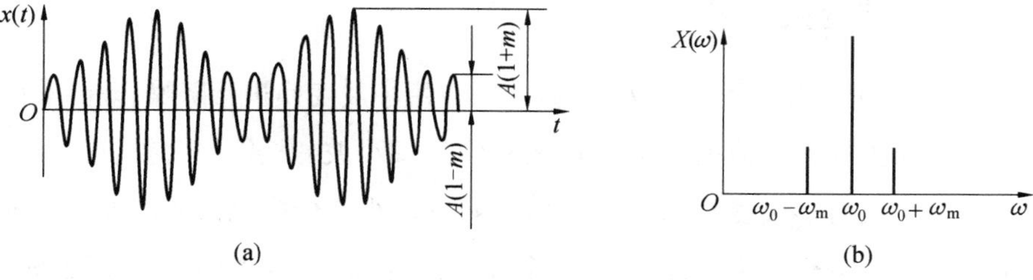

Figure 5.33　Beat weave phenomenon in gear meshing

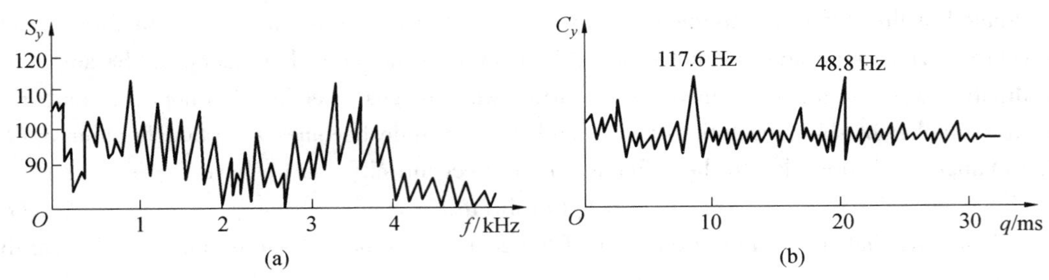

图 5.34　减速器的频谱图与倒频谱图

5.6　现代信号处理方法简介

目前,现代信号处理方法有多种,本节主要介绍以下几种常用的现代信号处理方法。

5.6.1　短时傅里叶变换

傅里叶变换的引入使频谱分析在故障诊断领域得到普及。它将振动信号由时域转换到频域,可以发现在时域内不易观察到的信号特征。然而,傅里叶变换只能处理线性平稳信号,而且只能应用在全局中,不能分析信号的局部变化,这在一定程度上限制了其应用。1946 年,Gabor 第一次提出了短时傅里叶变换的概念,即在傅里叶变换的框架下,将非平稳信号看作若干个短时平稳信号的叠加,短时性可通过在时域上加窗来实现,通过平移窗可以覆盖整个时域,这样对每个窗内的平稳信号进行傅里叶变换就可以得到信号的时频表示。短时傅里叶变换能实现对非平稳信号的时频局部化分析,但是其时频窗口是固定的。如果窗函数确定,那么时频窗的大小也就确定了,即其时频分辨率是固定的,对信号没有自适应性。

5.6.2　小波变换

小波变换是短时傅里叶变换的发展与创新,应用小波变换进行降噪,既能够保持傅里叶变换的优点,又能够弥补傅里叶变换本身的不足,即小波变换能同时提供信号时域和频域的局部化信息,具有多尺度和“数学显微镜”特性。小波变换提供的局部化分析是变化的。在高频端,频率分辨率不好,时域分辨率较好;而在低频端,频率分辨率较好,时域分辨率较差。小波变换的这种“变焦”性质用来处理突变的信号非常适合。但是,小波变换也存在着局限性,表现在以下方面。

(1) 基函数的选择。在小波变换中,可以根据不同的要求构造不同的小波基,但对某个信号来说,选择什么样的基函数是一个难点。而且在小波变换中,基函数一旦确定下

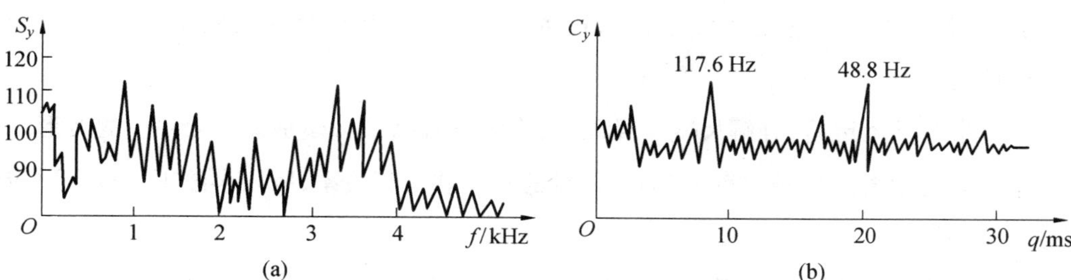

Figure 5.34　Spectrogram and Cepstrum of the reducer

5.6　Introduction to Modern Signal Processing Methods

At present, there are many modern signal processing methods. This section mainly introduces the following several commonly used modern signal processing methods.

5.6.1　Short-Time Fourier Transform

The introduction of Fourier transform has made frequency spectrum analysis popular in the field of fault diagnosis. It converts the vibration signal from the time domain to the frequency domain, and can find signal characteristics that are not easy to observe in the time domain. However, the Fourier transform that can only process linear and stationary signals can only be applied to the overall situation, and will be limited its application to a certain extent without analyzing the local changes of signals. In 1946, Gabor first proposed the concept of short-time Fourier transform, that is, under the framework of Fourier transform, non-stationary signals are regarded as the superposition of several short-time stationary signals, and the short-time performance can be achieved by adding windows in the time domain, and the entire time domain can be covered by the translation window, so that the Fourier transform of the stationary signal in each window can obtain the time-frequency representation of the signal. The short-time Fourier transform can realize the time-frequency localization analysis of non-stationary signals, but its time-frequency window is fixed. If the window function is determined, then the size of the time-frequency window is also determined, that is, the time-frequency resolution is fixed, and there is no adaptability to the signal.

5.6.2　Wavelet Transform

Wavelet transform is the development and innovation of short-time Fourier transform. The application of wavelet transform for noise reduction can not only maintain the advantages of Fourier transform, but also make up for the shortcomings of Fourier transform itself, that is, wavelet transform can provide localized information in the time domain and frequency domain of the signal at the same time, and has the characteristics of multi-scale and "mathematical microscope". The localized analysis provided by the wavelet transform is variable. The frequency resolution at the high frequency end is not good, and the resolution in the time domain is better; while the frequency resolution at the low frequency end is better and the resolution in

来,则整个分解和重构过程都无法更改,这将会出现小波基在全局上最佳,而在某一局部最差的情况,即小波基对信号没有适应性。

(2) 小波基固定后,分解尺度一经确定,小波分解结果必须是某个频率段的波形。这个频率段只与信号分析频率有关,而与信号的本身并没有关系,所以小波不具有自适应的信号分解特性。

(3) 虽然小波分析能实现多分辨率分析,但这只是基于不同的小波基函数实现的。一旦小波基确定,分辨率也就确定了,它并不会随着信号的变化而改变。

因此,小波变换在一定程度上限制了它的应用。

5.6.3 Hilbert - Huang 变换与经验模态分解

Norden E. Huang 等在深入研究了瞬时频率概念后,根据他们的研究成果,创造性地提出了本征模函数(MF) 的概念及任意信号分解为模式分量的新方法,即经验模态分解(EMD) 方法。将 EMD 方法和与之相应的 Hilbert 谱统称为 Hilbert - Huang 变换,其思想是:首先对信号进行 EMD 分解,得到一系列本征模函数(IMF);然后将各个 IMF 进行希尔伯特变换,得到希尔伯特谱;最后根据谱图来分析信号的特征。Huang 认为,对信号进行 EMD 分解,可以得到一系列具有一定物理特性的模态函数,它也是一种时频域分析方法,是根据振动信号所具有的局部时变特征进行自适应的时频域分解,能够得到极高的时频分辨率和良好的时频聚集性,对非线性、非平稳振动信号的分析是非常有效的。

经典的 EMD 分解方法存在抗混叠效应效果较差及模态混叠等问题。为克服这些缺陷, Huang 提出了 EEMD 分解方法。这是一种引入噪声辅助信号的处理方法,其实质是先将原信号叠加高斯白噪声,然后对叠加信号进行多次的 EMD 分解,利用附加的白噪声频谱均匀分布,均值为零,使用足够次数平均值处理后就会相互抵消的特性使得叠加的信号在不同尺度上具有连续性,使原始信号的极值点特性改变,有效地避免了混叠现象。

5.6.4 Wigner - Ville 分布

Wigner 在 1932 年首先提出了 Wigner 分布的概念,并把它应用到量子力学领域。1948 年,Ville 将其引入信号分析领域。因此,Wigner 分布又称 Wigner - Ville 分布,简称 WVD。与傅里叶谱不同,Wigner - Ville 谱是分析非平稳时变信号强有力的工具。利用解析信号,可极大地压缩多分量信号的交叉耦合项,时频图像和时域、频域信号图能相互对应,直观性强,它比短时傅里叶变换能够更好地描述信号的时变特征。Wigner - Ville 分布在信号的探测和故障诊断中应用非常广泛,但是 Wigner - Ville 分布存在着交叉干扰

the time domain is poor. This "zoom" property of wavelet transform is very suitable for processing abrupt signals. However, wavelet transform also has limitations, which are shown in following aspects.

(1) The choice of basis function. In wavelet transform, different wavelet bases can be constructed according to different requirements, but it is difficult to choose what base function for a certain signal. Moreover, the whole decomposition and reconstruction process cannot be changed once the basis function is determined in wavelet transform, which will lead to the situation that the wavelet basis is the best globally and the worst in a certain part, that is, the wavelet basis has no adaptability to the signal.

(2) The result of wavelet decomposition must be a waveform in a certain frequency range once the decomposition scale is determined after the wavelet basis is fixed. This frequency band is only related to the signal analysis frequency, and has nothing to do with the signal itself, so wavelet does not have adaptive signal decomposition characteristics.

(3) Although wavelet analysis can achieve multi-resolution analysis, it is only based on different wavelet basis functions. Once the wavelet basis is determined, the resolution is also determined, and it will not change with the change of the signal.

Therefore, the wavelet transform limits its application to a certain extent.

5.6.3 Hilbert-Huang Transformation and Empirical Mode Decomposition

After in-depth research on the concept of instantaneous frequency, Norden E. Huang, et al. creatively proposed the concept of intrinsic mode function (MF) and a new method of decomposing any signal into mode components based on their research results, that is, empirical mode decomposition (EMD) method. The idea that the EMD method and the corresponding Hilbert spectrum are collectively referred to as the Hilbert-Huang transform is: first decompose the signal by EMD to obtain a series of intrinsic mode function (IMF); and then perform the Hilbert transform of each IMF to obtain the Hilbert spectrum; finally analyze the characteristics of the signal according to the spectrum. Huang believes that EMD decomposition of the signal can obtain a series of modal functions with certain physical characteristics. It is also a time-frequency domain analysis method, which is based on the local time-varying characteristics of the vibration signal, and performs adaptive time-frequency domain. Decomposition can obtain extremely high time-frequency resolution and good time-frequency aggregation, which is very effective for the analysis of nonlinear and non-stationary vibration signals.

The classic EMD decomposition method has problems such as poor anti-aliasing effect and modal aliasing, so Huang proposed the EEMD decomposition method in order to overcome these shortcomings. This is a processing method that introduces a noise auxiliary signal. The essence is to first superimpose the original signal with Gaussian white noise, and then perform multiple EMD decomposition of the superimposed signal, using the additional white noise spectrum to be uniformly distributed with an average value of zero, and the characteristic of using enough times of average processing to cancel each other out makes the superimposed signal have continuity on different scales, changes the extreme point characteristics of the original signal, and effectively avoids the aliasing phenomenon.

项。交叉干扰项是指当信号含有多个成分时,信号的 Wigner - Ville 分布中将在两成分之间时频中心坐标的中点处存在振荡分量。它提供了虚假的能量分布,影响了 Wigner - Ville 分布的物理解释。

5.6.5 盲源分离方法

盲源分离(BSS)是20世纪90年代发展起来的一门技术,是研究在未知系统的传递函数、源信号的混合系数和概率分布的情况下,通过某种信号处理的方法,仅仅通过观测信号就可以恢复出原始信号和传输通道参数的过程,即在对源信号和其传输通道都是未知的情况下,只是根据多个传感器所能够观测到的信号来估计并且将各个源信号恢复出来的一种技术。

实现混合信号盲源分离的方法有很多种。其中,独立分量分析(ICA)是解决 BSS 问题的最为有效的方法之一。它是根据实际测得的混合信号之间具有统计独立的特性,将某一路或几路信号按统计独立的原则分离出来,并且对这些信号进行分析与处理,得到若干个独立分量成分,即从混合信号中能够分离出各自独立的源信号。在实际条件下,由于环境及机械设备自身结构组成的复杂性,通过传感器采集到的振动信号通常是多个源信号与噪声的叠加,而且各信号的混合方式也很复杂,因此盲源分离的模型很适合于描述机械振动信号。

5.6.6 LMD 分解

Jonathan Smith 提出了一种新的自适应时频分析方法,即局部均值分解(LMD),并将该方法应用到了脑电图(EE)信号的分析,结果表明其时频分析的效果要优于传统时频分析方法和 HHT 方法。LMD 方法的实质是将一个复杂的多分量信号分解为若干个乘积函数(PF)分量之和。它是基于信号的局部特征尺度参数的,是依据信号本身而进行的自适应分解,得到的每一个 PF 分量都具有一定的物理意义,反映了信号的内在本质,这样进一步获得的时频分布也必然能够准确地表现出原信号的真实特征。另外,在 LMD 方法中,每一个 PF 分量都是由一个包络信号和一个纯调频信号相乘得到的。因此,PF 分量实际上就是一个单分量的调幅调频信号,幅值调制信息和频率调制信息分别包含在包络信号和纯调频信号之中,能够很方便地实现解调。由此可以看出,LMD 方法是非常适合于处理非平稳和非线性信号的,特别是多分量的调幅 - 调频信号。

5.6.4 Wigner–Ville Distribution

Wigner first proposed the concept of Wigner distribution in 1932 and applied it to the field of quantum mechanics. In 1948, Ville introduced it to the field of signal analysis. Therefore, Wigner distribution is also called Wigner–Ville distribution, or WVD for short. Unlike Fourier spectrum, Wigner–Ville spectrum is a powerful tool for analyzing non-stationary time-varying signals. The use of analytical signals can greatly compress the cross-coupling terms of multi-component signals, and the time-frequency image and the time-domain and frequency-domain signal diagrams can correspond to each other, which is intuitive. It can better describe the time-varying characteristics of the signal than the short-time Fourier transform. The Wigner–Ville distribution is widely used in signal detection and fault diagnosis. But the Wigner–Ville distribution has cross-interference terms. The cross interference term means that when the signal contains multiple components, the Wigner–Ville distribution of the signal will have an oscillation component at the midpoint of the time-frequency center coordinates between the two components. It provided a false energy distribution, and affected the physical interpretation of Wigner–Ville distribution.

5.6.5 Blind Source Separation Method

Blind source separation (BSS) is a technology developed in the 1990s. It is to study the process of recovering the original signal and transmission channel parameters only by observing the signal through a certain signal processing method in the case of unknown system transfer function, mixing coefficient and probability distribution of source signal, that is, it's just a technique that estimates and recovers each source signal based on the signals that can be observed by multiple sensors when the source signal and its transmission channel are unknown.

There are many ways to achieve blind source separation of mixed signals. Among them, independent component analysis (ICA) is one of the most effective methods to solve the BSS problem. Based on the characteristics of statistical independence between the actually measured mixed signals, it separates a certain channel or several channels of signals according to the principle of statistical independence, and analyzes and processes these signals to obtain several independent component components, that is, independent source signals can be separated from the mixed signal. Under actual conditions, due to the complexity of the environment and the structural composition of the mechanical equipment itself, the vibration signal collected by the sensor is usually the superposition of multiple source signals and noise, and the mixing method of each signal is also very complicated. Therefore, the model of blind source separation is very suitable for describing mechanical vibration signals.

5.6.6 LMD Decomposition

Jonathan Smith proposed a new adaptive time-frequency analysis method, that is, local mean decomposition (LMD), and applied this method to the analysis of electro encephalogram (EE) signals, of which results showed the effect of time-frequency analysis is better than traditional time-frequency analysis methods and HHT methods. The essence of the LMD method

思考题与练习题

5.1　信号处理的目的是什么？信号处理有哪些主要方法？各个方法的主要内容是什么？

5.2　什么是采样定理？它在信号处理过程中有何作用？

5.3　栅栏效应对周期信号处理有何影响？如何避免？

5.4　求正弦信号 $x(t)=x_0\sin\omega t$ 的绝对值 $|\mu_x|$ 和方均根值 $x_{\rm rms}$。

5.5　求正弦信号 $x(t)=x_0\sin(\omega t+\varphi)$ 的均值 μ_x、均方值 ψ_x^2 和概率密度函数 $p(x)$。

5.6　考虑模拟信号

$$x_n(t)=3\cos 100\pi t$$

(1) 确定避免混叠所需要的最小采样率。

(2) 假设信号的采样率为 $F_{\rm s}=200$ Hz,求采样后得到的离散时间信号。

(3) 假设信号的采样率为 $F_{\rm s}=75$ Hz,求采样后得到的离散时间信号。

(4) 如果生成与(3) 得到的相同样本,相应的信号频率在 $0<F<F_{\rm s}/2$ 范围内为多少？

5.7　考虑模拟信号

$$x_n(t)=3\cos 50\pi t+10\sin 300\pi t-\cos 100\pi t$$

该信号的奈奎斯特频率为多少？

5.8　试述正弦信号、正弦加随机信号、窄带随机信号和宽带随机信号自相关函数的特点。

5.9　相关分析和功率谱分析在工程上各有哪些应用？

5.10　已知某信号的自相关函数 $R(\tau)=100\cos 100\pi\tau$,试求：

(1) 该信号的均值 μ_x；

(2) 均方值 ψ_x^2；

(3) 功率谱 $S_x(f)$。

5.11　已知某信号的自相关函数为 $R(\tau)=\frac{1}{4}{\rm e}^{-2\alpha|\tau|}(\alpha>0)$,求它的自谱 $S_x(f)$。

5.12　求信号 $x(t)$ 的自相关函数,其中

$$x(t)=\begin{cases}A{\rm e}^{-\alpha t}, & t\geqslant 0,\alpha>0\\ 0, & t<0\end{cases}$$

is to decompose a complex multi-component signal into the sum of several product function (PF) components. It is based on the local characteristic scale parameters of the signal, and is adaptively decomposed according to the signal itself. Each PF component obtained has a certain physical meaning with reflecting the intrinsic nature of the signal, so that the time-frequency distribution obtained in this way must also accurately show the true characteristics of the original signal. In addition, each PF component is obtained by multiplying an envelope signal and a pure FM signal in the LMD method. Therefore, the PF component is actually a single-component AM and FM signal. The amplitude modulation information and frequency modulation information are contained in the envelope signal and the pure FM signal respectively, which can be easily demodulated. It can be seen from this that the LMD method proves very suitable for processing non-stationary and non-linear signals, especially multi-component amplitude modulation–frequency modulation signals.

Questions and Exercises

5.1　What is the purpose of signal processing? What are the main methods of signal processing? What is the main content of each method?

5.2　What is the sampling theorem? What role does it play in signal processing?

5.3　What impact does the fence effect have on periodic signal processing? How to avoid it?

5.4　Find the absolute value $|\mu_x|$ and the root mean square value x_{rms} of the sine signal $x(t) = x_0 \sin \omega t$.

5.5　Find the mean value μ_x , mean square value ψ_x^2 and probability density function $p(x)$ of the sine signal $x(t) = x_0 \sin(\omega t + \varphi)$.

5.6　Consider analog signals

$$x_n(t) = 3\cos 100\pi t$$

(1) Determine the minimum sampling rate required to avoid aliasing.

(2) Assume that the sampling rate of the signal is $F_s = 200$ Hz , find the discrete time signal obtained after sampling.

(3) Assume that the sampling rate of the signal is $F_s = 75$ Hz , find the discrete time signal obtained after sampling.

(4) If the same sample as that obtained in (3) is generated, what is the corresponding signal frequency in the $0 < F < F_s/2$ range?

5.7　Consider analog signals

$$x_n(t) = 3\cos 50\pi t + 10\sin 300\pi t - \cos 100\pi t$$

What is the Nyquist frequency of this signal?

5.8　Try to describe the characteristics of autocorrelation function of sine signal, sine plus random signal, narrowband random signal and broadband random signal.

5.9　What are the applications of correlation analysis and power spectrum analysis in engineering?

5.10　Knowing the autocorrelation function $R(\tau) = 100\cos 100\pi\tau$ of a certain signal, try

5.13　信号 $x(t)$ 由两个频率和相位均不相等的余弦函数叠加而成，其数学表达式为 $x(t)=A_1\cos(\omega_1 t+\theta_1)+A_2\cos(\omega_2 t+\theta_2)$。求信号的自相关函数 $R_x(\tau)$。

5.14　已知信号的自相关函数为 $A\cos\omega t$，请确定该信号的均方值和方均根值 x_{rms}。

5.15　图5.35所示为两信号 $x(t)$ 和 $y(t)$。求当 $\tau=0$ 时，$x(t)$ 和 $y(t)$ 的互相关函数值 $R_{xy}(0)$，并说明理由。

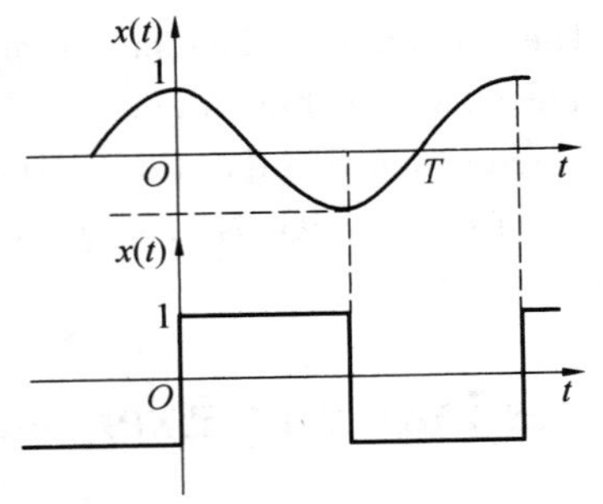

图5.35　题5.15图

5.16　某一系统的输入信号为 $x(t)$。若输出信号 $y(t)$ 与输入信号 $x(t)$ 波形相同，并且输入的自相关函数 $R_x(\tau)$ 和输入输出的互相关函数的关系为 $x_n(t)=3\cos 100\pi t$，请说明该系统的作用。

to find:

(1) the mean value of the signal μ_x ;

(2) the mean square value ψ_x^2 ;

(3) the power spectrum $S_x(f)$.

5.11 Knowing that the autocorrelation function of a certain signal is $R(\tau) = \frac{1}{4}e^{-2\alpha|\tau|}$ $(\alpha > 0)$, find its self-spectrum $S_x(f)$.

5.12 Find the autocorrelation function of signal $x(t)$, where

$$x(t) = \begin{cases} Ae^{-\alpha t} & t \leqslant 0, \alpha > 0 \\ 0, & t < 0 \end{cases}$$

5.13 Signal $x(t)$ is superimposed by two cosine functions with unequal frequency and phase, whose mathematical expression is $x(t) = A_1\cos(\omega_1 t + \theta_1) + A_2\cos(\omega_2 t + \theta_2)$. Find the autocorrelation function $R_x(\tau)$ of the signal.

5.14 The autocorrelation function of the known signal is $A\cos\ \omega t$, please determine the mean square value and root mean square value x_{rms} of this signal.

5.15 Figure 5.35 shows two signals $x(t)$ and $y(t)$. Find the cross-correlation function value $R_{xy}(0)$ of $x(t)$ and $y(t)$ when $\tau = 0$, and explain the reason.

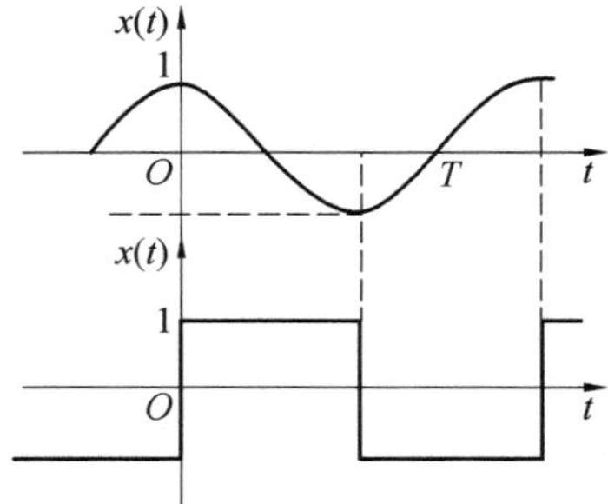

Figure 5.35 Figure of exercise 5.15

5.16 The input signal of a certain system is $x(t)$. If the output signal $y(t)$ has the same waveform as the input signal $x(t)$, and the relationship between the input auto-correlation function $R_x(\tau)$ and the input-output cross-correlation function is $x_n(t) = 3\cos\ 100\pi t$, please explain the function of the system.

第 6 章　工程应用实例

【本章学习目标】

1. 掌握静态测量工程实现过程；
2. 掌握动态测量工程实现过程。

Chapter 6 Engineering Application Examples

【Learning Objectives】

1. To be able to master the realization process of static measurement engineering;
2. To be able to master the realization process of dynamic measurement engineering.

6.1 静态测量工程应用实例——弹簧管刚度测量

6.1.1 被测对象概述

电液伺服系统主要应用在航空、航天、航海等领域，用来驱动各种各样的负载，进行位置、速度、力等控制。它的优点有很多，主要有功率大、精度高、响应快等。它的核心部件是伺服阀，伺服阀的主要作用是电与液的转换，一般由三部分组成：输入级、中间级和输出级。输入级输入电信号并输出机械信号，它主要由力矩马达、弹簧或扭簧等关键零部件组成；中间级是小型的滑阀结构，其主要作用是进行小功率放大；输出级是较大型的滑阀结构，它的输出功率很高。电液伺服阀原理图如图 6.1 所示。

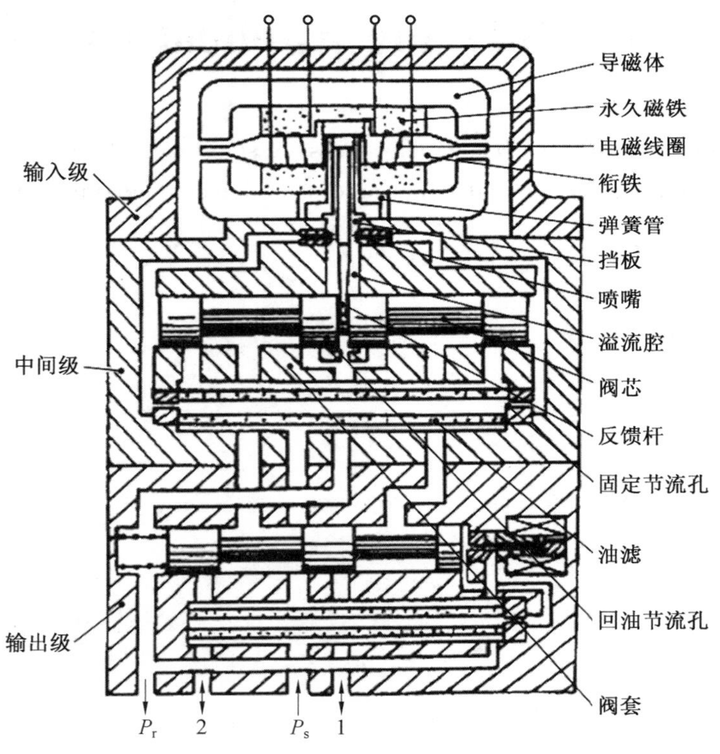

图 6.1　电液伺服阀原理图

从伺服阀的结构图中可以看出，伺服阀的上半部分是力矩马达。力矩马达主要由电磁线圈、导磁体、永久磁铁、衔铁组件（主要包括衔铁、弹簧管、喷嘴、挡板、反馈杆）组成。

6.1 Engineering Application Examples of Static Measurement Engineering—Spring Tube Stiffness Measurement

6.1.1 Overview of the Object

Electro-hydraulic servo systems are mainly used in aviation, aerospace, navigation and other fields to drive various loads, and control position, speed, and force. It has many advantages, mainly large power, high precision and fast response. Its core component is the servo valve. The main function of the servo valve is the conversion between electricity and liquid. It is generally composed of three parts: input stage, intermediate stage and output stage. The input stage, which is mainly composed of key components such as torque motors, springs or torsion springs, inputs electrical signals and outputs mechanical signals; the middle stage is a small slide valve structure, whose main function is to perform small power amplification; the output stage is a larger slide valve structure with its output power being very high. The principle diagram of electron-hydraulic servo valve is shown in Figure 6.1.

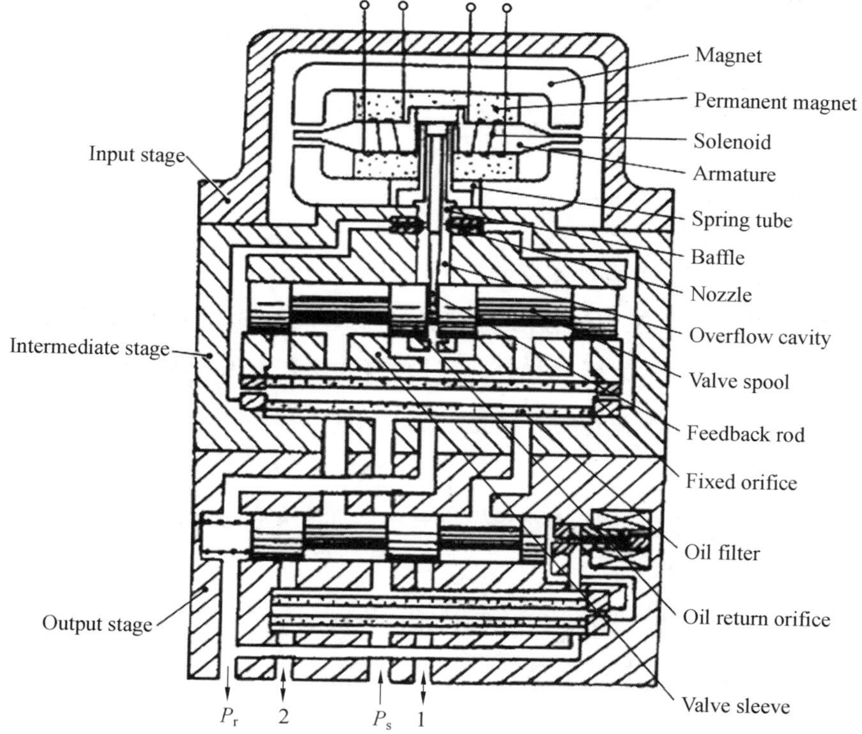

Figure 6.1　Principle diagram of electro-hydraulic servo valve

力矩马达的作用是将控制系统输入的电控信号转换成机械的运动，即控制通过电磁线圈的电流大小来对伺服系统进行控制。中间部分是中间级，它的作用是小功率放大。它由力反馈实现位置控制的四通滑阀机构，由喷嘴、挡板、衔铁组件和滑阀副组成，力反馈功能靠弹簧管和反馈杆来实现。伺服阀的下部分是输出级，它的作用是进行大功率输出。

电液伺服阀中的弹簧管是一种有固定机械特性的弹性元件，是一种广义的弹簧，它的二维结构图如 6.2(a) 所示，实物图如图 6.2(b) 所示。弹簧管的头部外表面与衔铁的内表面过盈配合，内表面与反馈杆的过盈段过盈配合。弹簧管的弹性段厚度最小可达到只有 0.06 mm，内外圆的圆柱度均小于 0.001 mm，并且要求非常均匀，不允许有任何轻微的划痕和表面缺陷。弹簧管的机械特性主要靠其薄壁部分来保证，所以弹簧管最重要的部分是其弹性段。

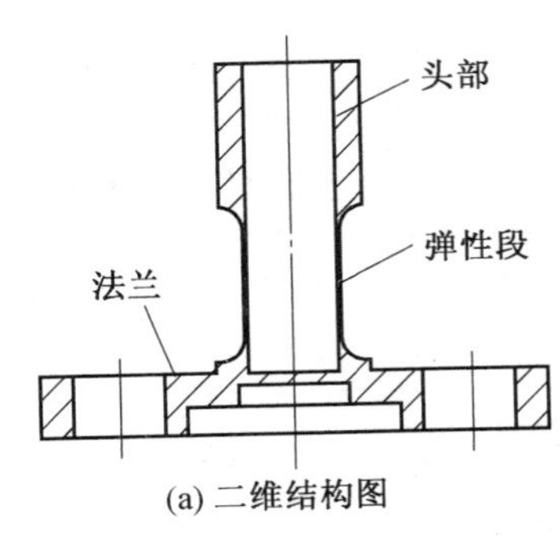

(a) 二维结构图

(b) 实物图

图 6.2　弹簧管结构

6.1.2　被测物理量分析

本实例的任务是测量图 6.2 所示弹簧管的刚度。

首先根据弹簧管的实际工作状态，提取弹簧管的实际受力模型，弹簧管实际工作时的受力分析图如图 6.3 所示。弹簧管的底部通过螺钉固定在电液伺服阀中间级壳体上表面，可以简化为固支模型。弹簧管的头部受弯矩作用，弯矩的施加是通过安装在其头部的衔铁实现的。

在此定义弹性元件刚度为:悬臂梁受载荷分析图如 6.4 所示，根据所受载荷不同，悬臂梁可以定义如下四种刚度。

当悬臂梁的一端所施加的载荷为力矩时，刚度定义如下。

(1) 抗力矩位移刚度 $K_{S.M} = M/S$。

(2) 抗力矩转角刚度 $K_{\theta.M} = M/\theta$。

当悬臂梁的一端所施加的载荷为集中力时，刚度定义如下。

It can be seen from the structure diagram of the servo valve that the upper part of the servo valve is a torque motor. The torque motor is mainly composed of solenoids, magnets, permanent magnets, and armature components (mainly including armature, spring tube, nozzle, baffle, feedback rod), whose role is to convert the electrical control signal input by thecontrol system into mechanical motion, that is, to control the servo system by controlling the current through the electromagnetic coil. The middle part is the middle stage of which role is to amplify small power. It is a four-way slide valve mechanism that realizes position control by force feedback. It consists of nozzle, baffle, armature component and slide valve pair. Force feedback function is realized by spring tube and feedback lever. The lower part of the servo valve is the output stage, which is used for high-power output.

The spring tube in the electro-hydraulic servo valve is an elastic element with fixed mechanical characteristics and a generalized spring. Its two-dimensional structure is shown in Figure 6.2(a), and the actual product is shown in Figure 6.2(b). The outer surface of the head of the spring tube is in interference fit with the inner surface of the armature, and the inner surface is in interference fit with the interference section of the feedback lever. The minimum thickness of the elastic section of the spring tube can reach only 0.06 mm, and the cylindricity of the inner and outer circles is less than 0.001 mm, and is required to be very uniform without allowing any slight scratches and surface defects. The mechanical characteristics of the spring tube mainly depend on its thin-walled part, so the most important part of the spring tube is its elastic section.

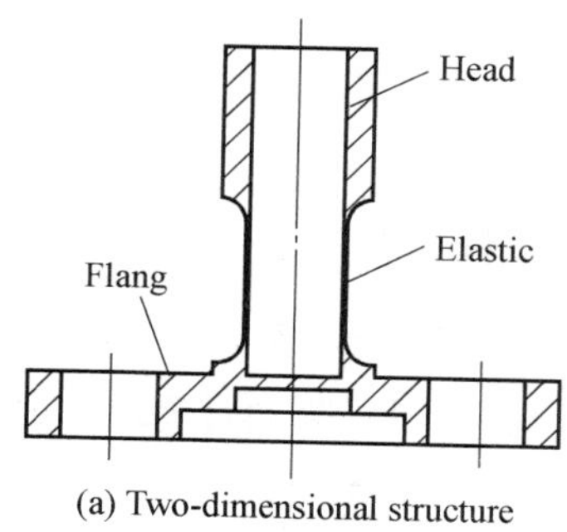

(a) Two-dimensional structure

(b) Physical image

Figure 6.2 Spring tube structure

6.1.2 Analysis of the Measured Physical Quantity

The task of this example is to measure the stiffness of the spring tube as shown in Figure 6.2.

First, extract the actual force model of the spring tube according to the actual working state of the spring tube, the force analysis diagram of the spring tube in actual work is shown in Figure 6.3. The bottom of the spring tube is fixed on the upper surface of the middle stage shell of the electro-hydraulic servo valve by screws, which can be simplified as a fixed support model. The head of the spring tube is subjected to the bending moment, and the application of the bending moment is realized by the armature mounted on the head.

The stiffness of the elastic element is defined here as: the load analysis diagram of

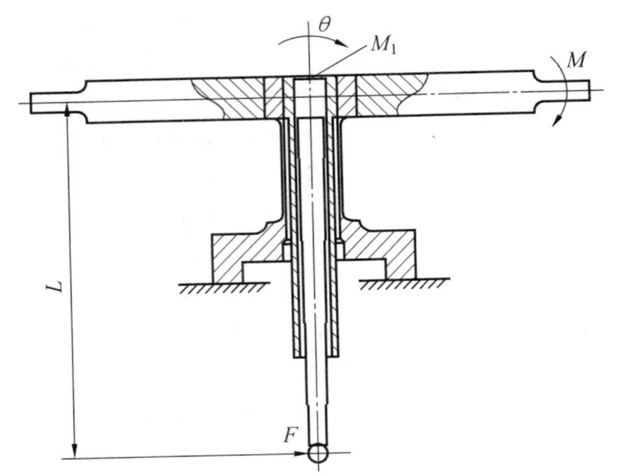

图 6.3　弹簧管实际工作时的受力分析图

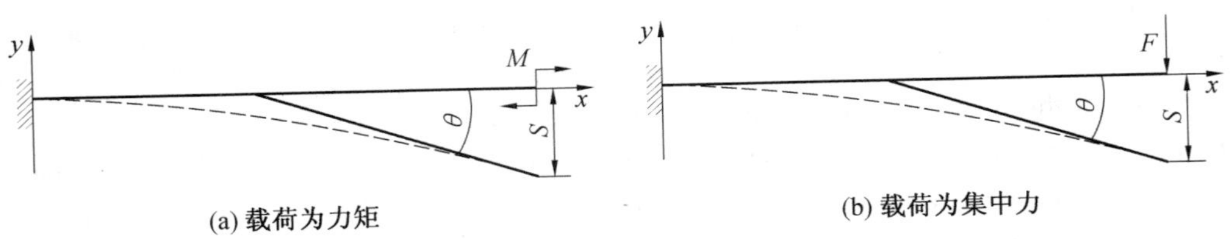

(a) 载荷为力矩　　(b) 载荷为集中力

图 6.4　悬臂梁受载荷分析图

(1) 抗力位移刚度 $K_{S.F} = F/S$；

(2) 抗力转角刚度 $K_{\theta.F} = F/\theta$。

根据以上刚度定义，弹簧管的实际使用刚度应为抗力矩转角刚度。但是，经过对测试装置进行初步结构设计就可以发现，力矩的施加和转角的测量在工程上会带来很大难度。因此，本实例采用抗力位移刚度进行设计和测量，测量结果通过简单的力学计算就可以转化为实际的抗力矩转角刚度。具体结构设计和转化方式不在本书中详述。

6.1.3　测量原理及误差分析

1. 测量原理

采用抗力位移刚度定义进行刚度测量时，弹簧管的实际加载方式可以简化为图 6.5(a)，此时相当于衔铁两端各施加有一大小相等、方向相反的电磁力 F，最终施加在弹簧管头部上的载荷是一个转矩，由双端力施加很难保证一致。这种不平衡必然在施加转矩的同时引入一个弯矩，而且其施加力的结构也很复杂。因此，简化起见，本实例中测量系统的加载方式采用图 6.5(b) 所示的方式，将双端的加载方式改为一个单向施力的测杆加载方式。

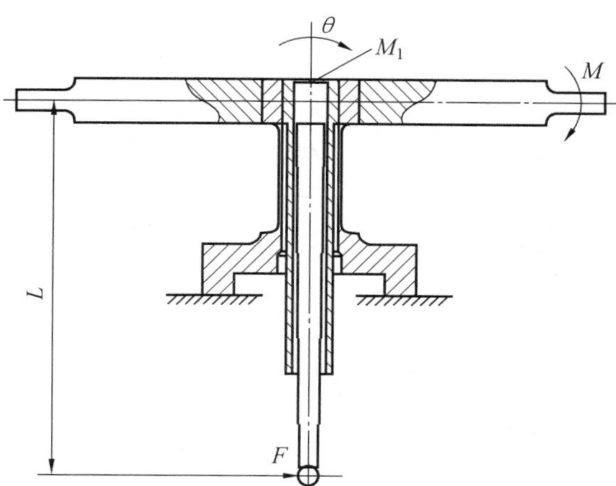

Figure 6.3　The force analysis diagram of the spring tube in actual work

cantilever beam is shown in Figure 6.4, the cantilever beam can be defined as the following four stiffnesses according to the different loads.

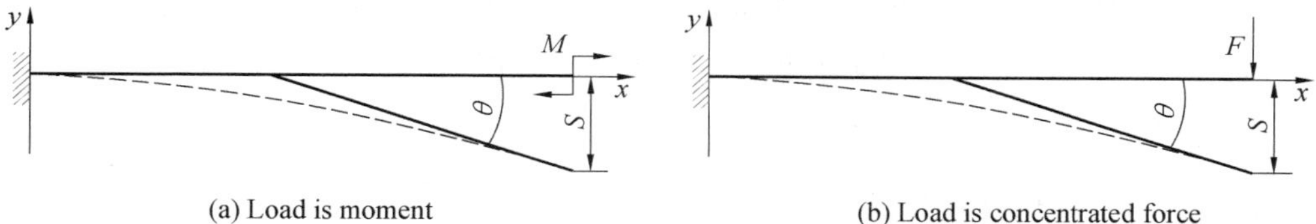

Figure 6.4　Load analysis diagram of cantilever beam

When the load applied by one end of the cantilever beam is moment, the stiffness is defined as followings.

(1) Moment-resistant displacement stiffness $K_{S.M} = M/S$.

(2) Moment-resistant corner stiffness $K_{\theta.M} = M/\theta$.

When the load applied by one end of the cantilever beam is a concentrated force, the stiffness is defined as followings.

(1) Resistance displacement stiffness $K_{S.F} = F/S$.

(2) Resistance corner stiffness $K_{\theta.F} = F/\theta$.

According to the above definition of stiffness, the actual stiffness of the spring tube should be the stiffness of the anti-moment corner. However, after the preliminary structural design of the device, it can be found that the application of torque and the measurement of the rotation angle will bring great difficulties in engineering. Therefore, this example uses the resistance displacement stiffness for design and measurement. The measurement results can be transformed into the actual torque resistance corner stiffness through simple mechanical calculations. The specific structure design and transformation methods are not detailed in this book.

6.1.3　Measurement Principle and Error Analysis

1. Measuring Principle

When using the definition of resistance displacement stiffness for stiffness measurement,

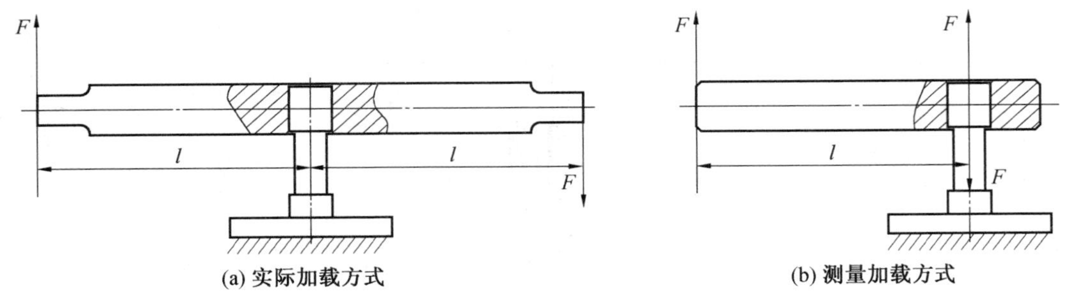

(a) 实际加载方式　　(b) 测量加载方式

图 6.5　加载方式简化原理图

通过给定力 F,测量测杆头部的位移 S。根据前文刚度定义,则可获得抗力位移刚度 $K_{S.F} = F/S$。

根据以上简化测量模型,可知测量方案在施力方式、位移计算方式上与实际工作状态有差别,会带来一定的误差,需要对该误差进行评估。另外,夹具本身的精度、夹持时接触刚度、传感器及测试系统都会带来一定的误差,理论上也应该一起考虑。方便起见,忽略了夹具制造误差。在只考虑接触刚度误差和测量系统误差时,一般可以认为二者的误差在测量总误差中的比重是一致的,即各占一半。

2. 施力方式简化误差分析

由于测量的过程相当于在力测杆的头部施加一个力,将力 F 平移至弹簧管处,因此根据力的平移定理,弹簧管头部会受到一个力和一个力矩的作用。力矩是有效的载荷,轴向力 F 则是附加载荷,会引起误差。假设在附加力的作用下弹簧管发生轴向位移 Δl_i,根据材料力学可得

$$\Delta l_i = \frac{F_i \times l}{E \times A} \tag{6.1}$$

式中　A —— 弹簧管弹性段的面积,m^2;

L—— 弹簧管弹性段的长度,m;

E—— 弹性模量,MPa。

由式(6.1)可以看出,在弹簧管的刚度测量过程中位移的测量引入了简化误差。弹簧管测量过程中受力分析图如图 6.6 所示。假设不考虑其他环节对位移测量的影响,弹簧管在测量力 ΔF 的作用下,由力矩产生的转角位移为 ΔS,其变形段产生轴向的拉伸位移为 Δl,位移传感器实际检测到的位移为 Δs。假设弹簧管的测量刚度为 K,理论刚度为 K',它可以表示为

$$K' = \frac{\Delta F}{\Delta S} = \frac{\Delta F}{\Delta s - \Delta l} L^2 \tag{6.2}$$

the actual loading method of the spring tube can be simplified as Figure 6.5(a). At this time, it is equivalent to applying an equal size and opposite direction electromagnetic force F at both ends of the armature. The load applied to the head of the spring tube is a torque, and it is difficult to ensure consistency by applying the double-ended force. This imbalance inevitably introduces a bending moment while applying a torque, and the structure of the applied force is also very complicated. Therefore, for the sake of simplification, the loading method of the measurement system in this example adopts the method shown in Figure 6.5(b), and the double-ended loading method is changed to a unidirectional force rod loading method.

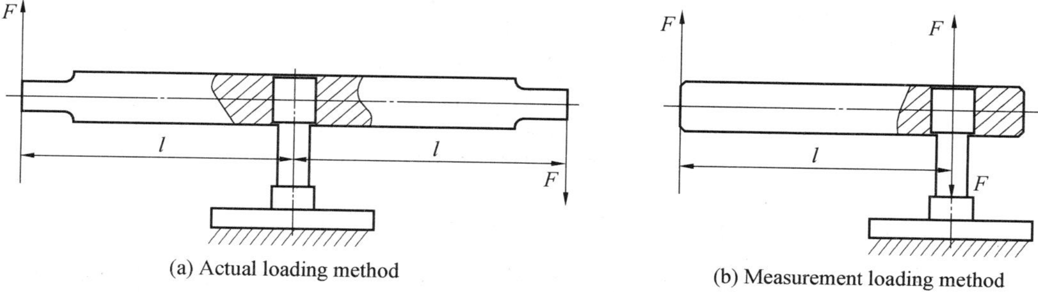

(a) Actual loading method　　(b) Measurement loading method

Figure 6.5　Simplified schematic diagram of loading mode

With a given force F, measure the displacement S of the head of the measuring rod. According to the above definition of stiffness, the resistance to displacement stiffness $K_{S.F} = F/S$ can be obtained.

According to the above simplified measurement model, it can be known that the measurement scheme is different from the actual working state in the force application method and the displacement calculation method, which will bring a certain error, which needs to be evaluated. In addition, the accuracy of the fixture itself, the contact stiffness during clamping, the sensor and the systems will all bring certain errors, which should also be considered together in theory. For convenience, the fixture manufacturing errors are ignored. When only considering the contact stiffness error and the measurement system error, it can generally be considered that the proportion of the two errors in the total measurement error is the same, that is, each accounts for half.

2. Simplify Error Analysis by Applying Force

Since the measurement process is equivalent to applying a force on the head of the force measuring rod, the force F is translated to the spring tube. According to the translation theorem of force, the head of the spring tube will be subjected to a force and a moment, the moment is effective load, and axial force F is an additional load that will cause errors. Assuming that the spring tube has an axial displacement Δl_i under the action of additional force, according to the material mechanics, it can be obtained that

$$\Delta l_i = \frac{F_i \times l}{E \times A} \tag{6.1}$$

where　A —the area of the elastic section of the spring tube, m^2;

L —the length of the elastic section of the spring tube, m;

因此,因简化加载方式而引起的刚度测量误差 ε 为

$$\varepsilon = \frac{|K' - K|}{K'} = \frac{lK}{EAL^2} = \frac{K}{K_l L^2} \tag{6.3}$$

式中　K_l—— 弹簧管弹性段的拉伸刚度,$K_l = EA/l$,N/m。

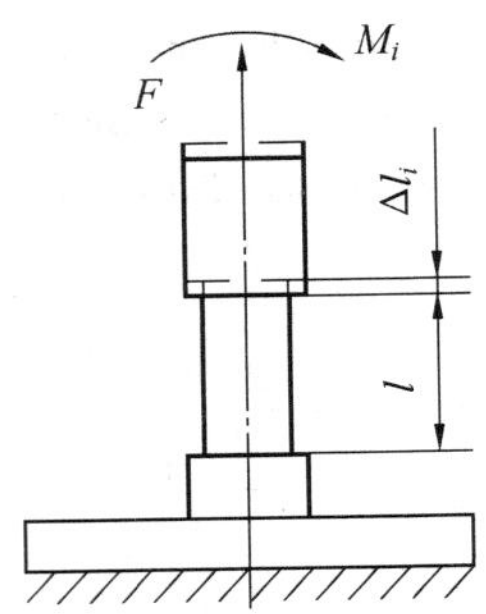

图 6.6　弹簧管测量过程中受力分析图

从式(6.3)中可以看出,在弹簧管刚度测量过程中,由简化加载方式而产生的附加力引起的误差与两个物理量成反比:一个是弹簧管弹性段的拉伸刚度 K_l;另一个则是弹簧管刚度测量中的力测杆长度,也就是施力力臂的长度。以刚度值为 $K = 90.865$ N · m/rad 的大刚度弹簧管为例,当加载力臂的长度为50 mm时,附加力引起的测量误差为0.08%。以刚度值 $K = 10.356$ N · m/rad 的弹簧管为例,当加载力臂长度为 35 mm 时,进行实际计算,附加力引起的测量误差为0.09%。从实际计算可以看出,附加力 F 引起的测量误差无论是对于大刚度弹簧管还是小刚度弹簧管都在合理的范围内,是可以忽略的。

3. 采用弦长代替弧长的计算误差分析

分析由弦长代替弧长引起的近似计算误差。在弹簧管刚度测量过程中,弹簧管头部的转角位移实际上是通过间接测量到的。位移传感器测量力测杆头部的位移为 ΔS,通过式(6.4)计算得到弹簧管头部的转角,即

$$\theta = \frac{L}{R} = \frac{\Delta S}{L} \tag{6.4}$$

由于弹簧管头部的转角是间接计算得到的,并且利用了力测杆头部的位移弦长来近似替代实际的弧长,因此这样的简化对于力臂 L 和转角 θ 就都引入了简化误差。弹簧管及力测杆转角分析图如图 6.7 所示。

可以看出,在实际测量过程中,实际的加载力臂随着弹簧管头部转角的增大在不断地减小,从而产生力臂误差,其大小可以表示为

$$\Delta L = L(1 - \cos \Delta\theta) \tag{6.5}$$

E —modulus of elasticity, MPa.

It can be seen from formula (6. 1) that the displacement measurement introduces a simplified error during the stiffness measurement of the spring tube. The force analysis diagram during the measurement of the spring tube is shown in Figure 6.6. Assuming that the influence of other links on the displacement measurement is not considered, under the action of the measuring force ΔF, the angular displacement of the spring tube generated by the moment is ΔS, and the axial tensile displacement produced by the deformation section is Δl, the displacement actually detected by the displacement sensor is Δs, assuming that the measured stiffness of the spring tube is K, and the theoretical stiffness is K', it can be expressed as

$$K' = \frac{\Delta F}{\Delta S} = \frac{\Delta F}{\Delta s - \Delta l} L^2 \tag{6.2}$$

Therefore, the stiffness measurement error ε caused by the simplified loading method is

$$\varepsilon = \frac{|K' - K|}{K'} = \frac{lK}{EAL^2} = \frac{K}{K_l L^2} \tag{6.3}$$

where　K_l —the tensile stiffness of the elastic section of the spring tube, $K_l = EA/l$, N/m.

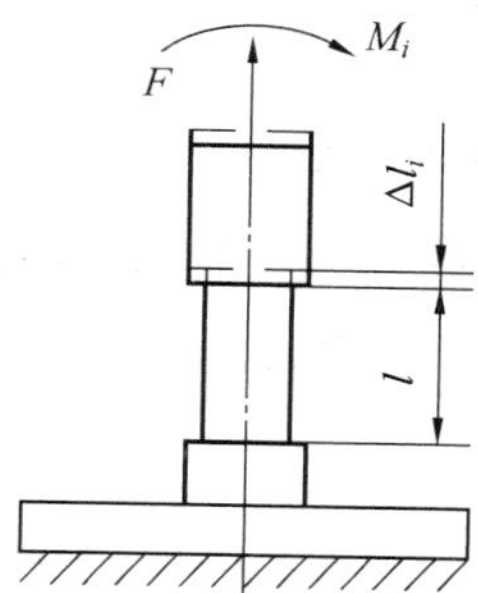

Figure 6.6　The force analysis diagram during the measurement of the spring tube

It can be seen from formula (6. 3) that in the process of measuring the stiffness of the spring tube, the error caused by the additional force generated by the simplified loading method is inversely proportional to two physical quantities: one is the tensile stiffness K_l of the elastic section of the spring tube; the other is the length of the force measuring rod in the spring tube stiffness measurement, that is, the length of the force arm. Take the high-rigidity spring tube with the stiffness value of $K = 90.865$ N · m/rad as an example. When the length of the loading arm is 50 mm, the measurement error caused by the additional force is 0.08%. Taking the spring tube with stiffness value $K = 10.356$ N · m/rad as an example. When the length of the loading arm is 35 mm, the measurement error caused by the additional force is 0.09% according to the actual calculation. It can be seen from the actual calculation that the measurement error caused by the additional force F is within a reasonable range for both large and small stiffness spring tubes and can be ignored.

3. The Calculation Error Analysis of Using Chord Length Instead of Arc Length

Analyze the approximate calculation error caused by the chord length instead of the arc length. In the process of measuring the stiffness of the spring tube, the angular displacement of the spring tube head is actually measured indirectly. The displacement sensor measures the

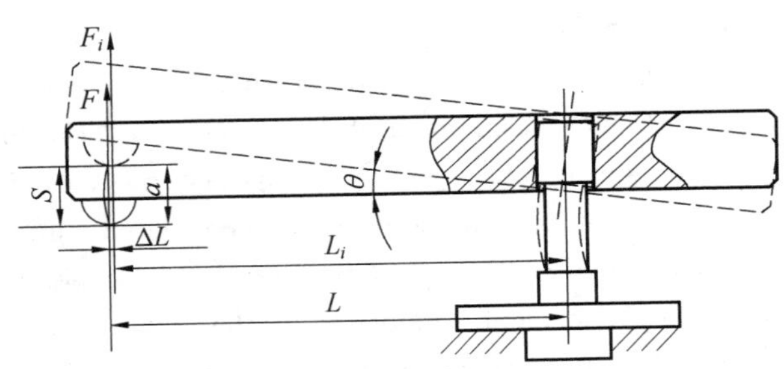

图 6.7　弹簧管及力测杆转角分析图

在刚度计算中使用的力臂依然是一个固定的力臂。力臂的改变就引起了刚度测量的误差,所以因测量力臂变化而引起的刚度测量误差为

$$\varepsilon = \frac{|K' - K|}{K'} = \frac{\left|\frac{\Delta F}{\Delta S}[(L - \Delta L)^2 - L^2]\right|}{\frac{\Delta F}{\Delta S}L^2} = \frac{|2\Delta LL - \Delta L^2|}{L^2} = 1 - \cos^2\Delta\theta \quad (6.6)$$

以实际力测杆长度为 35 mm,在不考虑其他引入误差的情况下,实际计算力臂改变引起的刚度测量误差最大不会超过 0.007 3%,在实际刚度测量过程中可以忽略。

下面分析由弦长代替弧长进行计算引入的舍入误差。从图 6.6 中可以看出,弧长偏差为

$$\Delta s = a - S = L(\theta - \sin\theta) \quad (6.7)$$

在实际测量过程中,由弦长代替弧长引起的刚度计算误差为

$$\varepsilon = \frac{|K' - K|}{K'} = \frac{\left|\frac{F}{S}L^2 - \frac{F}{a}L^2\right|}{\frac{F}{a}L^2} = \frac{\theta}{\sin\theta} - 1 \quad (6.8)$$

如果力测杆长度的实际长度为 35 mm,则在不考虑其他引入误差的情况下,计算由弧长代替弦长引入的舍入误差最大不会超过 0.004 9%,在实际刚度测量过程中可以忽略。

6.1.4　传感器的选用

(1) 传感器的精度。

在可以忽略测量原理误差的前提下,考虑测量系统误差,以便选择合适的传感器。假设被测弹簧管的刚度值设计值为(43 ±0.1)N·m/rad,则被测工件公差为

$$\varepsilon = \frac{\pm 0.1}{43} \times 100\% \approx \pm 0.23\%$$

为分辨被测元件的公差值,测量总误差一般应小于被测工件误差一个数量级,一般至

force of the measuring rod head as ΔS , and calculate the rotation angle of the spring tube head by formula (6.4), that is

$$\theta = \frac{L}{R} = \frac{\Delta S}{L} \tag{6.4}$$

Since the rotation angle of the spring tube head is calculated indirectly, and the displacement chord length of the force measuring rod head is used to approximate the actual arc length, this simplification introduces simplified errors for both the moment arm L and the rotation angle θ, of which the analysis diagram of the angle of the spring tube and the force measuring rod is shown in Figure 6.7.

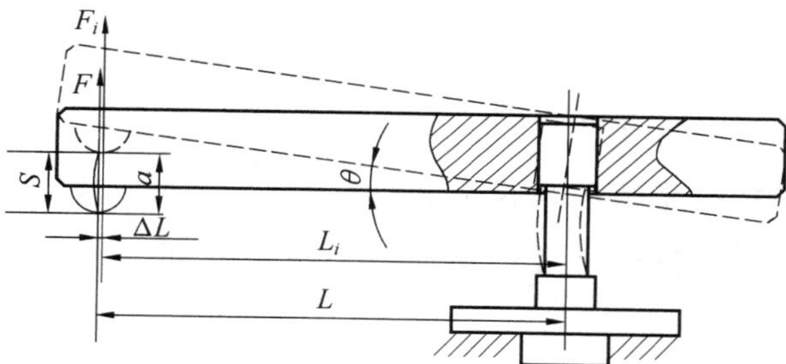

Figure 6.7 The analysis diagram of the angle of the spring tube and the force measuring rod

It can be seen that in the actual measurement process, the actual loading force arm is continuously reduced with the increase of the turning angle of the spring tube head, resulting in a force arm error, which can be expressed as

$$\Delta L = L(1 - \cos \Delta\theta) \tag{6.5}$$

The force arm used in the stiffness calculation is still a fixed force arm. The change of the force arm causes the stiffness measurement error, so the stiffness measurement error caused by the change of the measurement force arm is

$$\varepsilon = \frac{|K' - K|}{K'} = \frac{\left|\frac{\Delta F}{\Delta S}[(L - \Delta L)^2 - L^2]\right|}{\frac{\Delta F}{\Delta S}L^2} = \frac{|2\Delta LL - \Delta L^2|}{L^2} = 1 - \cos^2\Delta\theta \tag{6.6}$$

Taking the actual force measuring rod length as 35 mm, the stiffness measurement error caused by the actual calculation of the force arm change will not exceed 0.007 3% , which can be ignored in the actual stiffness measurement process.

The following analyzes the rounding error introduced by the calculation of the chord length instead of the arc length. From Figure 6.6, it can be seen that the arc length deviation is

$$\Delta s = a - S = L(\theta - \sin\theta) \tag{6.7}$$

In the actual measurement process, the stiffness calculation error caused by the chord length instead of the arc length is

$$\varepsilon = \frac{|K' - K|}{K'} = \frac{\left|\frac{F}{S}L^2 - \frac{F}{a}L^2\right|}{\frac{F}{a}L^2} = \frac{\theta}{\sin\theta} - 1 \tag{6.8}$$

If the actual length of the force measuring rod is 35 mm, the rounding error introduced by

少取被测刚度公差的 1/3。按照这一原则,测量的精度要求约为 ±0.08%,按照测试系统误差在总误差中占一半进行估算,测试系统的精度应为 ±0.04%。由于刚度测量主要涉及力和位移两个物理量的测量,而且力和位移测量误差对测试系统误差的贡献是在一个数量级上的,因此测量系统的精度应由力传感器和位移传感器共同分担,即传感器精度应为 ±0.02%。

(2) 传感器的量程。

选择传感器的量程。需根据弹簧管的实际受力范围和变形范围进行选择,量程一般应取实际受力和变形的 3 倍以上。

(3) 传感器的分辨率。

在确定量程和精度的前提下,即可确定传感器的分辨率。传感器的量程与精度之积即绝对误差值,分辨率的选择一般要比绝对误差提高一个数量级。如果分辨率过低,则会引入量化误差。

(4) 传感器之间分辨率的协调。

在本实例中,测杆长度的确定也是需要经过精心设计的,其长度的选取可以协调力和位移测量值变化量数值。理想情况使得力值每变化一个最小值,位移也相应地变化一个最小值,从而发挥传感器各自的最大效力。

经过以上过程,选择合适的力和位移传感器(图 6.8),在此基础上修订和设计测试装置方案和结构。本实例选择的传感器如下。

(1) 平行梁式力传感器。分辨力是 1 mN;量程是 0 ~ 10 000 mN;线性度和重复性均小于 0.02%(2 mN)。

(2) 激光位移传感器。分辨率为 0.03 μm,其量程为 2 mm,精度为 0.4 μm。

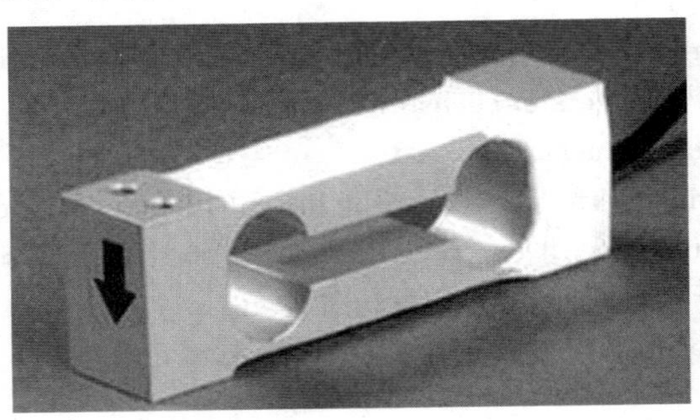

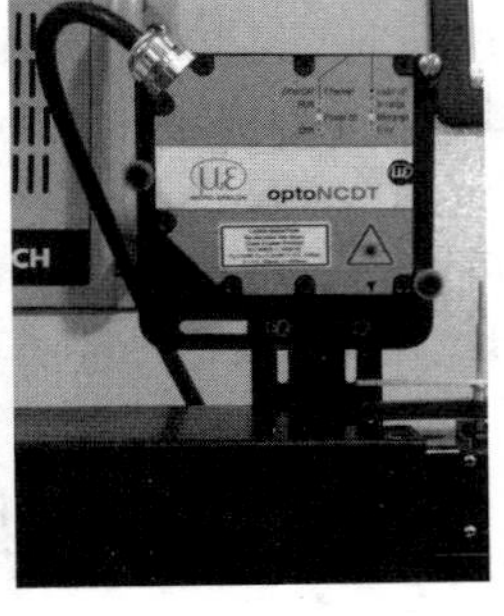

图 6.8　力传感器和位移传感器(此处为实际产品图片)

the arc length instead of the chord length will not exceed 0.004 9% without considering other errors introduced, which can be ignored in the actual stiffness measurement process.

6.1.4 The Selection of Sensors

(1) The accuracy of the sensor.

On the premise that the error of the measurement principle can be ignored, the error of the measurement system is considered in order to select the appropriate sensor. Suppose the design value of the stiffness of the spring tube is (43±0.1) N · m/rad, then the tolerance of the workpiece is

$$\varepsilon = \frac{\pm 0.1}{43} \times 100\% \approx \pm 0.23\%$$

In order to distinguish the tolerance value of the measured component, the total measurement error should generally be less than an order of magnitude of the measured workpiece error, generally at least 1/3 of the measured stiffness tolerance. According to this principle, the measurement accuracy requirement is about ±0.08% , and the accuracy of the systems should be ±0.04% according to the error of the systems accounts for half of the total error for estimation. Because stiffness measurement mainly involves the measurement of force and displacement these two physical quantities, and the contribution of force and displacement measurement error to the systems error is on an order of magnitude, the accuracy of the measurement system should be shared by the force sensor and the displacement sensor, that is, the accuracy of the sensor should be ±0.02% .

(2) The range of the sensor.

Select the range of the sensor. The range should generally be more than 3 times the actual force and deformation with being selected according to the actual force range and deformation range of the spring tube.

(3) The resolution of the sensor.

On the premise of determining the range and accuracy, the resolution of the sensor can be determined. The product of the sensor's range and accuracy is the absolute error value, and the choice of resolution is generally an order of magnitude higher than the absolute error. If the resolution is too low, quantization errors will be introduced.

(4) Coordination of resolution between sensors.

In this example, the determination of the length of the measuring rod also needs to be carefully designed. The selection of its length can coordinate the changes in the force and displacement measurement values. Ideally, every time the force value changes to a minimum, the displacement will change accordingly so as to give full play to the maximum effectiveness of the sensors.

After the above process, select the appropriate force and displacement sensors (Figure 6.8), and modify and design the device plan and structure on this basis. The sensors selected in this example are as follows.

(1) Parallel beam force sensor. Resolution is 1 mN; range is 0–10 000 mN; linearity and repeatability are both less than 0.02% (2 mN).

6.1.5 信号处理

1. 力信号的处理

力传感器在额定的工作电压下,其输出的信号非常微弱,一般在毫伏的级别。这样的信号是不易采集和进行 A/D 转换的,必须要经过放大电路的放大才能进行后续的处理。因此,要设计放大电路,传感器对放大电路也是有技术要求的:一是放大电路要是差动型,以排除温度对传感器的影响;二是放大倍数要高且稳定,并且可调以适应不同的 A/D 转换芯片;三是放大电路的零点漂移和噪声要低;四是放大电路自身的线性度要高。

综合上述技术指标,选择 AD524 作为放大电路的芯片。力传感器的供电形式有两种:一种是恒压源对电桥供电,其优点是结构简单,缺点则是在为电桥供电时不能排除温度的干扰;另一种是恒流源对电桥供电,其优点是在为电桥供电时可以根据差动原理,将温度的影响排除,缺点是结构复杂。综合考虑选择恒流源式电桥供电方式,并配合 AD524 放大电路芯片进行信号放大,其原理图如图6.9(a) 所示,实物图如图6.9(b) 所示。其工作原理是利用 LM399 做恒压源,然后利用 OP07 放大芯片和三极管配合工作,保证精密电阻 R_9 两端的电压差保持不变,从而保证输出电流不变,最后经过三极管的电流放大,达到给电桥供电电流的要求。应变电桥的输出端连接放大芯片 AD524 的输入端。当电桥的阻值发生变化时,电桥输出电压会发生变化。变化的电压经过放大后,就可以输出放大电路为后续信号处理做准备。

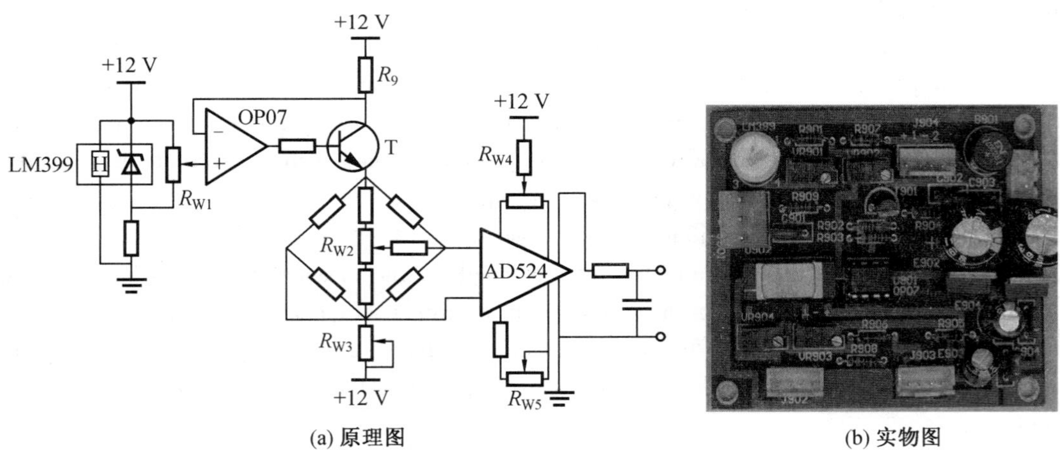

(a) 原理图　　(b) 实物图

图 6.9　力传感器放大电路

一般情况下,传感器信号经过放大电路放大之后输出的都是模拟电压信号,计算机所能处理的是数字量信号。由于模拟量的数据采集卡直接采集模拟信号时会有较大的干

(2) Laser displacement sensor. The resolution is 0.03 μm, the range is 2 mm, and the accuracy is 0.4 μm.

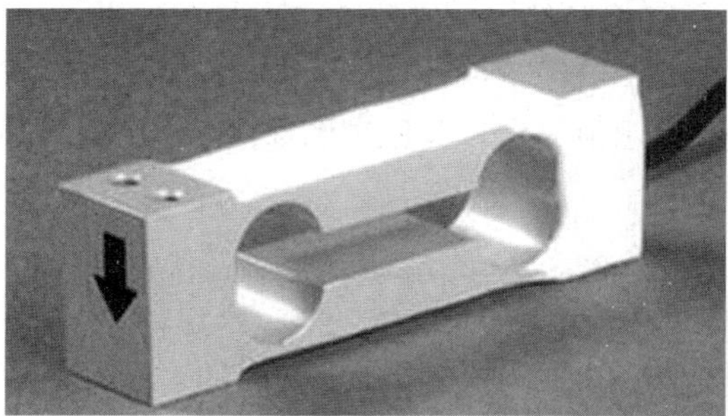

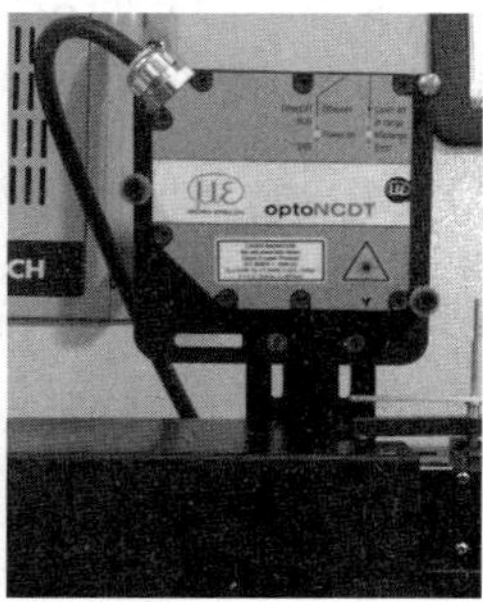

Figure 6.8　Force sensor and displacement sensor (here is the actual product picture)

6.1.5　Signal Processing

1. Force Signal Processing

Under the rated working voltage, the output signal of the force sensor is very weak, generally in the millivolt level. Such kind of a signal is not easy to collect and perform A/D conversion, and must be amplified by an amplifier circuit for subsequent processing. Therefore, the sensor also has technical requirements for the amplifying circuit to design amplifying circuit: first, if amplifying circuit proves a differential type, it can eliminate the influence of temperature on the sensor; second, the magnification should be high and stable, and adjustable to adapt to different A /D conversion chip; third, the zero drift and noise of the amplifying circuit should be low; fourth, the linearity of the amplifying circuit itself should be high.

Based on the above technical indicators, AD524 is selected as the chip of the amplifying circuit. There are two types of power supply for the force sensor: one is that the constant voltage source supplies power to the bridge, which has the advantage of simple structure, and the disadvantage is that the temperature interference cannot be eliminated when powering the bridge; the other is that the constant current source supplies power to the electric bridge. The advantage of bridge power supply is that the influence of temperature can be eliminated according to the differential principle when powering the bridge, while the complicated structure is its disadvantage. Consider the selection of the constant current source bridge power supply mode, and cooperate with the AD524 amplifying circuit chip to amplify the signal. The schematic diagram is shown in Figure 6.9(a), and the physical diagram is shown in Figure 6.9(b). Its working principle is to use LM399 as a constant voltage source, and then use OP07 amplifying chip and triode to work together to ensure that the voltage difference between the two ends of the precision resistor R_9 remains unchanged, thereby ensuring that the output current remains unchanged. Finally, the current through the transistor is amplified to meet the requirements of the power supply current for the bridge. The output end of the strain bridge is connected to the input end of the amplifier chip AD524. When the resistance of the bridge changes, the output voltage of the bridge will change. After the changed voltage is amplified, it can be output to the amplifier circuit to prepare for subsequent signal processing.

扰，因此本测量系统仍然选择数字量采集卡，A/D 转换电路自行设计。静态测量过程对数据的处理速度要求不高，所以对 A/D 转换芯片的响应速度要求较低。由于力测量值的显示精度和量程为 1 mN、10 000 mN，因此 A/D 转换芯片选择 4 位半的双积分转换芯片 ICL7135，其优点是输出值经过滤波，显示值稳定。其原理图如图 6.10(a) 所示，实物图如图 6.10(b) 所示。

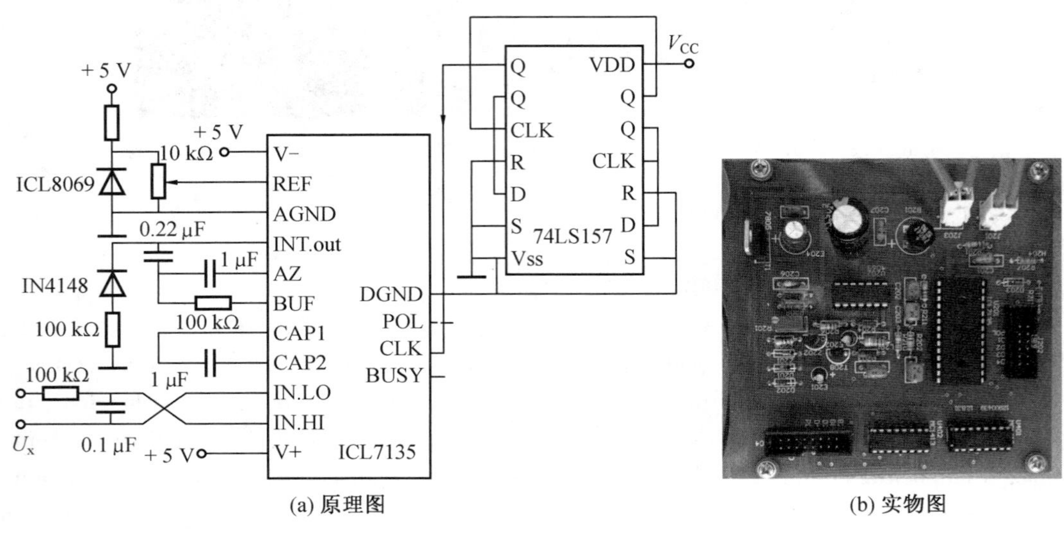

(a) 原理图　　(b) 实物图

图 6.10　A/D 转换电路

2. 位移信号的处理

选用的激光位移传感器内置放大电路和 A/D 转换电路，可直接利用数字量采集卡或串口进行数据采集。

6.1.6　测试过程及结果

1. 力传感器的标定

测量系统中安装的传感器在使用前都需要进行标定。对于本实例中使用的力传感器，具体标定方法是先调节力传感器放大电路的调零旋钮，使得力传感器的示值为 0 mN，然后将 10 000 mN 的标准块缓慢地放在力传感器的测头板上。此时，力传感器的示值应该是 10 000 mN。如果力传感器的显示不是此值，则需要调节放大电路的倍率旋钮，使得示值为 10 000 mN，然后缓慢地将标准块取下，看力传感器的示值是否为 0 mN。如果不是，则需要调整放大电路的调零旋钮，使示值为 0 mN。重复以上的步骤，直至力传感器满足要求即可。以上步骤结束之后，在 0 mN 与 10 000 mN 之间需再选择多点进行力传感器

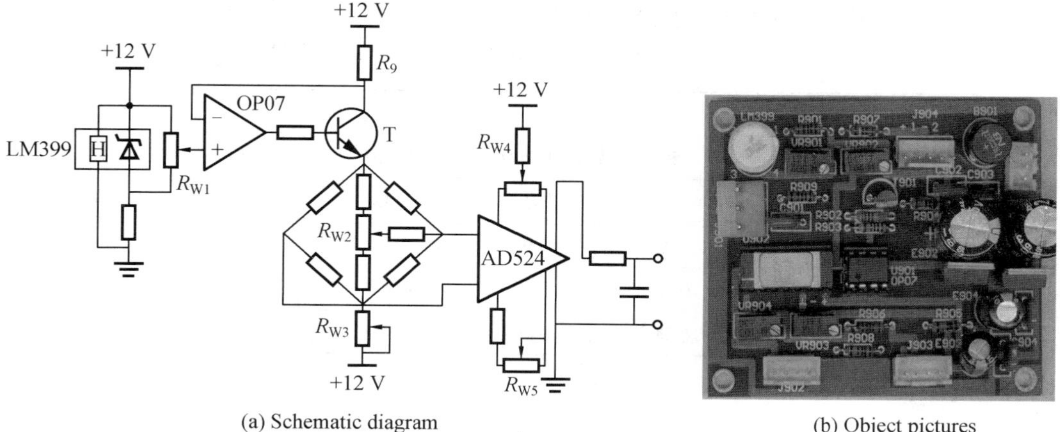

(a) Schematic diagram　　(b) Object pictures

Figure 6.9　Force sensor amplifier circuit

In general, the output of the sensor signal after being amplified by the amplifier circuit proves an analog voltage signal, and what the computer can handle is a digital signal. Since the analog data acquisition card directly collects the analog signal, there will be greater interference, so this measurement system still chooses the digital acquisition card, and the A/D conversion circuit is designed by itself. The static measurement process does not require high data processing speed, so the response speed of the A/D conversion chip is relatively low. Because the display accuracy and range of the force measurement value and the range are 1 mN, 10 000 mN, the A/D conversion chip chooses the 4 and a half double integral conversion chip ICL7135, whose advantage is that the output value is filtered and the display value is stable. The schematic diagram is shown in Figure 6.10(a), and the object picturesis shown in Figure 6.10(b).

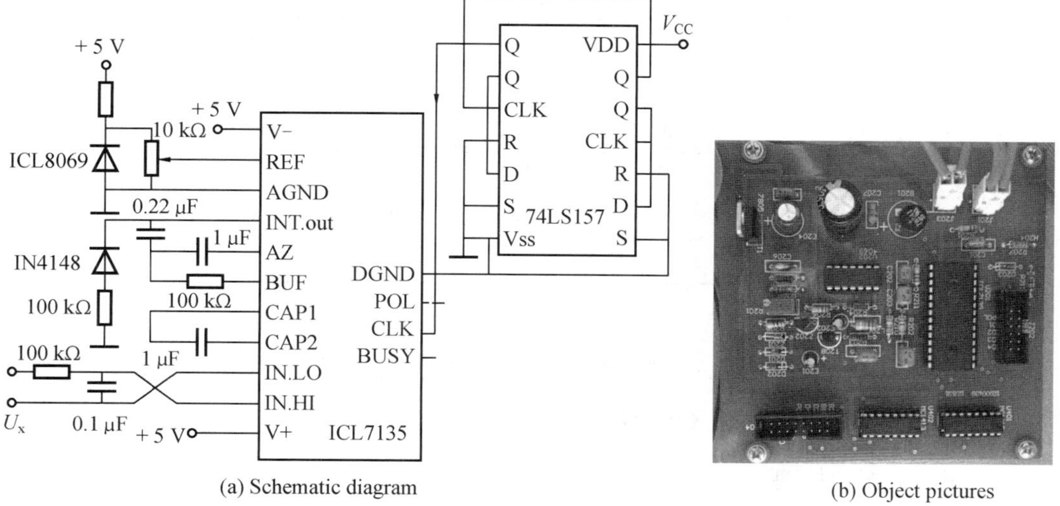

(a) Schematic diagram　　(b) Object pictures

Figure 6.10　A/D conversion circuit

的标定检验,来验证传感器的线性是否满足测量系统的要求,其标定数据见表6.1。

表6.1　力传感器标定数据　　mN

	1	2	3	4	5	6
理论值	0	500	1 000	2 000	3 000	4 000
测量值	0	501	1 000	2 000	3 003	4 003
	7	8	9	10	11	12
理论值	5 000	6 000	7 000	8 000	9 000	10 000
测量值	5 004	6 002	7 002	8 001	9 001	10 000

根据实际的标定的数据,可得力传感器的非线性误差为0.04%,满足刚度测量对力传感器线性度小于 ±0.1% 的要求。

2. 位移传感器的线性度测试

由于激光位移传感器在购买时已经自带放大电路,并且位移传感器在出厂前已经由厂家完成了标定,因此这里只进行激光位移传感器测量的线性度检验。激光位移传感器线性度检验的具体方法是:首先固定好激光位移传感器,并将它的测量距离调整合适;然后确定一个基准平面,并将1 mm厚的量块放在基准面上,调整激光位移传感器的位置,使得激光位移传感器的输出示数0 μm;最后以0.1 mm为间隔递增地更换量块,记录激光位移传感器的示数,计算得到激光位移传感器的线性度。位移传感器检定数据见表6.2。

表6.2　位移传感器检定数据　　mN

	1	2	3	4	5	6
理论值	0	100	200	300	400	500
测量值	0.1	100.2	200.2	300.4	400.6	500.8
	7	8	9	10	11	12
理论值	600	700	800	900	—	—
测量值	600.5	700.3	800.1	900.1	—	—

根据实际的标定的数据,可得激光位移传感器的非线性误差为0.07%,满足刚度测量对位移传感器线性度小于 ±0.1% 的要求。

3. 测试数据的处理

以测量装置作为对象进行试验。首先确定分别以刚度值为10 N·m/rad、50 N·m/rad、100 N·m/rad左右的弹簧管作为实验的对象;然后对不同型号的弹簧管进

2. Displacement Signal Processing

The selected laser displacement sensor has built-in amplifying circuit and A/D conversion circuit, which can directly use digital acquisition card or serial port for data acquisition.

6.1.6　Process and Results

1. Calibration of Force Sensor

The sensors installed in the measurement system need to be calibrated before use. The specific calibration method for the force sensor used in this example is to first adjust the zero adjustment knob of the force sensor amplifier circuit to make the force sensor display value 0 mN, and then slowly put the 10 000 mN standard block on the probe board of the force sensor. At this time, the display value of the force sensor should be 10 000 mN. If the display value of the force sensor is not this value, we need to adjust the magnification knob of the amplifier circuit to make the display value 10 000 mN, and then slowly remove the standard block to see if the indication value of the force sensor is 0 mN. If not, we need to adjust the zero adjustment knob of the amplifier circuit to make the indication value 0 mN. Repeat the above steps until the force sensor meets the requirements. After the above steps, it is necessary to select multiple points between 0 mN and 10 000 mN to perform the calibration of the force sensor in order to verify whether the linearity of the sensor meets the requirements of the measurement system. The calibration data is listed in Table 6.1 .

Table 6.1　Force sensor calibration data

mN

	1	2	3	4	5	6
Theoretical value	0	500	1 000	2 000	3 000	4 000
Measured value	0	501	1 000	2 000	3 003	4 003
	7	8	9	10	11	12
Theoretical value	5 000	6 000	7 000	8 000	9 000	10 000
Measured value	5 004	6 002	7 002	8 001	9 001	10 000

According to the actual calibrated data, the non-linear error of the force sensor is 0.04% , which satisfies the requirement of rigidity measurement for the linearity of the force sensor to be less than ±0.1%.

2. Linearity of displacement sensors

Since the laser displacement sensor has its own amplifier circuit when it is purchased, and the displacement sensor has been calibrated by the manufacturer before leaving the factory, we only perform the inspection of the linearity of the laser displacement sensor measurement. The specific method to check the linearity of the laser displacement sensor is to fix the laser displacement sensor first and adjust its measuring distance appropriately; then to determine a reference plane, and place a 1 mm thick gauge block on the reference surface, adjust the position of the laser displacement sensor, so that the output of the laser displacement sensor shows 0 μm; finally, we should replace the gauge block inerementlly at intervals of 0.1 mm,

行单次装夹多次测量、多次装夹测量试验；最后分别记录各个实验的数据，之后就可以根据实验数据的结果进行分析，判断弹簧管的刚度是否满足设计要求。各实验数据见表6.3 ~ 6.6。

表 6.3　弹簧管单次装夹多次刚度测量实验数据　　N · m/rad

	1	2	3	4	5	6	7	8
$K_n = 10$	10.319	10.315	10.312	10.325	10.339	10.335	10.319	10.326
$K_n = 50$	50.726	50.748	50.768	50.689	50.778	50.699	50.732	50.615
$K_n = 100$	99.895	99.758	99.687	99.856	99.998	99.836	99.528	99.482

表 6.4　弹簧管单次装夹多次刚度测量结果平均值和重复测量误差

平均值 /(N · m · rad^{-1})	重复性	平均值 /(N · m · rad^{-1})	重复性	平均值 /(N · m · rad^{-1})	重复性
10.32	0.26%	50.72	0.32%	99.76	0.52%

表 6.5　弹簧管多次装夹单次刚度测量实验数据　　N · m/rad

	1	2	3	4	5	6	7	8
$K_n = 10$	10.328	10.304	10.323	10.313	10.335	10.338	10.327	10.319
$K_n = 50$	50.625	50.736	50.732	50.701	50.768	50.683	50.698	50.543
$K_n = 100$	99.796	99.859	99.386	99.796	99.898	99.901	99.615	99.476

表 6.6　弹簧管新老装置多次装夹单次刚度测量结果平均值和重复测量误差

平均值 /(N · m · rad^{-1})	重复性	平均值 /(N · m · rad^{-1})	重复性	平均值 /(N · m · rad^{-1})	重复性
10.32	0.32%	50.69	0.44%	99.72	0.52%

至此，就完成了弹簧管刚度的测量。

注意：本实例主要是讲述工程实际应用过程，具体内容经过了大量的简化。为具有代表性，部分内容与实际装置相比也有较大区别。因此，不保证按照本实例进行装置设计能够实现相应的功能或达到相应的指标。

and record the laser displacement sensor's reading, and calculate the linearity of the laser displacement sensor. Verification data of displacement sensor is listed in Table 6.2.

Table 6.2　Verification data of displacement sensor

μm

	1	2	3	4	5	6
Theoretical value	0	100	200	300	400	500
Measured value	0.1	100.2	200.2	300.4	400.6	500.8
	7	8	9	10	11	12
Theoretical value	600	700	800	900	—	—
Measured value	600.5	700.3	800.1	900.1	—	—

According to the actual calibrated data, the nonlinear error of the laser displacement sensor is 0.07%, which meets the requirement of rigidity measurement for the linearity of the displacement sensor to be less than ±0.1%.

3. Processing of Data

The experiment was conducted with the measuring device as the object. Firstly, determine the spring tubes with stiffness values of 10 N·m/rad, 50 N·m/rad, and 100 N·m/rad respectively as the experimental objects; then perform a single clamping and multiple measurements on different types of spring tubes, multiple clamping measurement tests; finally, the data of each experiment are recorded separately, and then the results of the experimental data can be analyzed to determine whether the stiffness of the spring tube meets the design requirements. The experiment data is listed in Table 6.3–6.6.

Table 6.3　Experimental data of multiple stiffness measurement of spring tube with single clamping

N·m/rad

	1	2	3	4	5	6	7	8
$K_n = 10$	10.319	10.315	10.312	10.325	10.339	10.335	10.319	10.326
$K_n = 50$	50.726	50.748	50.768	50.689	50.778	50.699	50.732	50.615
$K_n = 100$	99.895	99.758	99.687	99.856	99.998	99.836	99.528	99.482

Table 6.4　Average value and repeated measurement error of multiple stiffness measurement results of the spring tube with a single clamping

Average /(N·m·rad^{-1})	Repeatability	Average /(N·m·rad^{-1})	Repeatability	Average /(N·m·rad^{-1})	Repeatability
10.32	0.26%	50.72	0.32%	99.76	0.52%

6.2 动态测量工程应用实例——磨削对刀检测

6.2.1 被测对象概述

上节中的滑阀阀芯的加工中有一道进行端面磨削的工艺，即需要对图6.11所示伺服阀阀芯实物图的a、b、c、d四个端面进行磨削，该项加工一般是在外圆磨床上进行的。由于磨削量在100 μm以内，在苛刻条件下的磨削量甚至小于10 μm，而且不允许超差，因此当磨削量较小时，磨削对刀误差的影响就较为明显。磨床的对刀误差是一个较为复杂的问题，即如何采用一种方法使得工件在对刀前后的误差控制在一定的范围内。若对刀问题没办法改善，则会直接影响整体的加工精度。以往的判断磨削对刀是操作者直接根据砂轮与工件接触时的声音变化来判断的，这种方法较为简单、操作方便、成本低，但对操作者的经验要求很高，而且对砂轮阀芯对刀点的判断具有主观性。不同的操作者对于同一工件的加工可能得到不同的对刀误差，并且同一操作者的对刀重复测量误差也无法控制，导致使用传统办法的对刀精度存在未预知性及不稳定性，有时甚至是直接使工件报废。

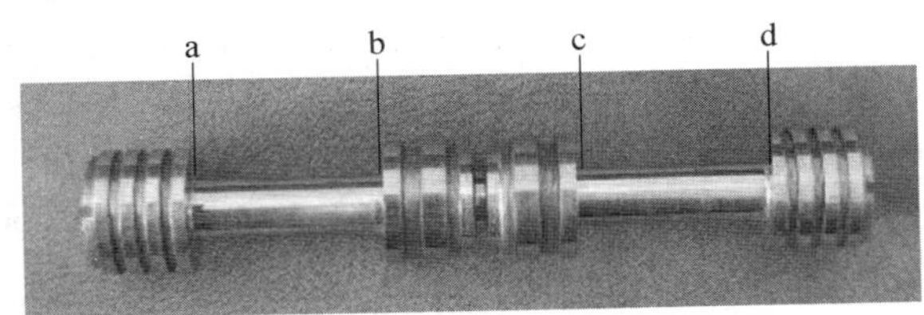

图6.11 伺服阀阀芯实物图

6.2.2 被测物理量分析

上述的人工对刀方法效率和精度都难以保证。因此，本实例测量砂轮与工件接触时的声音信号变化，由砂轮与工件接触前后的产生信号特征的改变判断对刀是否完成。这种信号称为声发射信号，是一种高频振动信号。因此，本实例的被测物理量是振动的位移量或者振动的能量，是与时间有关的量，属于动态测试范畴。

6.2.3 测量原理

如果想利用被加工件在加工时产生的声发射信号来检测砂轮与工件是否接触，以此来判断对刀状态，则有必要首先弄清声发射信号检测工件磨削过程。

磨削实际是利用砂轮表面上大量密布的微小磨粒切削刃进行的切削加工。磨削过程

Table 6.5　data of spring tube multiple clamping with a single stiffness measurement

N · m/rad

	1	2	3	4	5	6	7	8
$K_n = 10$	10.328	10.304	10.323	10.313	10.335	10.338	10.327	10.319
$K_n = 50$	50.625	50.736	50.732	50.701	50.768	50.683	50.698	50.543
$K_n = 100$	99.796	99.859	99.386	99.796	99.898	99.901	99.615	99.476

Table 6.6　Average value and repeated measurement error multiple clamping of new and old spring tube devices with single stiffness measurement results

Average /(N · m · rad^{-1})	Repeatability	Average /(N · m · rad^{-1})	Repeatability	Average /(N · m · rad^{-1})	Repeatability
10.32	0.32%	50.69	0.44%	99.72	0.52%

So far, the measurement of the stiffness of the spring tube is completed.

Note: this example mainly illustrates the actual application process of the project with the specific content having been simplified a lot. In order to be repersentative, some of the content is quite different from the actual device. Therefore, it is not guaranteed that the corresponding functions and indicators can be realized according to the device design of this example.

6.2　Engineering Application Examples of Dynamic Measurement—Grinding Tool Setting Inspection

6.2.1　Overview of the Object

In the processing of the spool valve core in the previous section, there is a process of end face grinding, that is, the four end faces of a, b, c and d in object pictures of servo valve spool as shown in Figure 6.11 need to be ground. The processing is generally carried out on a cylindrical grinder. Since the grinding amount is within 100 μm, the grinding amount under severe conditions is even less than 10 μm, and it is not allowed to exceed the tolerance, so when the grinding amount is small, the influence of grinding on the tool error proves more obvious. The tool setting error of the grinder is a more complicated problem, that is, how to use a method to control the error of the workpiece before and after the tool setting in a certain range. If the tool setting problem cannot be improved, it will directly affect the overall machining accuracy. The past judgment of grinding and tool setting is that the operator directly judges the sound changes when the grinding wheel is in contact with the workpiece. This method is relatively simple, convenient to operate, and low in cost, but it requires a high level of experience for the operator, and it also requires the grinding wheel valve core. The judgment of the tool setting point is subjective. Different operators may get different tool setting errors when processing the same workpiece, and the same operator ' s tool setting repeated measurement error cannot be controlled, resulting in the unpredictability and instability of the

的声发射信号来源如图 6.12 所示。可以首先从工件的磨削加工过程中对声发射信号的产生机理来分析。机械加工件在磨削过程中产生的声发射源可分为以下几种。

(1) 工件材料被加工时产生的剪切变形。

(2) 砂轮磨粒与工件加工表面之间的摩擦。

(3) 砂轮磨粒的破碎。

(4) 砂轮自身转动引起的噪声。

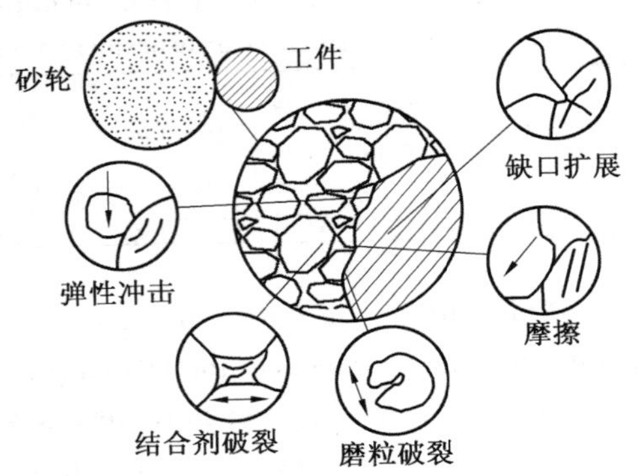

图 6.12　磨削过程的声发射信号来源

其中,磨削条件、工件材料、砂轮锋利程度等因素都均有影响。发生改变时,就会引起声发射信号的幅值、频谱等方面发生变化,这就使得可以通过检测声发射信号的变化来对磨削状态进行判别。通过分析可知,在一次磨削加工过程中,工件材料被加工时产生的剪切变形是声发射信号产生的主要原因,而这种变形又是金属材料内部的位错运动造成的,其声发射信号的频率带主要分布在 50 ~ 400 kHz。砂轮磨粒与工件加工表面之间的摩擦引起的噪声则频率很小,主要分布于 50 kHz 以下,这为分析磨削过程中的声发射信号可行性提供了依据。

声发射信号采集的原理是在各种实际工程或是实验项目中以声发射传感器为媒介,自动采集目标对象中声发射信号各种特征和信息的全过程,整个声发射采集过程离不开各类硬件采集设备与软件信号控制。通常声发射信号采集完整的过程有信号采集、信号预处理、A/D 转换和计算机数据分析等环节。

6.2.4　传感器的选择

声发射信号的频率主要集中在 50 ~ 400 kHz,中心频率在 100 ~ 200 kHz,采用的声发射传感器为基于压电陶瓷的振动传感器。振动传感器实物图如图 6.13 所示。该类型传感器对于声发射信号的灵敏度较高、多次采集信号的一致性较好、外部环境温度对其信

tool setting accuracy using traditional methods, and sometimes even directly scrapes the workpiece.

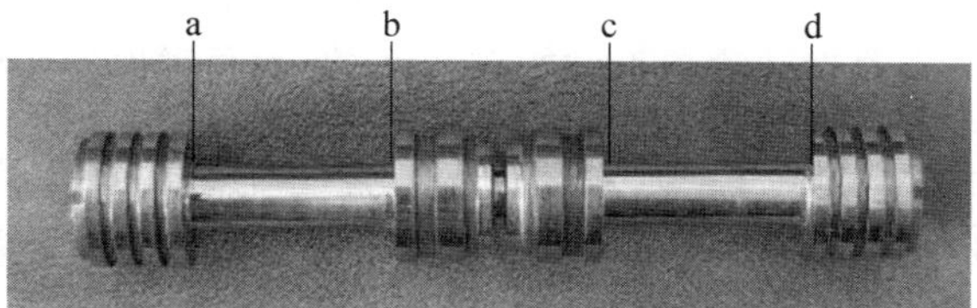

Figure. 6.11 Object pictures of servo valve spool

6.2.2 Analysis of the Measured Physical Quantity

The efficiency and accuracy of the manual tool setting method mentioned above are difficult to guarantee. Therefore, this example measures the change of the sound signal when the grinding wheel is in contact with the workpiece. The change in the signal characteristics before and after the contact between the grinding wheel and the workpiece is used to determine whether the toolsetting is completed. This signal is called an acoustic emission signal, which is a high-frequency vibration signal. Therefore, the measured physical quantity in this example is the displacement of vibration or the energy of vibration, which is a time-related quantity and belongs to the category of dynamic .

6.2.3 Measurement Principle

If we want to use the acoustic emission signal generated during the processing of the workpiece to detect whether the grinding wheel is in contact with the workpiece to determine the tool setting state, it is necessary to clarify the acoustic emission signal to detect the grinding process of the workpiece.

Grinding is actually a cutting process with a large number of densely distributed tiny abrasive cutting edges on the surface of the grinding wheel. The source of the acoustic emission signal during the grinding process is shown in Figure 6.12. We can first analyze the generation mechanism of the acoustic emission signal during the grinding process of the workpiece. The acoustic emission sources generated during the grinding of machined parts can be divided into the following types.

(1) The shear deformation generated when the workpiece material is processed.

(2) The friction between the abrasive grains of the grinding wheel and the machined surface of the workpiece.

(3) Breaking of abrasive grains on the grinding wheel.

(4) Noise caused by the rotation of the grinding wheel itself.

Among them, factors such as grinding conditions, workpiece material and sharpness of the grinding wheel all have an impact. When the change occurs, it will cause the amplitude and frequency spectrum of the acoustic emission signal to change, which makes it possible to judge the grinding state by detecting the change of the acoustic emission signal. Through analysis, it can be known that in a grinding process, the shear deformation of the workpiece material is the main reason for the acoustic emission signal, and this deformation is caused by the movement of

号采集结果影响较小且稳定性方面的性能较好。其采集声发射信号的中心频率大致在150 kHz,采集信号的频率范围为50 ~ 400 kHz,适用于声发射信号采集。传感器安装在临近震源的区域即可,震源为砂轮与工件的接触面,振动通过顶尖传导至顶尖尾座上。因此,在本实例中,将振动传感器安装在外圆磨床的顶尖尾座上,这一安装位置是最临近震源的平面区域。使用时,振动传感器通过磁性夹具固定在被测表面,传感器与被测表面之间需要涂敷声震耦合剂。

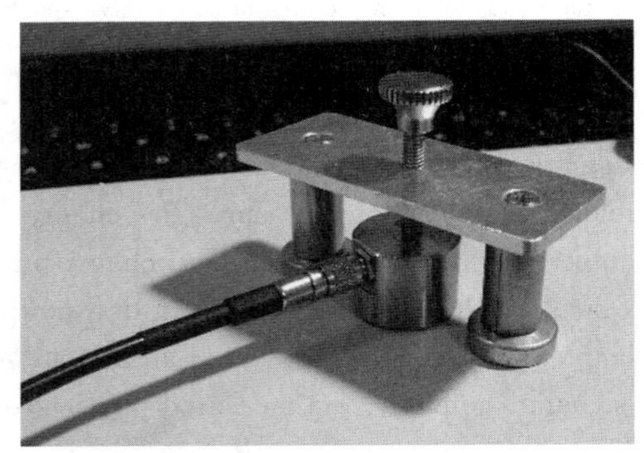

图 6.13　振动传感器实物图

6.2.5　信号处理

声发射信号的采集可以选用数据采集卡。本实例选用的数据采集卡主要参数为采样率 1 MHz,转换精度达到 12 位,即最小的电压分辨率达到

$$\gamma = \frac{V}{2^{\kappa}} = \frac{10}{2^{12}} = 2.44(\mathrm{mV}) \tag{6.9}$$

查阅资料可知,磨削过程中声发射对刀信号幅值一般在 0.5 ~ 1 V,所以该数据采集卡对振动幅值识别的分辨率在 0.2% ~ 0.4%,满足一般精度要求。根据采样定理,采集设备的采样频率f_s需要大于被检测信号中所关注频率成分最高频率f_{max}的 2 倍,才能反映所关注的信号特征。一般在实际的应用中,需要保证采集卡具有所关注频率成分最高频率的5 ~ 10 倍的采样频率。本实例选用的数据采集卡可以达到声发射对刀检测信号采集的要求。对刀时,采集卡采集信号会存在延迟,因为数据采集卡向上位机传送数据有两种工作模式:寄存器全满传送和寄存器半满传送,分别对于寄存器存储空间存满数据后统一传送和寄存器存储空间半满数据后传送。如果数据采集卡寄存器容量为 10 240 个数据,设置半满传送模式,根据采集卡的 1 MHz 采样频率,则相当于每 0.005 s 向工控机传送一次数据,每次5 120 个数据。在使用数据采集卡时,需要注意数据传输方式,切不可认为每个数据点是实时传送的。

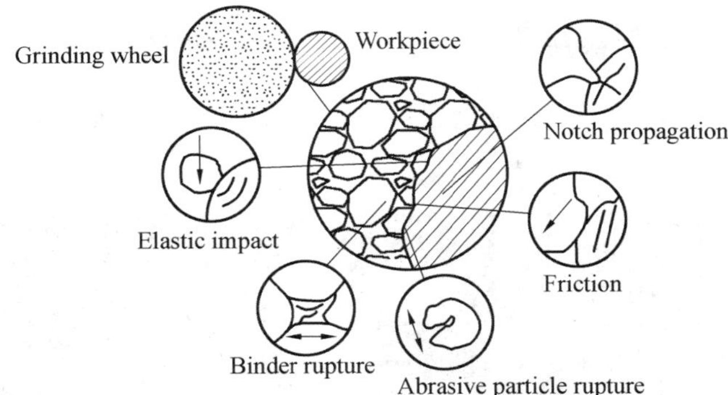

Figure 6.12　The source of the acoustic emission signal during the grinding process

the internal dislocation of the metal material. The frequency band of its acoustic emission signal is mainly distributed between 50 – 400 kHz. The noise caused by the friction between the abrasive particles of the grinding wheel and the workpiece surface is very small, mainly distributed below 50 kHz, which provides a basis for us to analyze the feasibility of the acoustic emission signal in the grinding process.

The principle of acoustic emission signal collection is the whole process of automatically collecting various characteristics and information of the acoustic emission signal from the target object through the acoustic emission sensor as a medium in various actual projects or experimental projects. The entire acoustic emission collection process is inseparable from various hardware collection equipment and software signal control. Usually the complete process of acoustic emission signal acquisition includes signal acquisition, signal preprocessing, A/D conversion and computer data analysis, etc.

6.2.4　The Choice of Sensors

The frequency of the acoustic emission signal is mainly concentrated in 50–400 kHz, and the center frequency is 100–200 kHz. The acoustic emission sensor used is a vibration sensor based on piezoelectric ceramics. The object pictures of the vibration sensor is shown in Figure 6.13. This type of sensor has high sensitivity to acoustic emission signals, good consistency of multiple collected signals, the external environment temperature has little influence on the signal collection results, and its stability performance is better. The center frequency of the collected acoustic emission signals roughly around 150 kHz, the frequency range of the collected signal is 50–400 kHz, which is suitable for acoustic emission signal collection. The sensor can be installed in the area close to the seismic source that is the contact surface between the grinding wheel and the workpiece. The vibration is transmitted to the top tailstock through the top. Therefore, in this example, the vibration sensor is installed on the top tailstock of the cylindrical grinder. An installation location is the plane area closest to the seismic source. The using vibration sensor is fixed on the surface to be measured by a magnetic clamp, and the acoustic shock coupling agent needs to be coated between the sensor and the surface to be measured.

6.2.6 测试过程及结果

安装好振动传感器及相应的辅助仪器后，就可以进行声发射振动测试。声发射对刀误差现场实验图如图 6.14 所示。

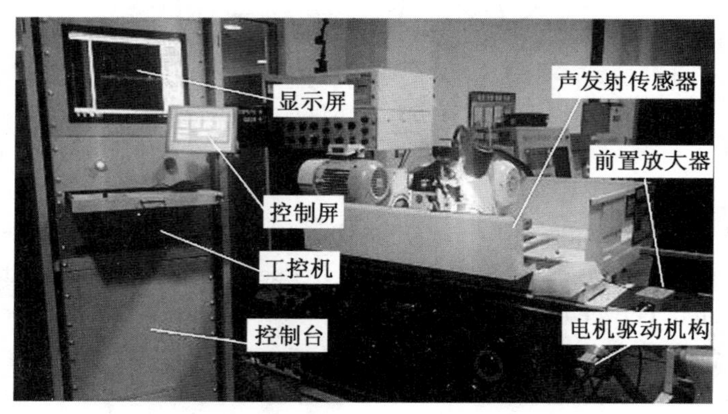

图 6.14　声发射对刀误差现场实验图

图 6.15 所示阀芯磨削声发射信号的滤波选择影响给出了滤波对振动测试信号的处理效果。采用滤波的目的是增加信噪比，尽量区分噪声和接触振动信号，有利于判断接触时刻。但是，值得注意的是，由于滤波是采用对数据点进行的数字滤波，处理后的得到的接触时刻将会发生滞后，影响对刀接触时刻的判断，因此为提高信噪比，理想的做法是在前置放大器处进行信号放大。本实例中的前置放大器为 60 dB，能够满足判断接触时刻的信噪比要求。

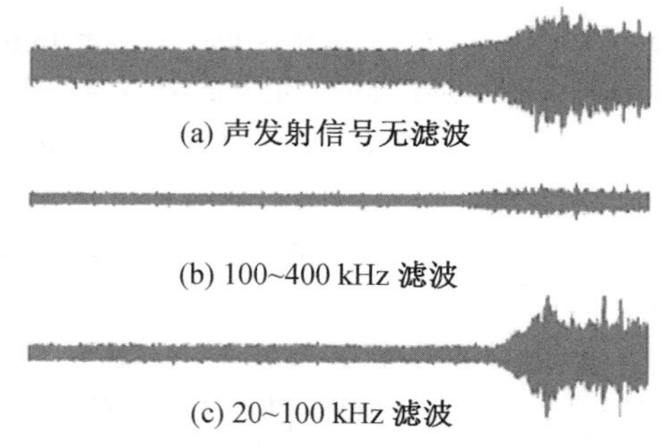

图 6.15　阀芯磨削声发射信号的滤波选择影响

传感器安置的位置不同也会对声发射信号产生影响，如图 6.16 所示。该试验有利于判断最优的测试位置。

通过实验可知，当传感器安置在顶尖或床头架处时，声发射信号很强，但噪音也很大，不利于声发射信号的分析。同时，现场试验知道传感器贴于尾座处效果较佳，放置于尾座

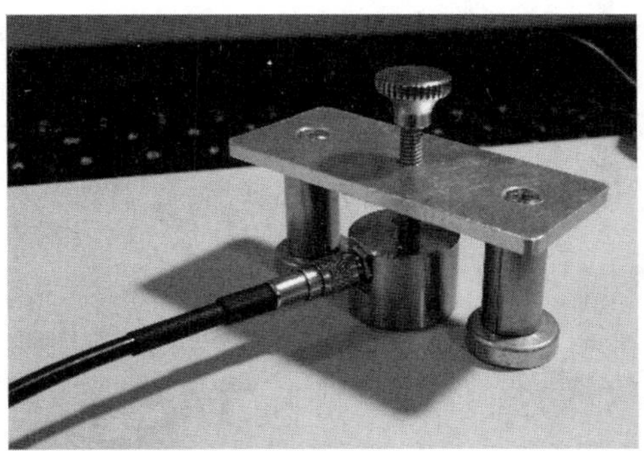

Figure 6.13　Object pictures of the vibration sensor

6.2.5　Signal Processing

A data acquisition card can be used for the acquisition of acoustic emission signals. The main parameters of the data acquisition card used in this example are the sampling rate of 1 MHz and the conversion accuracy of 12 bit, that is, the minimum voltage resolution is

$$\gamma = \frac{V}{2^{K}} = \frac{10}{2^{12}} = 2.44(\mathrm{mV}) \tag{6.9}$$

According to the information, in the grinding process, the amplitude of the acoustic emission tool setting signal is generally at 0.5 – 1 V. The resolution of vibration amplitude recognition of the data acquisition card is between 0.2% –0.4%, which meets the general accuracy requirements. According to the sampling theorem, when the sampling frequency f_s of the acquisition device is the highest frequency f_{max} with greater than 2 times in the spectrum component of the interest detected signal, reflecting concerned signal characteristics. Generally, in practical applications, it is necessary to ensure that the acquisition card can reach a sampling frequency that is 5–10 times the highest frequency of the detected signal component. The selected data acquisition card can meet the requirements of acoustic emission tool setting detection signal acquisition. During tool calibration, there is a delay in the acquisition signal, because there are two working modes for the data acquisition card to transfer data to the upper computer: register full transfer and register half-full transfer, unified transfer after the register storage space is full of data and transfer after the register storage space is half full. If the register capacity of the data acquisition card is 10 240 data, and set the half-full transmission mode, According to the 1 MHz sampling frequency of the acquisition card, it is equivalent to transmitting data to the industrial computer every 0.005 s, 5 120 data each time according to the 1 MHz sampling frequency of the acquisition card. When using the data acquisition card, pay attention to the data transmission mode, and do not think that each data point is transmitted in real time.

6.2.6　Process and Results

After installing the vibration sensor and the corresponding auxiliary equipment, the acoustic emission vibration can be carried out. Field experiment diagram of acoustic emission tool setting error is shown in Figure 6.14.

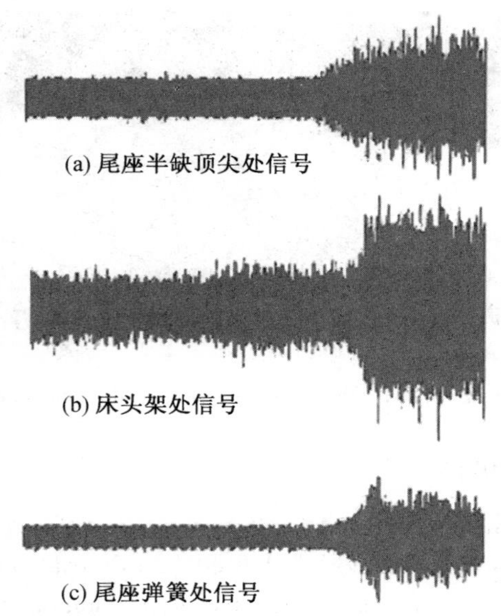

图 6.16　传感器位置对声发射信号的影响

弹簧处效果最佳。

选用加工时正常的进给速度，以保证与平时的加工状态类似。图 6.17 所示为对刀过程声发射信号图。整个过程分为四个阶段：砂轮未接触工件阶段、对刀阶段、磨削加工阶段和光整磨削阶段。这四阶段在图 6.17 中得到了很好的展示。第一个阶段主要是加工环境的噪音，其信号幅值较小；第二个阶段对刀开始，声发射信号幅值开始迅速增加；第三个阶段加工时声发射信号保持在较高幅值；第四个阶段不再进给，声发射信号逐步减弱。整个过程的对刀接触位置发生在声发射电压幅值迅速增大的区域，且幅值增大幅度很明显。

图 6.17　对刀过程声发射信号图

根据实际测试结果可知，在调整进给速度并保证进给速度平稳的前提下，最后对刀磨削量可以控制在 0.5 ~ 1.5 μm。

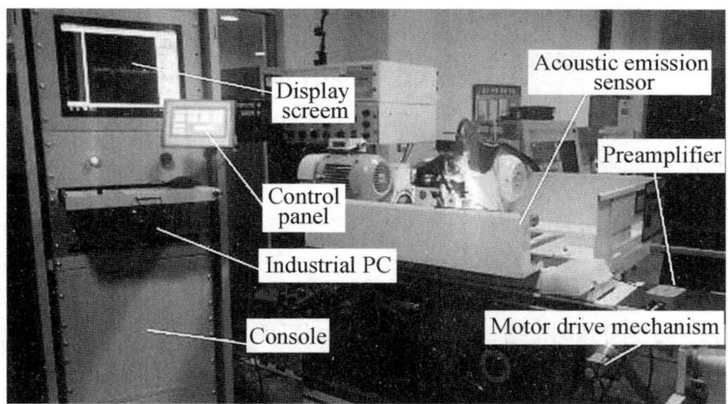

Figure 6. 14 Field experiment diagram of acoustic emission tool setting error

The influence of filter selection of spool with grinding acoustic emission signals shown in Figure 6. 15 shows the processing effect of filtering on vibration signals. The purpose of filtering is to increase the signal-to-noise ratio and try to distinguish between noise and contact vibration signals, which is helpful to judge the contact time. However, it is worth noting that the filtering is using digital filtering of data points, the contact time obtained after processing will lag, which affects the judgment of the knife contact time. Therefore, the ideal approach is to develop the signal at the preamplifier in order to improve the signal-to-noise ratio. The preamplifier in this example is 60 dB, which can meet the signal-to-noise ratio requirement for judging the contact moment.

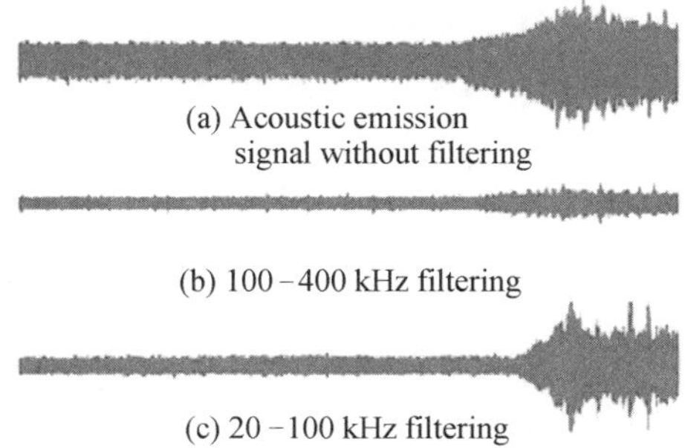

Figure 6. 15 The influence of filter selection of spool with grinding acoustic emission signals

With different placement, the sensor will also affect the acoustic emission signal, as shown in Figure 6. 16. This experiement is conducive to judging the optimal location.

Through experiments, it can be known that when the sensor is placed at the top or the bedside frame, the acoustic emission signal is very strong, but the noise is also great, which is not conducive to the analysis of the acoustic emission signal. At the same time, the sensor is better when it is attached to the tailstock, and the effect is best if it is placed on the tailstock spring through the field.

Select the normal feed speed during processing to ensure that it is similar to the usual processing state. Figure 6. 17 shows the influence of sensor position on acoustic emission signal. The whole process is divided into four stages: the stage with the grinding wheel not touching the workpiece, the tool setting stage, the grinding processing stage and the finishing

思考题与练习题

6.1　静态测试与动态测试的关注点有什么不同?

6.2　静态测试实例中,还可以选择哪些类型传感器?选择传感器需要关注的问题是什么?

6.3　动态测试实例中如果给出一组时域下的离散数据点,如何进行频谱分析?频谱分析的结果有什么用途?

6.4　动态测试中的传感器是否需要标定?如果需要,如何标定?

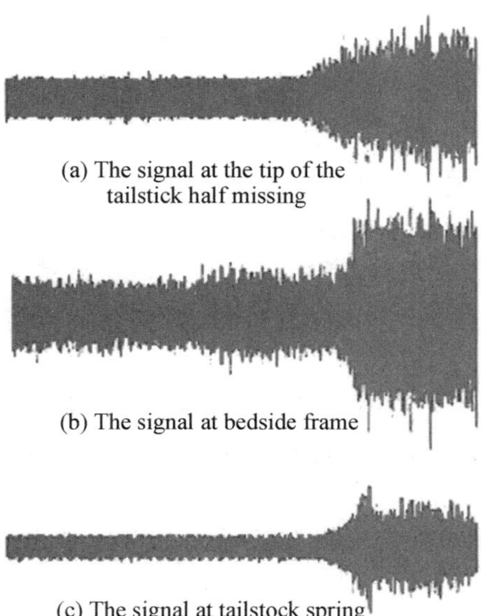

(a) The signal at the tip of the tailstick half missing
(b) The signal at bedside frame
(c) The signal at tailstock spring

Figure 6.16　The influence of sensor position on acoustic emission signal

grinding stage. These four stages are well illustrated in Figure 6.17. The first stage is mainly the noise of the processing environment, and the signal amplitude is small; the second stage starts the tool setting, and the acoustic emission signal amplitude begins to increase rapidly; the third stage is processing the acoustic emission signal to maintain a higher amplitude; the fourth stage is no more feed, and the acoustic emission signal gradually weakens. The tool setting contact position of the whole process occurs in the area where the amplitude of the acoustic emission voltage increases rapidly, and the amplitude increase is obvious.

Figure 6.17　Acoustic emission signal diagram during tool setting

According to the actual results, the final tool grinding amount can be controlled within 0.5–1.5 μm by adjusting the feed rate and ensuring a stable feed rate.

Questions and Exercises

6.1　What is the difference between static and dynamic?

6.2　In the static example, what other types of sensors can be selected? What are the issues that need to be paid attention to when choosing a sensor?

6.3　If a set of discrete data points in the time domain is given in the dynamic example, how to perform spectrum analysis? What is the function of the results of spectrum analysis?

6.4　Does the sensor in the dynamic need to be calibrated? If necessary, how to calibrate?

参 考 文 献

[1] 黄长艺,严普强. 机械工程测试技术基础[M]. 2 版. 北京:机械工业出版社,1995.
[2] 熊诗波. 液压测试技术[M]. 北京:清华大学出版社,1982.
[3] 熊诗波. 机械工程测试技术基础[M]. 4 版. 北京:机械工业出版社,2018.
[4] 李玮华. 机械工程测试技术基础学习指导、典型题解析与习题解答 [M] . 北京:机械工业出版社,2013.
[5] 祝海林. 机械工程测试技术[M]. 北京:机械工业出版社,2017.
[6] 李云峰. 电液伺服阀中弹簧管刚度精密测量技术的研究 [D]. 哈尔滨:哈尔滨工业大学,2007.
[7] 潘旭东. 伺服阀滑阀叠合量液动测量系统及其关键技术的研究 [D]. 哈尔滨:哈尔滨工业大学,2008.
[8] 张耀祖. 伺服阀阀芯节流工作边亚微米级精密磨削装置的研究 [D]. 哈尔滨:哈尔滨工业大学,2008.
[9] HORST C. Measurement testing and sensor technology[M]. Berlin:Springer, 2018.
[10] 黄文涛. 传感与测试技术[M]. 哈尔滨:哈尔滨工业大学出版社,2014.
[11] 祝海林. 机械工程测试技术[M]. 2 版. 北京:机械工业出版社,2017.
[12] 黄长艺,严普强. 机械工程测试技术基础[M]. 2 版. 北京:机械工业出版社,1995.
[13] 马怀祥 王艳颖. 工程测试技术[M]. 武汉:华中科技大学出版社, 2014.
[14] 许同乐. 机械工程测试技术[M]. 2 版. 北京:机械工业出版社,2017.
[15] 秦树人. 机械工程测试原理与技术[M]. 重庆:重庆大学出版社,2017.
[16] THOMAS G B. Mechanical measurements[M]. New Jersey:Prentice Hall,2006.
[17] 王振成,张雪松. 工程测试技术及应用[M]. 重庆: 重庆大学出版社,2014.
[18] 王化祥. 传感器原理与应用技术[M]. 北京: 化学工业出版社,2018.
[19] 于彤. 传感器原理及应用[M]. 北京:机械工业出版社,2015
[20] 吴大正. 信号与线性系统分析[M]. 北京:高等教育出版社,2005.
[21] 郑君里, 应启珩, 杨为理. 信号与系统[M]. 北京:高等教育出版社,2000.
[22] 华成英, 童诗白. 模拟电子技术基础[M]. 北京:高等教育出版社,2005
[23] 李立华. 模拟电子技术[M]. 2 版. 北京:电子工业出版社,2016

名 词 索 引

A

B

C

D

E

F

G

H

J

K

L

M

N

O

P

Q

R

S

T

W

X

Y

Z